Cell Motility and Cancer

Cell Motility and Cancer

Editors

José I. López
Ildefonso M. de la Fuente

MDPI • Basel • Beijing • Wuhan • Barcelona • Belgrade • Manchester • Tokyo • Cluj • Tianjin

Editors
José I. López
Pathology
Cruces University Hospital
Barakaldo
Spain

Ildefonso M. de la Fuente
Department of Nutrition
CEBAS-CSIC Institute
Murcia, and
Department of Mathematics
University of the Basque Country
Leioa
Spain

Editorial Office
MDPI
St. Alban-Anlage 66
4052 Basel, Switzerland

This is a reprint of articles from the Special Issue published online in the open access journal *Cancers* (ISSN 2072-6694) (available at: www.mdpi.com/journal/cancers/special_issues/Motility).

For citation purposes, cite each article independently as indicated on the article page online and as indicated below:

LastName, A.A.; LastName, B.B.; LastName, C.C. Article Title. *Journal Name* **Year**, *Volume Number*, Page Range.

ISBN 978-3-0365-2054-4 (Hbk)
ISBN 978-3-0365-2053-7 (PDF)

Contents

About the Editors

José I. López

José I. López is head of Department of Pathology at the Hospital Universitario Cruces and principal investigator of the Biomarker in Cancer Unit at the Biocruces-Bizkaia Health Research Institute. He graduated at the Faculty of Medicine, University of the Basque Country, Leioa, Spain, and trained in Pathology at the Hospital Universitario 12 de Octubre, Madrid, Spain. He received his PhD degree at the Universidad Complutense of Madrid, Spain. Dr. López has served as pathologist for more than 30 years in several hospitals in Spain and is subspecialized in Uropathology. Dr. López is interested in translational uropathology in general and in renal cancer in particular, and collaborates with several international research groups unveiling the genomic landscape of urological cancer. Intratumor heterogeneity, tumor sampling, tumor microenvironment, tumor ecology, immunotherapy, and basic mechanisms of carcinogenesis are his main topics of interest.

Ildefonso M. de la Fuente

Ildefonso M. de la Fuente is a tenured scientist at Spanish National Research Council (CSIC), an academic researcher of the Department of Mathematics at University of the Basque Country, and a postgraduate master's degree professor on "Cellular and Molecular Advances in Cancer" at the School of Medicine in the same University. He has been the scientific director of the Biocruces Health Research Institute, and a pioneer in Systems Biology in the Basque Country. Currently, he is also the team leader of the multicenter "Quantitative Systems Biology Group". Dr. De la Fuente's research focuses mainly on the field of enzymatic self-organization and systemic metabolic functionality of cells. His fascination for the complex dynamical organization of the cellular metabolism goes throughout his scientific career. From his university studies to the present, the aim of his studies is to understand the origin of the molecular order that underlies the complex functional architecture of celular life.

Preface to "Cell Motility and Cancer"

Cell motility is a crucial systemic behavior essential for a plethora of fundamental biological processes and human diseases. Migration is an intrinsic key property of cells necessary for embryogenesis, tissue repair, inflammation, autoimmunity, and other fundamental physiological activities. There has been notable progress in the understanding of biochemical mechanisms involved in cell migration, however, how unicellular organisms efficiently regulate their locomotion system at a systemic level is a topic that still remains unresolved.

Cancer is a leading human disease with persistent high mortality rates that poses a serious economical concern for health systems in Western societies. Malignant tumors are now understood as communities of billions of individuals (cells) characterized by a variable tendency to invade locally and to metastasize to distant organs. Both local invasion and metastases have received much attention in recent years, and both take place through tumor cell migration.

This Special Issue is conceived as a forum for basic, translational, and clinical research related to cell directional motility mechanisms and tumor cell migration. Under such a generic umbrella, basic researchers in biology, systems biology, and other quantitative sciences, as well as in physiology and pharmacology, will have the opportunity to join clinical specialties in medical oncology, biochemistry, immunology, and pathology. Such unique convergence of disciplines will enrich the panorama of a central characteristic of malignant tumors, cell motility.

José I. López, Ildefonso M. de la Fuente
Editors

cancers

MDPI

Editorial

An Approach to Cell Motility as a Key Mechanism in Oncology

José I. López [1,2,*] and Ildefonso M. De la Fuente [3,4,*]

1 Department of Pathology, Cruces University Hospital, 48903 Barakaldo, Spain
2 Biocruces-Bizkaia Health Research Institute, 48903 Barakaldo, Spain
3 Department of Nutrition, CEBAS-CSIC Institute, Espinardo University Campus, 30100 Murcia, Spain
4 Department of Mathematics, Faculty of Science and Technology, University of the Basque Country, 48940 Leioa, Spain
* Correspondence: joseignacio.lopez@osakidetza.eus (J.I.L.); mtpmadei@ehu.eus (I.M.D.l.F.)

check for updates

Citation: López, J.I.; De la Fuente, I.M. An Approach to Cell Motility as a Key Mechanism in Oncology. *Cancers* **2021**, *13*, 3576. https://doi.org/10.3390/cancers13143576

Received: 5 July 2021
Accepted: 8 July 2021
Published: 16 July 2021

Motility is an inherent characteristic of living cells manifesting cell migration, a fundamental mechanism of survival and development. In unicellular organisms, cell migration is needed to prey and escape from predators. In multicellular individuals, however, embryogenesis, tissue repair, and adaptation to external changes do happen through cell migration. Cancer cells also display motile abilities; actually, aggressiveness in most malignant tumors depends fundamentally on two properties related to cell motility: local invasion and metastases.

This Special Issue contains up to fourteen contributions focusing on the cell motility/cancer binomial from very different approaches and tries to serve as a showcase/sample book of the enormous possibilities still pending to be analyzed and discovered in the field. Eight articles and six reviews are in this issue. An international cast of contributors has deepened in a broad spectrum of specific processes related to cell migration in breast and colorectal cancers, as well as in rhabdomyosarcoma, melanoma, and Leydig cell tumor of the testis. The reviews revisit several basic mechanisms related to drug resistance, epithelial-to-mesenchymal transition processes, and the transfer of knowledge related to motility from single organisms to cancer cells. Finally, an ecological approach to cancer biology highlights the benefit obtained to sum on the oncology board allied scientific disciplines.

Rhabdomyosarcoma is a malignant mesenchymal neoplasm more frequently diagnosed in childhood and adolescence, where it pursues an aggressive clinical course with 3-year survival rates of only 25%. Skrzypek et al. [1] demonstrate for the first time that Snail, a transcription factor linked to E-cadherin regulation in epithelial to mesenchymal transition processes, also regulates the metastatic behavior of rhabdomyosarcoma cells, both in vivo and in vitro, promoting cell motility, invasion, and chemotaxis. This effect is accomplished by upregulating the protein expression of Ezrin and Akt. Besides, the authors have shown that the Snail-miRNA axis regulates motility of rhabdomyosarcoma cells, especially miR-28-3p through indirect modulation of Ezrin levels. The authors conclude that this new regulatory mechanism of cell motility in this type of sarcoma could be shared by other mesenchymal neoplasms and propose to consider Snail a potential new target in future therapy modalities.

Colorectal adenocarcinoma is a common neoplasm in Western countries and a paradigmatic example of intratumor heterogeneity. Kryczka et al. [2] have observed that ABCC4, a protein belonging to a superfamily of ATP-binding cassette proteins, can regulate cell migration in colorectal adenocarcinomas through a cAMP-dependent way. Since the inhibition of ABCC4 seems to increase the migratory and invasive capacities of these neoplasms, the authors call attention to such a pathway as a potentially actionable therapeutic target.

Two contributions in this Special Issue deal with breast cancer [3,4], the leading cause of cancer-related death in European women. In their work, Panzetta et al. [3] study how the extracellular matrix stiffness interferes in the adhesion and migration properties of two different mammary cell lines under the exposure of two different X-ray doses. The authors conclude that the microenvironment simulating healthy tissue has a radioprotective role in preventing cell motility and invasion. Instead, a supraphysiological matrix stiffness promotes cell motility. This cellular response, called durotaxis and originally described in fibroblasts, observed in mammary cell lines reproduces the results obtained in previous experiments performed in unicellular organisms (*Dictyostelium discoideum* and *Caenorhabditis elegans*, among others) and melanoma cell lines, as it will be mentioned elsewhere in this collection [5]. On the other hand, Levine and Ogunwobi [4] focus their work on triple-negative breast cancer, a subset of around 10% of mammary carcinomas which pursues an especially aggressive clinical course. More specifically, they have centered their research on one of the six subtypes of this tumor variant, that is, the so-called claudin-low triple-negative breast carcinoma. Such spectrum of molecular variants of triple-negative breast cancer has been identified based on their specific genomic profiles. In brief, they have found that targeting the plasmacytoma variant translocation 1 (PTV1) exon 9 results in the re-expression in these tumors of the claudin 4 protein, this way inhibiting tumor cell migration. The authors stress that this finding could have important clinical implications in this specific subset of patients.

Although rare, Leydig cell tumor is probably the most frequent non-germ cell tumor in the human testis [6]. The majority of them are benign, but a small percentage (<10%) pursue a malignant course. Here, Ponikwicka-Tyszko et al. [7] have analyzed the effect of mifepristone, the selective progesterone receptor modulator, in a transgenic mouse model and two Leydig tumor cell lines. They conclude that mifepristone acts as a membrane progesterone agonist promoting Leydig cell tumor progression.

Melanoma is a classic model to analyze tumor cell migration in clinics and research [8,9] and this collection of Cell Motility and Cancer includes three contributions [10–12]. An Australian clinical study of 306 metastatic melanomas has found that *BRAF* + *NRAS* mutations were associated to the central nervous system and liver metastases, while *BRAF* mutation was to lymph node metastases and *NRAS* mutation with lung metastases [8], so tumor mutation status may advise to direct specifically to these sites the clinical surveillance of these patients. A recent review describes the last advances in the remodeling of melanoma cell metabolism, e.g., glycosylation and oxidative phosphorylation, along with its temporal development from nevus to metastases [9]. El-Kharbili et al. [10] delineate how keratinocytes cooperate with melanoma cells in dermal colonization through dermal-epidermal junction proteolysis induced by the Tspan8 action. Moreover, the same author has shown in other studies that its encoding gene, *TSPAN8*, acts not only in reducing matrix adherence [13] but also in promoting invasion [14]. The authors conclude that using Tspan8-blocking antibodies would prevent early melanoma from spreading [10]. Naffa et al. [11] show that PMCA4b, a plasma membrane Ca^{2+} key pump in the regulation of cytosolic Ca^{2+} concentration, regulates melanoma cell migration via remodeling the actin cytoskeleton. PMCA4b plays a key role in regulating cell polarity through F-actin rearrangement resulting in a less aggressive phenotype. Interestingly, the same group has previously shown that PMCA4b inhibits cell migration and metastatic capacities in BRAF mutant melanoma cells [15]. Kwan et al. [12] analyze the role of LRG1, a leucine-rich alpha 2 glycoprotein, in melanoma and conclude that this protein is required for metastatic dissemination but not for cell growth.

Heissig et al. [16] have contributed to this Special Issue with a review of the functional role of the epidermal growth factor-like protein-7 (EGFL7) in cancer and drug resistance. This protein is involved in cell migration and neoangiogenesis thus contributing to tumor metastases. The review includes a detailed description of the protein, its contribution to the development of a pathological tumor vessel phenotype, its role enhancing tumor immune escape, its regulation of the extracellular matrix stiffness, and its contribution to drug resistance.

The cytoskeletal dynamics involved in the epithelial–mesenchymal transition processes and their role as potential targets for cancer metastases have been reviewed by Datta et al. [17]. The key role of the cytoskeleton in cell motility is deeply analyzed in this review, from its structure and functions to its implication in epithelial–mesenchymal transition processes and its importance in multidrug resistance. The authors conclude that the interplay between cytoskeleton dynamics and epithelial–mesenchymal transition should be utilized to identify potential biomarkers.

Cheng et al. [18] review the role of peroxisome proliferator-activated receptors (PPAR) as metabolic regulators in neoplasms. PPARs are essential in reprogramming cancer-associated fibroblasts and adipocytes and regulate the paracrine and autocrine signaling of cancer-associated fibroblasts and tumor-associated macrophages/immune cells. The authors conclude that PPAR-based anticancer treatment could be achieved by modulating its physiological activity.

Readers interested in knowing how motility mechanisms of unicellular organisms can be translated to human cancer cells, and how the analysis of motility properties of cells in a wide variety of protists, worms, insects, etc., have helped to understand cell motility mechanisms in mammals, have an excellent opportunity reading the review by De la Fuente and López [5] included in this collection. The authors explain why simple organism models are necessary to understand human cell behavior. Additionally, they focus on the similarities between the locomotion system in unicellular eukaryotic organisms and human cells, the connection between external stimuli (galvanotaxis, chemotaxis, haptotaxis, barotaxis, durotaxis, etc.), migration, and cancer. They have verified that a cell migratory behavior can be modified by changes in the signals coming from the external medium (cellular associative conditioning) [19]. Likewise, the role of the nucleus in cell migration analyzed from a quantitative perspective is also a special topic in this review [5], linked to cancer, and in which the authors reflect their own previous experience [20].

Capp et al. [21] connect essential points in cancer biology such as the development of metastases with strict ecological principles. They conjecture if the so-called Parrondo's paradox [22] may play any role in cancer biology. The paradox, defined as how combinations of losing strategies produce a winning outcome, may help to explain some particular behaviors in biological collectivities. The authors hypothesize if stability is a losing strategy for malignant cells, why should cell populations with high stochasticity be needed for long term survival and proliferation, how dormancy can be considered as a losing strategy in Parrondo's dynamics, and if the metastatic behavior is a strategy under the paradox [21]. In the authors' opinion, this perspective may have therapeutic implications. Alternating two losing strategies, i.e., to treat aggressively a tumor thus promoting the development of resistant clones and not to treat it at all, the cost of the investment needed to acquire drug resistance by tumor cells would need to be shared with the cost to maintain tumor cell proliferation. As proposed by Kam et al. [23], alternating a fake drug (called ersatzdroges) with a real drug may allow keeping the tumor size constant, without resistance selection. Interestingly, a close strategy promoting tumor containment has been recently proposed by Viossat and Noble [24] to avoid, or delay, tumor resistances.

Last but not least, Keller-Pinter et al. [25] review the role of syndecan-4, a transmembrane proteoglycan, in cell motility of several tumors, including melanoma, breast and lung cancers, among others, and point to this protein as a potential therapeutic target.

In conclusion, cell motility is an essential characteristic of cells, from unicellular organisms to cancer, and this Special Issue has intended to put together very different viewpoints of the intricated mechanisms involucred on it. Since cell migration is a constant event in cancer and is responsible for tumor invasion and metastases, the two dismal effects of neoplasia, this collection of articles and reviews aims to serve as a translational bridge between basic researchers and clinicians promoting interdisciplinary collaboration. We hope to meet the objective.

Author Contributions: J.I.L. and I.M.D.l.F. conceived, designed, and wrote the manuscript. Both authors have read and agreed to the published version of the manuscript.

Funding: This study received no external funding.

Conflicts of Interest: The authors declare no conflict of interest.

References

1. Skrzypek, K.; Kot, M.; Konieczny, P.; Nieszporek, A.; Kusienicka, A.; Lasota, M.; Bobela, W.; Jankowska, U.; Kedracka-Krok, S.; Kajka, M. SNAIL promotes metastatic behavior of rhabdomyosarcoma by increasing EZRIN and AKT expression and regulating microRNA networks. *Cancers* **2020**, *12*, 1870. [CrossRef]
2. Kryczka, J.; Sochacka, E.; Papiewska-Pajak, I.; Boncela, J. Implications of ABCC4-mediated cAMP efflux for CRC migration. *Cancers* **2020**, *12*, 3547. [CrossRef]
3. Panzetta, V.; La Verde, G.; Pugliese, M.; Artiola, V.; Arrichiello, C.; Muto, P.; La Commara, M.; Netti, P.A.; Fusco, S. Adhesion and migration response to radiation therapy of mammary epithelial and adenocarcinoma cells interacting with different stiffness substrates. *Cancers* **2020**, *12*, 1170. [CrossRef]
4. Levine, F.; Ogunwobi, O.O. Targeting PVT1 exon 9 re-expresses claudin 4 protein and inhibits migration by claudin-low triple negative breast cancer cells. *Cancers* **2021**, *13*, 1046. [CrossRef]
5. De la Fuente, I.M.; López, J.I. Cell motility and cancer. *Cancers* **2020**, *12*, 2177. [CrossRef] [PubMed]
6. Kim, I.; Young, R.H.; Scully, R.E. Leydig cell tumors of the testis. A clinicopathological analysis of 40 cases and review of the literature. *Am. J. Surg. Pathol.* **1985**, *9*, 177–192. [CrossRef] [PubMed]
7. Ponikwicka-Tyszko, D.; Chrusciel, M.; Pulawska, K.; Bernaczyk, P.; Sztacheklska, M.; Guo, P.; Li, X.; Toppari, J.; Huhtaniemi, I.T.; Wolczynski, S.; et al. Mifepristone treatment promotes testicular Leydig cell tumor progression in transgenic mice. *Cancers* **2020**, *12*, 3263. [CrossRef] [PubMed]
8. Adler, N.R.; Wolfe, R.; Kelly, J.W.; Haydon, A.; McArthur, G.A.; McLean, C.A.; Mar, V.J. Tumor mutation status and sites of metastasis in patients with cutaneous melanoma. *Br. J. Cancer* **2017**, *11*, 1026–1035. [CrossRef] [PubMed]
9. Sobiepanek, A.; Paone, A.; Cutruzzolà, F.; Kobiela, T. Biophysical characterization of melanoma cell phenotype markers during metastatic progression. *Eur. Biophys. J.* **2021**, *50*, 523–542. [CrossRef]
10. El-Kharbili, M.; Cario, M.; Béchetoille, N.; Pain, C.; Boucheix, C.; Degoul, F.; Masse, I.; Berthier-Vergnes, O. Tspan8 drives melanoma dermal invasion by promoting proMMP-9 activation and basement membrane proteolysis in a keratinocyte-dependent manner. *Cancers* **2020**, *12*, 1297. [CrossRef]
11. Naffa, R.; Padányi, R.; Ignácz, A.; Hegyi, Z.; Jezsó, B.; Tóth, S.; Varga, K.; Homolya, L.; Hegedüs, L.; Schlett, K.; et al. The plasma membrane Ca^{2+} pump PMCA4b regulates melanoma cell migration through remodeling of the actin cytoskeleton. *Cancers* **2021**, *13*, 1354. [CrossRef]
12. Kwan, Y.P.; Teo, M.H.Y.; Lim, J.C.W.; Tan, M.S.; Rosellinny, G.; Wahli, W.; Wang, X. LRG1 promotes metastatic dissemination of melanoma through regulating EGFR/STAT3 signalling. *Cancers* **2021**, *13*, 3279. [CrossRef] [PubMed]
13. El-Kharbili, M.; Robert, C.; Witkowski, T.; Danty-Berger, E.; Barbollat-Boutrand, L.; Masse, I.; Gadot, N.; de la Fouchardière, A.; McDonald, P.C.; Dedhar, S.; et al. Tetraspanin 8 is a novel regulator of ILK-driven beta1 integrin adhesion and signaling in invasive melanoma cells. *Oncotarget* **2017**, *8*, 17140–17155. [CrossRef] [PubMed]
14. El-Kharbili, M.; Agaësse, G.; Barbollat-Boutrand, L.; Pommier, R.M.; de la Fouchardière, A.; Larue, L.; Caramel, J.; Puisieux, A.; Berthier-Vergnes, O.; Masse, I. Tspan8-β-catenin positive feedback loop promotes melanoma invasion. *Oncogene* **2019**, *38*, 3781–3793. [CrossRef] [PubMed]
15. Hegedũs, L.; Garay, T.; Molnár, E.; Varga, K.; Bilecz, Á.; Török, S.; Padányi, R.; Pászty, K.; Wolf, M.; Grusch, M.; et al. The plasma membrane Ca^{2+} pump PMCA4b inhibits the migratory and metastatic activity of BRAF mutant melanoma cells. *Int. J. Cancer* **2017**, *140*, 2758–2770. [CrossRef]
16. Heissig, B.; Salama, Y.; Takahashi, S.; Okumura, K.; Hattori, K. The multifaceted roles of EGFL7 in cancer and drug resistance. *Cancers* **2021**, *13*, 1014. [CrossRef]
17. Datta, A.; Deng, S.; Gopal, V.; Yap, K.C.H.; Halim, C.E.; Lye, M.L.; Ong, M.S.; Tan, T.Z.; Sethi, G.; Hooi, S.C.; et al. Cytoskeletal dynamics in epithelial-mesenchymal transition: Insights into therapeutic targets for cancer metastasis. *Cancers* **2021**, *13*, 1882. [CrossRef]
18. Cheng, H.S.; Yip, Y.S.; Lim, E.K.Y.; Wahli, W.; Tan, N.S. PPARs and tumor microenvironment; the emerging roles of the metabolic master regulators in tumor stromal-epithelial crosstalk and carcinogenesis. *Cancers* **2021**, *13*, 2153. [CrossRef]
19. De la Fuente, I.M.; Bringas, C.; Malaina, I.; Fedetz, M.; Carrasco-Pujante, J.; Morales, M.; Knafo, S.; Martinez, L.; Pérez-Samartin, A.; López, J.I.; et al. Evidence of conditioned behavior in amoebae. *Nat. Commun.* **2019**, *10*, 3690. [CrossRef]
20. De la Fuente, I.M.; Bringas, C.; Malaina, I.; Regner, B.; Pérez-Samartín, A.; Boyano, M.D.; Fedetz, M.; López, J.I.; Pérez-Yarza, G.; Cortés, J.M.; et al. The nucleus does not significantly affect the migratory trajectories of amoeba in two-dimensional environments. *Sci. Rep.* **2019**, *9*, 16369. [CrossRef]
21. Capp, J.P.; Nedelcu, A.M.; Dujon, A.M.; Roche, B.; Catania, F.; Ujvari, B.; Alix-Panabières, C.; Thomas, F. Does cancer biology rely on Parrondo's principles? *Cancers* **2021**, *13*, 2197. [CrossRef] [PubMed]
22. Cheong, K.H.; Koh, J.M.; Jones, M.C. Paradoxical survival: Examining the Parrondo effect across biology. *BioEssays* **2019**, *41*, 1900027. [CrossRef] [PubMed]
23. Kam, Y.; Das, T.; Tian, H.; Foroutan, P.; Ruiz, E.; Martinez, G.; Minton, S.; Gillies, R.J.; Gatenby, R.A. Sweat but not gain: Inhibiting proliferation to multidrug resistant cancer cells with "ersatzdroges". *Int. J. Cancer* **2015**, *136*, E188–E196. [CrossRef]
24. Viossat, Y.; Noble, R. A theoretical analysis of tumor containment. *Nat. Ecol. Evol.* **2021**, *5*, 826–835. [CrossRef] [PubMed]
25. Keller-Pinter, A.; Gyulai-Nagy, S.; Becsky, D.; Dux, L.; Rovo, L. Syndecan-4 in tumor cell motility. *Cancers* **2021**, *13*, 3322. [CrossRef]

cancers

MDPI

Review

Syndecan-4 in Tumor Cell Motility

Aniko Keller-Pinter [1,*], Szuzina Gyulai-Nagy [1], Daniel Becsky [1], Laszlo Dux [1] and Laszlo Rovo [2]

1 Department of Biochemistry, Faculty of Medicine, University of Szeged, H-6720 Szeged, Hungary; gyulai-nagy.szuzina@med.u-szeged.hu (S.G.-N.); becsky.daniel@med.u-szeged.hu (D.B.); dux.laszlo@med.u-szeged.hu (L.D.)
2 Department of Oto-Rhino-Laryngology and Head-Neck Surgery, University of Szeged, H-6725 Szeged, Hungary; office.orl@med.u-szeged.hu
* Correspondence: keller.aniko@med.u-szeged.hu

Simple Summary: Cell migration is crucial fReaor metastasis formation and a hallmark of malignancy. The primary cause of high mortality among oncology patients is the ability of cancer cells to metastasize. To form metastasis, primary tumor cells must be intrinsically able to move. The transmembrane, heparan sulfate proteoglycan syndecan-4 (SDC4) exhibits multiple functions in signal transduction by regulating Rac1 GTPase activity and consequently actin remodeling, as well as regulating focal adhesion kinase, protein kinase C-alpha and the level of intracellular calcium. By affecting several signaling pathways and biological processes, SDC4 is involved in cell migration under physiological and pathological conditions as well. In this review, we discuss the SDC4-mediated cell migration focusing on the role of SDC4 in tumor cell movement.

Abstract: Syndecan-4 (SDC4) is a ubiquitously expressed, transmembrane proteoglycan bearing heparan sulfate chains. SDC4 is involved in numerous inside-out and outside-in signaling processes, such as binding and sequestration of growth factors and extracellular matrix components, regulation of the activity of the small GTPase Rac1, protein kinase C-alpha, the level of intracellular calcium, or the phosphorylation of focal adhesion kinase. The ability of this proteoglycan to link the extracellular matrix and actin cytoskeleton enables SDC4 to contribute to biological functions like cell adhesion and migration, cell proliferation, cytokinesis, cellular polarity, or mechanotransduction. The multiple roles of SDC4 in tumor pathogenesis and progression has already been demonstrated; therefore, the expression and signaling of SDC4 was investigated in several tumor types. SDC4 influences tumor progression by regulating cell proliferation as well as cell migration by affecting cell-matrix adhesion and several signaling pathways. Here, we summarize the general role of SDC4 in cell migration and tumor cell motility.

Keywords: syndecan-4; proteoglycan; migration; EMT; metastasis; cancer; cell polarity; extracellular matrix; actin; calcium; centrosome

check for updates

Citation: Keller-Pinter, A.; Gyulai-Nagy, S.; Becsky, D.; Dux, L.; Rovo, L. Syndecan-4 in Tumor Cell Motility. *Cancers* **2021**, *13*, 3322. https://doi.org/10.3390/cancers13133322

Academic Editors: José I. López and Ildefonso M. de la Fuente

Received: 30 April 2021
Accepted: 27 June 2021
Published: 1 July 2021

Publisher's Note: MDPI stays neutral with regard to jurisdictional claims in published maps and institutional affiliations.

1. Introduction

Cell migration is a hallmark of tumor cell malignancy and essential for the multistep process of metastasis formation. The capability of invasion and metastasis enables cancer cells to escape the primary tumor mass and colonize new terrain in the body [1]. Beyond its role in metastasis formation and tumor progression, cell motility is essential in a variety of physiological and pathological tasks, such as tissue regeneration, wound healing, angiogenesis, embryonic development, as well as immune cell responses [2]. To form metastasis, primary tumor cells must be intrinsically able to move. These motility mechanisms do not differ from the normal motility cycles [3].

Epithelial-mesenchymal transition (EMT) is defined as the transdifferentiation of epithelial cells into motile mesenchymal cells. EMT occurs during different biological processes, such as embryonic development, tissue regeneration or cancer progression.

During EMT, cells acquire enhanced invasion ability, escape from apoptotic signals [4] and gain drug resistance [5]. The epithelial cells maintain cell-to-cell junctions and apico-basal polarity, whereas mesenchymal cells display a motile phenotype and front-rear polarity. The loss of apico-basal polarization and the development of front-rear polarity are characteristic features of EMT. The individual or collective migration of cancer cells require several steps of EMT. For effective single-cell migration, cells must acquire a complete EMT. In contrast, collective cell migration requires a wide spectrum of EMT states: the leader cells gain mesenchymal phenotype, but the follower cells keep the connection with their neighbors with intact cell-cell junctions [6,7].

EMT is controlled by complex signaling pathways, including transcriptional regulation, epigenetic modifications, alternative splicing and modulated by miRNAs, other non-coding RNAs, translational control and post-translational modifications [6,8]. The transforming growth factor beta (TGFβ) signaling is crucial for the induction of EMT, as well as other signaling pathways, including tyrosine kinase receptor signaling [9,10].

The front-rear polarity of migrating cells is developed during the early stages of movement. In 2D environment, migrating cells display flattened morphology, while protrusions of the plasma membrane (i.e., sheet-like lamellipodia and finger-like filopodia) are formed at the cell's leading edge (Figure 1) [11]. During migration, cell front defines the direction of movement as the tail region forms (Figure 1), causing the morphology of cells to change, forcing them to elongate as a result of actin-cytoskeleton and cell-matrix rearrangement [12]. The shaped tail region is known as the trailing edge, while the front region as the leading edge [13].

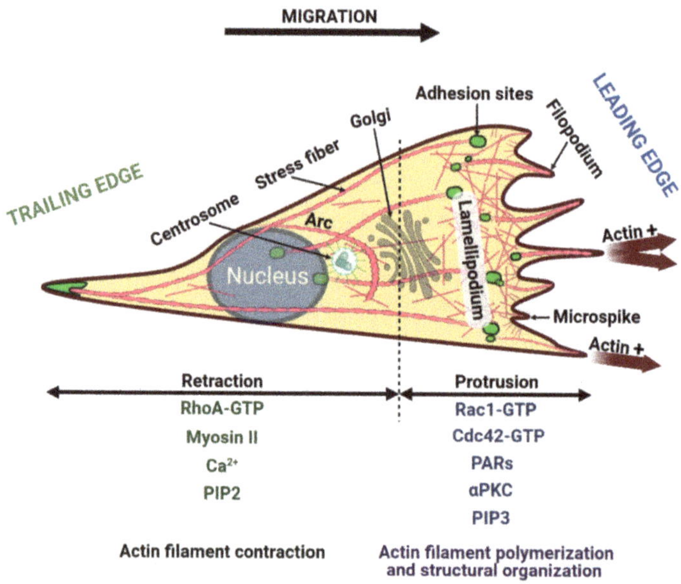

Figure 1. Schematic structure and polarity of a migrating cell in 2D environment during mesenchymal cell migration. Contractile actin bundles (stress fibers) in a migrating cell are represented. Arc-shaped bundles are also observed to move inward under the dorsal cell surface (Arc). At the cell front, in lamellipodia and filopodia, actin filaments are all polarized with their fast-polymerizing ends forwards (for pushing); in the body of the cytoskeleton, actin filaments form bipolar assemblies with myosin to form contractile arrays (for retracting). RhoA: Ras homolog family member A; Ca^{2+}: Calcium; Rac1: Ras-related C3 botulinum toxin substrate 1; Cdc42: Cell division control protein 42 homolog; PAR: Partitioning-defective (polarity complex); aPKC: Atypical protein kinase C; PIP2: Phosphatidylinositol 4,5-bisphosphate; PIP3: Phosphatidylinositol 3,4,5-trisphosphate. Image was created with BioRender.com.

Focal adhesions, the cell-matrix contact points, are dynamic, multi-protein structures composed of over 150 proteins [14]. Cell migration requires the continuous assembly and disassembly of focal adhesions, formation of new focal adhesions at the front and disruption at the tail, causing the cell to move [15].

Importantly, migrating cells exhibit different morphologies in vitro in 2D, 3D collagen or 3D cell derived matrix environments [16]. In contrast to the flattened morphology of the cells in 2D, in 3D collagen matrix the migrating cells display a spindle-like phenotype and exhibit multiple small lamellipodia at the leading edge. Moreover, the physical properties of the collagen substrate also affect cancer cell shape both in 2D and 3D [17]. In 3D cell-derived matrix, cells use lobopodial migration and exhibit a more tubular shape with lateral blebs and a leading edge that lacks lamellipodia [18]. The in vitro 1D systems containing matrix fibrils, which usually have a width of 1–2 μm, can closely mimic the biological characteristics of cell migration in 3D matrix, but not on flat 2D substrates [18]. In the 3D living environments, cells exhibit multiple types of migration, such as mesenchymal, amoeboid, lobopodial and collective migration, depending on the local matrix environment [7,19]. All these migration modes are regulated by the local extracellular microenvironment, Rho GTPase signaling and non-muscle myosin contractility [19].

2. Cytoskeletal System during Cell Migration

The cytoskeletal system of the mammalian cells is composed of actin network (microfilaments), intermediate filaments and microtubules. The continuous crosstalk between actin, microtubules and intermediate filaments provides their coordinated dynamics to facilitate cell migration [7]. Recently, the septin network was also described as the fourth component of the cytoskeleton. Evidence indicates that all these cytoskeletal systems participate in mammalian cell motility. The roles of the actin network and microtubules in cell motility are well characterized, while less is known about the roles of intermediate filaments and septins.

2.1. Rearrangement of the Actin Cytoskeleton during Migration

The dynamic rearrangement of the actin cytoskeleton and cell-matrix interactions is a prerequisite for cell migration [20]. Actin stress fibers play a critical role in in cell adhesion, cell contractility and movement and they are also crucial for preserving and changing the cell's shape, as well as determining the mechanical properties of the cell surface [21,22]. The main components of these actomyosin contractile stress fibers are the actin microfilaments, myosin II (mechanochemical enzyme) and α-actinin (actin-binding protein) [23]. There are three main types of stress fibers in a migrating cell: ventral stress fibers, transverse arcs and dorsal stress fibers [24]. Ventral stress fibers are associated with focal adhesions at both ends and are located in the tail region of the cells [15]. Transverse arcs are not connected directly to focal adhesions and usually stream back from the anterior edge of the cell toward the center. The dorsal stress fibers are located in the front edge of the cell. They are attached to focal adhesions on the ventral surface of the leading edge and extend dorsally, towards the cell center to bind to transverse arcs (Figure 1). During cell migration, the actin fibers are recycled by a retrograde actin flow process, creating a dynamically active cyclic system [25,26].

Actin polymerization, retrograde actin flow and myosin II-based contractility are all essential for cell migration [27,28]. Cells move by repeating cycles of cell front protrusion and attachment, followed by rear decoupling and retraction. Coordinated polymerization of multiple actin filaments generates protrusive forces that drive plasma membrane protrusion to the cell's leading edge [29]. Contractile force is generated by myosin motors. Not only the active fibers, but also the cells' posterior ends are pulled back by this force. Muscle contraction is identical to this process [30,31].

The "dendritic nucleation" is a mechanism of actin turnover in lamellipodia that involves Arp2/3 complex continuously nucleating new actin filaments alongside the preexisting "primordial" filaments [32,33]. After that, the newly formed filaments elongate and push against the plasma membrane. The diameter of the actin filaments is ~7 nm. They are polar structures, with a plus end (also known as barbed end), where the actin monomers assemble and a minus and (also known as pointed end), where monomers disassemble. The barbed end of the filaments is "capped" after a brief period of elongation; thus, elongation is terminated. Disassembly of the network occurs through a combination of debranching and severing of actin filaments, followed by depolymerization of filament fragments [29]. Overall, the array of branched filaments in lamellipodia undergoes treadmilling by assembling at the front and disassembling throughout its body. Cadherin complexes regulate actin dynamics mainly via α-catenin, which inhibits Arp2/3-mediated branching polymerization [34] and recruits formin, an actin nucleator, to adherent junctions. In addition to their role in providing junctional stability, β-catenin and p120-catenin can act as transcriptional regulators [35]. The key organizers of the actin cytoskeleton dynamics are the members of Rho family of small GTPases [36].

2.2. The Role of Intermediate Filaments in Cell Motility

The intermediate filaments are non-polar components of the cytoskeleton with a diameter of 11 nm. The expression of intermediate filaments is tissue specific. During tumor development, changes in intermediate filament expression and composition, such as increases in vimentin levels, are associated with increased invasive capacities [37–39]. Vimentin can act as a scaffold for signaling molecules involved in cell motility [40], as well as interact with cell-matrix adhesions [41]. Moreover, vimentin organization modulates the formation of lamellipodia [42]. Keratin intermediate filaments are associated with cell-cell (desmosomal) and cell-matrix (hemidesmosomal) junctions, thereby regulating cell shape, cell adhesion and mechanotransduction [41]. Intermediate filaments exhibit a role in collective migration as well, as keratin filaments control traction forces during collective migration [43,44].

2.3. The Complex Function of Microtubules in Cell Migration

Microtubules are dynamic components of the cytoskeleton coordinating cellular migration. They are wider than actin and intermediate filaments with a diameter of 25 nm and composed of α-tubulin and β-tubulin heterodimers. Microtubule assembly is a polarized process and starts from microtubule organizing centers (MTOCs). In most cell the centrosomes serve as a major MTOCs stabilizing the minus ends of microtubules; however, the Golgi complex also participates in microtubule network organization in some cell types [45].

Microtubules are involved in intracellular transport processes, which are crucial for delivery of new membrane components and signaling molecules to the leading edges of migrating cells and the recycling of adhesion receptors (intergrins) [46,47]. The delivery of membranes, mRNAs and polarity factors to the leading edge of a migrating cell supports the formation of protrusions [47,48]. Microtubules also contribute to cell motility through their ability to resist high compressive loads and generate pushing forces to support the formation and maintenance of cell protrusions [46,49]. Moreover, microtubule cytoskeleton controls the formation and maturation of focal adhesions [50] and is also essential for the disassembly of focal adhesions [51].

The microtubule cytoskeleton is an essential regulator of the polarized organization of migrating cells. During cell motility, microtubules display an asymmetric organization, thereby creating a front-rear polarity. By providing pulling forces, they move the nucleus forwards and determine the position of centrosomes [52].

2.4. The Role of Septins in Cell Migration

Septins are guanine nucleotide-binding proteins that are highly conserved in eukaryotes and polymerize into hetero-oligomeric complexes, filaments, bundles and rings [53,54]. Septins are recognized as novel components of the cytoskeleton; however, they remain relatively poorly understood compared with other cytoskeletal systems [54]. The septin filaments are formed at the cell cortex or in association with other cytoskeletal components, such as actin or microtubules. By directly associating with cellular membranes, septins are implicated in providing membrane stability, organization of plasma membrane by serving as diffusion barriers for membrane proteins and orientation of cell polarity [54]. Moreover, septins have been shown to function as multimolecular scaffolds by recruiting components of signaling pathway. Growing evidence indicates the role of septins in cell migration. It was reported that septin filaments crosslink actin stress fibers, thereby promoting focal adhesion maturation and cell migration [55].

3. Multiple Functions of Rho GTPases in Cell Motility

The Rho family of small GTPases including Rac1 (Ras-related C3 botulinum toxin substrate 1), Cdc42 (Cell division control protein 42 homolog) and RhoA (Ras homolog family member A) are evolutionarily conserved regulators of cell polarity and the actin cytoskeleton [56]. Rho GTPases function as molecular switches alternating between inactive GDP-bound and active GTP-bound forms. The GTP-bound form binds and activates downstream effector proteins, thereby regulating different signaling pathways [57]. The two-state cycle is regulated by three sets of proteins: the guanine nucleotide exchange factors (GEFs), the GTPase-activating proteins (GAPs) and the guanine dissociation inhibitors (GDIs). The GEFs catalyze the exchange of GDP for GTP. GAPs are able to increase intrinsic GTP hydrolysis and are responsible for switching between the active and inactive forms of Rho GTPases. Alternating between GDP- and GTP-bound states may involve cytosol-membrane translocation, as GDIs prevent Rho GTPases from membrane-targeting and activation [58].

Rho GTPases are crucial molecules in the establishment and sustenance of front-rear polarity in migrating cells [59]. Moreover, they play a role in cell division, cell morphology, differentiation and cell migration [60]. Activated Rac1 is enriched along the leading edge (Figure 1), thereby increasing actin polymerization and the formation of lamellipodial membrane protrusions [61]. Rac1 activity decreases towards the tail region of the cell [29,62]. In contrast, RhoA activity is the highest in the tail region (Figure 1) leading to the appearance of contractile actin bundles (stress fibers). RhoA activity also influences the development of mature focal adhesions [63].

The formation of filopodia is regulated by the activation of Cdc42 [63,64]. Both Rac1 and Cdc42 are able to activate the Arp2/3 complex, leading to actin polymerization and the formation of a branched lamellipodial actin network. Cdc42 and Rac1 regulate the polymerization of cortical actin through the members of the Wiskott–Aldrich syndrome protein (WASP)/Scar1 superfamily [65]. The interaction of Cdc42/Rac1 with WASP/Scar proteins unmasks the C-terminal region, thereby mediating the binding of WASP/Scar to the Arp2/3 complex [66]. Arp2/3 complex binds to the sides of preexisting actin filaments and stimulates new filament formation to create branched actin networks [32]. Actin nucleation is induced by Arp2/3 and enhanced by binding of WASP-family carboxyl-terminal domains to the Arp2/3 complex [66]; therefore, the Arp2/3 and WASP proteins act as molecular links for Cdc42 and Rac1 induced cortical actin polymerization [67,68]. Beyond the role of Rho GTPases in the regulation of actin polymerization, they are involved in actin depolymerization as well. Rac1 and RhoA also regulate cofilin activity, thereby affecting actin depolymerization [69].

4. Front-Rear Polarity of Migrating Cells

The existence of asymmetry within a cell is referred to cell polarity. The polarization of migrating cells, such as the formation of front-rear edges and the proper orientation of cellular components, is one of the most remarkable conditions for cell movement [15]. Polarity lipids, such as PIP2 (phosphatidylinositol 4,5-bisphosphate) and PIP3 (phosphatidylinositol 3,4,5-trisphosphate) and 3 sets of polarity protein complexes, such as Par (partition defective), Crumbs and Scribble complexes are responsible for the establishment and maintenance of cellular polarity [59,70]. During intracellular polarization of migrating cells, the leading edge is determined by the presence of PIP3, whilst the tail region is determined by PIP2 [59,71] (Figure 1).

The Par polarity complex, composed of Par3, Par6 and atypical protein kinase C (PKC), can determine the front of a migrating cell and the accumulation of Rac1 and Cdc42 [72]. Because these proteins are missing in the back of the cell, the formation of protrusions is inhibited in the rear resulting in directional migration of the cell [15].

During cell migration, actin accumulates in the lamellipodium, thereby creating a front-rear asymmetry within the cell [73]. Polarization of a migrating cell is also defined by the positioning of the cell nucleus and reorientation of the Golgi network and microtubule organizing center towards the leading edge [74,75]. The activity of Rho GTPases is also asymmetrical during migration creating a gradient between the front and the rear of the cell [58,61,63,76,77]. The Rac1 and Cdc42 GTPases exhibit high activity at the front which decreases towards the rear. In contrast, the activity of RhoA is lower at the front and gradually increases towards the trailing edge [63].

In addition, Tiam1, along with the Par polarity complex, facilitates persistent migration through the stabilization of anterior-posterior cell polarization [78]. Par3 interacts with Tiam1, leading to localized Rac1 activation and consequently creating the front-rear gradient of Rac1 and RhoA GTPases in migrating cells [79]. Because Tiam1-mediated Rac1 signaling is required for establishing and maintaining cell polarity [80], the impaired Tiam1 signaling inhibits the formation of front-rear polarization in migrating cells, thereby inhibiting persistent migration.

Migrating cells also create a front-to-back calcium (Ca^{2+}) gradient that is essential for cell migration and serves as a coordinator for polarized distribution of molecules [81]. Both Ca^{2+} influx from the extracellular space through different Ca^{2+} channels of the plasma membrane [82] and Ca^{2+} release from intracellular stores (primarily the endoplasmic reticulum) contribute to cytosolic Ca^{2+} concentration [83]. The increasing front–rear Ca^{2+} gradient is involved in the disassembly of focal adhesions and, consequently, the rear end retraction and the movement of the cell. The Ca^{2+} gradient is required to maintain the front–rear polarization of migrating cells by restricting spontaneous lamellipodia formation in the trailing edges [84]. In addition to contractility, changes in intracellular Ca^{2+} affect the activity of calmodulin-dependent enzymes and actin-crosslinking proteins, thus playing a key role in the assembly of adhesions and multilevel junctions [77,85]. High levels of RhoA activity and subsequent actomyosin contractility define the rear of a migrating cell as well as an increased Ca^{2+} concentration and the activation of Ca^{2+}-dependent proteases is required to cleave focal adhesion proteins. It was suggested by Tsai et al. that the crosstalk between Ca^{2+} signaling and Rho GTPases would coordinate the oscillations of these factors in the leading edges of migrating cells [86].

5. Syndecan Family of Transmembrane Proteoglycans

Syndecans (SDCs) are transmembrane proteoglycans and four family members are distinguished in vertebrates [87]. Due to their transmembrane structure, the most important task of SDCs is to participate in the physical connection and signaling between the extracellular matrix and the cell. SDCs are major mediators of cellular interactions with the pericellular environment, thereby contributing critical functions to cell adhesion receptors. Moreover, they also participate in cell signaling events and numerous biological processes. The expression of SDCs is cell-, tissue- and development-specific. Syndecan-1 (SDC1), also known as CD-138, is expressed in endothelial, epithelial, smooth muscle and plasma cells. Syndecan-2 (SDC2), also known as fibroglycan, is presented mainly in fibroblasts, mesenchymal tissues, whilst syndecan-3 (SDC3) (N-syndecan) is expressed in neurons and developing musculoskeletal system. Syndecan-4 (SDC4, ryudocan), unlike other members of the family, is universally expressed in virtually all cell types in a development state specific manner [87–89].

General Structure of Syndecans

SDCs consist of three domains (Figure 2), an N-terminal, variable extracellular domain (ectodomain), the highly conserved transmembrane domain and the C-terminal intracellular domain [87,90]. Glycosaminoglycan (GAG) side chains are attached to the core protein extracellularly [87,88,91]. Near the N-terminus, heparan sulfate (HS) chains are linked via a tetrasaccharide linker to one of the serine (Ser) residues of the ectodomain by an O-glycosidic bond [92] (Figure 2). Chondroitin sulfate (CS) side chains are also present for SDC1 and SDC3 and bind closer to the transmembrane region [92,93] (Figure 2). The repeating disaccharide of HS is N-acetylglucosamine and uronic acid, which is modified by sulfate and uronic acid epimerization to iduronic acid. The HS chains contain 2-O-sulfated iduronic acid and N-, 6-O, or (rarely) 3-O-sulfated glucosamine subunits. In CS chains, N-, 6-O or 4-O-sulfated acetylgalactosamine subunits are present [94,95].

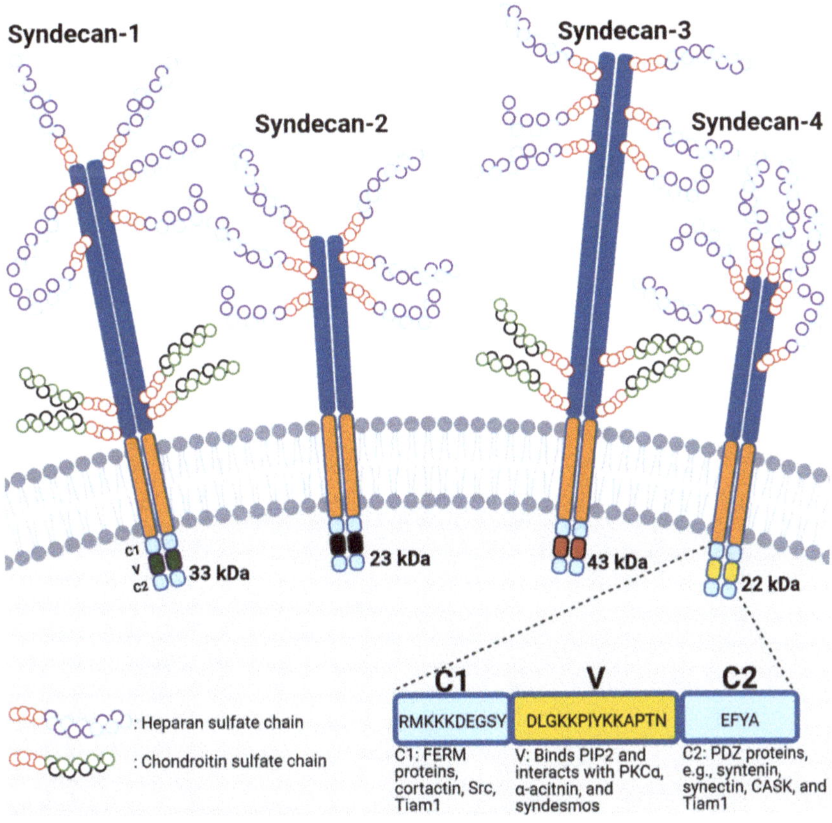

Figure 2. The four-member family of vertebrate syndecans (SDCs). The core proteins of SDC1 and SDC3 are larger than that of SDC2 and SDC4 and can carry both heparan and chondroitin sulfate chains. The glycosaminoglycan chains are attached to the serine residues of the core protein. The cytoplasmic domains are composed of two strongly conserved regions (C1 and C2) separated by an SDC-specific variable (V) region. The main interacting partners of the cytoplasmic domain of SDC4 are shown. PIP2: Phosphatidylinositol-4,5-bisphosphate; PKC: Protein kinase C; PDZ: Postsynaptic density protein; CASK: Calcium/calmodulin-dependent serine protein kinase; Tiam1: T-lymphoma invasion and metastasis-inducing protein 1. Image was created with BioRender.com.

The extracellular domain has plenty of interacting partners, such as matrix proteins, e.g., fibronectin, matrix metalloproteinases (MMPs), growth factors and cytokines. SDCs can recruit soluble ligands, thereby increasing their local concentration and they can also modulate the ligand-dependent activation of primary signaling receptors at the cell surface [89,96,97], or can protect growth factor precursors from activation [98]. The role of SDCs in tumor cell proliferation was reported in numerous cases. SDC1 drives proliferation through the Wnt/β-catenin pathway in multiple myeloma, but defeats cell growth in colorectal carcinoma via the inhibition of JAK1/STAT and Ras/Raf/MEK/ERK pathways [99]. Moreover, SDC1 is the key mediator of the reactive stromal response that promotes the proliferation of breast cancer cells [100]. SDC2 promotes tumorigenic activity in colon carcinoma cells [101]. SDC4 regulates autotaxin-β induced proliferation in osteosarcoma [102]. The SDC4-α5β1 integrin mediated cell adhesion to fibronectin reduces tumor cell proliferation, whilst the tenascin-C-mediated inhibition of SDC4-fibronectin interaction and consequently the impaired fibronectin-induced signaling enhances the proliferation of glioblastoma cells [103]. The ectodomain of SDCs can also promote the adhesion and penetration of bacteria and viruses [104–107], the uptake of positively charged cell-penetrating peptides [108], the cell surface binding of cationic poly- and lipoplexes [109], as well as the cellular internalization of lipoplexes [110]. The ectodomain of SDCs can be cleaved by proteolytic enzymes (secretases), such as members of ADAM (disintegrin and metalloproteinase) family and MMPs. This ectodomain shedding also plays a role in pathophysiological processes, including inflammation and tissue regeneration [94,105,111,112].

The transmembrane domain is the most conserved part of the molecule and also shows high similarity within the family. It contains a GXXXG motif that strongly influences the formation of SDS (sodium dodecyl sulfate) resistant dimers [113,114].

The cytoplasmic domain is short and comprises a variable (V) region that is unique for each member of the SDC family and two conserved regions preceding (C1) and following (C2) the V region [91,115]. The C1 region can bind to the members of the FERM (four-point-one, ezrin, radixin, moezin) family, which are membrane- and actin-associated proteins and also binds Src kinase and cortactin [116]. The EFYA motif of the C2 region binds PDZ (postsynaptic density protein) domain containing proteins, such as syntenin, synectin, synbindin, CASK (calcium/calmodulin-dependent protein kinase) or Tiam1 (T-lymphoma invasion and metastasis-inducing protein 1) [91,117,118].

6. Structure, Interacting Partners and Signaling of Syndecan-4

SDC4, similarly to other members of the family, is involved in signal transduction processes across the cell membrane. Unlike other SDCs, it is universally expressed and present in virtually all cell types. SDC4 plays a major role in cell proliferation, migration, cell adhesion and it is also involved in cytokinesis, endocytosis and mechanotransduction [88,97,102,119–123]. The extracellular domain binds several growth factors, such as FGF2 (fibroblast growth factor-2) [124], HGF (hepatocyte growth factor) [125], VEGF (vascular endothelial growth factor), PDGF (platelet derived growth factor) [117,126], or the myostatin precursor promyostatin [98] and also different cytokines, like MCP-1 (Monocyte chemoattractant protein-1) [127], or SDF-1 (Stromal cell-derived factor-1, also known as CXCL12) [128]. In addition, extracellular matrix components (e.g., fibronectin), proteases, protease inhibitors are interacting partners, as well. By directly binding to fibronectin, SDC4 is involved in cell adhesion [129], thereby also influencing cell migration.

SDC4 participates in several signaling pathways and functions as a structural protein (Figure 2). The V region of the cytoplasmic domain of SDC4 also binds to PIP2 and activates PKCα [130–133]. SDC4 dimer forms a tetramer with 2 PIP2 molecules, which binds to the catalytic subunit of PKCα. The resulting activation complex is regulated by the phosphorylation of the cytoplasmic Ser179 (human Ser179, rat Ser183) of SDC4 [134], which alters the conformation of the C2 region of the cytoplasmic domain, leading to loss of PIP2 binding and consequently the lack of PKCα activation [93,135]. PKCα is a Ca^{2+}-dependent conventional PKC isoform, but its activation through SDC4 is independent of changes in intracellular Ca^{2+} levels and consequently it is active in the presence of EDTA (ethylenediaminetetraacetic acid) [81].

Moreover, the roles of SDC4 in the regulation of intracellular Ca^{2+} levels were also reported. SDC4 regulates transient receptor potential canonical (TRPCs) channels to control cytosolic Ca^{2+} equilibria, thus consequently cell behavior. SDC4 can recruit PKCα to target serine714 of TRPC7 increasing intracellular Ca^{2+} concentration with a subsequent control of the cytoskeleton in fibroblasts [136]. However, a direct interaction between SDC4 and TRPC7 has not been reported. In contrast, in podocytes, SDC4 knockdown reduced the cell surface expression of TRPC6 channel and reduced the Ca^{2+} concentration [137]. Furthermore, knocking down of SDC4 expression in HaCaT keratinocytes did not affect intracellular Ca^{2+} level, whereas silencing the expression of both SDC1 and SDC4 decreased it by modulating TRPC4 channels [136]. Moreover, the development of intracellular front-to-rear Ca^{2+} gradient is also determined by SDC4 in migrating cells [138]. Knocking down of SDC4 expression decreased cell motility and abrogated Ca^{2+} gradient and centrosome reorientation during migration [138].

SDC4 also establishes contact with the actin cytoskeleton through the biding of SDC4 cytoplasmic domain to α-actinin, a cross-linking protein between actin filaments [131]. SDC4 expression affects the nanoscale structure of the lamellipodial actin network during cell migration. SDC4 knockdown decreased the number of branches as well as the length of branches of the lamellipodial actin cytoskeleton in migrating cells [138].

6.1. Syndecan-4 and the Regulation of Rac1/RhoA Activity

SDC4 affects Rac1 activation and accumulates active Rac1 at the leading edges of migrating cells, thus ensuring the formation of membrane extensions [139,140]. The polarized distribution of active Rac1 is essential for directional cell movement. SDC4 knockout fibroblasts migrate randomly as a result of high delocalized Rac1 activity [140].

Tiam1 is a GEF acting as a specific activator for Rac1 [141]. Tiam1 is involved in essential biological processes such as cell migration [57] and cell polarization [141]. Via its relationship with the Arp2/3 complex, Tiam1 regulates actin polymerization and actin cytoskeleton rearrangement [142]. The direct interaction between SDC4 and Tiam1 has been previously demonstrated. SDC4 binds Tiam1 via C2 region of the cytoplasmic domain and the cytoplasmic Ser of SDC4 is also involved in Tiam1 binding [118]. Consequently, SDC4 regulates Tiam1 binding and Rac1 activity in a Ser179 phosphorylation-dependent manner [118]. Moreover, SDC4 also affects the expression and distribution of Tiam1 and influences the persistence of the cell movement in myoblasts [119].

SDC4-dependent binding and activation of PKCα guide PKCα activity to SDC4-regulated membrane microdomains, where PKCα can phosphorylate specific substrates locally. Regulators of the small GTPase RhoA, which facilitates focal adhesion and stress fiber assembly, are potential candidates. RhoA-GTP is necessary for signaling after SDC4 engagement at the cell surface [143], where there is an increase in GTP load and, thus, activity [144]. RhoGDIα (also known as RhoGDI1), which is considered to be phosphorylated, seems to be one of the substrates [144]. The SDC4-dependent activation of RhoA is mediated by PKCα during focal adhesion formation [144]. Moreover, SDC4-mediated Rac1 activation is also controlled by the RhoG activation pathway [145]. SDC4 clustering activates PKCα, which phosphorylates RhoGDI1 at Ser96, thereby triggering the release of RhoG and leading to polarized activation of Rac1 [145].

6.2. Syndecan-4 and Focal Adhesion Formation

The formation of α5β1 integrin-dependent focal adhesions requires SDC4 enrichment in focal adhesions [115,143,146,147]. The heparin binding domain of fibronectin binds to the HS side chains of SDC4 [146,148], thereby fibronectin forms a bridge between SDC4 and α5β1 integrins. The binding of fibronectin to HS chains of SDC4 is essential for focal adhesion formation [146–148].

During the accumulation of integrins in focal adhesions, focal adhesion kinase (FAK) is autophosphorylated at Tyr397 to serve as a binding site for Src kinase and subsequently phosphorylated on additional tyrosine side chains [149]. Because syndecan-4 regulates the phosphorylation of FAK, the phosphorylation levels of FAK Tyr397 were lower in SDC4 knockout fibroblasts [150].

PKCα activity is required for the formation of mature focal adhesions. PKCα is directly linked to β1 integrins [151]. In this way, the cytoplasmic domain of SDC4 binds to β1 integrin indirectly via PKCα [147,151]. The cytoplasmic domain of SDC4 can also bind to integrin receptors through focal adhesion proteins. The cytoplasmic domain of SDC4 interacts with paxillin through syndesmos [152], which coordinates the organization of focal adhesions. Paxillin can bind to α4 or α9 and β1 integrins directly or indirectly via other focal adhesion proteins such as vinculin and talin [153].

SDC4 interacts directly with α-actinin [130,131]. Because α-actinin binds focal adhesion proteins, such as vinculin and zyxin, the α-actinin binding serves as a link between SDC4, focal adhesions and the cytoskeleton [154]. Moreover, knocking down of SDC4 expression was reported to induce the decoupling of vinculin from F-actin filaments [155]. SDC4 has been identified as a binding partner of dynamin II GTPase via its PH domain and the interaction between dynamin II and SDC4 is important in mediating focal adhesion and stress-fiber formation [156]. Therefore, SDC4 serves as a central mediator in focal adhesion formation by bridging the interactions between integrins, fibronectin and intracellular molecules.

7. SDC4 and Tumor Cell Migration

SDC4 contributes to the development and progression of tumors by affecting cell proliferation, invasive growth, migration, metastases formation, or angiogenesis [157–159]. SDC4 functions at the cell surface as a signaling interface to affect these processes serving as a co-receptor for soluble ligands, such as growth factors and chemokines and interacting with integrins and growth factor receptors [160].

SDC4 expression is dysregulated in several tumor types, in most cases the tumor cells exhibit SDC4 overexpression [160]. However, it has been also demonstrated, that SDC4 has the potential to act as an anti-migratory/anti-invasive tumor suppressor [161]. SDC4 expression is downregulated in colon carcinoma cells [162] and it is upregulated in normal breast tissue compared to malignant breast tissue [163]. However, SDC4 is overexpressed in melanoma, liver cancer [160], ovarian carcinoma [164], mesothelioma and fibrosarcoma [165]. SDC4 has previously been linked to a high histological grade and a negative estrogen receptor status [166], implying that it may be a predictor of poor prognosis in breast cancer. SDC4-silenced breast carcinoma cells show decreased ability to form bone metastasis in mice [102] and reduced SDC4 expression is associated with reduced metastatic potential in testicular germ cell tumors [167]. Increased SDC4 expression is related to the existence of distant metastasis and increased size of the tumor mass in osteosarcoma [168], but increased patient survival in renal cell carcinoma [169].

Several studies discuss the role of SDCs in EMT. SDC1 is known to inhibit EMT in human oral cancer cells [170]. In contrast, SDC1 mediates EMT in prostate cancer [171] and the expression of SDC1 (and also SDC2) is correlated with EMT markers (E-cadherin, β-catenin) in prostate cancer [172]. SDC2 has a tumorigenic role by promoting EMT in colorectal cancer [173]. Less is known about the role of SDC4 in EMT. SDC4 is known to positively regulate TGFβ1-induced EMT (via Snail) in lung adenocarcinoma cells [174], whilst SDC4-signalling negatively regulates the production of TGFβ1 (reported in the kidneys of SDC4 KO mice) [175]. Moreover, SDC4 silencing is shown to repress EMT in papillary thyroid cancer cells [176].

SDC4 contributes to the regulation of cell motility in various cancer cell types, such as melanoma, breast cancer, lung, or cervical cancer cells (Table 1).

7.1. Melanoma

SDC4 has a tumor suppressor property in melanoma. SDC4 silencing increases the migration, whilst SDC4 overexpression decreases the migration of melanoma cells [177–180]. The tumor suppressor role of SDC4 was also shown in vivo as the overexpression of SDC4 resulted in decreased pulmonary metastatic potential and decreased lymph node metastasis of B16F10 melanoma cells in mice [179]. Similarly, it has been recently shown, that lumican, a small leucine-rich proteoglycan, inhibits in vivo metastasis formation of melanoma [181]. Moreover, syntenin-1 negatively regulates cell migration and SDC4-mediated cytoskeletal organization [179]. FGF2 is essential for the migration of M5 melanoma cells by downregulating FAK Tyr397 phosphorylation during fibronectin-mediated cell adhesion and, thereby promoting cell migration [177]. FGF2 also decreased SDC4 expression in M5 melanoma cells [177]. The matricellular protein cysteine-rich angiogenic inducer 61 (Cyr61) interacts with SDC4, activates integrins and induces metastasis formation, migration and tumorigenicity in MV3 human melanoma cells [178]. Lysophosphatidylcholine (LysoPC) C18:0 decreased the metastatic spread of murine melanoma cells, the cell membrane rigidification by LysoPC C18:0 appears to prevent the formation of focal adhesion [180], which is required for migration and tumor metastasis. Saturated LysoPC activates PKCδ to phosphorylate SDC4 thereby deactivating PKCα and reducing FAK activity [180].

7.2. Breast Cancer

The role of SDC4 in breast cancers has not clearly been understood, as we have controversial data regarding the correlation of SDC4 expression and breast cancer prognosis [160,182]. MMPs cleave the extracellular domains of SDCs, which may have a significant role in tumor progression. ADAMTS (a disintegrin and metalloproteinase with thrombospondin motifs)s, a family of secreted proteinases, are involved in the cleavage of proteoglycans. Overexpression of ADAMTSs in cancer cells might be a possible invasive mechanism in order to degrade proteoglycans [183]. ADAMTS-15 decreases the migration of MDA-MB-231 and MCF-7 breast cancer cells in association with the increased cell surface expression of SDC4 [184]. This effect of ADAMTS-15 is not linked to its metalloproteinase function [184]. Moreover, silencing of SDC4 expression rescued the effect of ADAMTS-15 on cell motility in breast cancer cells [184]. SDC4 silencing decreases EGF-mediated chemokinesis and human epidermal growth factor receptor 1 (HER1, also known as EGFR)-induced migration of MCF10A human mammary gland epithelial cells [185]. Overexpression of SDC4 decreases the invasion of breast adenocarcinoma cells into 3D collagen matrix, whilst SDC4 silencing increases the invasiveness. SDC4 inhibits cell invasion, whilst K-Ras-induced α2β1 integrin and membrane type-1 matrix metalloproteinase (MT1-MMP) promote this function. The mutational activation of K-Ras increases the expression of all these proteins suggesting a complex regulatory mechanism of tumor cell invasiveness and metastasis formation [186].

The antimicrobial peptide LL-37 promotes the migration of breast cancer cells via PI3K/AKT signaling and increases intracellular Ca^{2+} levels via Transient Receptor Potential Cation Channel Subfamily V Member 2 (TRPV2). Because the silencing of SDC4 expression decreased LL-37-induced migration and decreased Ca^{2+} influx, SDC4 is essential for both functions of LL-37 [187]. Moreover, by its GAG chains, SDC4 is crucial for LL-37 binding to the cell surface [187]. The Ca^{2+}-binding protein S100A4 and its interacting partner, Ca^{2+}-dependent protein crosslinking enzyme tissue transglutaminase (TG2), promote tumor cell migration. S100A4 directly interacts with SDC4 and increases the expression of SDC4, whilst recombinant SDC4 administration inhibits the migration of R37 rat mammary cells by competing with the cell surface SDC4 [188]. The SDC4-α5β1 integrin signaling through PKCα participates in TG2/S100A4-mediated tumor cell migration [188]. The branched peptide NT4 exhibits antagonist binding to the GAG chains of HS proteoglycans. NT4 binds SDC4, thereby target cancer cells and inhibit their migration and FGF-induced invasion [189].

Estrogen receptor signaling plays a critical role in the development and progression of hormone-dependent breast cancer. Estradiol (E2) decreases the expression of SDC4 and insulin-like growth factor receptor (IGFR) regulates the expression of SDC4 in the presence, as well as in the absence of E2 [190]. The proteoglycan lumican is known to play a role in estrogen-mediated functions of breast cancer cells, including EMT. Lumican downregulates integrin signaling (FAK, Erk1/2, AKT) [191] and inhibits EMT and the formation of lamellipodia in breast cancer cells [192].

7.3. Lung Cancer

SDC4 participates in tumor growth as the size of lung carcinoma tumors was reduced in SDC4 KO mice [193]. Increased levels of SDC4 expression were found in response to lung injury [194], as well as after tumor cell seeding [195]. Moreover, the cell surface expression of SDC4 is regulated by ADAMTS-1 via MMP9 and SDC4 (together with ADAMTS-1) inhibits migration of lung endothelial cells [126]. Cell migration is also inhibited by the interaction of SDC4 and the antifibrotic chemokine CXCL10 in primary lung fibroblasts [194]. In contrast, SDC4 promotes cell migration and invasion of A549 lung adenocarcinoma cells both in wound healing and chemotaxis assays and SDC4 positively regulates TGFβ1-mediated EMT via Snail [174]. The proteolytic shedding of SDCs leads to the release of the soluble N-terminal ectodomain from a transmembrane C-terminal fragment (tCTF). The transmembrane C-terminal fragment (tCTF) of SDC4 increased in vitro migration (examined in wound scratch assay) of SDC1-deficient A459 cells equivalently to that of SDC1 tCTF, whilst the presence of the tCTF of SDC1 was sufficient for the lung metastasis formation in vivo [196].

7.4. Other Tumor Types

SDC4 contributes to the regulation of cell migration in numerous cancer cell types and several extracellular modulators of this process are identified. The chemokine SDF-1, also known as C-X-C motif chemokine 12 (CXCL12), binds to SDC4 thereby regulating migration and invasion of choriocarcinoma cells [197]. Moreover, SDC4 is essential for CXCL12-induced migration and invasion of hepatoma cells [198] and human cervix carcinoma (HeLa) cells [199]. The extracellular calumenin decreases HeLa cell migration via SDC4 and α5β1-integrin-dependent suppression of ERK1/2 signaling [200]. Human epidermal receptor 1 (HER1), also known as epidermal growth factor receptor (EGFR), induces cell invasion of skin squamous cancer cells via SDC4-dependent activation of α6β4 integrins [185]. Silencing of SDC4 expression decreases migration and invasion of papillary thyroid cancer cells and inhibits epithelial-mesenchymal transition via Wnt/beta-catenin pathway [176]. SDC4 is also involved in the RANTES/CCL5 signaling and is necessary in RANTES/CCL5-induced invasion and migration of hepatoma cell lines [201].

Table 1. Overview of SDC4-dependent migration and SDC4 expression in different tumor cell models.

Cell Type	Migration Assay	Signaling Pathway	Biological Effect	Citation
4T1 and MDA-MB-231 breast cancer cells	-	-	SDC4 has an anti-migratory, anti-invasive tumor suppressor role.	[161]
Colon carcinoma cells	-	SDC4 expression	SDC4 is downregulated in colon carcinoma cells.	[162]
Infiltrating breast carcinoma tissues	-	SDC4 expression	SDC4 is upregulated in normal breast tissue compared to malignant breast tissue	[202]
Human ovarian carcinoma cell line NIH:OVCAR5	Modified Boyden chamber chemotaxis, Matrigel invasion assay	Carbohydrate modifications	The migration, invasion and tumor growth of ovarian carcinoma is mediated by the carbohydrate modifications of proteoglycans.SDC4 is upregulated in ovarian carcinoma.	[164]
Mesothelioma, fibrosarcoma	-	SDC4 expression	SDC4 is upregulated in mesothelioma and fibrosarcoma.	[165]

Table 1. *Cont.*

Cell Type	Migration Assay	Signaling Pathway	Biological Effect	Citation
Breast carcinoma samples from patients	-	SDC4 expression	SDC4 is associated with high histological grade and a negative estrogen receptor status in breast carcinoma.	[166]
4T1 mouse breast cancer cells	-	bone metastasis formation	SDC4-silenced breast carcinoma cells have decreased ability to form bone metastasis in mice.	[102]
JKT-1 human seminoma cell line, NTERA-2 human embryonal carcinoma cell line, NCCIT teratocarcinoma cell line	-	SDC4 expression—metastatic potential	Reduced SDC4 expression is associated with reduced metastatic potential in testicular germ cell tumors.	[167]
Patients with primary high grade intramedullary osteosarcoma, with low grade central osteosarcoma, with osteoid osteoma and normal bone tissues	-	SDC4 expression—metastasis formation, tumor size	Increased SDC4 expression is associated with the formation of distant metastasis and increased tumor size in osteosarcoma.	[168]
Renca (mouse), 786-O and Caki-2 (human) renal carcinoma cells	Wound scratch assay, Transwell assay	High SDC4 expression in renal cell carcinoma	High SDC4 expression determines increased patient survival in renal cell carcinoma.	[169]
M5 human metastatic melanoma cells	Chemotaxis assay, wound scratch assay	FGF-2/SDC4	FGF-2 regulates melanoma cell migration in a SDC4-dependent manner.	[177]
MV3 human melanoma cell line	Wound scratch assay	Cyr61/SDC4	Cyr61 is exocytosed by binding to SDC4. Cyr61 binds to and activates integrins, thus induce migration, metastasis formation and tumorigenicity.	[178]
Rat embryonic fibroblasts (REFs), A375 melanoma cells, B16F10 melanoma cells, C57BL/6 mice	Transwell migration assay, lung metastasis model	Syntenin-1/SDC4 SDC4—inhibition of cancer-associated melanoma migration	SDC4 overexpression decreases melanoma cell migration in vitro and reduces the metastatic potential of melanoma in vivo. Syntenin-1 negatively regulates SDC4-mediated inhibition of cell migration and SDC4-mediated tumor suppression in melanoma.	[179]
B16.F10 murine melanoma cells	Wound scratch assay	LysoPC/PKCδ/SDC4/PKCα/FAK	LysoPC C18:0 decreases the metastatic spread of melanoma cells. LysoPC activates PKCδ to phosphorylate SDC4 thereby deactivating PKCα and reducing FAK activity.	[180]
MDA-MB-231 and MCF7 human breast cancer cells	2D: wound scratch assay 3D: Matrigel and Collagen Type I	ADAMTS-15/SDC4	Inhibition of mammary cancer cell migration by ADAMTS-15 requires SDC4.	[184]
Human HaCat keratinocytes, A431 (human squamous skin epithelial) carcinoma cells, MCF10A (human mammary gland epithelial) cells	Wound scratch assay	HER1(EGFR)/α6β4 integrin/SDC4	HER1-dependent activation of α6β4 integrin and α6β4 integrin-mediated cell invasion require SDC4.	[185]
MDA-MB-231 breast adenocarcinoma cells	Cell invasion into 3D collagen gel	Integrin α2β1/MT1-MMP/SDCs–K-Ras mutant cell invasion	K-Ras mutant cells show increased expression of SDC1 and SDC4. MT1-MMP and α2β1 integrin promote invasive phenotype, SDCs reduce invasion into collagen matrices.	[186]
MCF7, MDA-MB-435s and MDA-MB-231 breast cancer cells	Migration chamber (insert with polyethylene filter with 8 μM pores)	LL-37/SDC4LL-37/TRPV2/ic. Ca^{2+} LL-37/PI3K/AKT/motility	SDC4 is a receptor for LL-37 increasing Ca2+ levels via TRPV2 channels and increasing the motility of breast cancer cells via PI3K/AKT signaling.	[187]

Table 1. *Cont.*

Cell Type	Migration Assay	Signaling Pathway	Biological Effect	Citation
Non-metastatic rat mammary R37 cells, highly metastatic KP1 cells (R37 cells transfected with S100A4)	Wound scratch assay	SDC4/$\alpha5\beta1$ integrin/PKCα—TG2 and S100A4-mediated cell migration	S100A4 mediates migration of tumor cells via SDC4 and $\alpha5\beta1$ integrin-mediated PKCα activation.	[188]
PANC-1 human pancreas adenocarcinoma cells, HT-29 human colon adenocarcinoma cells, MCF-7 and MDA-MB-231 human breast adenocarcinoma cells	-	NT4—SDC4	The branched peptide NT4 inhibits cancer cell migration and FGF-induced invasion. NT4 binds to SDC4, the expression of SDC4 is upregulated breast cancer cells.	[189]
MCF-7 (low metastatic ERa+), MDA-MB-231 (highly invasive ERa-) breast cancer cells	Wound scratch assay	IGFR/SDC4 expression	IGFR regulates the expression of SDC4 both in the presence and in the absence of E2 in breast cancer cells.IGFR inhibitors reduced the migration of MCF-7 cells but did not have a significant effect on MDA-MB-231 cells.	[190]
C57Bl/6 mouse primary lymphatic endothelial cells, Lewis lung carcinoma cells, bone marrow–derived DCs (BMDCs)	Transwell migration assay, in vivo migration assay (BMDCs migration into lymph node), tumor growth studies	SDC4—dendritic cell maturation	SDC4-deficient mice exhibit impaired tumor growth and increased infiltration by mature dendritic cells. SDC4 is the dominant proteoglycan on dendritic cells.	[193]
Primary lung fibroblasts	Boyden chamber, chemotaxis assay	CXCL10—SDC4	In response to lung injury, the expression of SDC4 is increased. SDC4 directly interacts with CXCL10 and they inhibit the migration of fibroblasts. SDC4 is required for the inhibitory effect of CXCL10 during fibrosis.	[194]
Human blood–derived monocytes, primary pulmonary endothelial cells, Lewis lung carcinoma cells (LLC1)	Boyden chamber, Transwell assay, spontaneous metastasis in mice	-	Increased expression of SDC4 is observed in endothelial cells after tumor cell seeding to the lungs.	[195]
Mouse lung endothelial cells	Random migration assay; ex vivo C57BL/6 mice aortic ring assay	ADAMTS-1—MMP9—SDC4	ADAMTS-1 modulates the cell surface expression of SDC4 via MMP9. ADAMTS-1 and SDC4 inhibit cell migration, whilst their inhibition increase angiogenesis.	[126]
A549 human lung adenocarcinoma cells	Wound scratch assay, transwell chemotaxis assay	SDC4/Snail/TGFβ1-induced EMT	SDC4 promotes migration and invasion of lung adenocarcinoma cells. SDC4 positively regulates TGFβ1-induced EMT (via Snail), consequently promoting a more motile phenotype.	[174]
A549 lung tumor epithelial cells	Wound scratch assay, matrigel invasion assay, in vivo lung tumor metastasis	ADAM17–SDC4 cleavage;SDC1—in vivo lung tumor metastasis	SDC1 tCFT was sufficient to induce lung metastasis formation in SCID mice, whilst SDC4 tCFT achieved as efficient wound closure as SDC1 tCFT. (tCTF = transmembrane C-terminal fragment)	[196]
JAR choriocarcinoma cells	Modified Boyden-chamber chemotactic assay	CXCL12/SDC4	SDC4 binds to CXCL12 and regulates CXCL12-mediated cell migration and invasion. SDC4 plays a role in the invasiveness of extravillous cytotrophoblast in moles.	[197]

Table 1. *Cont.*

Cell Type	Migration Assay	Signaling Pathway	Biological Effect	Citation
Huh7 human hepatoma cells	Bio-coat cell migration chambers, Matrigel invasion assay	SDF-1 (CXCL12)/CXCR4/SDC4	SDC4 is essential for SDF-1 (CXCL12) induced migration and invasion of hepatoma cells.	[198]
Human cervix epitheloid carcinoma (HeLa) cells	Bio-coat cell migration chambers, Matrigel invasion assay	SDC4–SDF-1/CXCL12–PKCδ, JNK/SAPK	SDC4 plays a role in SDF-1/CXCL12-mediated cell invasion and chemotaxis. PKCδ and c-jun NH2-terminal kinase/stress-activated protein kinase (JNK/SAPK) are involved in the SDF-1/CXCL12-induced cell invasion.	[199]
Human cervix epitheloid carcinoma (HeLa) cells	Wound scratch assay, Transwell assays	Calumenin–FN, SDC4, α5β1 integrin–ERK1/2	Calumenin inhibits cell migration and tumor metastasis through FN, SDC4 and α5β1-integrin by the suppression of ERK1/2 signaling.	[200]
Papillary thyroid cancer cells K1, BCPAP, TPC-1 and IHH-4, normal thyroid Nthy-ori3-1 cells	Transwell assay, wound scratch assay	SDC4—Wnt/β-catenin signaling pathway	SDC4-silencing decreased papillary thyroid cancer cell migration and invasion and represses EMT. Furthermore, SDC4-silencing suppresses Wnt/βcatenin signaling, thus promoting apoptosis.	[176]
Huh7, HepG2and Hep3B human hepatoma cells	Bio-coat migration chambers, Matrigel invasion assay	RANTES/CCL5—SDC4	SDC4 is essential in RANTES/CCL5-mediated hepatoma cell invasion and migration and its binding to the cell plasma membrane.	[201]

8. Syndecan-4 and Non-Cancer Cell Migration

Beyond the role of SDC4 in tumor cell migration, SDC4 was shown previously to affect migration in various non-cancerous cell types as well, including fibroblasts [140], myoblasts [119,138], endothelial cells [203], or hepatic stellate cells [204]. SDC4 may also contribute to arthritis development by affecting the migration of B-cells [205] and the pathogenesis of preeclampsia by modulating trophoblast migration [206]. SDC4 is necessary for the maturation of dendritic cells, which requires a switch in SDC expression and the elevated level of SDC4 ensures the increased motility of the cells and their relocation to the lymphoid tissues [207]. The monitoring of intestinal wound healing in SDC4 KO mice revealed that SDC4 is necessary for wound closure both in vitro and in vivo [208]. Moreover, in vivo wound healing assays of myofibroblasts indicate that SDC4 is important for the proper cardiac functions after myocardial infarction as it is a crucial mediator of granulation tissue formation thereby preventing cardiac rupture [209]. Because SDC4 KO mice also exhibit impaired angiogenesis, SDC4 may affect angiogenesis by the modulation of endothelial cell migration [210]. Administration of SDC4 proteoliposomes intensified the proliferation, migration and angiogenic tube formation of endothelial cells [211].

Shin et al. reported that SDC4 overexpression increased the migration of turkey satellite cells and increased the activation of RhoA GTPase and these phenomena required the cytoplasmic domain of SDC4 [212]. Other studies observed reduced motility after SDC4 knockdown in different cell types (hepatic stellate cells [204], lens epithelial cells [213], human umbilical vein endothelial cells (HUVECs) [210] and dendritic cells [207]), consistent with our observations [119,138], whereas high SDC4 level promoted migration [174,204,206].

9. Conclusions

The metastasis formation is a key cause of mortality and the failure of cancer therapy. For the development of metastases, the migratory ability of cancer cells is required. The identification of key molecules in cancer cell migration can open new therapeutic perspectives for successful cancer treatment. In this review, we highlighted the numerous functions of SDC4, a transmembrane proteoglycan, in cell motility and we summarized the recent knowledge about the role of SDC4 in cancer cell movement. Changes in SDC4 expression contribute to cancer growth and progression and have diagnostic and prognostic significance in numerous tumor types. SDC4 modulates several steps in the development and progression of tumors, such as uncontrolled cell proliferation, invasive growth, migration, metastases formation, angiogenesis, as well as tumor-associated inflammation.

Given the ubiquitous expression of SDC4, the summarized SDC4-mediated signaling pathways are likely applicable to several cell types. Importantly, a couple of anticancer drugs modulate SDC4 expression. Because SDC4 has multiple roles in tumor development and progression, targeting SDC4-mediated signaling may be a promising possibility for cancer treatment and drug development; however, the ubiquitous expression of SDC4 would require cancer cell specific targeting.

Author Contributions: Conceptualization, A.K.-P.; writing—original draft preparation, D.B., S.G.-N., A.K.-P.; writing—review and editing, A.K.-P., S.G.-N., D.B., L.D., L.R.; visualization, D.B., S.G.-N., A.K.-P.; supervision, A.K.-P. All authors have read and agreed to the published version of the manuscript.

Funding: This work was funded by the National Research, Development and Innovation Office of Hungary [grant numbers: GINOP-2.3.2-15-2016-00040 (MYOTeam), NKFI FK 134684 and NKFI K 132446], New National Excellence Program of the Ministry for Innovation and Technology Sciences (grant number: UNKP-20-5-SZTE-162), and was further supported by the János Bolyai Research Scholarship of the Hungarian Academy of Sciences (to A.K.-P.) and by the Szeged Scientists Academy under the sponsorship of the Hungarian Ministry of Innovation and Technology (FEIF/433-4/2020-ITM_SZERZ; to S.G.-N.). The work was also funded by EFOP 3.6.3 VEKOP-16-2017 00009 (to D.B.).

Conflicts of Interest: The authors declare no conflict of interest.

References

1. Hanahan, D.; Weinberg, R.A. The Hallmarks of Cancer. *Cell* **2000**, *100*, 57–70. [CrossRef]
2. Stuelten, C.H.; Parent, C.A.; Montell, D.J. Cell Motility in Cancer Invasion and Metastasis: Insights from Simple Model Organisms. *Nat. Rev. Cancer* **2018**, *18*, 296–312. [CrossRef]
3. Welch, D.R.; Hurst, D.R. Defining the Hallmarks of Metastasis. *Cancer Res.* **2019**, *79*, 3011–3027. [CrossRef]
4. Mittal, V. Epithelial Mesenchymal Transition in Tumor Metastasis. *Annu. Rev. Pathol.* **2018**, *13*, 395–412. [CrossRef]
5. Zhang, Y.; Weinberg, R.A. Epithelial-to-Mesenchymal Transition in Cancer: Complexity and Opportunities. *Front. Med.* **2018**, *12*, 361–373. [CrossRef]
6. Lu, W.; Kang, Y. Epithelial-Mesenchymal Plasticity in Cancer Progression and Metastasis. *Dev. Cell* **2019**, *49*, 361–374. [CrossRef]
7. Seetharaman, S.; Etienne-Manneville, S. Cytoskeletal Crosstalk in Cell Migration. *Trends Cell Biol.* **2020**, *30*, 720–735. [CrossRef] [PubMed]
8. Ribatti, D.; Tamma, R.; Annese, T. Epithelial-Mesenchymal Transition in Cancer: A Historical Overview. *Transl. Oncol.* **2020**, *13*, 100773. [CrossRef]
9. Lindsey, S.; Langhans, S.A. Crosstalk of Oncogenic Signaling Pathways During Epithelial-Mesenchymal Transition. *Front. Oncol.* **2014**, *4*, 358. [CrossRef] [PubMed]
10. Lamouille, S.; Xu, J.; Derynck, R. Molecular Mechanisms of Epithelial-Mesenchymal Transition. *Nat. Rev. Mol. Cell Biol.* **2014**, *15*, 178–196. [CrossRef]
11. Chhabra, E.S.; Higgs, H.N. The Many Faces of Actin: Matching Assembly Factors with Cellular Structures. *Nat. Cell Biol.* **2007**, *9*, 1110–1121. [CrossRef]
12. Svitkina, T. The Actin Cytoskeleton and Actin-Based Motility. *Cold Spring Harb. Perspect. Biol.* **2018**, *10*, a018267. [CrossRef]
13. Sackmann, E.; Tanaka, M. Critical Role of Lipid Membranes in Polarization and Migration of Cells: A Biophysical View. *Biophys. Rev.* **2021**, *13*, 123–138. [CrossRef]
14. Zaidel-Bar, R.; Itzkovitz, S.; Ma'ayan, A.; Iyengar, R.; Geiger, B. Functional Atlas of the Integrin Adhesome. *Nat. Cell Biol.* **2007**, *9*, 858–867. [CrossRef] [PubMed]
15. Ridley, A.J.; Schwartz, M.A.; Burridge, K.; Firtel, R.A.; Ginsberg, M.H.; Borisy, G.; Parsons, J.T.; Horwitz, A.R. Cell Migration: Integrating Signals from Front to Back. *Science* **2003**, *302*, 1704–1709. [CrossRef]
16. Petrie, R.J.; Gavara, N.; Chadwick, R.S.; Yamada, K.M. Nonpolarized Signaling Reveals Two Distinct Modes of 3d Cell Migration. *J. Cell Biol.* **2012**, *197*, 439–455. [CrossRef]
17. Franchi, M.; Masola, V.; Bellin, G.; Onisto, M.; Karamanos, K.A.; Piperigkou, Z. Collagen Fiber Array of Peritumoral Stroma Influences Epithelial-to-Mesenchymal Transition and Invasive Potential of Mammary Cancer Cells. *J. Clin. Med.* **2019**, *8*, 213. [CrossRef] [PubMed]

18. Yamada, K.M.; Collins, J.W.; Walma, D.A.C.; Doyle, A.D.; Morales, S.G.; Lu, J.; Matsumoto, K.; Nazari, S.S.; Sekiguchi, R.; Shinsato, Y.; et al. Extracellular Matrix Dynamics in Cell Migration, Invasion and Tissue Morphogenesis. *Int. J. Exp. Pathol.* **2019**, *100*, 144–152. [CrossRef] [PubMed]
19. Yamada, K.M.; Sixt, M. Mechanisms of 3d Cell Migration. *Nat. Rev. Mol. Cell Biol.* **2019**, *20*, 738–752. [CrossRef]
20. Yamaguchi, H.; Condeelis, J. Regulation of the Actin Cytoskeleton in Cancer Cell Migration and Invasion. *Biochim. Biophys. Acta* **2007**, *1773*, 642–652. [CrossRef]
21. Bachir, A.I.; Horwitz, A.R.; Nelson, W.J.; Bianchini, J.M. Actin-Based Adhesion Modules Mediate Cell Interactions with the Extracellular Matrix and Neighboring Cells. *Cold Spring Harb. Perspect. Biol.* **2017**, *9*, a023234. [CrossRef]
22. Burridge, K.; Guilluy, C. Focal Adhesions, Stress Fibers and Mechanical Tension. *Exp. Cell Res.* **2016**, *343*, 14–20. [CrossRef]
23. Kemp, J.P., Jr.; Brieher, W.M. The Actin Filament Bundling Protein Alpha-Actinin-4 Actually Suppresses Actin Stress Fibers by Permitting Actin Turnover. *J. Biol. Chem.* **2018**, *293*, 14520–14533. [CrossRef] [PubMed]
24. Byers, H.R.; White, G.E.; Fujiwara, K. Organization and Function of Stress Fibers in Cells In Vitro and In Situ. A Review. *Cell Muscle Motil.* **1984**, *5*, 83–137. [PubMed]
25. Hu, Y.L.; Li, S.; Miao, H.; Tsou, T.C.; del Pozo, M.A.; Chien, S. Roles of Microtubule Dynamics and Small Gtpase Rac in Endothelial Cell Migration and Lamellipodium Formation under Flow. *J. Vasc. Res.* **2002**, *39*, 465–476. [CrossRef] [PubMed]
26. Burnette, D.T.; Manley, S.; Sengupta, P.; Sougrat, R.; Davidson, M.W.; Kachar, B.; Lippincott-Schwartz, J. A Role for Actin Arcs in the Leading-Edge Advance of Migrating Cells. *Nat. Cell Biol.* **2011**, *13*, 371–381. [CrossRef] [PubMed]
27. Small, J.V.; Herzog, M.; Anderson, K. Actin Filament Organization in the Fish Keratocyte Lamellipodium. *J. Cell Biol.* **1995**, *129*, 1275–1286. [CrossRef]
28. Svitkina, T.M.; Verkhovsky, A.B.; McQuade, K.M.; Borisy, G.G. Analysis of the Actin-Myosin Ii System in Fish Epidermal Keratocytes: Mechanism of Cell Body Translocation. *J. Cell Biol.* **1997**, *139*, 397–415. [CrossRef]
29. Pollard, T.D.; Borisy, G.G. Cellular Motility Driven by Assembly and Disassembly of Actin Filaments. *Cell* **2003**, *112*, 453–465. [CrossRef]
30. Huxley, A.F.; Niedergerke, R. Measurement of Muscle Striations in Stretch and Contraction. *J. Physiol.* **1954**, *124*, 46–47.
31. Huxley, H.; Hanson, J. Changes in the Cross-Striations of Muscle During Contraction and Stretch and Their Structural Interpretation. *Nature* **1954**, *173*, 973–976. [CrossRef]
32. Mullins, R.D.; Heuser, J.A.; Pollard, T.D. The Interaction of Arp2/3 Complex with Actin: Nucleation, High Affinity Pointed End Capping, and Formation of Branching Networks of Filaments. *Proc. Natl. Acad. Sci. USA* **1998**, *95*, 6181–6186. [CrossRef]
33. Mullins, R.D.; Kelleher, J.F.; Xu, J.; Pollard, T.D. Arp2/3 Complex from Acanthamoeba Binds Profilin and Cross-Links Actin Filaments. *Mol. Biol. Cell* **1998**, *9*, 841–852. [CrossRef]
34. Drees, F.; Pokutta, S.; Yamada, S.; Nelson, W.J.; Weis, W.I. Alpha-Catenin Is a Molecular Switch That Binds E-Cadherin-Beta-Catenin and Regulates Actin-Filament Assembly. *Cell* **2005**, *123*, 903–915. [CrossRef]
35. Krause, M.; Dent, E.W.; Bear, J.E.; Loureiro, J.J.; Gertler, F.B. Ena/Vasp Proteins: Regulators of the Actin Cytoskeleton and Cell Migration. *Annu. Rev. Cell Dev. Biol.* **2003**, *19*, 541–564. [CrossRef] [PubMed]
36. Spiering, D.; Hodgson, L. Dynamics of the Rho-Family Small Gtpases in Actin Regulation and Motility. *Cell Adh. Migr.* **2011**, *5*, 170–180. [CrossRef] [PubMed]
37. Eriksson, J.E.; Dechat, T.; Grin, B.; Helfand, B.; Mendez, M.; Pallari, H.M.; Goldman, R.D. Introducing Intermediate Filaments: From Discovery to Disease. *J. Clin. Invest.* **2009**, *119*, 1763–1771. [CrossRef] [PubMed]
38. Chung, B.M.; Rotty, J.D.; Coulombe, P.A. Networking Galore: Intermediate Filaments and Cell Migration. *Curr. Opin. Cell Biol.* **2013**, *25*, 600–612. [CrossRef] [PubMed]
39. Leduc, C.; Etienne-Manneville, S. Intermediate Filaments in Cell Migration and Invasion: The Unusual Suspects. *Curr. Opin. Cell Biol.* **2015**, *32*, 102–112. [CrossRef]
40. Chernoivanenko, I.S.; Minin, A.A.; Minin, A.A. Role of Vimentin in Cell Migration. *Ontogenez* **2013**, *44*, 186–202. [CrossRef] [PubMed]
41. Cheng, F.; Eriksson, J.E. Intermediate Filaments and the Regulation of Cell Motility During Regeneration and Wound Healing. *Cold Spring Harb. Perspect. Biol.* **2017**, *9*, a022046. [CrossRef] [PubMed]
42. Helfand, B.T.; Mendez, M.G.; Murthy, S.N.; Shumaker, D.K.; Grin, B.; Mahammad, S.; Aebi, U.; Wedig, T.; Wu, Y.I.; Hahn, K.M.; et al. Vimentin Organization Modulates the Formation of Lamellipodia. *Mol. Biol. Cell* **2011**, *22*, 1274–1289. [CrossRef] [PubMed]
43. Sonavane, P.R.; Wang, C.; Dzamba, B.; Weber, G.F.; Periasamy, A.; DeSimone, D.W. Mechanical and Signaling Roles for Keratin Intermediate Filaments in the Assembly and Morphogenesis of Xenopus Mesendoderm Tissue at Gastrulation. *Development* **2017**, *144*, 4363–4376.
44. De Pascalis, C.; Pérez-González, C.; Seetharaman, S.; Boëda, B.; Vianay, B.; Burute, M.; Leduc, C.; Borghi, N.; Trepat, X.; Etienne-Manneville, S. Intermediate Filaments Control Collective Migration by Restricting Traction Forces and Sustaining Cell-Cell Contacts. *J. Cell Biol.* **2018**, *217*, 3031–3044. [CrossRef] [PubMed]
45. Vinogradova, T.; Miller, P.M.; Kaverina, I. Microtubule Network Asymmetry in Motile Cells: Role of Golgi-Derived Array. *Cell Cycle* **2009**, *8*, 2168–2174. [CrossRef] [PubMed]
46. Etienne-Manneville, S. Microtubules in Cell Migration. *Annu. Rev. Cell Dev. Biol.* **2013**, *29*, 471–499. [CrossRef]
47. Garcin, C.; Straube, A. Microtubules in Cell Migration. *Essays Biochem.* **2019**, *63*, 509–520.

48. Liao, G.; Mingle, L.; van de Water, L.; Liu, G. Control of Cell Migration through Mrna Localization and Local Translation. *Wiley Interdiscip. Rev. RNA* **2015**, *6*, 1–15. [CrossRef]
49. Laan, L.; Husson, J.; Munteanu, E.L.; Kerssemakers, J.W.; Dogterom, M. Force-Generation and Dynamic Instability of Microtubule Bundles. *Proc. Natl. Acad. Sci. USA* **2008**, *105*, 8920–8925. [CrossRef] [PubMed]
50. Rooney, C.; White, G.; Nazgiewicz, A.; Woodcock, S.A.; Anderson, K.I.; Ballestrem, C.; Malliri, A. The Rac Activator Stef (Tiam2) Regulates Cell Migration by Microtubule-Mediated Focal Adhesion Disassembly. *EMBO Rep.* **2010**, *11*, 292–298. [CrossRef]
51. Krylyshkina, O.; Anderson, K.I.; Kaverina, I.; Upmann, I.; Manstein, D.J.; Small, J.V.; Toomre, D.K. Nanometer Targeting of Microtubules to Focal Adhesions. *J. Cell Biol.* **2003**, *161*, 853–859. [CrossRef]
52. Letort, G.; Nedelec, F.; Blanchoin, L.; Théry, M. Centrosome Centering and Decentering by Microtubule Network Rearrangement. *Mol. Biol. Cell* **2016**, *27*, 2833–2843. [CrossRef]
53. Mostowy, S.; Cossart, P. Septins: The Fourth Component of the Cytoskeleton. *Nat. Rev. Mol. Cell Biol.* **2012**, *13*, 183–194. [CrossRef]
54. Woods, B.L.; Gladfelter, A.S. The State of the Septin Cytoskeleton from Assembly to Function. *Curr. Opin. Cell Biol.* **2021**, *68*, 105–112. [CrossRef]
55. Dolat, L.; Hunyara, J.L.; Bowen, J.R.; Karasmanis, E.P.; Elgawly, M.; Galkin, V.E.; Spiliotis, E.T. Septins Promote Stress Fiber-Mediated Maturation of Focal Adhesions and Renal Epithelial Motility. *J. Cell Biol.* **2014**, *207*, 225–235. [CrossRef] [PubMed]
56. Hall, A. Rho Gtpases and the Control of Cell Behaviour. *Biochem. Soc. Trans.* **2005**, *33*, 891–895. [CrossRef]
57. Hall, A. Rho Family Gtpases. *Biochem. Soc. Trans.* **2012**, *40*, 1378–1382. [CrossRef]
58. Hodge, R.G.; Ridley, A.J. Regulating Rho Gtpases and Their Regulators. *Nat. Rev. Mol. Cell Biol.* **2016**, *17*, 496–510. [CrossRef] [PubMed]
59. Iden, S.; Collard, J.G. Crosstalk between Small Gtpases and Polarity Proteins in Cell Polarization. *Nat. Rev. Mol. Cell Biol.* **2008**, *9*, 846–859. [CrossRef]
60. Bos, J.L.; Rehmann, H.; Wittinghofer, A. Gefs and Gaps: Critical Elements in the Control of Small G Proteins. *Cell* **2007**, *129*, 865–877. [CrossRef]
61. Nobes, C.D.; Hall, A. Rho, Rac, and Cdc42 Gtpases Regulate the Assembly of Multimolecular Focal Complexes Associated with Actin Stress Fibers, Lamellipodia, and Filopodia. *Cell* **1995**, *81*, 53–62. [CrossRef]
62. Itoh, R.E.; Kurokawa, K.; Ohba, Y.; Yoshizaki, H.; Mochizuki, N.; Matsuda, M. Activation of Rac and Cdc42 Video Imaged by Fluorescent Resonance Energy Transfer-Based Single-Molecule Probes in the Membrane of Living Cells. *Mol. Cell Biol.* **2002**, *22*, 6582–6591. [CrossRef]
63. Ridley, A.J. Rho Gtpases and Actin Dynamics in Membrane Protrusions and Vesicle Trafficking. *Trends Cell Biol.* **2006**, *16*, 522–529. [CrossRef]
64. Nobes, C.D.; Hawkins, P.; Stephens, L.; Hall, A. Activation of the Small Gtp-Binding Proteins Rho and Rac by Growth Factor Receptors. *J. Cell Sci.* **1995**, *108*, 225–233. [CrossRef]
65. Mullins, R.D. How Wasp-Family Proteins and the Arp2/3 Complex Convert Intracellular Signals into Cytoskeletal Structures. *Curr. Opin. Cell Biol.* **2000**, *12*, 91–96. [CrossRef]
66. Machesky, L.M.; Gould, K.L. The Arp2/3 Complex: A Multifunctional Actin Organizer. *Curr. Opin. Cell Biol.* **1999**, *11*, 117–121. [CrossRef]
67. Machesky, L.M.; Mullins, R.D.; Higgs, H.N.; Kaiser, D.A.; Blanchoin, L.; May, R.C.; Hall, M.E.; Pollard, T.D. Scar, a Wasp-Related Protein, Activates Nucleation of Actin Filaments by the Arp2/3 Complex. *Proc. Natl. Acad. Sci. USA* **1999**, *96*, 3739–3744. [CrossRef]
68. Rohatgi, R.; Ma, L.; Miki, H.; Lopez, M.; Kirchhausen, T.; Takenawa, T.; Kirschner, M.W. The Interaction between N-Wasp and the Arp2/3 Complex Links Cdc42-Dependent Signals to Actin Assembly. *Cell* **1999**, *97*, 221–231. [CrossRef]
69. Oleinik, N.V.; Helke, K.L.; Kistner-Griffin, E.; Krupenko, N.I.; Krupenko, S.A. Rho Gtpases Rhoa and Rac1 Mediate Effects of Dietary Folate on Metastatic Potential of A549 Cancer Cells through the Control of Cofilin Phosphorylation. *J. Biol. Chem.* **2014**, *289*, 26383–26394. [CrossRef]
70. Assemat, E.; Bazellieres, E.; Pallesi-Pocachard, E.; le Bivic, A.; Massey-Harroche, D. Polarity Complex Proteins. *Biochim. Biophys. Acta* **2008**, *1778*, 614–630. [CrossRef]
71. Insall, R.H.; Weiner, O.D. Pip3, Pip2, and Cell Movement—Similar Messages, Different Meanings? *Dev. Cell* **2001**, *1*, 743–747. [CrossRef]
72. Mack, N.A.; Georgiou, M. The Interdependence of the Rho Gtpases and Apicobasal Cell Polarity. *Small GTPases* **2014**, *5*, 10. [CrossRef]
73. Krause, M.; Gautreau, A. Steering Cell Migration: Lamellipodium Dynamics and the Regulation of Directional Persistence. *Nat. Rev. Mol. Cell Biol.* **2014**, *15*, 577–590. [CrossRef]
74. Vicente-Manzanares, M.; Webb, D.J.; Horwitz, A.R. Cell Migration at a Glance. *J. Cell Sci.* **2005**, *118*, 4917–4919. [CrossRef]
75. Zhang, J.; Wang, Y.L. Centrosome Defines the Rear of Cells During Mesenchymal Migration. *Mol. Biol. Cell* **2017**, *28*, 3240–3251. [CrossRef] [PubMed]
76. Burridge, K. Crosstalk between Rac and Rho. *Science* **1999**, *283*, 2028–2029. [CrossRef]
77. Xiang, B.; Liu, Y.; Zhao, W.; Zhao, H.; Yu, H. Extracellular Calcium Regulates the Adhesion and Migration of Osteoclasts Via Integrin Alphav Beta 3/Rho a/Cytoskeleton Signaling. *Cell Biol. Int.* **2019**, *43*, 1125–1136. [CrossRef]

78. Pegtel, D.M.; Ellenbroek, S.I.; Mertens, A.E.; van der Kammen, R.A.; de Rooij, J.; Collard, J.G. The Par-Tiam1 Complex Controls Persistent Migration by Stabilizing Microtubule-Dependent Front-Rear Polarity. *Curr. Biol.* **2007**, *17*, 1623–1634. [CrossRef]
79. Wang, S.; Watanabe, T.; Matsuzawa, K.; Katsumi, A.; Kakeno, M.; Matsui, T.; Ye, F.; Sato, K.; Murase, K.; Sugiyama, I.; et al. Tiam1 Interaction with the Par Complex Promotes Talin-Mediated Rac1 Activation During Polarized Cell Migration. *J. Cell Biol.* **2012**, *199*, 331–345. [CrossRef]
80. Pasten, C.; Cerda, J.; Jausoro, I.; Court, F.A.; Caceres, A.; Marzolo, M.P. Apoer2 and Reelin Are Expressed in Regenerating Peripheral Nerve and Regulate Schwann Cell Migration by Activating the Rac1 Gef Protein, Tiam1. *Mol. Cell Neurosci.* **2015**, *69*, 1–11. [CrossRef]
81. Oh, E.S.; Woods, A.; Couchman, J.R. Syndecan-4 Proteoglycan Regulates the Distribution and Activity of Protein Kinase C. *J. Biol. Chem.* **1997**, *272*, 8133–8136. [CrossRef]
82. Putney, J.W. Capacitative Calcium Entry: From Concept to Molecules. *Immunol. Rev.* **2009**, *231*, 10–22. [CrossRef]
83. Tsai, F.C.; Kuo, G.H.; Chang, S.W.; Tsai, P.J. Ca^{2+} Signaling in Cytoskeletal Reorganization, Cell Migration, and Cancer Metastasis. *BioMed Res. Int.* **2015**, *2015*, 409245. [CrossRef]
84. Machaca, K. Ca^{2+} Signaling and Lipid Transfer 'Pas a Deux' at Er-Pm Contact Sites Orchestrate Cell Migration. *Cell Calcium* **2020**, *89*, 102226. [CrossRef]
85. Schulman, H. Activity-Dependent Regulation of Calcium/Calmodulin-Dependent Protein Kinase Ii Localization. *J. Neurosci.* **2004**, *24*, 8399–8403. [CrossRef] [PubMed]
86. Saneyoshi, T.; Hayashi, Y. The Ca^{2+} and Rho Gtpase Signaling Pathways Underlying Activity-Dependent Actin Remodeling at Dendritic Spines. *Cytoskeleton* **2012**, *69*, 545–554. [CrossRef]
87. Bernfield, M.; Kokenyesi, R.; Kato, M.; Hinkes, M.T.; Spring, J.; Gallo, R.L.; Lose, E.J. Biology of the Syndecans: A Family of Transmembrane Heparan Sulfate Proteoglycans. *Annu. Rev. Cell Biol.* **1992**, *8*, 365–393. [CrossRef]
88. Kim, C.W.; Goldberger, O.A.; Gallo, R.L.; Bernfield, M. Members of the Syndecan Family of Heparan Sulfate Proteoglycans Are Expressed in Distinct Cell-, Tissue-, and Development-Specific Patterns. *Mol. Biol. Cell* **1994**, *5*, 797–805. [CrossRef] [PubMed]
89. Carey, D.J. Syndecans: Multifunctional Cell-Surface Co-Receptors. *Biochem. J.* **1997**, *327*, 1–16. [CrossRef]
90. Zimmermann, P.; David, G. The Syndecans, Tuners of Transmembrane Signaling. *FASEB J.* **1999**, *13*, S91–S100. [CrossRef]
91. Couchman, J.R. Syndecans: Proteoglycan Regulators of Cell-Surface Microdomains? *Nat. Rev. Mol. Cell Biol.* **2003**, *4*, 926–937. [CrossRef]
92. Couchman, J.R.; Woods, A. Syndecans, Signaling, and Cell Adhesion. *J. Cell Biochem.* **1996**, *61*, 578–584. [CrossRef]
93. Couchman, J.R.; Vogt, S.; Lim, S.T.; Lim, Y.; Oh, E.S.; Prestwich, G.D.; Theibert, A.; Lee, W.; Woods, A. Regulation of Inositol Phospholipid Binding and Signaling through Syndecan-4. *J. Biol. Chem.* **2002**, *277*, 49296–49303. [CrossRef]
94. Manon-Jensen, T.; Itoh, Y.; Couchman, J.R. Proteoglycans in Health and Disease: The Multiple Roles of Syndecan Shedding. *FEBS J.* **2010**, *277*, 3876–3889. [CrossRef]
95. Deepa, S.S.; Yamada, S.; Zako, M.; Goldberger, O.; Sugahara, K. Chondroitin Sulfate Chains on Syndecan-1 and Syndecan-4 from Normal Murine Mammary Gland Epithelial Cells Are Structurally and Functionally Distinct and Cooperate with Heparan Sulfate Chains to Bind Growth Factors. A Novel Function to Control Binding of Midkine, Pleiotrophin, and Basic Fibroblast Growth Factor. *J. Biol. Chem.* **2004**, *279*, 37368–37376.
96. Bernfield, M.; Götte, M.; Park, P.W.; Reizes, O.; Fitzgerald, M.L.; Lincecum, J.; Zako, M. Functions of Cell Surface Heparan Sulfate Proteoglycans. *Annu. Rev. Biochem.* **1999**, *68*, 729–777. [CrossRef] [PubMed]
97. Elfenbein, A.; Simons, M. Syndecan-4 Signaling at a Glance. *J. Cell Sci.* **2013**, *126*, 3799–3804. [CrossRef] [PubMed]
98. Keller-Pinter, A.; Szabo, K.; Kocsis, T.; Deak, F.; Ocsovszki, I.; Zvara, A.; Puskas, L.; Szilak, L.; Dux, L. Syndecan-4 Influences Mammalian Myoblast Proliferation by Modulating Myostatin Signalling and G1/S Transition. *FEBS Lett.* **2018**, *592*, 3139–3151. [CrossRef]
99. Wang, S.; Zhang, X.; Wang, G.; Cao, B.; Yang, H.; Jin, L.; Cui, M.; Mao, Y. Syndecan-1 Suppresses Cell Growth and Migration Via Blocking Jak1/Stat3 and Ras/Raf/Mek/Erk Pathways in Human Colorectal Carcinoma Cells. *BMC Cancer* **2019**, *19*, 1160. [CrossRef]
100. Maeda, T.; Alexander, C.M.; Friedl, A. Induction of Syndecan-1 Expression in Stromal Fibroblasts Promotes Proliferation of Human Breast Cancer Cells. *Cancer Res.* **2004**, *64*, 612–621. [CrossRef]
101. Park, H.; Kim, Y.; Lim, Y.; Han, I.; Oh, E.S. Syndecan-2 Mediates Adhesion and Proliferation of Colon Carcinoma Cells. *J. Biol. Chem.* **2002**, *277*, 29730–29736. [CrossRef] [PubMed]
102. Leblanc, R.; Sahay, D.; Houssin, A.; Machuca-Gayet, I.; Peyruchaud, O. Autotaxin-Beta Interaction with the Cell Surface Via Syndecan-4 Impacts on Cancer Cell Proliferation and Metastasis. *Oncotarget* **2018**, *9*, 33170–33185. [CrossRef]
103. Huang, W.; Chiquet-Ehrismann, R.; Moyano, J.V.; Garcia-Pardo, A.; Orend, G. Interference of Tenascin-C with Syndecan-4 Binding to Fibronectin Blocks Cell Adhesion and Stimulates Tumor Cell Proliferation. *Cancer Res.* **2001**, *61*, 8586–8594.
104. Fitzgerald, M.L.; Wang, Z.; Park, P.W.; Murphy, G.; Bernfield, M. Shedding of Syndecan-1 and -4 Ectodomains Is Regulated by Multiple Signaling Pathways and Mediated by a Timp-3-Sensitive Metalloproteinase. *J. Cell Biol.* **2000**, *148*, 811–824. [CrossRef]
105. Park, P.W.; Foster, T.J.; Nishi, E.; Duncan, S.J.; Klagsbrun, M.; Chen, Y. Activation of Syndecan-1 Ectodomain Shedding by Staphylococcus Aureus Alpha-Toxin and Beta-Toxin. *J. Biol. Chem.* **2004**, *279*, 251–258. [CrossRef]
106. Yuan, K.; Hong, T.M.; Chen, J.J.; Tsai, W.H.; Lin, M.T. Syndecan-1 up-Regulated by Ephrinb2/Ephb4 Plays Dual Roles in Inflammatory Angiogenesis. *Blood* **2004**, *104*, 1025–1033. [CrossRef]

107. Chen, D.; Adenekan, B.; Chen, L.; Vaughan, E.D.; Gerald, W.; Feng, Z.; Knudsen, B.S. Syndecan-1 Expression in Locally Invasive and Metastatic Prostate Cancer. *Urology* **2004**, *63*, 402–407. [CrossRef]
108. Letoha, T.; Keller-Pinter, A.; Kusz, E.; Kolozsi, C.; Bozso, Z.; Toth, G.; Vizler, C.; Olah, Z.; Szilak, L. Cell-Penetrating Peptide Exploited Syndecans. *Biochim. Biophys. Acta* **2010**, *1798*, 2258–2265. [CrossRef] [PubMed]
109. Rehman, Z.U.; Sjollema, K.A.; Kuipers, J.; Hoekstra, D.; Zuhorn, I.S. Nonviral Gene Delivery Vectors Use Syndecan-Dependent Transport Mechanisms in Filopodia to Reach the Cell Surface. *ACS Nano* **2012**, *6*, 7521–7532. [CrossRef]
110. Letoha, T.; Kolozsi, C.; Ekes, C.; Keller-Pinter, A.; Kusz, E.; Szakonyi, G.; Duda, E.; Szilak, L. Contribution of Syndecans to Lipoplex-Mediated Gene Delivery. *Eur. J. Pharm. Sci.* **2013**, *49*, 550–555. [CrossRef] [PubMed]
111. Adepu, S.; Rosman, C.W.; Dam, W.; van Dijk, M.C.; Navis, G.; van Goor, H.; Bakker, S.J.; van den Born, J. Incipient Renal Transplant Dysfunction Associates with Tubular Syndecan-1 Expression and Shedding. *Am. J. Physiol. Renal Physiol.* **2015**, *309*, F137–F145. [CrossRef] [PubMed]
112. Vuong, T.T.; Reine, T.M.; Sudworth, A.; Jenssen, T.G.; Kolset, S.O. Syndecan-4 Is a Major Syndecan in Primary Human Endothelial Cells in Vitro, Modulated by Inflammatory Stimuli and Involved in Wound Healing. *J. Histochem. Cytochem.* **2015**, *63*, 280–292. [CrossRef] [PubMed]
113. Choi, S.; Lee, E.; Kwon, S.; Park, H.; Yi, J.Y.; Kim, S.; Han, I.O.; Yun, Y.; Oh, E.S. Transmembrane Domain-Induced Oligomerization Is Crucial for the Functions of Syndecan-2 and Syndecan-4. *J. Biol. Chem.* **2005**, *280*, 42573–42579. [CrossRef]
114. Dews, I.C.; Mackenzie, K.R. Transmembrane Domains of the Syndecan Family of Growth Factor Coreceptors Display a Hierarchy of Homotypic and Heterotypic Interactions. *Proc. Natl. Acad. Sci. USA* **2007**, *104*, 20782–20787. [CrossRef]
115. Couchman, J.R.; Chen, L.; Woods, A. Syndecans and Cell Adhesion. *Int. Rev. Cytol.* **2001**, *207*, 113–150.
116. Granes, F.; Berndt, C.; Roy, C.; Mangeat, P.; Reina, M.; Vilaro, S. Identification of a Novel Ezrin-Binding Site in Syndecan-2 Cytoplasmic Domain. *FEBS Lett.* **2003**, *547*, 212–216. [CrossRef]
117. Multhaupt, H.A.; Yoneda, A.; Whiteford, J.R.; Oh, E.S.; Lee, W.; Couchman, J.R. Syndecan Signaling: When, Where and Why? *J. Physiol. Pharmacol.* **2009**, *60*, 31–38.
118. Keller-Pinter, A.; Ughy, B.; Domoki, M.; Pettko-Szandtner, A.; Letoha, T.; Tovari, J.; Timar, J.; Szilak, L. The Phosphomimetic Mutation of Syndecan-4 Binds and Inhibits Tiam1 Modulating Rac1 Activity in Pdz Interaction-Dependent Manner. *PLoS ONE* **2017**, *12*, e0187094. [CrossRef] [PubMed]
119. Becsky, D.; Gyulai-Nagy, S.; Balind, A.; Horvath, P.; Dux, L.; Keller-Pinter, A. Myoblast Migration and Directional Persistence Affected by Syndecan-4-Mediated Tiam-1 Expression and Distribution. *Int. J. Mol. Sci.* **2020**, *21*, 823. [CrossRef]
120. Gopal, S.; Multhaupt, H.A.B.; Pocock, R.; Couchman, J.R. Cell-Extracellular Matrix and Cell-Cell Adhesion Are Linked by Syndecan-4. *Matrix Biol.* **2017**, *60–61*, 57–69. [CrossRef]
121. Kawaguchi, Y.; Takeuchi, T.; Kuwata, K.; Chiba, J.; Hatanaka, Y.; Nakase, I.; Futaki, S. Syndecan-4 Is a Receptor for Clathrin-Mediated Endocytosis of Arginine-Rich Cell-Penetrating Peptides. *Bioconjug. Chem.* **2016**, *27*, 1119–1130. [CrossRef]
122. Bellin, R.M.; Kubicek, J.D.; Frigault, M.J.; Kamien, A.J.; Steward, R.L., Jr.; Barnes, H.M.; Digiacomo, M.B.; Duncan, L.J.; Edgerly, C.K.; Morse, E.M.; et al. Defining the Role of Syndecan-4 in Mechanotransduction Using Surface-Modification Approaches. *Proc. Natl. Acad. Sci. USA* **2009**, *106*, 22102–22107. [CrossRef]
123. Keller-Pinter, A.; Bottka, S.; Timar, J.; Kulka, J.; Katona, R.; Dux, L.; Deak, F.; Szilak, L. Syndecan-4 Promotes Cytokinesis in a Phosphorylation-Dependent Manner. *Cell. Mol. Life Sci.* **2010**, *67*, 1881–1894. [CrossRef] [PubMed]
124. Zhang, Y.; Li, J.; Partovian, C.; Sellke, F.W.; Simons, M. Syndecan-4 Modulates Basic Fibroblast Growth Factor 2 Signaling in Vivo. *Am. J. Physiol. Heart Circ. Physiol.* **2003**, *284*, H2078–H2082. [CrossRef]
125. Cornelison, D.D.; Wilcox-Adelman, S.A.; Goetinck, P.F.; Rauvala, H.; Rapraeger, A.C.; Olwin, B.B. Essential and Separable Roles for Syndecan-3 and Syndecan-4 in Skeletal Muscle Development and Regeneration. *Genes Dev.* **2004**, *18*, 2231–2236. [CrossRef] [PubMed]
126. Lambert, J.; Makin, K.; Akbareian, S.; Johnson, R.; Alghamdi, A.A.A.; Robinson, S.D.; Edwards, D.R. Adamts-1 and Syndecan-4 Intersect in the Regulation of Cell Migration and Angiogenesis. *J. Cell Sci.* **2020**, *133*, jcs235762. [CrossRef]
127. Slimani, H.; Charnaux, N.; Mbemba, E.; Saffar, L.; Vassy, R.; Vita, C.; Gattegno, L. Interaction of Rantes with Syndecan-1 and Syndecan-4 Expressed by Human Primary Macrophages. *Biochim. Biophys. Acta* **2003**, *1617*, 80–88. [CrossRef] [PubMed]
128. Charnaux, N.; Brule, S.; Hamon, M.; Chaigneau, T.; Saffar, L.; Prost, C.; Lievre, N.; Gattegno, L. Syndecan-4 Is a Signaling Molecule for Stromal Cell-Derived Factor-1 (Sdf-1)/Cxcl12. *FEBS J.* **2005**, *272*, 1937–1951. [CrossRef]
129. Tumova, S.; Woods, A.; Couchman, J.R. Heparan Sulfate Chains from Glypican and Syndecans Bind the Hep Ii Domain of Fibronectin Similarly Despite Minor Structural Differences. *J. Biol. Chem.* **2000**, *275*, 9410–9417. [CrossRef]
130. Choi, Y.; Kim, S.; Lee, J.; Ko, S.G.; Lee, W.; Han, I.O.; Woods, A.; Oh, E.S. The Oligomeric Status of Syndecan-4 Regulates Syndecan-4 Interaction with Alpha-Actinin. *Eur. J. Cell Biol.* **2008**, *87*, 807–815. [CrossRef]
131. Greene, D.K.; Tumova, S.; Couchman, J.R.; Woods, A. Syndecan-4 Associates with Alpha-Actinin. *J. Biol. Chem.* **2003**, *278*, 7617–7623. [CrossRef] [PubMed]
132. Tkachenko, E.; Simons, M. Clustering Induces Redistribution of Syndecan-4 Core Protein into Raft Membrane Domains. *J. Biol. Chem.* **2002**, *277*, 19946–19951. [CrossRef]
133. Baciu, P.C.; Goetinck, P.F. Protein Kinase C Regulates the Recruitment of Syndecan-4 into Focal Contacts. *Mol. Biol. Cell* **1995**, *6*, 1503–1513. [CrossRef] [PubMed]
134. Horowitz, A.; Simons, M. Regulation of Syndecan-4 Phosphorylation In Vivo. *J. Biol. Chem.* **1998**, *273*, 10914–10918. [CrossRef]

135. Koo, B.K.; Jung, Y.S.; Shin, J.; Han, I.; Mortier, E.; Zimmermann, P.; Whiteford, J.R.; Couchman, J.R.; Oh, E.S.; Lee, W. Structural Basis of Syndecan-4 Phosphorylation as a Molecular Switch to Regulate Signaling. *J. Mol. Biol.* **2006**, *355*, 651–663. [CrossRef] [PubMed]
136. Gopal, S.; Sogaard, P.; Multhaupt, H.A.; Pataki, C.; Okina, E.; Xian, X.; Pedersen, M.E.; Stevens, T.; Griesbeck, O.; Park, P.W.; et al. Transmembrane Proteoglycans Control Stretch-Activated Channels to Set Cytosolic Calcium Levels. *J. Cell Biol.* **2015**, *210*, 1199–1211. [CrossRef]
137. Liu, Y.; Echtermeyer, F.; Thilo, F.; Theilmeier, G.; Schmidt, A.; Schulein, R.; Jensen, B.L.; Loddenkemper, C.; Jankowski, V.; Marcussen, N.; et al. The Proteoglycan Syndecan 4 Regulates Transient Receptor Potential Canonical 6 Channels Via Rhoa/Rho-Associated Protein Kinase Signaling. *Arterioscler. Thromb. Vasc. Biol.* **2012**, *32*, 378–385. [CrossRef]
138. Becsky, D.; Szabo, K.; Gyulai-Nagy, S.; Gajdos, T.; Bartos, Z.; Balind, A.; Dux, L.; Horvath, P.; Erdelyi, M.; Homolya, L.; et al. Syndecan-4 Modulates Cell Polarity and Migration by Influencing Centrosome Positioning and Intracellular Calcium Distribution. *Front. Cell Dev. Biol.* **2020**, *8*, 575227. [CrossRef]
139. Tkachenko, E.; Elfenbein, A.; Tirziu, D.; Simons, M. Syndecan-4 Clustering Induces Cell Migration in a Pdz-Dependent Manner. *Circ. Res.* **2006**, *98*, 1398–1404. [CrossRef]
140. Bass, M.D.; Morgan, M.R.; Humphries, M.J. Integrins and Syndecan-4 Make Distinct, but Critical, Contributions to Adhesion Contact Formation. *Soft Matter* **2007**, *3*, 372–376. [CrossRef]
141. Mertens, A.E.; Pegtel, D.M.; Collard, J.G. Tiam1 Takes Part in Cell Polarity. *Trends Cell Biol.* **2006**, *16*, 308–316. [CrossRef]
142. Ten Klooster, J.P.; Evers, E.E.; Janssen, L.; Machesky, L.M.; Michiels, F.; Hordijk, P.; Collard, J.G. Interaction between Tiam1 and the Arp2/3 Complex Links Activation of Rac to Actin Polymerization. *Biochem. J.* **2006**, *397*, 39–45. [CrossRef]
143. Saoncella, S.; Echtermeyer, F.; Denhez, F.; Nowlen, J.K.; Mosher, D.F.; Robinson, S.D.; Hynes, R.O.; Goetinck, P.F. Syndecan-4 Signals Cooperatively with Integrins in a Rho-Dependent Manner in the Assembly of Focal Adhesions and Actin Stress Fibers. *Proc. Natl. Acad. Sci. USA* **1999**, *96*, 2805–2810. [CrossRef] [PubMed]
144. Dovas, A.; Yoneda, A.; Couchman, J.R. Pkcbeta-Dependent Activation of Rhoa by Syndecan-4 During Focal Adhesion Formation. *J. Cell Sci.* **2006**, *119*, 2837–2846. [CrossRef]
145. Elfenbein, A.; Rhodes, J.M.; Meller, J.; Schwartz, M.A.; Matsuda, M.; Simons, M. Suppression of Rhog Activity Is Mediated by a Syndecan 4-Synectin-Rhogdi1 Complex and Is Reversed by Pkcalpha in a Rac1 Activation Pathway. *J. Cell Biol.* **2009**, *186*, 75–83. [CrossRef]
146. Bass, M.D.; Humphries, M.J. Cytoplasmic Interactions of Syndecan-4 Orchestrate Adhesion Receptor and Growth Factor Receptor Signalling. *Biochem. J.* **2002**, *368*, 1–15. [CrossRef]
147. Mostafavi-Pour, Z.; Askari, J.A.; Parkinson, S.J.; Parker, P.J.; Ng, T.T.; Humphries, M.J. Integrin-Specific Signaling Pathways Controlling Focal Adhesion Formation and Cell Migration. *J. Cell Biol.* **2003**, *161*, 155–167. [CrossRef]
148. Woods, A.; Couchman, J.R. Syndecan-4 and Focal Adhesion Function. *Curr. Opin. Cell Biol.* **2001**, *13*, 578–583. [CrossRef]
149. Brunton, V.G.; Avizienyte, E.; Fincham, V.J.; Serrels, B.; Metcalf, C.A., 3rd; Sawyer, T.K.; Frame, M.C. Identification of Src-Specific Phosphorylation Site on Focal Adhesion Kinase: Dissection of the Role of Src Sh2 and Catalytic Functions and Their Consequences for Tumor Cell Behavior. *Cancer Res.* **2005**, *65*, 1335–1342. [CrossRef]
150. Wilcox-Adelman, S.A.; Denhez, F.; Goetinck, P.F. Syndecan-4 Modulates Focal Adhesion Kinase Phosphorylation. *J. Biol. Chem.* **2002**, *277*, 32970–32977. [CrossRef]
151. Parsons, M.; Keppler, M.D.; Kline, A.; Messent, A.; Humphries, M.J.; Gilchrist, R.; Hart, I.R.; Quittau-Prevostel, C.; Hughes, W.E.; Parker, P.J.; et al. Site-Directed Perturbation of Protein Kinase C- Integrin Interaction Blocks Carcinoma Cell Chemotaxis. *Mol. Cell Biol.* **2002**, *22*, 5897–5911. [CrossRef]
152. Denhez, F.; Wilcox-Adelman, S.A.; Baciu, P.C.; Saoncella, S.; Lee, S.; French, B.; Neveu, W.; Goetinck, P.F. Syndesmos, a Syndecan-4 Cytoplasmic Domain Interactor, Binds to the Focal Adhesion Adaptor Proteins Paxillin and Hic-5. *J. Biol. Chem.* **2002**, *277*, 12270–12274. [CrossRef]
153. Brown, M.C.; Turner, C.E. Paxillin: Adapting to Change. *Physiol. Rev.* **2004**, *84*, 1315–1339. [CrossRef]
154. Pataki, C.A.; Couchman, J.R.; Brábek, J. Wnt Signaling Cascades and the Roles of Syndecan Proteoglycans. *J. Histochem. Cytochem.* **2015**, *63*, 465–480. [CrossRef]
155. Cavalheiro, R.P.; Lima, M.A.; Jarrouge-Boucas, T.R.; Viana, G.M.; Lopes, C.C.; Coulson-Thomas, V.J.; Dreyfuss, J.L.; Yates, E.A.; Tersariol, I.L.S.; Nader, H.B. Coupling of Vinculin to F-Actin Demands Syndecan-4 Proteoglycan. *Matrix Biol.* **2017**, *63*, 23–37. [CrossRef]
156. Yoo, J.; Jeong, M.J.; Cho, H.J.; Oh, E.S.; Han, M.Y. Dynamin Ii Interacts with Syndecan-4, a Regulator of Focal Adhesion and Stress-Fiber Formation. *Biochem. Biophys. Res. Commun.* **2005**, *328*, 424–431. [CrossRef]
157. Yip, G.W.; Smollich, M.; Gotte, M. Therapeutic Value of Glycosaminoglycans in Cancer. *Mol. Cancer Ther.* **2006**, *5*, 2139–2148. [CrossRef]
158. Espinoza-Sanchez, N.A.; Gotte, M. Role of Cell Surface Proteoglycans in Cancer Immunotherapy. *Semin. Cancer Biol.* **2020**, *62*, 48–67. [CrossRef]
159. Hassan, N.; Greve, B.; Espinoza-Sanchez, N.A.; Gotte, M. Cell-Surface Heparan Sulfate Proteoglycans as Multifunctional Integrators of Signaling in Cancer. *Cell. Signal.* **2021**, *77*, 109822. [CrossRef]
160. Onyeisi, J.O.S.; Lopes, C.C.; Gotte, M. Syndecan-4 as a Pathogenesis Factor and Therapeutic Target in Cancer. *Biomolecules* **2021**, *11*, 503. [CrossRef]

161. Liao, W.C.; Yen, H.R.; Chen, C.H.; Chu, Y.H.; Song, Y.C.; Tseng, T.J.; Liu, C.H. Chpf Promotes Malignancy of Breast Cancer Cells by Modifying Syndecan-4 and the Tumor Microenvironment. *Am. J. Cancer Res.* **2021**, *11*, 812–826.
162. Jayson, G.C.; Vives, C.; Paraskeva, C.; Schofield, K.; Coutts, J.; Fleetwood, A.; Gallagher, J.T. Coordinated Modulation of the Fibroblast Growth Factor Dual Receptor Mechanism During Transformation from Human Colon Adenoma to Carcinoma. *Int. J. Cancer* **1999**, *82*, 298–304. [CrossRef]
163. Warner, T.F.; Wrone, D.A.; Williams, E.C.; Cripps, D.J.; Mundhenke, C.; Friedl, A. Heparan Sulphate Proteoglycan in Scleromyxedema Promotes Fgf-2 Activity. *Pathol. Res. Pract.* **2002**, *198*, 701–707. [CrossRef]
164. Casey, R.C.; Oegema, T.R., Jr.; Skubitz, K.M.; Pambuccian, S.E.; Grindle, S.M.; Skubitz, A.P. Cell Membrane Glycosylation Mediates the Adhesion, Migration, and Invasion of Ovarian Carcinoma Cells. *Clin. Exp. Metastasis* **2003**, *20*, 143–152. [CrossRef] [PubMed]
165. Szatmari, T.; Dobra, K. The Role of Syndecan-1 in Cellular Signaling and Its Effects on Heparan Sulfate Biosynthesis in Mesenchymal Tumors. *Front. Oncol.* **2013**, *3*, 310. [CrossRef]
166. Baba, F.; Swartz, K.; van Buren, R.; Eickhoff, J.; Zhang, Y.; Wolberg, W.; Friedl, A. Syndecan-1 and Syndecan-4 Are Overexpressed in an Estrogen Receptor-Negative, Highly Proliferative Breast Carcinoma Subtype. *Breast Cancer Res. Treat.* **2006**, *98*, 91–98. [CrossRef] [PubMed]
167. Labropoulou, V.T.; Skandalis, S.S.; Ravazoula, P.; Perimenis, P.; Karamanos, N.K.; Kalofonos, H.P.; Theocharis, A.D. Expression of Syndecan-4 and Correlation with Metastatic Potential in Testicular Germ Cell Tumours. *BioMed Res. Int.* **2013**, *2013*, 214864. [CrossRef] [PubMed]
168. Na, K.Y.; Bacchini, P.; Bertoni, F.; Kim, Y.W.; Park, Y.K. Syndecan-4 and Fibronectin in Osteosarcoma. *Pathology* **2012**, *44*, 325–330. [CrossRef] [PubMed]
169. Majo, S.; Courtois, S.; Souleyreau, W.; Bikfalvi, A.; Auguste, P. Impact of Extracellular Matrix Components to Renal Cell Carcinoma Behavior. *Front. Oncol.* **2020**, *10*, 625. [CrossRef]
170. Wang, X.; He, J.; Zhao, X.; Qi, T.; Zhang, T.; Kong, C. Syndecan-1 Suppresses Epithelial-Mesenchymal Transition and Migration in Human Oral Cancer Cells. *Oncol. Rep.* **2018**, *39*, 1835–1842. [CrossRef]
171. Fujii, T.; Shimada, K.; Tatsumi, Y.; Tanaka, N.; Fujimoto, K.; Konishi, N. Syndecan-1 up-Regulates Microrna-331-3p and Mediates Epithelial-to-Mesenchymal Transition in Prostate Cancer. *Mol. Carcinog.* **2016**, *55*, 1378–1386. [CrossRef]
172. Contreras, H.R.; Ledezma, R.A.; Vergara, J.; Cifuentes, F.; Barra, C.; Cabello, P.; Gallegos, I.; Morales, B.; Huidobro, C.; Castellón, E.A. The Expression of Syndecan-1 and -2 Is Associated with Gleason Score and Epithelial-Mesenchymal Transition Markers, E-Cadherin and Beta-Catenin, in Prostate Cancer. *Urol. Oncol.* **2010**, *28*, 534–540. [CrossRef]
173. Hua, R.; Yu, J.; Yan, X.; Ni, Q.; Zhi, X.; Li, X.; Jiang, B.; Zhu, J. Syndecan-2 in Colorectal Cancer Plays Oncogenic Role Via Epithelial-Mesenchymal Transition and Mapk Pathway. *Biomed. Pharmacother.* **2020**, *121*, 109630. [CrossRef] [PubMed]
174. Toba-Ichihashi, Y.; Yamaoka, T.; Ohmori, T.; Ohba, M. Up-Regulation of Syndecan-4 Contributes to Tgf-Beta1-Induced Epithelial to Mesenchymal Transition in Lung Adenocarcinoma A549 Cells. *Biochem. Biophys. Rep.* **2016**, *5*, 1–7. [PubMed]
175. Cevikbas, F.; Schaefer, L.; Uhlig, P.; Robenek, H.; Theilmeier, G.; Echtermeyer, F.; Bruckner, P. Unilateral Nephrectomy Leads to up-Regulation of Syndecan-2- and Tgf-Beta-Mediated Glomerulosclerosis in Syndecan-4 Deficient Male Mice. *Matrix Biol.* **2008**, *27*, 42–52. [CrossRef] [PubMed]
176. Chen, L.L.; Gao, G.X.; Shen, F.X.; Chen, X.; Gong, X.H.; Wu, W.J. Sdc4 Gene Silencing Favors Human Papillary Thyroid Carcinoma Cell Apoptosis and Inhibits Epithelial Mesenchymal Transition Via Wnt/Beta-Catenin Pathway. *Mol. Cells* **2018**, *41*, 853–867.
177. Chalkiadaki, G.; Nikitovic, D.; Berdiaki, A.; Sifaki, M.; Krasagakis, K.; Katonis, P.; Karamanos, N.K.; Tzanakakis, G.N. Fibroblast Growth Factor-2 Modulates Melanoma Adhesion and Migration through a Syndecan-4-Dependent Mechanism. *Int. J. Biochem. Cell Biol.* **2009**, *41*, 1323–1331. [CrossRef] [PubMed]
178. Schmitz, P.; Gerber, U.; Schutze, N.; Jungel, E.; Blaheta, R.; Naggi, A.; Torri, G.; Bendas, G. Cyr61 Is a Target for Heparin in Reducing Mv3 Melanoma Cell Adhesion and Migration Via the Integrin Vla-4. *Thromb. Haemost.* **2013**, *110*, 1046–1054.
179. Choi, Y.; Yun, J.H.; Yoo, J.; Lee, I.; Kim, H.; Son, H.N.; Kim, I.S.; Yoon, H.S.; Zimmermann, P.; Couchman, J.R.; et al. New Structural Insight of C-Terminal Region of Syntenin-1, Enhancing the Molecular Dimerization and Inhibitory Function Related on Syndecan-4 Signaling. *Sci. Rep.* **2016**, *6*, 36818. [CrossRef]
180. Ross, T.; Jakubzig, B.; Grundmann, M.; Massing, U.; Kostenis, E.; Schlesinger, M.; Bendas, G. The Molecular Mechanism by Which Saturated Lysophosphatidylcholine Attenuates the Metastatic Capacity of Melanoma Cells. *FEBS Open Bio* **2016**, *6*, 1297–1309. [CrossRef]
181. Karamanou, K.; Franchi, M.; Proult, I.; Rivet, R.; Vynios, D.; Brézillon, S. Lumican Inhibits in Vivo Melanoma Metastasis by Altering Matrix-Effectors and Invadopodia Markers. *Cells* **2021**, *10*, 841. [CrossRef] [PubMed]
182. Afratis, N.A.; Karamanou, K.; Piperigkou, Z.; Vynios, D.H.; Theocharis, A.D. The Role of Heparins and Nano-Heparins as Therapeutic Tool in Breast Cancer. *Glycoconj. J.* **2017**, *34*, 299–307. [CrossRef]
183. Filou, S.; Korpetinou, A.; Kyriakopoulou, D.; Bounias, D.; Stavropoulos, M.; Ravazoula, P.; Papachristou, D.J.; Theocharis, A.D.; Vynios, D.H. Adamts Expression in Colorectal Cancer. *PLoS ONE* **2015**, *10*, e0121209. [CrossRef]
184. Kelwick, R.; Wagstaff, L.; Decock, J.; Roghi, C.; Cooley, L.S.; Robinson, S.D.; Arnold, H.; Gavrilovic, J.; Jaworski, D.M.; Yamamoto, K.; et al. Metalloproteinase-Dependent and -Independent Processes Contribute to Inhibition of Breast Cancer Cell Migration, Angiogenesis and Liver Metastasis by a Disintegrin and Metalloproteinase with Thrombospondin Motifs-15. *Int. J. Cancer* **2015**, *136*, E14–E26. [CrossRef]

185. Wang, H.; Jin, H.; Beauvais, D.M.; Rapraeger, A.C. Cytoplasmic Domain Interactions of Syndecan-1 and Syndecan-4 with Alpha6beta4 Integrin Mediate Human Epidermal Growth Factor Receptor (Her1 and Her2)-Dependent Motility and Survival. *J. Biol. Chem.* **2014**, *289*, 30318–30332. [CrossRef] [PubMed]
186. Vuoriluoto, K.; Hognas, G.; Meller, P.; Lehti, K.; Ivaska, J. Syndecan-1 and -4 Differentially Regulate Oncogenic K-Ras Dependent Cell Invasion into Collagen through Alpha2beta1 Integrin and Mt1-Mmp. *Matrix Biol.* **2011**, *30*, 207–217. [CrossRef]
187. Habes, C.; Weber, G.; Goupille, C. Sulfated Glycoaminoglycans and Proteoglycan Syndecan-4 Are Involved in Membrane Fixation of Ll-37 and Its Pro-Migratory Effect in Breast Cancer Cells. *Biomolecules* **2019**, *9*, 481. [CrossRef]
188. Wang, Z.; Griffin, M. The Role of Tg2 in Regulating S100a4-Mediated Mammary Tumour Cell Migration. *PLoS ONE* **2013**, *8*, e57017. [CrossRef] [PubMed]
189. Brunetti, J.; Riolo, G.; Depau, L.; Mandarini, E.; Bernini, A.; Karousou, E.; Passi, A.; Pini, A.; Bracci, L.; Falciani, C. Unraveling Heparan Sulfate Proteoglycan Binding Motif for Cancer Cell Selectivity. *Front. Oncol.* **2019**, *9*, 843. [CrossRef] [PubMed]
190. Tsonis, A.I.; Afratis, N.; Gialeli, C.; Ellina, M.I.; Piperigkou, Z.; Skandalis, S.S.; Theocharis, A.D.; Tzanakakis, G.N.; Karamanos, N.K. Evaluation of the Coordinated Actions of Estrogen Receptors with Epidermal Growth Factor Receptor and Insulin-Like Growth Factor Receptor in the Expression of Cell Surface Heparan Sulfate Proteoglycans and Cell Motility in Breast Cancer Cells. *FEBS J.* **2013**, *280*, 2248–2259. [CrossRef]
191. Karamanou, K.; Franchi, M.; Onisto, M.; Passi, A.; Vynios, D.H.; Brézillon, S. Evaluation of Lumican Effects on Morphology of Invading Breast Cancer Cells, Expression of Integrins and Downstream Signaling. *FEBS J.* **2020**, *287*, 4862–4880. [CrossRef]
192. Karamanou, K.; Franchi, M.; Vynios, D.; Brézillon, S. Epithelial-to-Mesenchymal Transition and Invadopodia Markers in Breast Cancer: Lumican a Key Regulator. *Semin. Cancer Biol.* **2020**, *62*, 125–133. [CrossRef]
193. El Ghazal, R.; Yin, X.; Johns, S.C.; Swanson, L.; Macal, M.; Ghosh, P.; Zuniga, E.I.; Fuster, M.M. Glycan Sulfation Modulates Dendritic Cell Biology and Tumor Growth. *Neoplasia* **2016**, *18*, 294–306. [CrossRef] [PubMed]
194. Jiang, D.; Liang, J.; Campanella, G.S.; Guo, R.; Yu, S.; Xie, T.; Liu, N.; Jung, Y.; Homer, R.; Meltzer, E.B.; et al. Inhibition of Pulmonary Fibrosis in Mice by Cxcl10 Requires Glycosaminoglycan Binding and Syndecan-4. *J. Clin. Invest.* **2010**, *120*, 2049–2057. [CrossRef]
195. Roblek, M.; Strutzmann, E.; Zankl, C.; Adage, T.; Heikenwalder, M.; Atlic, A.; Weis, R.; Kungl, A.; Borsig, L. Targeting of Ccl2-Ccr2-Glycosaminoglycan Axis Using a Ccl2 Decoy Protein Attenuates Metastasis through Inhibition of Tumor Cell Seeding. *Neoplasia* **2016**, *18*, 49–59. [CrossRef] [PubMed]
196. Pasqualon, T.; Pruessmeyer, J.; Weidenfeld, S.; Babendreyer, A.; Groth, E.; Schumacher, J.; Schwarz, N.; Denecke, B.; Jahr, H.; Zimmermann, P.; et al. A Transmembrane C-Terminal Fragment of Syndecan-1 Is Generated by the Metalloproteinase Adam17 and Promotes Lung Epithelial Tumor Cell Migration and Lung Metastasis Formation. *Cell Mol. Life Sci.* **2015**, *72*, 3783–3801. [CrossRef] [PubMed]
197. Schanz, A.; Baston-Bust, D.; Krussel, J.S.; Heiss, C.; Janni, W.; Hess, A.P. Cxcr7 and Syndecan-4 Are Potential Receptors for Cxcl12 in Human Cytotrophoblasts. *J. Reprod. Immunol.* **2011**, *89*, 18–25. [CrossRef]
198. Sutton, A.; Friand, V.; Brule-Donneger, S.; Chaigneau, T.; Ziol, M.; Sainte-Catherine, O.; Poire, A.; Saffar, L.; Kraemer, M.; Vassy, J.; et al. Stromal Cell-Derived Factor-1/Chemokine (C-X-C Motif) Ligand 12 Stimulates Human Hepatoma Cell Growth, Migration, and Invasion. *Mol. Cancer Res.* **2007**, *5*, 21–33. [CrossRef]
199. Brule, S.; Friand, V.; Sutton, A.; Baleux, F.; Gattegno, L.; Charnaux, N. Glycosaminoglycans and Syndecan-4 Are Involved in Sdf-1/Cxcl12-Mediated Invasion of Human Epitheloid Carcinoma Hela Cells. *Biochim. Biophys. Acta* **2009**, *1790*, 1643–1650. [CrossRef]
200. Wang, Q.; Shen, B.; Chen, L.; Zheng, P.; Feng, H.; Hao, Q.; Liu, X.; Liu, L.; Xu, S.; Chen, J.; et al. Extracellular Calumenin Suppresses Erk1/2 Signaling and Cell Migration by Protecting Fibulin-1 from Mmp-13-Mediated Proteolysis. *Oncogene* **2015**, *34*, 1006–1018. [CrossRef]
201. Charni, F.; Friand, V.; Haddad, O.; Hlawaty, H.; Martin, L.; Vassy, R.; Oudar, O.; Gattegno, L.; Charnaux, N.; Sutton, A. Syndecan-1 and Syndecan-4 Are Involved in Rantes/Ccl5-Induced Migration and Invasion of Human Hepatoma Cells. *Biochim. Biophys. Acta* **2009**, *1790*, 1314–1326. [CrossRef]
202. Mundhenke, C.; Meyer, K.; Drew, S.; Friedl, A. Heparan Sulfate Proteoglycans as Regulators of Fibroblast Growth Factor-2 Receptor Binding in Breast Carcinomas. *Am. J. Pathol.* **2002**, *160*, 185–194. [CrossRef]
203. Chaudhuri, P.; Colles, S.M.; Fox, P.L.; Graham, L.M. Protein Kinase Cdelta-Dependent Phosphorylation of Syndecan-4 Regulates Cell Migration. *Circ. Res.* **2005**, *97*, 674–681. [CrossRef]
204. Yin, L.; Qi, Y.; Xu, Y.; Xu, L.; Han, X.; Tao, X.; Song, S.; Peng, J. Dioscin Inhibits Hsc-T6 Cell Migration Via Adjusting Sdc-4 Expression: Insights from Itraq-Based Quantitative Proteomics. *Front. Pharmacol.* **2017**, *8*, 665. [CrossRef]
205. Endo, T.; Ito, K.; Morimoto, J.; Kanayama, M.; Ota, D.; Ikesue, M.; Kon, S.; Takahashi, D.; Onodera, T.; Iwasaki, N.; et al. Syndecan 4 Regulation of the Development of Autoimmune Arthritis in Mice by Modulating B Cell Migration and Germinal Center Formation. *Arthritis Rheumatol.* **2015**, *67*, 2512–2522. [CrossRef] [PubMed]
206. Jeyarajah, M.J.; Bhattad, G.J.; Kops, B.F.; Renaud, S.J. Syndecan-4 Regulates Extravillous Trophoblast Migration by Coordinating Protein Kinase C Activation. *Sci. Rep.* **2019**, *9*, 10175. [CrossRef]
207. Averbeck, M.; Gebhardt, C.; Anderegg, U.; Termeer, C.; Sleeman, J.P.; Simon, J.C. Switch in Syndecan-1 and Syndecan-4 Expression Controls Maturation Associated Dendritic Cell Motility. *Exp. Dermatol.* **2007**, *16*, 580–589. [CrossRef]

208. Frohling, M.; Tepasse, P.; Intemann, J.; Sambale, M.; Sherwood, J.; Paruzel, P.; Tiemeyer, N.M.; Nowacki, T.M.; Bruckner, M.; Mennigen, R.; et al. Syndecan-4 Modulates Epithelial Gut Barrier Function and Epithelial Regeneration in Experimental Colitis. *Inflamm. Bowel Dis.* **2018**, *24*, 2579–2589. [CrossRef] [PubMed]
209. Matsui, Y.; Ikesue, M.; Danzaki, K.; Morimoto, J.; Sato, M.; Tanaka, S.; Kojima, T.; Tsutsui, H.; Uede, T. Syndecan-4 Prevents Cardiac Rupture and Dysfunction after Myocardial Infarction. *Circ. Res.* **2011**, *108*, 1328–1339. [CrossRef]
210. Li, R.; Wu, H.; Xie, J.; Li, G.; Gu, R.; Kang, L.; Wang, L.; Xu, B. Syndecan-4 Regulates the Bfgf-Induced Chemotactic Migration of Endothelial Cells. *J. Mol. Histol.* **2016**, *47*, 503–509. [CrossRef]
211. Jang, E.; Albadawi, H.; Watkins, M.T.; Edelman, E.R.; Baker, A.B. Syndecan-4 Proteoliposomes Enhance Fibroblast Growth Factor-2 (Fgf-2)-Induced Proliferation, Migration, and Neovascularization of Ischemic Muscle. *Proc. Natl. Acad. Sci. USA* **2012**, *109*, 1679–1684. [CrossRef] [PubMed]
212. Shin, J.; McFarland, D.C.; Velleman, S.G. Migration of Turkey Muscle Satellite Cells Is Enhanced by the Syndecan-4 Cytoplasmic Domain through the Activation of Rhoa. *Mol. Cell. Biochem.* **2013**, *375*, 115–130. [CrossRef]
213. Qin, Y.; Zhu, Y.; Luo, F.; Chen, C.; Chen, X.; Wu, M. Killing Two Birds with One Stone: Dual Blockade of Integrin and Fgf Signaling through Targeting Syndecan-4 in Postoperative Capsular Opacification. *Cell Death Dis.* **2017**, *8*, e2920. [CrossRef] [PubMed]

cancers

MDPI

Article

LRG1 Promotes Metastatic Dissemination of Melanoma through Regulating EGFR/STAT3 Signalling

Yuet Ping Kwan [1,2], Melissa Hui Yen Teo [1,2], Jonathan Chee Woei Lim [3], Michelle Siying Tan [4], Graciella Rosellinny [1,2], Walter Wahli [5,6,7] and Xiaomeng Wang [1,2,8,*]

1 Centre for Vision Research, Duke NUS Medical School, 8 College Road, Singapore 169857, Singapore; joyce.kwan@duke-nus.edu.sg (Y.P.K.); melissateo@duke-nus.edu.sg (M.H.Y.T.); graciella.rosellinny@duke-nus.edu.sg (G.R.)
2 Singapore Eye Research Institute (SERI) The Academia, 20 College Road, Level 6 Discovery Tower, Singapore 169856, Singapore
3 Pharmacotherapeutics Unit, Department of Medicine, Faculty of Medicine and Health Sciences, Universiti Putra Malaysia, Serdang 43400, Selangor, Malaysia; cheewoei@upm.edu.my
4 Department of Surgery, Yong Yoo Lin School of Medicine, National University of Singapore, MD6, 14 Medical Drive, Singapore 117599, Singapore; surmts@nus.edu.sg
5 Center for Integrative Genomics, Université de Lausanne, Le Génopode, CH-1015 Lausanne, Switzerland; walter.wahli@unil.ch
6 Lee Kong Chian School of Medicine, Nanyang Technological University Singapore, Clinical Sciences Building, 11 Mandalay Road, Singapore 308232, Singapore
7 Toxalim (Research Center in Food Toxicology), INRAE, ENVT, INP-PURPAN, UMR 1331, UPS, Université de Toulouse, F-31027 Toulouse, France
8 Institute of Molecular and Cell Biology (IMCB), Agency for Science, Technology and Research (A*STAR), Proteos, 61 Biopolis Dr, Singapore 138673, Singapore
* Correspondence: xiaomeng.wang@duke-nus.edu.sg

Simple Summary: Melanoma is a highly metastatic and lethal form of skin cancer. Currently, there is no effective treatment available once melanoma cells spread to other parts of the body. Our study demonstrated that LRG1 regulates multiple aspects of melanoma metastasis through modulating EGFR/STAT3 signalling. Targeting LRG1 may offer an alternative way to control the metastatic spread of melanoma cells.

Abstract: Although less common, melanoma is the deadliest form of skin cancer largely due to its highly metastatic nature. Currently, there are limited treatment options for metastatic melanoma and many of them could cause serious side effects. A better understanding of the molecular mechanisms underlying the complex disease pathophysiology of metastatic melanoma may lead to the identification of novel therapeutic targets and facilitate the development of targeted therapeutics. In this study, we investigated the role of leucine-rich α-2-glycoprotein 1 (LRG1) in melanoma development and progression. We first established the association between LRG1 and melanoma in both human patient biopsies and mouse melanoma cell lines and revealed a significant induction of LRG1 expression in metastatic melanoma cells. We then showed no change in tumour cell growth, proliferation, and angiogenesis in the absence of the host *Lrg1*. On the other hand, there was reduced melanoma cell metastasis to the lungs in *Lrg1*-deficient mice. This observation was supported by the promoting effect of LRG1 in melanoma cell migration, invasion, and adhesion. Mechanistically, LRG1 mediates melanoma cell invasiveness in an EGFR/STAT3-dependent manner. Taken together, our studies provided compelling evidence that LRG1 is required for melanoma metastasis but not growth. Targeting LRG1 may offer an alternative strategy to control malignant melanoma.

Keywords: leucine-rich α-2-glycoprotein-1; melanoma; metastasis; STAT3; EGFR

check for updates

Citation: Kwan, Y.P.; Teo, M.H.Y.; Lim, J.C.W.; Tan, M.S.; Rosellinny, G.; Wahli, W.; Wang, X. LRG1 Promotes Metastatic Dissemination of Melanoma through Regulating EGFR/STAT3 Signalling. *Cancers* **2021**, *13*, 3279. https://doi.org/10.3390/cancers13133279

Academic Editors: José I. López and Ildefonso M. de la Fuente

Received: 28 May 2021
Accepted: 23 June 2021
Published: 30 June 2021

1. Introduction

Malignant melanoma is the most aggressive form of skin cancer [1,2] and the death rate of melanoma skin cancer patients is significantly higher than that of non-melanoma skin cancer patients [3]. Furthermore, the WHO predicts that death from melanoma will increase to 20% by 2025 [4]. Owing to the increase in life expectancy, ozone layer depletion, and low awareness of sun exposure, the number of death cases may escalate by 74% in 2040 [5]. Currently, there are only limited treatment options available for metastatic melanoma. Skin and gastrointestinal toxicity, as well as reduced efficacy due to resistance to immune and chemotherapies, are key challenges facing current melanoma drugs [6]. Therefore, there is an urgent need to identify novel therapeutic targets to improve the survival and quality of life of patients suffering from metastatic melanoma.

Our earlier study led to the identification of leucine-rich alpha 2 glycoprotein 1 (LRG1), a novel regulator of TGFβ1 signalling [7]. In endothelial cells (EC), LRG1 promotes angiogenesis by tipping the balance of TGFβ1 signalling toward the ALK1/Smad1,5,8 pathway, which is dependent on the presence of the type III TGF β1 receptor endoglin [7]. Besides its role in retinal angiogenesis [7], LRG1 has been linked to abnormal angiogenesis in glomerular [8], ischemic brain [9], cornea [10], and diabetic wounds [11]. Neovascularization plays an essential role in tumour expansion [12] and tumour vasculature provides a route of transportation for tumour cell dissemination [13]. Indeed, altered LRG1 expression is associated with various cancers and LRG1 has been proposed as a prognosis/diagnosis marker for hepatocellular carcinoma, gastric cancer, pancreatic cancer, leukaemia, ovarian cancer, breast cancer, prostate cancer, biliary tract cancer, bladder cancer, and non-small cell lung cancer [14–21]. LRG1 regulates tumour angiogenesis by inducing VEGFA through HIF1α activation [22]. Recently, angiogenesis-independent roles of LRG1 in tumourigenesis have also been reported. For instance, LRG1 regulates epithelial–mesenchymal transition (EMT) by activating RUNX1 [23] and TGFβ/Smad2,3 signalling [24–26]. However, the biological function of LRG1 in melanoma development and progression remains to be elucidated.

In the present study, we first investigated the expression of LRG1 in human melanoma biopsies and murine melanoma cell lines and established an association between LRG1 and melanoma. We then determined the role of LRG1 in melanoma growth in a tumour xenograft model as well as in vitro cell-based tumour cell viability and proliferation assays. Despite the important role of LRG1 in angiogenesis, there was no change in tumour angiogenesis and tumour growth in Lrg1-deficient mice. Instead, we found that LRG1 promotes melanoma cell migration, invasion, adhesion, and lung metastasis. Furthermore, both tumour cell and endothelial cell-derived LRG1 were important for tumour cell extravasation. Mechanistically, LRG1 exerts its function by regulating EGFR/STAT3 signalling, a central pathway involved in tumour metastasis. Stattic, a Stat3 inhibitor, completely abolishes the promoting effects of LRG1 on tumour cell activation. Targeting LRG1, therefore, may offer an alternative strategy to control STAT3-mediated melanoma metastasis.

2. Materials and Methods

2.1. Mice

Lrg1$^{-/-}$ mice used in this study were originally generated by the University of California Davies Knockout Mouse Project (KOMP) repository (https://www.mmrrc.org/catalog/sds.php?mmrrc_id=48463) and were a generous gift from Professors John Greenwood and Steven Moss at UCL Institute of Ophthalmology. Animal care and procedures were performed under the guidelines of the Institutional Animal Care and Use Committee (IACUC, Protocol number: A0269) of the Nanyang Technological University in Singapore and the Guide for Care and Use of Laboratory Animals from the US National Institutes of Health. All mice were housed in an environmentally controlled room (22 °C, 40–60% humidity, and a 12-h light cycle).

2.2. Tissue Microarray and Immunohistochemistry

Human skin cancer and normal tissue arrays (cat#SK721) were purchased from US Biomax (Rockville, MD, USA). The paraffin-embedded slides were deparaffinized and rehydrated before being subjected to antigen retrieval in a 10 mM citrate buffer (pH 9.0) under boiling conditions for 25 min. The slides were then incubated with 3% hydrogen peroxide (Sigma Aldrich, Burlington, MA, USA) for 10 min followed by blocking with 10% blocking buffer containing horse serum for 30 min before being incubated with anti-LRG1 antibodies (1:100, Proteintech, Rosemont, IL, USA) overnight at room temperature. The next day, the unbound primary antibodies were washed off and the section was incubated with HRP-conjugated secondary antibodies followed by treatment with a substrate reagent containing diaminobenzidine (DAB) for 5 min (Dako Real Envision Detection Kit). The section was counterstained with hematoxylin, dehydrated, and mounted with Leica Ultra CV mounting media (Leica, Wetzlar, Germany).

2.3. Cell Lines and Cell Culture Conditions

Mouse melanoma cell lines B16F0 and B16F10 and the human melanoma cell line A375 were obtained from the American Type Culture Collection (Manassas, VA, USA). Cells were cultured in Dulbecco's modified Eagle's medium (DMEM; Gibco, USA) supplemented with 10% fetal bovine serum (FBS; Gibco, USA), 2 mM of l-glutamine (Gibco, USA), 100 U/mL of penicillin, and 100 μg/mL of streptomycin (Nacalai Tesque, Kyoto, Japan). Human pulmonary microvascular endothelial cells (HPMEC) were obtained from Promocell (Heidelberg, Germany) and cultured in Endothelial Cell Medium-2 supplemented with endothelial cell growth medium bullet kits (Lonza, Basel, Switzerland). All cell lines were maintained at 37 °C in a humidified atmosphere of 95% air and 5% CO_2.

2.4. Chemicals

The chemicals used were erlotinib (Sigma-Aldrich), FAK inhibitor 14 (Sigma-Aldrich), Src-I1 (Sigma-Aldrich), and stattic (Sigma-Aldrich). Erlotinib, Src-I1, and stattic were dissolved in dimethylsulfoxide (DMSO), while FAK inhibitor 14 was dissolved in water at the desired concentrations and stored at −20 °C.

2.5. Molecular Biology Methods

The coding sequence of human LRG1 (NM_052972) carrying a 6xHis-tag at the 3′ end and a Kozak consensus sequence at the 5′ end was cloned into pcDNA3.1 (Invitrogen, Waltham, MA, USA) at the HindIII/XhoI sites to form pcDNA-hLRG1. The coding sequence of mouse Lrg1 (NM_029796) carrying a 6xHis-tag at the 3′ end and a Kozak consensus sequence at the 5′ end was cloned into pcDNA3.1 (Invitrogen, Waltham, MA, SA) at the HindIII/XbaI sites to form pcDNA-mLrg1. Cells were transfected with pcDNA-hLRG1 or pcDNA-mLrg1 plasmid (2500 ng) using Lipofectamine 2000 (Invitrogen, Waltham, MA, USA) according to the manufacturer's protocol. Small interfering RNA against Lrg1 (siLrg1; L-015179-01-0010; ON-TARGETplus SMARTpool human LRG1 siRNA) and non-targeting siRNA (negative control, siScr: D-001810-01-20; ON-TARGETplus Nontargeting siRNA#1) were purchased from Dharmacon (Lafayette, LA, USA). Cells were transfected with the siRNA oligonucleotides (25 nM) using Lipofectamine RNAiMAX transfection reagent (Life Technologies, Carlsbad, CA, USA), based on the manufacturer's protocol.

2.6. Cell Viability Assay

The CellTiter 96® AQueous One Solution Cell Proliferation Assay (MTS) (Promega, Madison, WI, USA) was used according to the manufacturer's instructions. In brief, cells were seeded at 5×10^3 cells/well in a 96-well flat-bottom plate. The next day, cells were starved for 6 h to synchronize cell growth before being cultured in full growth media. At each time point, 20 uL of MTS reagent was added into each well and the plate was incubated at 37 °C for 2 h in a humified, 5% CO_2 atmosphere. The absorbance at 490 nm was recorded using a microplate reader (Tecan, Mannedorf, Switzerland).

2.7. Cell Proliferation Assay

Cells were seeded at a density of 4.5×10^3 cells/well in a 48-well plate. The next day, cells were serum-starved in DMEM supplemented with 0.5% FBS for 6 h before being cultured in full growth media for another 72 h. Cells were then washed with PBS and fixed with 4% paraformaldehyde for 20 min at room temperature before being washed again in PBS and blocked with staining buffer containing 1% BSA, 1% Tween 20, and 3% Triton-X in PBS. One hour later, cells were incubated with Ki67 antibodies (Abcam, UK) at 1:500 dilution overnight. Cells were washed with staining buffer before being incubated with Goat anti-Rabbit IgG (H+L) Cross-Adsorbed Secondary Antibody, Alexa Fluor 594 (Thermo Fischer Scientific, Waltham, MA, USA, cat#A11012) at 1:500 dilution and 4,6-diamidino-2-phenylindole (DAPI). Five random fields of 10× objective images were taken using a Nikon Ti-E fluorescence microscope (Nikon, Tokyo, Japan). Cell proliferation rate was calculated as the percentage of Ki67 positive cells of the total cell number per well as determined using the cell counter plugin of the Image J software (National Institutes of Health, Bethesda, MD, USA).

2.8. Cell Migration and Invasion Assays

Cell migration and invasion assays were performed using 24-well plates with 8 μm pore size transwell inserts (Corning Life Science, Corning, NY, USA). For migration assays, 8×10^4 cells in serum-free DMEM medium were seeded into the upper chamber of the transwell, while NIH-3T3 cell-conditioned medium was applied to the lower chamber of the transwell as a chemoattractant. For invasion assays, the inserts were coated with 1:10 diluted Matrigel (50 μL/well) (BD Biosciences, Franklin Lakes, NJ, USA) and kept at 37 °C for 2 h to allow polymerisation to occur before seeding cells into the transwell. Cells were fixed in 1% paraformaldehyde (PFA), permeabilised in 0.5% Triton X after 4 h for migration assays or 18 h for invasion assays. Non-migrating or invading cells on the top of the transwell membrane were gently removed using cotton swabs while the migrated/invaded cells at the bottom of the membrane were stained with DAPI. Migrated/invaded cells were visualised under a Nikon Ti-E fluorescence microscope (Nikon, Tokyo, Japan) Five fields from each insert were captured and quantified using the Image J software.

2.9. Transendothelial Migration Assay

One $\times 10^5$ HPMEC cells were plated in the upper chamber of a collagen-coated transwell insert with 8 μm pore size (Corning Costar, Cambridge, MA, USA) and grown in the complete endothelial medium for 4 days to reach 100% confluence. Melanoma cells were pulsed with 25 μM CellTracker Green CMFDA (Invitrogen, Waltham, MA, USA) dye for 30 min before being trypsinised and plated on top of the HPMEC monolayer. Cells were allowed to migrate for 24 h toward the complete DMEM in the lower chamber of the transwell. Transwell inserts were fixed with 1% PFA and permeabilised by 0.5% Triton X-100. Non-migrated cancer cells were removed from the upper side of the filter using cotton buds. Migrated cells on the lower side of the filter were visualised under a Nikon Ti-E fluorescence microscope (Nikon, Tokyo, Japan). Five fields from each insert were captured and quantified using the Image J software.

2.10. Xenograft Tumour Model

B16F10 cells at a density of 2×10^6 cells were injected subcutaneously into the left flank of six- to eight-week-old wild-type and *Lrg1*$^{-/-}$ mice. Mice were monitored daily. Once tumours became visible, they were measured using a caliper for two weeks. Mice were sacrificed once the tumour size reached 250 mm^3.

2.11. Lung Metastasis Model

Five $\times 10^5$ of B16F10 cells in 200 μL of PBS were intravenously inoculated into 6- to 8-week-old wild-type and *Lrg1*$^{-/-}$ mice. The mice were sacrificed two weeks after the inoculations.

2.12. Extravasation Assay

One × 10^6 B16F10 cells labelled with CellTracker Green CMFDA (Invitrogen, Waltham, MA, USA) were intravenously inoculated into 6- to 8-week-old wild-type and *Lrg1*$^{-/-}$ mice. Mice were sacrificed and their lungs were harvested 24 h post-inoculation. Lung tissues were histologically processed and the number of labelled B16F10 cells were determined using a Nikon Ti-E fluorescence microscope (Nikon, Tokyo, Japan).

2.13. Western Blot

Cells were lysed in RIPA buffer containing protease inhibitor cocktail-EDTA free (Nacalai Tesque, Kyoto, Japan, cat#03969-21) and phosphotase inhibitor cocktail-EDTA free (Nacalai Tesque, Kyoto, Japan, Cat#07575-51). Protein concentration was determined via the Bradford method. Eighty micrograms of protein was separated via SDS-PAGE and transferred to PVDF membranes (Milipore, Burlington, MA, USA). The membranes were then incubated with p-STAT3 (Tyr705) antibodies (Cell Signaling Technology, Danvers, MA, USA, cat#9145), STAT3 antibodies (Cell Signaling Technology, Danvers, MA, USA, cat#30835), phospho-Src (Tyr527) antibodies (Cell Signaling Technology, Danvers, MA, USA, cat# 2105), Src antibodies (Cell Signaling Technology, Danvers, MA, USA, cat#2108), and GAPDH antibodies (Santa Cruz Biotechnology, Santa Cruz, CA, USA, cat#32233) at 4 °C overnight followed by horseradish peroxidase (HRP)-labelled secondary antibodies (Santa Cruz Biotechnology, Santa Cruz, CA, USA) at room temperature for 2 h.

2.14. Quantitative Real-Time PCR (qRT-PCR)

Total RNA was extracted using RNAzol RT (Molecular Research Center, Albany, NY, USA). RNA concentrations were determined using a Nanodrop 2000C Spectrophotometer 19 (Thermo Fisher Scientific, Waltham, MA, USA). The cDNA was synthesized using qScript cDNA Supermix (Quantabio, Beverly, MA, USA) according to manufacturer protocols. The qRT-PCR was conducted with SYBR Green (PrimerDesign Precision, Chandler's Ford, UK) using a QuantStudio 6 Flex Real-Time PCR System (Life Technologies, Carlsbad, CA, USA). Data were analysed using the 2 (-Delta Delta C(T)) method. The primers used in this study are listed in Table 1.

Table 1. Sequences of the forward and reverse primers utilized for gene expression analysis.

Target (Mouse)	Forward Sequence (5′–3′)	Reverse Sequence (5′–3′)
Lrg1	TGCACCTCTCGAGCAATCG	AGAGCATTGCGGGTCAGATC
Endoglin	CGATAGCAGCACTGGATGAC	AGAATGGTGCCTTTGGGTCT
Alk1	CTTGGGGAGCTTCAGAAGGGG	GGTGGCCTCCAGCATCAGAGA
Alk5	AAATTGCTCGACGCTGTTCT	GGTACAAGATCATAATAAGGCAACTG
Gapdh	ACTGAGGACCAGGTTGTCTCC	CTGTAGCCGTATTCATTGTCATACC

2.15. Histology and Immunofluorescence Staining

Resected tumours and lungs were fixed in 4% PFA for 24 h, washed with PBS, and gradually transferred to 15% sucrose followed by 30% sucrose before being embedded in O.C.T. compound (Thermo Fisher Scientific, Waltham, MA, USA). Six-micrometre cryosections of tumour samples were dehydrated and blocked with a blocking buffer containing 2% BSA, 1% Tween 20, 3% Triton X, and horse serum for an hour before being incubated with primary antibodies followed by a washing step and then incubated with secondary antibodies. The primary antibodies used were Ki67 (Abcam, Cambridge, UK, cat#ab15580) and CD31 (BD Biosciences, Franklin Lakes, NJ, USA, cat#553370). The secondary antibodies employed were Goat anti-Rabbit IgG (H+L) Cross-Adsorbed Secondary Antibody, Alexa Fluor 594 (Thermo Fischer Scientific, Waltham, MA, USA, cat#A11012), and Goat anti-Rat IgG (H+L) Cross-Adsorbed Secondary Antibody, Alexa Fluor 488 (Thermo Fischer Scientific, Waltham, MA, USA, cat#A11006). The nuclei were stained with DAPI. The slides were subsequently washed with PBS, mounted with Mowiol, and visualised under a Zeiss LSM 800 inverted confocal microscope or a Nikon Ti-E fluorescence microscope (Nikon,

Tokyo, Japan). Images were captured at 20× magnification. Haematoxylin-eosin staining was performed to quantify the total area and the number of metastases in each lung using the ImageJ software (National Institutes of Health, Bethesda, MD, USA).

2.16. Statistical Analyses

Data are represented as mean ± standard error of mean (SEM). Statistical comparison of results was performed via two-tailed, unpaired Student's t-test using Prism 8.0 (GraphPad, La Jolla, CA, USA). Statistical significance is denoted with asterisks as follows: * $p \leq 0.05$; ** $p \leq 0.01$; *** $p \leq 0.001$ and **** $p < 0.0001$.

3. Results

3.1. LRG1 Levels Are Highly Enhanced in Melanoma Cells

To establish the association between LRG1 and melanoma, we investigated the expression of LRG1 in human biopsy specimens via immunohistochemistry staining using an LRG1-specific antibody. First, our study showed that LRG1 is expressed in melanocytes of normal human skin tissues but at a slightly lower level compared to neighbouring keratinocytes (Figure 1A). Second, LRG1 is much enhanced in melanoma cells of malignant melanoma tissue as compared to cells in the stroma (Figure 1A). Consistent with this observation, data from The Human Protein Atlas (https://www.proteinatlas.org/, accessed on 23 April 2021) show that a higher expression of *LRG1* is associated with poor three-year clinical outcomes in patients with melanoma (Figure 1B). We further tested Lrg1 expression in two different murine melanoma cell lines, B16F0 and B16F10. B16F10 cells are more aggressive and highly metastatic compared to B16F0 cells [27–30]. As demonstrated via qRT-PCR, higher *Lrg1* mRNA levels are observed in B16F10 cells compared to B16F0 cells (Figure 1C). Our previous study showed that LRG1 exerts its function in endothelial cells (EC) through interaction with the TGFβ type III receptor, Endoglin [7]. Similar to *Lrg1*, *Endoglin* is highly expressed in metastatic B16F10 cells (Figure 1C). Interestingly, the expression of TGFβ type I receptor *Alk1* is also significantly higher in B16F10 cells (Figure 1C). It is worth noting that the expression levels of TGFβ type I receptor *Alk5* is comparable in B16F10 and B16F0 cells (Supplementary Figure S1).

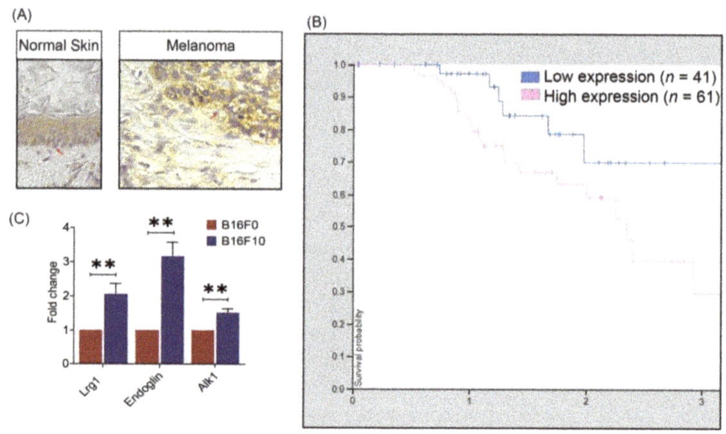

Figure 1. LRG1 expression in human melanoma biopsies and mouse melanoma cell lines. (**A**) Representative photomicrographs of immunohistochemistry staining demonstrating the expression pattern of LRG1 in normal human skin and malignant melanoma tissue. (**B**) Kaplan–Meier curves showing stratification of three-year survival probability as a function of *LRG1* RNA expression (adapted from The Human Protein Atlas). (**C**) Real-time quantitative PCR analysis of *Lrg1*, *endoglin*, and *Alk 1* expression. Data are presented as the mean ± S.E.M of three independent experiments. Statistical analysis was performed via two-tailed, unpaired Student's t-test; ** $p < 0.01$.

3.2. Host Lrg1 Deficiency Has No Impact on Melanoma Growth

Murine melanoma cell line B16F10 is highly malignant and has been widely used to study melanoma growth and metastasis. In this project, B16F10 cells were inoculated subcutaneously into the flank of wild-type and *Lrg1*$^{-/-}$ mice. Once tumours became visible, tumour volume was measured daily for a continuous 14 days using a caliper. Surprisingly, no significant differences in tumour volume (Figure 2A,B) and growth rate (Figure 2C) were observed between wild-type and *Lrg1*$^{-/-}$ mice. Additionally, the Kaplan–Meier survival curves did not show any difference between mice from the two experimental groups (Figure 2D). This was further confirmed by immunofluorescence staining with the cell proliferation marker Ki67. The percentage of Ki67+ cells in tumour tissues collected from *Lrg1*-deficient mice was comparable to that in wild-type controls (Figure 2E). To complement these in vivo observations, the role of LRG1 in tumour cell proliferation was investigated in vitro. The MTS assay was used to evaluate the viability of B16F10 cells transfected with pCDNA3.1-mLrg1, which resulted in Lrg1 overexpression, or pcDNA3.1 control plasmid; no difference was observed between the two experimental groups (Figure 2F). Furthermore, plasmid-mediated Lrg1 overexpression did not affect B16F10 cell proliferation as demonstrated by the percentage of Ki67+ cells (Figure 2G). Similar observations were made in the human melanoma cell line A375 (Supplementary Figure S2A,B).

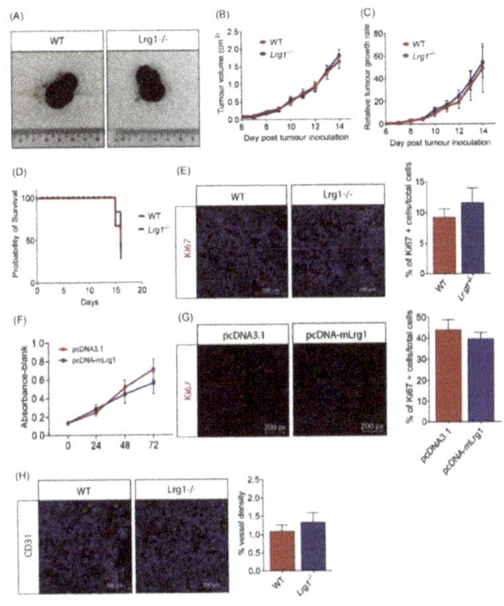

Figure 2. Host *Lrg1*-deficiency does not affect tumour growth in a tumour xenograft model. (**A**) Representative images of B16F10 tumours dissected from wild-type and *Lrg1*$^{-/-}$ mice. (**B**) Tumour volume was plotted over time in both experimental groups. (**C**) Tumour growth rates were determined based on changes in tumour volume over time in both experimental groups. (**D**) Kaplan–Meier survival curves of wild-type or *Lrg1*$^{-/-}$ tumour-bearing mice. (**E**) Representative images (left) and quantification (right) of the percentage of Ki67+ cells (red) in tumour samples collected from wild-type and *Lrg1*$^{-/-}$ mice. Nuclei were labelled with DAPI (blue). (**F**) MTS assay was performed on B16F10 cells transfected with pcDNA-mLrg1 or pcDNA3.1 control plasmid at 24, 48, and 72 h post-transfection. (**G**) Representative images (left) and quantification (right) of Ki67+ cells (red) in Lrg1 overexpressing or pcDNA3.1 plasmid transfected control B16F10 cells. (**H**) Representative images (left) and quantification (right) of CD31+ vessel (Green) in tumour samples collected from wild-type or *Lrg1*$^{-/-}$ mice. All images are representative. Data are presented as the mean ± SEM. of three independent experiments. Statistical analysis was performed via two-tailed, unpaired Student's *t*-test.

Tumours require access to blood vessels to grow beyond 2 mm^3 [31]. Considering the role of LRG1 in ocular angiogenesis [7], we next investigated whether tumour vascularization is affected in wild-type and *Lrg1*$^{-/-}$ mice. To our surprise, there was no change in tumour vessel density in the absence of host Lrg1 compared to wild-type mice as demonstrated by similar CD31 positive areas (Figure 2H). Together, these data suggest that host Lrg1-deficiency does not affect tumour cell viability, proliferation, and tumour angiogenesis.

3.3. Host Lrg1 Deficiency Leads to Reduced Pulmonary Metastasis of Melanoma In Vivo

As melanoma is a highly metastatic disease, we then explored the potential role of Lrg1 in melanoma metastasis. Lung metastasis was induced in wild-type and *Lrg1*$^{-/-}$ mice via intravenous delivery of B16F10 cells. Pulmonary metastases can be visualised under the dissection microscope. There is a clear difference in the number of metastatic nodules with black pigmentation between wild-type and *Lrg1*$^{-/-}$ mice (Figure 3A). The number of pulmonary metastatic nodules was then counted and represented as metastasis frequency following a scoring system by denoting 0 for no metastases, 1 for within 10 metastases, and 2 for more than 10 metastases. Our study revealed a significantly lower metastasis frequency score in *Lrg1*$^{-/-}$ mice as compared to that in wild-type mice (Figure 3B). Concomitant with this observation, haematoxylin-eosin staining of excised lungs showed a similar reduction in the percentage of tumour nodule area in the lungs of *Lrg1*$^{-/-}$ mice following the histopathological analysis (Figure 3C). Further analysis of the percentage of Ki67+ cells in the lung samples of wild-type and *Lrg1*$^{-/-}$ mice showed no significant changes, suggesting that post-extravasation proliferation was not affected (Figure 3D). Finally, the extravasation capability of melanoma cells was studied by inoculating CMFDA Green-labelled B16F10 cells through the tail vein of wild-type and *Lrg1*$^{-/-}$ mice. The number of extravasated B16F10 cells was visualised under epifluorescence microscopy 24 h post-inoculation and quantified. There was a significant reduction in the number of extravasated B16F10 cells in *Lrg1*$^{-/-}$ mice compared to that in wild-type mice (Figure 3E). Taken together, these data provided compiling evidence that Lrg1 is required for melanoma metastasis into the lungs in vivo.

3.4. Lrg1 Promotes B16F10 Cell Invasiveness In Vitro

Having established a role for Lrg1 in melanoma metastasis, especially the extravasation step, in vivo, we next investigated how Lrg1 modulates melanoma cell function in vitro. The ability of circulating tumour cells to tether to the vasculature is a prerequisite step for the extravasation [32]. Consistent with the increased invasiveness, there was a significantly higher number of Lrg1 overexpressing B16F10 cells that adhered to the human pulmonary microvascular endothelial cell monolayer in the in vitro assay (Figure 4A). The adhered tumour cells would then migrate through the endothelium to eventually settle at secondary sites. Our study showed that Lrg1 overexpressing B16F10 cells were more prone to transmigrate through the HPMEC monolayer (Figure 4B). To mimic the previous observations in vivo, HPMECs were subjected to siRNA-mediated LRG1 knockdown (Supplementary Figure S3). Our study showed that the ability of B16F10 cells to adhere to (Figure 4C) or migrate (Figure 4D) through the siLrg1-treated HPMEC monolayer was significantly compromised.

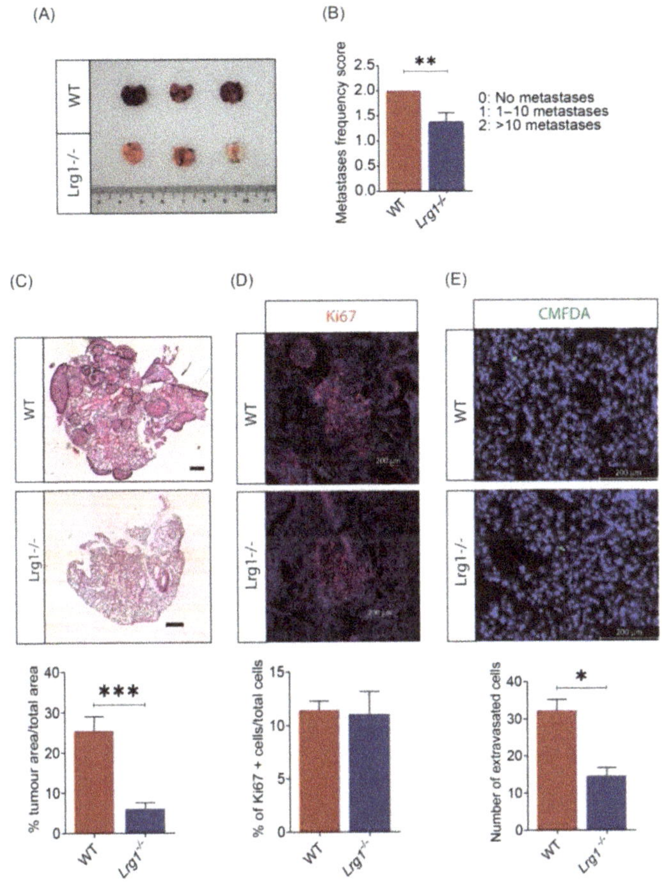

Figure 3. Host *Lrg1* deficiency reduces pulmonary metastasis of melanoma in vivo. (**A**) Representative images of lung metastases in wide-type and $Lrg1^{-/-}$ mice. Nodules are easily identifiable due to their black pigmentation. (**B**) Quantification of lung metastases frequency in wide-type and $Lrg1^{-/-}$ mice. (**C**) Representative images of haematoxylin-eosin staining (top) and quantification (below) of the metastatic burden of lung tissues in wide-type and $Lrg1^{-/-}$ mice. The percentage of tumour nodule area over the total lung area was measured. (**D**) Representative images (top) and quantification (below) of Ki67+ cells (red) in lung tissues of wide-type and $Lrg1^{-/-}$ mice. (**E**) Representative images of lung sections (top) to visualise extravasated melanoma cells identified via CMFDA Green signal (green) and quantification (below) of the number of extravasated cells out from blood vessel 24 h following intravenous injection of B16F10. Nuclei were labelled with DAPI (blue). Data are presented as the mean ± SEM. Statistical analysis was performed via two-tailed, unpaired Student's *t*-test; * $p < 0.05$, ** $p < 0.01$; *** $p < 0.001$. Wildtype group: $n = 9$, $Lrg1^{-/-}$ group: $n = 8$.

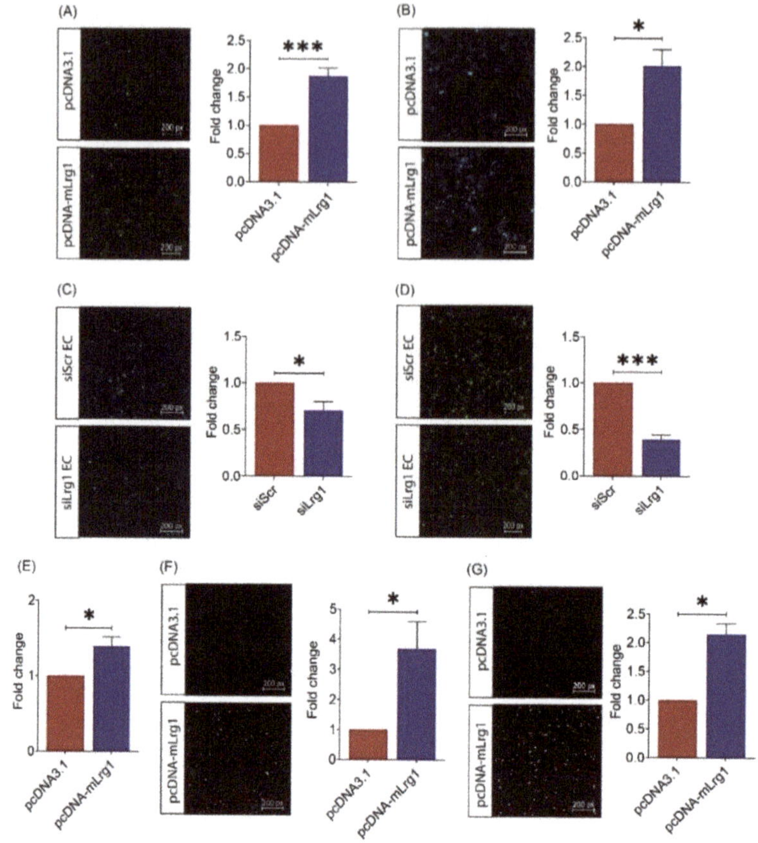

Figure 4. Lrg1 promotes B16F10 cell invasiveness in vitro. (**A**) Representative images and quantitative analysis of the number of CMFDA-positive Lrg1 overexpressing or control B16F10 cells adhered to the HPMEC monolayer. (**B**) Representative images and quantitative analysis of the number of CMFDA-positive Lrg1 overexpressing B16F10 cells that migrated through the HPMEC monolayer. (**C**) Representative images and quantitative analysis of the number of CMFDA-positive B16F10 cells adhered to the HPMECs subjected to siRNA-mediated LRG1 knockdown. (**D**) Representative images and quantitative analysis of the number of CMFDA-positive B16F10 cells that migrated through the HPMECs subjected to siRNA-mediated LRG1 knockdown. (**E**) Quantitative analysis of Lrg1 overexpressing or control B16F10 cells adhered to fibronectin. (**F**) Representative images and quantitative analysis of invaded Lrg1 overexpressing and control B16F10 cells using the Matrigel invasion assay. (**G**) Representative images and quantitative analysis of migrated Lrg1 overexpressing or control B16F10 cells in Transwell. Migrated and invaded cells are labelled by DAPI. All images are representative. Data are presented as mean ± SEM. of three independent experiments. Statistical analyses were performed via two-tailed, unpaired Student's t-test. * $p < 0.05$; *** $p < 0.001$.

The success of colonization at the secondary tumour site following extravasation depends on the ability of tumour cells to adhere to the extracellular matrix (ECM), degrade ECM components, and move around [33]. This prompted us to test the impact of Lrg1 on melanoma cell adhesion to ECM. Fibronectin is a major component of the tumour ECM and plays key regulatory roles in the tumour matrisome [34]. As shown in Figure 4E, there was an increased number of Lrg1 overexpressing B16F10 cells that adhered to fibronectin. Next, we performed the Matrigel invasion assay to test the role of Lrg1 in melanoma cell invasion. Consistent with the results presented above, Lrg1 overexpression significantly increased the invasiveness of B16F10 cells (Figure 4F). Furthermore, a transwell migration assay was performed using pcDNA3.1-mLrg1 or pcDNA3.1 control plasmid transfected B16F10 cells. As compared to the pcDNA3.1 transfected controls, Lrg1 overexpressing B16F10 cells showed increased motility (Figure 4G). We further confirmed LRG1's role in tumour cell adhesion to the endothelium (Supplementary Figure S2C), transendothelial migration (Supplementary Figure S2D), invasion (Supplementary Figure S2E), and migration (Supplementary Figure S2F) in a human melanoma cell line A375. Together, these data demonstrated that Lrg1 promotes tumour cell dissemination by affecting various properties of melanoma cells in both mice and humans.

3.5. Lrg1-Induced Activation of the EGFR/STAT3 Pathway Is Required for Melanoma Cell Invasiveness

To understand the mechanism of action for Lrg1-mediated melanoma cell activation, Western blot analysis was performed to determine the signalling pathways regulated by Lrg1 in B16F10 cells. Lrg1 was previously reported to regulate EGFR/STAT3 signalling in regenerating corneal epithelium [35]. Our study showed that the level of the phosphorylated form of STAT3 was significantly increased in Lrg1 overexpressing B16F10 cells compared to pcDNA3.1 plasmid transfected control cells, whereas the total STAT3 level remained unchanged (Figure 5A). STAT3 is a transcription factor that is responsible for relaying signals from various activated receptors of cytokines and growth factors, including focal adhesion kinases (FAK) and epidermal growth factor receptors (EGFR) [36]. To identify whether Lrg1-mediated activation of STAT3 occurs through EGFR or FAK, EGFR- and FAK-specific inhibitors erlotinib and FAK inhibitor 14, were used to treat pcDNA3.1 or pcDNA-mLrg1 transfected B16F10 cells. Interestingly, the Lrg1 induced-increase in STAT3 phosphorylation was not affected by the presence of FAK inhibitor 14 but was significantly attenuated in the presence of erlotinib (Figure 5B). Src, a non-receptor protein tyrosine kinase associated with EGFR within lipid rafts [37], was reported to activate STATs directly [38]. To figure out whether Src mediates Lrg1-regulated STAT3 activation, pcDNA3.1 or pcDNA-mLrg1 transfected B16F10 cells were treated with the Src-specific inhibitor Src-I1. Interestingly, Src-I1 did not affect the Lrg1-induced STAT3 phosphorylation (Figure 5C). On the other hand, the STAT3 specific inhibitor significantly inhibited the Lrg1-induced STAT3 phosphorylation (Figure 5C). To further explore whether Lrg1-induced activation of STAT3 signalling is required for Lrg1 regulated melanoma cell invasiveness, Lrg1 overexpressing B16F10 cells with the presence or absence of stattic were subjected to cell migration and invasion assays as described earlier. Lrg1-induced B16F10 cell migration (Figure 5D) and invasion (Figure 5E) were significantly suppressed by stattic. Together, these results suggest that Lrg1 promotes melanoma cell migration and invasion by activating the EGFR/STAT3 signalling pathway in a Src-independent manner.

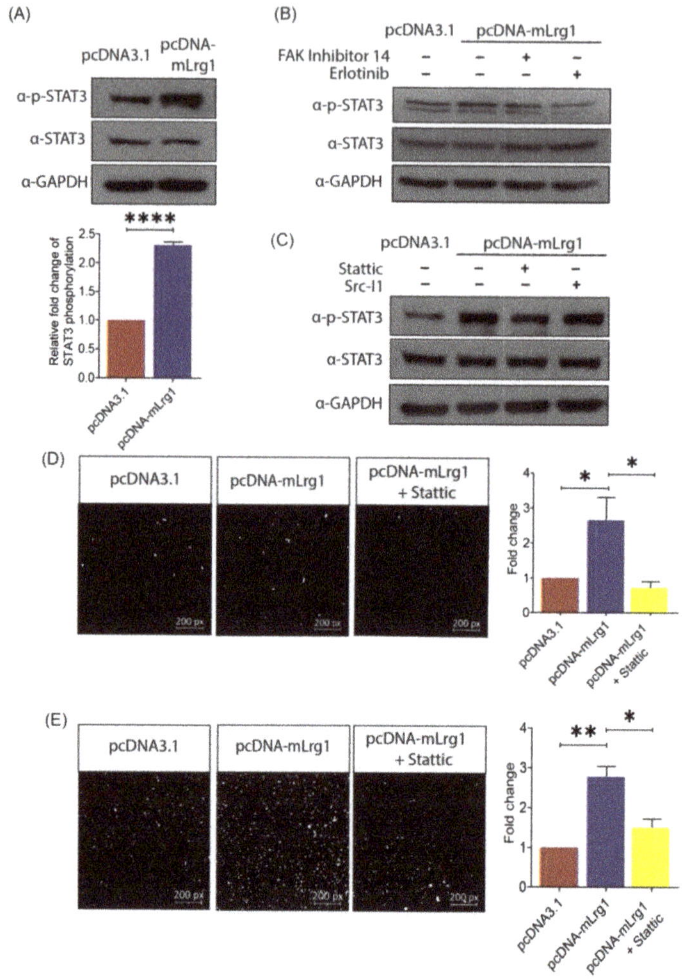

Figure 5. Lrg1-induced activation of the EGFR/STAT3 pathway is required for melanoma cell invasiveness. (**A**) Representative Western blot (left) and densitometry (right) analyses (right) showing the levels of phospho-STAT3 (Tyr705) and total STAT3 in Lrg1 overexpressing B16F10 cells. GAPDH was used as a loading control. (**B**) Representative images of Western blot analysis showing the levels of phospho-STAT3 (Tyr705), STAT3, and GAPDH in Lrg1 overexpressing B16F10 cells subjected to treatment with FAK inhibitor 14 (FAK inhibitor) or erlotinib (EGFR inhibitor). (**C**) Representative images of Western blot analysis showing the levels of phospho-STAT3 (Tyr705) and total STAT3 in Lrg1 overexpressing B16F10 cells subjected to treatment with Src-I1 (Src inhibitor) and stattic (Stat3 inhibitor). GAPDH was used as a loading control. (**D**) Representative images and quantification of migrated Lrg1 overexpressing B16F10 cells subjected to stattic treatment. (**E**) Representative images and quantitative analysis of invaded Lrg1 overexpressing B16F10 cells subjected to stattic treatment. All images are representative. Data are presented as mean ± SEM of three independent experiments. Statistical analyses were performed via two-tailed, unpaired Student's t-test. * $p < 0.05$; ** $p < 0.01$; **** $p < 0.0001$.

4. Discussion

Malignant melanoma is characterized by its high resistance to chemotherapy and the ability to rapidly metastasize to distant organs. To date, limited treatments are available to control malignant melanoma effectively. A better understanding of melanoma pathogenesis may facilitate the development of new therapeutic modalities. LRG1 is a novel angiogenic factor [7] that was previously associated with a variety of cancers, including endometrial carcinoma, gastric, colorectal, and pancreatic cancers [39–42]. However, the role of LRG1 in melanoma development and progression has not been established.

Although expressed at low levels in normal melanocytes, we show for the first time that LRG1 is significantly enhanced in malignant melanoma cells of human melanoma tissues. Furthermore, Lrg1 expression levels in the metastatic mouse melanoma cell line B16F10 are significantly higher than those in the parental melanoma cell line B16F0. This observation is consistent with the data from The Human Protein Atlas (https://www.proteinatlas.org/, accessed on 23 April 2021), which show that high LRG1 expression levels are associated with a poor three-year prognosis of patients with melanoma. However, these data do not inform whether the upregulated expression of LRG1 is the cause or consequence of melanoma development and progression.

We used the xenograft tumour model to establish the cause–effect relationship between Lrg1 and melanoma development. Interestingly, the growth of xenografted tumours, the mouse survival rate, and melanoma cell proliferation were not affected in the absence of host *Lrg1*. Considering the large amount of Lrg1 produced by B16F10 cells, we further validated the role of Lrg1 on melanoma cell growth using an in vitro cell-based assay, which again showed no impact of Lrg1 on melanoma cell viability and proliferation. To grow beyond 2 mm^3, a tumour requires blood vessels to supply oxygen and nutrients [32]. Despite the important role of Lrg1 in ocular angiogenesis, no changes were observed in tumour vessel density in xenograft tumours of wild-type and *Lrg1*$^{-/-}$ mice. It is possible that Lrg1 derived from implanted B16F10 cells is able to compensate for the loss of Lrg1 in Lrg1-deficient mice, therefore, leaving tumour angiogenesis unaffected. To further elucidate the role of Lrg1 in tumour angiogenesis, future work should compare tumour angiogenesis in Lrg1 knockout mice inoculated with control B16F10 cells, Lrg1 overexpressing B16F10 cells, and B16F10 cells subjected to siRNA-mediated Lrg1 knockdown.

As melanoma is highly metastatic, we next investigated the role of Lrg1 in melanoma dissemination to the lungs. Our finding revealed a significant reduction in tumour burden, total tumour nodule area, and melanoma cell extravasation in the lungs of *Lrg1*$^{-/-}$ mice as compared to wild-type controls. Successful extravasation depends on the ability of tumour cells to adhere to the endothelial cell and migrate across the endothelium through a process termed transendothelial migration [38,39]. To support in vivo observations, we showed an increased capability B16F10 cells overexpressing Lrg1 to adhere to and migrate across the HPMEC monolayer. Consistent with the in vivo observations, the ability of parental B16F10 cells to adhere to and transmigrate across the HPMEC subjected to siRNA-mediated Lrg1 knockdown was significantly lower, suggesting that both tumour cell and endothelial cell-derived Lrg1 affect melanoma cell extravasation. We further showed that Lrg1 overexpressing B16F10 cells are more migratory and invasive and show increased adhesion to fibronectin. Similar findings were made for glioma cells [43], colorectal cancer cells [22], and thyroid carcinoma cells [44].

STAT-family proteins are latent cytoplasmic transcription factors [45]. Upon phosphorylation, STATs form homodimers and translocate into the cell nucleus to regulate gene expression [34]. Numerous oncogenic signalling pathways converge on STATs proteins, particularly to STAT3 [46,47]. Hyperactivation of STAT3 has been associated with poor prognosis in various malignancies [48,49], including melanoma [34,50,51]. STAT3 exerts its function by promoting tumour metastasis through mediating tumour cell proliferation, migration, and invasion as well as tumour angiogenesis [52]. As such, targeting STAT3 has been considered a promising therapeutic strategy for highly metastatic melanoma. To date, several STAT3 inhibitors have been tested and demonstrated promising results in early-phase clinical trials, but none of them has been approved for melanoma treatment due to adverse side effects and toxicity [53,54]. Therefore, the development of a safer and more effective way to control STAT3 signalling is highly desired. Here, we demonstrated that Lrg1 promotes STAT3 phosphorylation in an EGFR-dependent manner. Furthermore, STAT3 activation is required for the promoting effect of LRG1 on melanoma cell migration and invasion. Since *Lrg1*$^{-/-}$ mice are viable and show no obvious abnormality, targeting Lrg1 may offer an alternative way to control STAT3-mediated metastasis.

5. Conclusions

Overall, our study provided strong evidence of Lrg1's role in melanoma metastasis and Lrg1 exerting its function through activation of the EGFR/STAT3 signalling pathway. As *Lrg1*$^{-/-}$ mice are healthy and have a normal life span, unlike mice treated with current STAT3 inhibitors, targeting LRG1 may cause fewer unwanted side effects and offers an alternative strategy to control STAT3-mediated melanoma metastasis.

Supplementary Materials: The following are available online at https://www.mdpi.com/article/10.3390/cancers13133279/s1, Figure S1: Real-time quantitative PCR analysis of Alk5 gene expression, Figure S2: LRG1 promotes A375 cell invasiveness in vitro, Figure S3: Gene silencing efficiency of LRG1 siRNA in HPMEC.

Author Contributions: Conceptualization, X.W. and W.W.; methodology, X.W., Y.P.K., J.C.W.L., M.S.T., M.H.Y.T. and G.R.; investigation, Y.P.K., J.C.W.L., M.S.T., M.H.Y.T. and G.R., data curation, X.W. and Y.P.K.; formal analysis, X.W. and Y.P.K.; funding acquisition, X.W. and W.W.; project administration, X.W.; supervision, X.W.; validation, X.W. and Y.P.K.; writing—original draft, Y.P.K.; writing—review and editing, X.W. and W.W. All authors have read and agreed to the published version of the manuscript.

Funding: This project was supported by the MOE Academic Research Fund Tier 2 (MOE2014-T2-1-036) to X.W. and W.W., the Duke-NUS Start-Up Grant to X.W. and the Singapore National Medical Research Council DYNAMO NMRC/OFLCG/001/2017 and TAPP NMRC/OFLCG/004/2018 to X.W., and the Lee Kong Chian School of Medicine, Nanyang Technological University Singapore Start-up Grant to W.W.

Institutional Review Board Statement: Animal care and procedures were performed under the guidelines of the Institutional Animal Care and Use Committee (IACUC, Protocol number: A0269) of the Nanyang Technological University in Singapore and the Guide for Care and Use of Laboratory Animals from the US National Institutes of Health.

Conflicts of Interest: The authors declare no conflict of interest.

References

1. Crowson, A.N.; Magro, C.; Miller, A.; Mihm, M.C., Jr. The molecular basis of melanomagenesis and the metastatic phenotype. *Semin. Oncol.* **2007**, *34*, 476–490. [CrossRef]
2. Miller, A.J.; Mihm, M.C., Jr. Melanoma. *New Engl. J. Med.* **2006**, *355*, 51–65. [CrossRef] [PubMed]
3. Bray, F.; Ferlay, J.; Soerjomataram, I.; Siegel, R.L.; Torre, L.A.; Jemal, A. Global cancer statistics 2018: GLOBOCAN estimates of incidence and mortality worldwide for 36 cancers in 185 countries. *CA Cancer J. Clin.* **2018**, *68*, 394–424. [CrossRef] [PubMed]
4. Global Cancer Observatory—Projected 2025 Data from 'Cancer Tomorrow'. Available online: https://gco.iarc.fr/tomorrow/en/dataviz/isotype?cancers=16&single_unit=10000&years=2025&types=1 (accessed on 2 February 2021).
5. Global Cancer Observatory—Projected 2040 Data from 'Cancer Tomorrow'. Available online: https://gco.iarc.fr/tomorrow/en/dataviz/isotype?cancers=16&single_unit=10000&years=2040&types=1 (accessed on 2 February 2021).
6. Domingues, B.; Lopes, J.M.; Soares, P.; Pópulo, H. Melanoma treatment in review. *Immunotargets Ther.* **2018**, *7*, 35–49. [CrossRef] [PubMed]
7. Wang, X.; Abraham, S.; McKenzie, J.; Jeffs, N.; Swire, M.; Tripathi, V.B.; Luhmann, U.; Lange, C.A.K.; Zhai, Z.; Arthur, H.M.; et al. LRG1 promotes angiogenesis by modulating endothelial TGF-β signalling. *Nature* **2013**, *499*, 306–311. [CrossRef]
8. Zhang, A.; Fang, H.; Chen, J.; He, L.; Chen, Y. Role of VEGF-A and LRG1 in abnormal angiogenesis associated with diabetic nephropathy. *Front. Physiol.* **2020**, *11*, 1064. [CrossRef]
9. Meng, H.; Song, Y.; Zhu, J.; Liu, Q.; Lu, P.; Ye, N.; Zhang, Z.; Pang, Y.; Qi, J.; Wu, H. LRG1 promotes angiogenesis through upregulating the TGF-β1 pathway in ischemic rat brain. *Mol. Med. Rep.* **2016**, *14*, 5535–5543. [CrossRef]

10. Song, S.; Cheng, J.; Yu, B.J.; Zhou, L.; Xu, H.F.; Yang, L.L. LRG1 promotes corneal angiogenesis and lymphangiogenesis in a corneal alkali burn mouse model. *Int. J. Ophthalmol.* **2020**, *13*, 365–373. [CrossRef]
11. Liu, C.; Teo, M.H.Y.; Pek, S.L.T.; Wu, X.; Leong, M.L.; Tay, H.M.; Hou, H.W.; Ruedl, C.; Moss, S.E.; Greenwood, J.; et al. A Multifunctional Role of Leucine-Rich α-2-Glycoprotein 1 in Cutaneous Wound Healing Under Normal and Diabetic Conditions. *Diabetes* **2020**, *69*, 2467–2480. [CrossRef] [PubMed]
12. Ziyad, S.; Iruela-Arispe, M.L. Molecular mechanisms of tumor angiogenesis. *Genes Cancer* **2011**, *12*, 1085–1096. [CrossRef] [PubMed]
13. Katt, M.E.; Wong, A.D.; Searson, P.C. Dissemination from a solid tumor: Examining the multiple parallel pathways. *Trends Cancer* **2018**, *4*, 20–37. [CrossRef]
14. Uen, Y.H.; Lin, K.Y.; Sun, D.P.; Liao, C.C.; Hsieh, M.S.; Huang, Y.K.; Chen, Y.W.; Huang, P.H.; Chen, W.J.; Tai, C.C.; et al. Comparative proteomics, network analysis and post-translational modification identification reveal differential profiles of plasma con a-bound glycoprotein biomarkers in gastric cancer. *J. Proteom.* **2013**, *83*, 197–213. [CrossRef] [PubMed]
15. Wu, J.; Xie, X.; Nie, S.; Buckanovich, R.J.; Lubman, D.M. Altered expression of sialylated glycoproteins in ovarian cancer sera using lectinbased elisa assay and quantitative glycoproteomics analysis. *J. Proteome Res.* **2013**, *12*, 3342–3352. [CrossRef] [PubMed]
16. Lindén, M.; Lind, S.B.; Mayrhofer, C.; Segersten, U.; Wester, K.; Lyutvinskiy, Y.; Zubarev, R.; Malmström, P.U.; Pettersson, U. Proteomic analysis of urinary biomarker candidates for nonmuscle invasive bladder cancer. *Proteomics* **2012**, *12*, 135–144. [CrossRef] [PubMed]
17. Liu, Y.; Luo, X.; Hu, H.; Wang, R.; Sun, Y.; Zeng, R.; Chen, H. Integrative proteomics and tissue microarray profiling indicate the association between overexpressed serum proteins and non-small cell lung cancer. *PLoS ONE* **2012**, *7*. [CrossRef]
18. Furukawa, K.; Kawamoto, K.; Eguchi, H.; Tanemura, M.; Tanida, T.; Tomimaru, Y.; Akita, H.; Hama, N.; Wada, H.; Kobayashi, S.; et al. Clinicopathological Significance of Leucine-Rich α2-Glycoprotein-1 in Sera of Patients with Pancreatic Cancer. *Pancreas* **2015**, *44*, 93–98. [CrossRef] [PubMed]
19. Zhang, Y.S.; Han, L.; Yang, C.; Liu, Y.J.; Zhang, X.M. Prognostic Value of LRG1 in Breast Cancer: A Retrospective Study. *Oncol. Res. Treat.* **2021**, *44*, 36–41. [CrossRef]
20. Guldvik, I.J.; Zuber, V.; Braadland, P.R.; Grytli, H.H.; Ramberg, H.; Lilleby, W.; Thiede, B.; Zucknick, M.; Saatcioglu, F.; Gislefoss, R.; et al. Identification and Validation of Leucine-rich α-2-glycoprotein 1 as a Noninvasive Biomarker for Improved Precision in Prostate Cancer Risk Stratification. *Eur. Urol. Open Sci.* **2020**, *21*, 51–60. [CrossRef]
21. Sandanayake, N.; Sinclair, J.; Andreola, F.; Chapman, M.H.; Xue, A.; Webster, G.J.; Clarkson, A.; Gill, A.; Norton, I.D.; Smith, R.C.; et al. A combination of serum leucine-rich α-2-glycoprotein 1, CA19-9 and interleukin-6 differentiate biliary tract cancer from benign biliary strictures. *Br. J. Cancer.* **2011**, *105*, 1370–1378. [CrossRef]
22. Zhang, J.; Zhu, L.; Fang, J.; Ge, Z.; Li, X. LRG1 modulates epithelial-mesenchymal transition and angiogenesis in colorectal cancer via HIF-1α activation. *J. Exp. Clin. Cancer Res.* **2016**, *35*, 29. [CrossRef] [PubMed]
23. Lu, C.; Yang, Z.; Yu, D.; Lin, J.; Cai, W. RUNX1 regulates TGF-β induced migration and EMT in colorectal cancer. *Pathol. Res. Pract.* **2020**, *216*, 153142. [CrossRef] [PubMed]
24. Zhang, N.; Ren, Y.; Wang, Y.; Zhao, L.; Wang, B.; Ma, N.; Gao, Z.; Cao, B. LRG1 Suppresses migration and invasion of esophageal squamous cell carcinoma by modulating epithelial to mesenchymal transition. *J. Cancer* **2020**, *11*, 1486–1494. [CrossRef] [PubMed]
25. Honda, H.; Fujimoto, M.; Serada, S.; Urushima, H.; Mishima, T.; Lee, H.; Ohkawara, T.; Kohno, N.; Hattori, N.; Yokoyama, A. Leucine-rich α-2 glycoprotein promotes lung fibrosis by modulating TGF-β signaling in fibroblasts. *Physiol. Rep.* **2017**, *5*, e13556. [CrossRef]
26. Liu, C.; Lim, S.T.; Teo, M.H.Y.; Tan, M.S.Y.; Kulkarni, M.D.; Qiu, B.; Li, A.; Lal, S.; dos Remedios, C.G.; Tan, N.S.; et al. Collaborative Regulation of LRG1 by TGF-β1 and PPAR-β/δ Modulates Chronic Pressure Overload–Induced Cardiac Fibrosis. *Circ. Heart Fail.* **2019**, *12*, e005962. [CrossRef] [PubMed]
27. Mathieu, V.; Le Mercier, M.; De Neve, N.; Sauvage, S.; Gras, T.; Roland, I.; Lefranc., F.; Kiss, R. Galectin-1 Knockdown Increases Sensitivity to Temozolomide in a B16F10 Mouse Metastatic Melanoma Model. *J. Invest. Dermatol.* **2007**, *127*, 2399–2410. [CrossRef] [PubMed]
28. Mathieu, V.; de Lassalle, E.M.; Toelen, J.; Mohr, T.; Bellahcene, A.; Goietsenoven, G.V.; Verschuere, T.; Bouzin, C.; Debyser, Z.; De Vleeschouwer, S.; et al. Galectin-1 in Melanoma Biology and Related Neo-Angiogenesis Processes. *J. Invest. Dermatol.* **2012**, *132*, 2245–2254. [CrossRef]
29. Fidler, I.J.; Gersten, D.M.; Budmen, M.B. Characterization in vivo and in vitro of tumor cells selected for resistance to syngeneic lymphocyte-mediated cytotoxicity. *Cancer Res.* **1976**, *36*, 3160–3165.
30. Overwijk, W.W.; Restifo, N.P. B16 as a mouse model for human melanoma. *Curr. Protoc. Immunol.* **2001**, *20*. [CrossRef]
31. Nishida, N.; Yano, H.; Nishida, T.; Kamura, T.; Kojiro, M. Angiogenesis in cancer. *Vasc. Health Risk Manag.* **2006**, *2*, 213–219. [CrossRef]
32. Azevedo, A.S.; Follain, G.; Patthabhiraman, S.; Harlepp, S.; Goetz, J.G. Metastasis of circulating tumor cells: Favorable soil or suitable biomechanics, or both? *Cell Adh. Migr.* **2015**, *9*, 345–356. [CrossRef]
33. Fares, J.; Fares, M.Y.; Khachfe, H.H.; Salhan, H.A.; Fares, Y. Molecular principles of metastasis: A hallmark of cancer revisited. *Sig. Transduct. Target Ther.* **2020**, *5*, 28. [CrossRef]
34. Spada, S.; Tocci, A.; Di Modugno, F.; Nistico, P. Fibronectin as a multiregulatory molecule crucial in tumor matrisome: From structural and functional features to clinical practice in oncology. *J. Exp. Clin. Cancer Res.* **2021**, *40*, 102. [CrossRef]

35. Li, W.; Wang, X.; Cheng, J.; Li, J.; Wang, Q.; Zhou, Q.; Li, H.; Xue, J.; Zhang, Y.; Yang, L. Leucine-rich α-2-glycoprotein-1 promotes diabetic corneal epithelial wound healing and nerve regeneration via regulation of matrix metalloproteinases. *Exp. Eye Res.* **2020**, *196*, 108060. [CrossRef] [PubMed]
36. Yu, H.; Lee, H.; Herrmann, A.; Buettner, R.; Jove, R. Revisiting STAT3 signalling in cancer: New and unexpected biological functions. *Nat. Rev. Cancer.* **2014**, *14*, 736–746. [CrossRef]
37. Irwin, M.E.; Bohin, N.; Boerner, J.L. Src family kinases mediate epidermal growth factor receptor signaling from lipid rafts in breast cancer cells. *Cancer Biol. Ther.* **2011**, *12*, 718–726. [CrossRef]
38. Silva, C.M. Role of STATs as downstream signal transducers in Src family kinase-mediated tumorigenesis. *Oncogene* **2004**, *23*, 8017–8023. [CrossRef]
39. Yamamoto, M.; Takahashi, T.; Serada, S.; Sugase, T.; Tanaka, K.; Miyazaki, Y.; Makino, T.; Kurokawa, Y.; Yamasaki, M.; Nakajima, K.; et al. Overexpression of leucine-rich alpha2-glycoprotein-1 is a prognostic marker and enhances tumor migration in gastric cancer. *Cancer Sci.* **2017**, *108*, 2052–2060. [CrossRef] [PubMed]
40. Ladd, J.J.; Busald, T.; Johnson, M.M.; Zhang, Q.; Pitteri, S.J.; Wang, H.; Brenner, D.E.; Lampe, P.D.; Kucherlapati, R.; Feng, Z.; et al. Increased plasma levels of the APC-interacting protein MAPRE1, LRG1, and IGFBP2 preceding a diagnosis of colorectal cancer in women. *Cancer Prev. Res.* **2012**, *5*, 655–664. [CrossRef]
41. Wen, S.Y.; Zhang, L.N.; Yang, X.M.; Zhang, Y.L.; Ma, L.; Ge, Q.L.; Jiang, S.H.; Zhu, X.L.; Xu, W.; Ding, W.J. LRG1 is an independent prognostic factor for endometrial carcinoma. *Tumor Biol.* **2014**, *35*, 7125–7133. [CrossRef] [PubMed]
42. Xie, Z.B.; Zhang, Y.F.; Jin, C.; Mao, Y.S.; Fu, D.L. LRG-1 promotes pancreatic cancer growth and metastasis via modulation of the EGFR/p38 signaling. *J. Exp. Clin. Cancer Res.* **2019**, *38*, 75. [CrossRef] [PubMed]
43. Zhong, D.; He, G.; Zhao, S.; Li, J.; Lang, Y.; Ye, W.; Li, Y.; Jiang, C.; Li, X. LRG1 modulates invasion and migration of glioma cell lines through TGF-β signaling pathway. *Acta Histochem.* **2015**, *117*, 551–558. [CrossRef] [PubMed]
44. Ban, Z.; He, J.; Tang, Z.; Zhang, L.; Xu, Z. LRG-1 enhances the migration of thyroid carcinoma cells through promotion of the epithelial-mesenchymal transition by activating MAPK/p38 signaling. *Oncol. Rep.* **2019**, *41*, 3270–3280. [CrossRef] [PubMed]
45. Yu, H.; Jove, R. The stats of cancer—New molecular targets come of age. *Nat. Rev. Cancer.* **2004**, *4*, 97–105. [CrossRef]
46. Wang, X.; Crowe, P.J.; Goldstein, D.; Yang, J.L. STAT3 inhibition, a novel approach to enhancing targeted therapy in human cancers (review). *Int. J. Oncol.* **2012**, *41*, 1181–1191. [CrossRef] [PubMed]
47. Johnston, P.A.; Grandis, J.R. STAT3 signaling: Anticancer strategies and challenges. *Mol. Interv.* **2011**, *11*, 18–26. [CrossRef] [PubMed]
48. Liu, Y.; Huang, J.; Li, W.; Chen, Y.; Liu, X.; Wang, J. Meta-analysis of STAT3 and phospho-STAT3 expression and survival of patients with breast cancer. *Oncotarget* **2018**, *9*, 13060–13067. [CrossRef]
49. Zou, S.; Tong, Q.; Liu, B.; Huang, W.; Tian, Y. Targeting STAT3 in Cancer Immunotherapy. *Mol. Cancer* **2020**, *19*, 145. [CrossRef] [PubMed]
50. Laudisi, F.; Cherubini, F.; Monteleone, G.; Stolfi, C. STAT3 Interactors as potential therapeutic targets for cancer treatment. *Int. J. Mol. Sci.* **2018**, *19*, 1787. [CrossRef] [PubMed]
51. Niu, G.; Bowman, T.; Huang, M.; Shivers, S.; Reintgen, D.; Daud, A.; Chang, A.; Kraker, A.; Jove, R.; Yu, H. Roles of activated Src and Stat3 signaling in melanoma tumor cell growth. *Oncogene* **2002**, *21*, 7001–7010. [CrossRef] [PubMed]
52. Lee, H.; Jeong, A.J.; Ye, S.K. Highlighted STAT3 as a potential drug target for cancer therapy. *BMB Rep.* **2019**, *52*, 415–423. [CrossRef]
53. Yang, L.; Lin, S.; Xu, L.; Lin, J.; Zhao, C.; Huang, X. Novel activators and small-molecule inhibitors of STAT3 in cancer. *Cytokine Growth Factor Rev.* **2019**, *49*, 10–22. [CrossRef] [PubMed]
54. Huynh, J.; Chand, A.; Gough, D.; Ernst, M. Therapeutically exploiting STAT3 activity in cancer—using tissue repair as a road map. *Nat. Rev. Cancer* **2019**, *19*, 82–96. [CrossRef] [PubMed]

cancers

MDPI

Review

Does Cancer Biology Rely on Parrondo's Principles?

Jean-Pascal Capp [1,*], Aurora M Nedelcu [2], Antoine M Dujon [3], Benjamin Roche [4], Francesco Catania [5], Beata Ujvari [3], Catherine Alix-Panabières [4,5,6] and Frédéric Thomas [4,*]

1 Toulouse Biotechnology Institute, University of Toulouse, INSA, CNRS, INRAE, 31400 Toulouse, France
2 Department of Biology, University of New Brunswick, P.O. Box 4400, Fredericton, NB E3B 5A3, Canada; anedelcu@unb.ca
3 Centre for Integrative Ecology, School of Life and Environmental Sciences, Deakin University, Deakin, VIC 3216, Australia; a.dujon@deakin.edu.au (A.M.D.); beata.ujvari@deakin.edu.au (B.U.)
4 CREEC/CANECEV, MIVEGEC (CREES), Centre de Recherches Ecologiques et Evolutives sur le Cancer, University of Montpellier, CNRS, IRD, 34000 Montpellier, France; benjamin.roche@ird.fr (B.R.); c-panabieres@chu-montpellier.fr (C.A.-P.)
5 Institute for Evolution and Biodiversity, University of Münster, Hüfferstrasse 1, 48149 Münster, Germany; fcata_01@uni-muenster.de
6 Laboratory of Rare Human Circulating Cells (LCCRH), University Medical Centre of Montpellier, 34093 Montpellier, France
* Correspondence: capp@insa-toulouse.fr (J.-P.C.); frederic.thomas2@ird.fr (F.T.)

Simple Summary: Parrondo's paradox, whereby losing strategies or deleterious effects can combine to provide a winning outcome, has been increasingly applied by biologists to explain complex adaptations in many living systems. Here, we suggest that considering this paradox in oncology, particularly in relation to the phenotypic diversity of malignant cells, could also be a promising approach to understand several puzzling aspects of cancer biology. For example, the high genetic and epigenetic instability of cancer cells, their metastatic behavior and their capacity to enter dormancy could be explained by Parrondo's theory. We also discuss the relevance of Parrondo's paradox in a therapeutical framework using different examples. This work provides a compelling argument that the traditional separation between medicine and other disciplines remains a fundamental limitation that needs to be overcome if complex processes, such as oncogenesis, are to be completely understood.

Abstract: Many aspects of cancer biology remain puzzling, including the proliferative and survival success of malignant cells in spite of their high genetic and epigenetic instability as well as their ability to express migrating phenotypes and/or enter dormancy despite possible fitness loss. Understanding the potential adaptive value of these phenotypic traits is confounded by the fact that, when considered separately, they seem to be rather detrimental at the cell level, at least in the short term. Here, we argue that cancer's biology and success could frequently be governed by processes underlying Parrondo's paradox, whereby combinations of intrinsically losing strategies may result in winning outcomes. Oncogenic selection would favor Parrondo's dynamics because, given the environmental adversity in which malignant cells emerge and evolve, alternating between various less optimal strategies would represent the sole viable option to counteract the changing and deleterious environments cells are exposed to during tumorigenesis. We suggest that malignant processes could be viewed through this lens, and we discuss how Parrondo's principles are also important when designing therapies against cancer.

Keywords: cancer; dormancy; metastasis; Parrondo's paradox; therapy

Citation: Capp, J.-P.; Nedelcu, A.M; Dujon, A.M; Roche, B.; Catania, F.; Ujvari, B.; Alix-Panabières, C.; Thomas, F. Does Cancer Biology Rely on Parrondo's Principles?. *Cancers* **2021**, *13*, 2197. https://doi.org/10.3390/cancers13092197

Academic Editors: José I. López and Ildefonso M. de la Fuente

Received: 9 April 2021
Accepted: 29 April 2021
Published: 3 May 2021

1. Introduction

One of the most intriguing features of cancer cell populations is their high levels of stochasticity and plasticity states, especially in advanced cancers. It is increasingly evident that the associated non-genetic intratumoral heterogeneity (ITH) poses a significant challenge to cancer prognosis and treatment. For instance, phenotypic plasticity [1] and the interplay between genetic and non-genetic phenomena [2,3] have been recognized as very important factors in the emergence of resistant cells. It is then essential to fully understand both the proximate and the ultimate causes for the observed increased cellular stochasticity in cancer.

Many studies have provided evidence that cancer evolution and tumor dynamics are characterized by a progressive increase in epigenetic and gene expression diversity [4–6] (see below). In fact, high levels of non-genetic ITH coupled with plasticity might be general features of solid and hematological cancers. This global increase in diversity indicates that the contribution of cells with high plasticity and stem-like states increases during progression, and that their progeny harbor more diverse fates, with less defined gene expression patterns. Such a scenario can also easily explain the progressive dedifferentiation associated with increased aggressiveness that is usually observed in advanced cancers.

The emerging question is why do cancer cell lineages experience such a global increase in stochasticity and transient fluctuations between infrequent expression patterns. In spite of detailed molecular studies providing proximate explanations for the observed increased epigenetic and gene expression diversity [7], ultimate causes and evolutionary explanations are still obscure. High levels of phenotypic heterogeneity associated with stochastic gene expression (even among genetically identical individuals) are also known in microbial populations [8]. Such heterogeneity allows populations to survive in fluctuating environments and promotes interactions among distinct phenotypic subpopulations leading to cell specialization (e.g., [8]). We have previously discussed the parallel between cancer cells and microbial populations from the point of view of cellular stochasticity (related to gene expression variability) and the way such stochasticity can be exploited to produce subpopulations better adapted to a given environment [9]. Specifically, we proposed that oncogenic processes rely on the initial increase in cellular stochasticity associated with cell de-differentiation, followed by the specialization of some cancer sub-populations while maintaining a less specialized lineage (cancer stem cells) with high stochasticity levels [9]. This can explain the co-existence of less specialized cells with higher stochasticity at the epigenetic and transcriptional levels able to diversify into many phenotypes, and more specialized cells with more stable epigenetic and transcriptional profiles that maximize exploitation of available resources in the surrounding environment. In this scenario, such unstable cells would be free to explore all the possible combinations of phenotypes thanks to their high plasticity.

Here, we go a step further by arguing that the observed increased cellular stochasticity and the presence of high plasticity states in advanced cancers could reflect the so-called "Parrondo's paradox", in which combinations of intrinsically losing strategies may result in wining outcomes (Figure 1, Boxes 1 and 2). While high plasticity and transient/fluctuating levels of gene expression could appear detrimental for growth, such traits (hereinafter also referred to as strategies) might be favorable in a highly disrupted and rapidly changing microenvironment, because they can prompt phenotypic diversification and thus enhance the tumor's chance of success. In fact, maintaining a combination of highly plastic cells with transient gene expression and a set of more stable differentiated cells could constitute the only evolutionary strategy able to withstand the changing and deleterious environments that characterize advanced tumorigenesis. Gene expression noise and phenotypic diversification strategies have also been considered survival strategies for microbial populations in stressful environments [10]. We also argue that important characteristics of malignant cells, like their capacity to become dormant or to leave the primary tumor and migrate to other locations in the body, could be strategies that can be understood in the light of Parrondo's paradox.

2. The Origins of the Increased Epigenetic and Transcriptional Heterogeneity in Cancer

The increased epigenetic and transcriptional diversity that characterizes cancer progression could reflect an increased number of discrete cancer states among which cells can randomly fluctuate, resulting in increased phenotypic plasticity. The random fluctuations between distinct phenotypic states in cancer cell populations observed a decade ago argue for such a phenomenon [11]. The most usual hypothesis to explain this dynamic heterogeneity in cancer is based on the notion of cancer attractor state [12,13]. Indeed, cancer cells would explore parts of the global regulatory network that are not accessible to normal cells thanks to a reconfiguration of the epigenetic landscape (as defined by Conrad Waddington [14,15]) with the appearance of new valleys that would correspond to cancer attractors. Together with disruptive factors such as the abnormal microenvironment and the increased stochastic gene expression that would help cells to switch between attractor states, this cancer landscape would allow cells to experience gene expression patterns that are not observed in normal tissues.

The increased epigenetic and transcriptional heterogeneity could also be, at least in part, the result of a global loss of chromatin coordination that is translated into single-cell phenotypic instability. The mutually exclusive activating and repressing histone modifications that co-map in single chronic lymphocytic leukemia cells [6] indicate that cancer cells lose at the epigenomic level the defined hierarchy of normal tissue, and can be characterized by co-occurrence of normally exclusive phenotypic markers. Previous studies already suggested that cancer cells revert to a 'pseudo-primitive' epigenetic status that combines features of embryonic stem cells and of different developing lineages [16], and that stochasticity of gene expression appears to be increased in cancer cells at higher levels than in normal stem cells because of a less organized and less stable chromatin structure [17]. Epigenetic regulators such as such as KDM5 family members have a key role in the generation of higher transcriptomic heterogeneity [7].

As recently shown in lung adenocarcinoma [4], cancer cells are characterized by a continuum of epigenetic states representing loss of cellular identities rather than by discrete and distinct states. This is consistent with previous data revealing that cancer should be conceived as a continuum of heterogeneous phenotype states because gradients of marker expression are observed rather than distinct subpopulations [18]. More generally, instead of being organized into well-categorized and discrete mature cell types carrying out specified functions as in normal tissues, cancer cells harbour increased plasticity and are distributed across a dynamic continuum of states from normal-like states with skewed differentiation to abnormal states [19].

However, some subpopulations are more plastic than others as recently shown in lung cancers [5]. In this case, it is proposed that highly plastic cells can give rise to more diverse fates, and are responsible for the emergence and maintenance of cellular heterogeneity. Thus, only certain subpopulations would really correspond to a state of high plasticity/instability that could be assimilated to a stem-like state with aggressive features, including robust potential for differentiation and proliferation as well as drug resistance, while the other subpopulations would harbor more discrete and stable states. Nevertheless, the highly plastic state does not overlap with the classical normal and cancer stem cell gene expression signatures [5], suggesting that high plasticity does not necessarily imply expression of stemness markers as classically defined.

Epigenetic and transcriptional diversity could originate from cellular reprogramming that results in cells with stemness or plasticity, through pathological cell reprogramming processes involving abnormal stem cell signal activation and suppressor gene inactivation [20]. Especially, epithelial-to-mesenchymal transition (EMT) is expected to be a source of tumor heterogeneity [21] and to contribute to stemness and cell plasticity [22,23]. EMT is a hallmark of many different carcinomas, known to be associated with the initiation of metastasis [22,23]. However, it was recently shown that in breast tumors EMT is an inherent feature of most clones (which were found to harbor a major population of epithelial cells and a smaller and variable subpopulation of mesenchymal cells) [24], suggesting that the differentiation state of tumor cells is inherently unstable or plastic. Finally, since stochastic gene expression is normally controlled by cell-cell interactions [25], tissue disruption could produce a global destabilization of gene expression resulting in increased cellular stochasticity, and especially in high epigenetic and transcriptional diversity [26–28].

3. Parrondo's Paradox in Cancer?

Based on the findings reviewed above, can Parrondo's logics—that is, 'losing/chaos + losing/chaos = winning/order' underlie the persistence and/or progression of malignant tumors? Starting from the observation of a global increase in epigenetic and transcriptional diversity in cancer, the first cancer trait that should be explored through this angle is cellular instability/stochasticity (unstable and variable epigenetic landscape and gene expression pattern).

3.1. Are Both Stability and Instability Losing Strategies for Malignant Cells?

Building on Parrondo's framework, we posit that, by itself, adopting a stable cellular state is a losing strategy because it can result in the death of the malignant cells. However, the association of this strategy with an alternation of transient states (which could be losing strategies on their own) can provide a winning outcome (e.g., by allowing persistence and/or proliferation in a changing environment). Is there evidence to support this possibility?

First indication about a possible involvement of Parrondo's paradox in cancer comes from studies analyzing the role of gene expression variability in drug response. Specifically, it was shown that rare and transient gene expression patterns, which results from the inherent instability of cancer cells, fortuitously confer resistance to chemotherapeutic drugs [29]. In this pioneering work, the authors showed that the rare and transient transcription of a number of resistance markers at high levels in a very small percentage of single melanoma cells is at the origin of non-genetic resistance. Thus, a strategy consisting in increasing cellular stochasticity probably would allow a myriad of subpopulations that transiently harbor rare combinations of expression levels, ultimately leading to survival of a few resistant cells in a highly selective environment. More recently, the same authors found that groups of genes co-fluctuate in "coordinated rare-cell expression programs" that are heritable for several generations but ultimately transient [30]. One can generalize this observation by postulating that the stochastic appearance of a myriad of transient rare subpopulations expressing distinct gene expression patterns could be a general feature of dynamically fluctuating cancer cell populations.

Hinohara et al. also found a prominent role of gene expression variability in the emergence of resistant cells [7]. When the activity of members of the KDM5 demethylase family is inhibited so as to diminish gene expression stochasticity and reduce transcriptomic and phenotypic heterogeneity among estrogen receptor-positive breast cancers, resistance to endocrine therapies is reduced because fewer cells acquire resistance [7]. Thus, there is a clear relationship between the level of cellular stochasticity and the acquisition of drug resistance.

To extend these observations, we propose that this high stochasticity is also globally required for survival in the highly selective conditions found in advanced tumors because low or no variability would make cancer cells unable to deal with the changing disrupted environments that themselves and surrounding cells continuously modify. Cancer cell survival might rely, at least in part, on transient non-genetic changes that, although unfavorable in the current environment, may favor an adaptive response to future conditions (with some cells harboring the 'right' combination of expressed genes at every moment in every place), in line with the dynamic nature of the tumor microenvironment.

Nevertheless, high instability by itself is likely a losing strategy. Indeed, recent single-cell analyses at the epigenomic level revealed no evidence that only highly unstable cells are present in advanced cancers; rather, cells with various levels of plasticity at various stages of tumor progression were found to co-exist [4]. These observations suggest that the co-existence of highly unstable and more stable subpopulations [5,9] should still be required for tumor maintenance and/or progression, even in advanced cancers. Furthermore, there is constant switch between the two states, as cells enriched for stem-like properties (i.e., unstable) known as tumor-initiating cells, can generate non-tumor-initiating cells, and the opposite [11,31], suggesting that the ability to change states (and the ratio between subpopulations expressing different states) is a necessary evolutionary strategy for tumor growth.

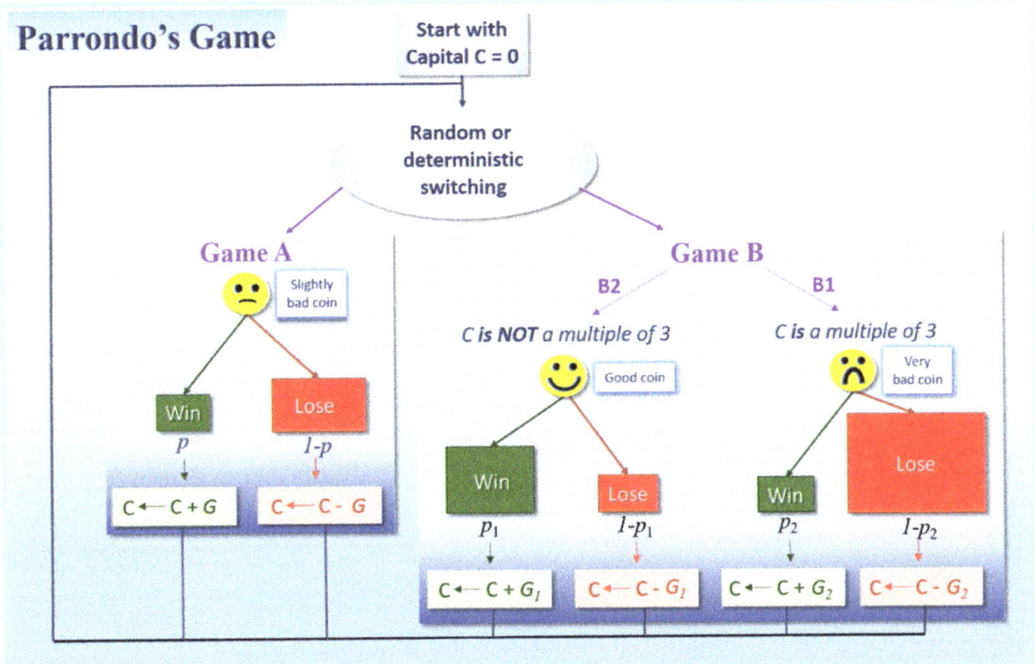

Figure 1. Parrondo's original game, involving two games, A and B (adapted from [32]). See Box 1 for details.

Mathematically speaking, for Parrondo's paradox to be at work, both a stable loosing strategy (Game A in Figure 1, see Box 1 for explanations) and a more stochastic loosing strategy (Game B in Figure 1, see Box 1 for explanations) are required, so that a ratchet effect producing the winning outcome can occur (it does not work without the ratchet). Thus, both a relatively stable cell population and a cell population exhibiting stochasticity are necessary for the Parrondo's paradox to explain a tumor's success (Figure 2a). Since the benefits from cell heterogeneity would depend on population size, the relative proportions between these stable and unstable subpopulations would vary depending on cancer stage, with the most advantageous ratio changing over time (Figure 2b). For instance, relatively more cells exhibiting elevated levels of variability might be needed at later stages to cope with an increasingly unpredictable and unstable micro-environment.

Box 1. Parrondo's paradox.

Can the combination of two individually-losing strategies yield a winning outcome? Since the pioneering work of Harmer et al. [32,33], it has been accepted that this counterintuitive phenomenon, called "Parrondo's paradox", exists in a large variety of contexts [34,35]. A classical way to introduce Parrondo's paradox considers a pair of losing games whose alternation provides a player with a winning outcome (Figure 1). Briefly, the basic principle is as follow: a player has some capital, which is increased by one unit when he/she wins, and decreased by one when he/she loses. In Game A, the player always loses in the long term because the game relies on a biased coin that slightly increases the player's probability to lose each time the game is played (i.e., the loosing probability 1-p is slightly above 0.5). Game B is slightly more complex since there are two options depending on the player's capital value: (i) if the capital is divisible by an integer (let's say 3), the player uses Game B1 relying on a biased coin which is strongly unfavorable (9/10 chance to lose); (ii) if the capital is not divisible by 3, the player uses Game B2 relying on a biased coin that this time is favorable (3/4 chance to win). Despite this latter beneficial option B2, it can be demonstrated with discrete-time Markov analysis that Game B remains also on average a losing game [33,36,37]. Nevertheless, provided the unfavorable biasing parameter in Game A remains small, computer simulations (Brownian ratchet and discrete-time Markov chain) show that a player alternating the two losing games in a random or a deterministic manner will on average yield a winning game [33,36,37]. As it can be intuitively perceived, the construction of game B is a critical factor for paradoxical phenomena to emerge in Parrondo's games (see [38]).

Parrondo's principle has generated significant multidisciplinary interest in the literature, becoming paradigmatic for all situations (including in many biological contexts, see text) where losing strategies or deleterious effects can combine to provide a winning outcome (e.g., [39–41]). Interestingly, by exploiting Parrondo's rule of alternating strategies, it can also be demonstrated that the periodic mixing of two chaotic dynamics can, in certain circumstances, result in ordered dynamics, illustrating a different Parrondian paradoxical phenomenon: "chaos + chaos = order" [42–45]. Although many aspects of cancer biology, like ITH and cellular stochasticity/instability are intriguing and somewhat counterintuitive, until now little attention has been devoted to exploring the extent to which cancer's success could, at least partially, rely on Parrondo's principles. Nevertheless, some applications of the Parrondo's game to cancer in a theoretical modeling and chaos control framework can be found in the literature [46,47], especially showing that switching control parameters can make a previously chaotic tumor growth trajectory nonchaotic, and inversely [47].

Box 2. Parrondo's effects in biology.

Although Parrondo's paradox has initially received a lot of attention from mathematicians and physicists, it is increasingly recognized that the genetic, ecological and evolutionary dynamics of numerous living systems (from genes to populations) can also, to some extent, be influenced by Parrondo's effects. Few examples are presented below (but see [48] for a recent review).

In genetics, Reed [49] used Parrondo's logic to explain how, in a sexually reproducing species, an autosomal allele that is, on average, deleterious for each sex (compared to an alternative allele) can nevertheless increase in frequency, persist in the population, and even continue to fixation. This situation is possible when the detrimental autosomal allele enhances fitness in combination with an allosomal allele in females only (i.e., a positive epistatic interaction coupled with a sexually antagonistic selection). This optimal context, equivalent to Game B2 in Figure 1 (see also Box 1), inevitably occurs in alternation because of genetic inheritance processes arising in sexual reproduction, allowing individually-losing strategies to be temporally intercalated to yield winning outcomes.

In bacteria, random phase variation (RPV; that is, unpredictable transitions between alternative states) is a strategy often favored in environments that are rapidly changing in time or space, even if some of the resulting phenotypes are likely to be at any time maladapted to the current environment. This suboptimal strategy implies that bacteria evolve imperfect ways to detect environmental transitions, with selection actually favoring individuals with sensors of lower accuracy and/or with enhanced signal transduction delays. Because RPV populations display on average a reduced growth rate variance, they may also become in the long term vulnerable to extinction when all extrinsic factor variations are considered. Interestingly, Wolf et al. [50] demonstrated with a game-theory model that a mixed stochastic/deterministic strategy can emerge as an evolutionarily stable strategy. Thus, random alternations between losing strategies likely to produce bacteria with the wrong phase variation or sequence of variations, can result in a winning outcome, especially when the regimes of bacterial switching rates match the rate of environmental instability (see also [51–53]).

Box 2. *Cont.*

Parrondo's effects have also been proposed as a possible explanation for the enigma of populations able to persist in environments exclusively composed of sink habitats. Jansen and Yoshimura [54] showed that such a persistence becomes possible when habitats' quality fluctuates through time and offspring produced can disperse between habitats. In a similar vein, Cheong et al. [55–57] showed that a mix of two losing lifestyles, called nomads and colonists, can result in a winning outcome. In their example, nomads are independent and neither compete nor cooperate, and rely on relatively low levels of resources. Being a nomad is a losing strategy because in the long term, individuals do not reproduce at a rate that compensates losses in death associated with this life-style. Being a colonist is also a losing strategy because even if individuals in this state have access to more resources and interact (cooperation but also competition), sooner or later they deplete their habitat and cannot persist. The switching between these two losing strategies not only ensures that population extinction is avoided, it can even facilitate proliferation if, for instance, colonists (i) strategically switch to a nomadic lifestyle when the over-exploited resources of their habitat become scarce, and (ii) switch back to a colonist lifestyle once novel resources are found (e.g., a novel habitat or once their habitat is replete). The cellular slime mold *Dictyostelium discoideum* may provide a possible example. These amoebae have a life cycle including a single-cellular stage and a multicellular one, when individual amoebae aggregate. Selection for aggregation occurs only in environments where food is slow to replenish, otherwise unicellularity is most of the time favored. While each strategy is intrinsically a losing one in the long term, alternating the two lifestyles depending on the food availability is a winning one [58].

Since research on Parrondo's paradox has been extended into ecology and evolutionary biology, it is increasingly apparent that this phenomenon is an important, if not a major process to consider when attempting to explain numerous general features of life, including adaptive as well as apparently maladaptive traits. For instance, its significance is illustrated by the recent provocative suggestion that the evolution of alternating unicellular and multicellular life history stages that enabled the success of multicellular lineages involved Parrondo's dynamics [59,60]. Surprisingly, the possibility that aspects of cancer's biology and success could, at least partially, rely on Parrondo's dynamics has received little attention.

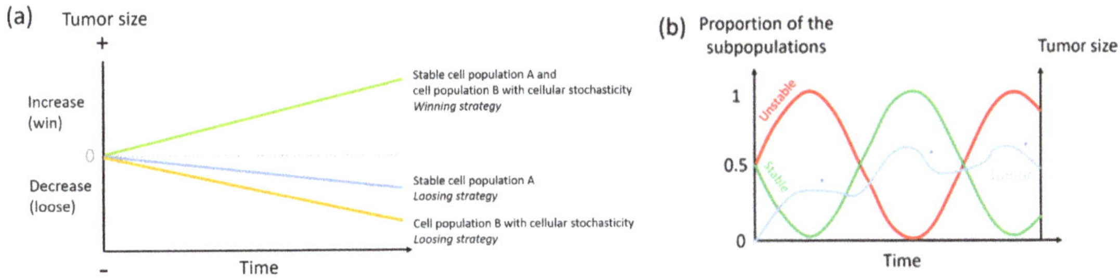

Figure 2. Hypothetical dynamics illustrating that (**a**) the co-existence of relatively stable cell populations and cell populations exhibiting stochasticity rely on the Parrondo's paradox (adapted from [34]), (**b**) the co-existence of a relatively stable cell subpopulation (game A) and a cell subpopulation exhibiting stochasticity (game B) is necessary for tumoral growth (i.e., relying on Parrondo's paradox). Tumor growth is impaired when the size of one subpopulation exceeds (or goes below) an optimal threshold of representation. The slope of the decrease depends on which unique strategy is overrepresented (small arrows). Also, the optimal ratio between the two strategies may vary during tumorigenesis, explaining the different slopes for the same ratios at different times.

3.2. Why Should Cell Populations with High Instability/Stochasticity Be Needed for Long Term Survival and Proliferation?

To date, all theoretical and empirical examples of Parrondo's paradox in biology have considered situations involving the mix of only a few losing strategies, usually two. Given the large diversity of cellular phenotypes expressed through time by unstable and stochastic cells, one might question if these phenotypes are indeed manifestations of different losing strategies, and why Parrondo's dynamics in malignant cells needs so many alternative strategies.

A first possible explanation comes from the fact that a major difference between malignant cells and other living forms is that the former do not possess adaptations finely-tuned by selection over millions of years. Except for transmissible cancers, each cancer must 'reinvent the wheel' as their evolutionary products die with the host. In the same way, the bet-hedging of malignant cells may appear rudimentary/primitive compared to, for example, that of desert plants for which only the seed germination timing is variable. The fine tuning of the potentially adaptive combination of strategies underlying the Parrondo's logic in malignant cells is not possible in such a short time. Differently put, the genesis of a biological substrate underlying a Parrondo's logic in malignant cells could lack sophisticated calibration. If this hypothesis is correct, then Parrondo's dynamics in transmissible cancers, which have the opportunity to evolve over longer period of time (beyond the lifespan of their hosts), should rely on fewer options, resulting in a reduced cellular instability/stochasticity compared to normal cancers. Similarly, examples of Parrondo's dynamics in micro-organisms like bacteria should rely on fewer, better adjusted, losing strategies [50].

A second possible explanation for the elevated number of "losing/suboptimal strategies" in cancer cells is that the environmental conditions experienced by malignant cells are so unstable, diversified and adverse, that any viable Parrondo's dynamics could only rely on a myriad of strategies to be sustainable. From an initially normal cell's perspective, the tumor environment is indeed characterized by a high level of adversity and instability [61,62]. These unfavorable ecological conditions may originate from the host (microenvironment, immune system), but also from the malignant cells themselves since their activities and proliferation largely contribute to altering, in a non-predictable manner, numerous variables in the tumor environment. Because malignant cells typically die with their hosts, these unprecedented ecological conditions result each time in a novel habitat for which no specific (cross-generational) adaptation could have evolved. In this context, a cellular instability/stochasticity exploited within a Parrondo's logic may be selected as a viable option.

Another non-mutually exclusive explanation could be a run-away process, initiated by external ecological factors and then maintained and amplified by internal cell factors. Cellular instability/stochasticity would not only permit malignant cells to cope with unprecedented environmental conditions (see above), but it would also impair the cells' normal functioning by introducing genetic/epigenetic abnormalities. While in response to such abnormalities healthy cells usually activate apoptosis, oncogenic selection in malignant cells is likely to favor increased cellular instability/stochasticity; this diversification will favor many possible survival strategies. However, an increase of cellular instability/stochasticity is likely to introduce additional internal cell impairments that will subsequently reinforce the selection for an exacerbated Parrondo's compensatory response via a novel enhancement of the cellular instability/stochasticity etc. In this runaway scenario, it is expected that the levels of cellular impairment and instability/stochasticity are positively correlated and should increase during tumorigenesis, until a level for which cell instability/stochasticity probably becomes insufficient to compensate, via Parrondo's effects, all cellular dysfunctions.

3.3. *Dormancy: A Losing Strategy in Parrondo's Dynamics?*

Dormancy and quiescence are frequently observed in malignant cells [63]. Although advantageous in adverse conditions, dormancy and quiescence may be considered a losing strategy because cells that would permanently switch toward this lifestyle are exposed to the risk of dying without producing offspring. A switch toward these states of suspended growth seems even more surprising when it occurs in environmental conditions that are a priori not adverse for malignant proliferation [64,65]. Alternating dormancy with other malignant lifestyles could however be interpreted as a mixed strategy shaped by Parrondo's dynamics. Malignant cells that are highly proliferative have the advantage of reproducing rapidly, but experience a higher mortality rate if conditions become unfavorable, e.g., because of changes in the microenvironment and of therapies (i.e., the major focus of therapy development remains on directly targeting viability or proliferation of tumor cells). Conversely, dormant/quiescent malignant cells exhibit a lower reproduction rate, but they benefit from increased survival under adverse conditions, as illustrated by the relapse they cause years, even decades, later. These two phenotypes (i.e., high proliferation and suspended growth) correspond to fast and slow life-history strategies, respectively. In unstable environments, organisms frequently adjust by adopting different phenotypes in response to different external conditions [66]. However, when reliable cues for predicting environmental changes are lacking, and/or when populations did not evolve the capacity to exploit them (as expected for malignant cells that have at best few years of evolution [67]), individual organisms are often constrained to develop a strategy based on a stochastic switching between different phenotypes/stages (i.e., a bet-hedging strategy [68]). Because malignant cells can only respond to direct selection forces without the possibility to anticipate future conditions and dynamics of the ecosystem, dormancy/quiescence may be a part of a generalized bet-hedging strategy. In summary, stochastic alternation between strategies through the entry in and exit from dormancy/quiescence does not maximize fitness within a generation, but it reduces fitness variance and hence maximizes tumor's success under environmental unpredictability in the long term. This ultimately represents a winning outcome based on the combination of two losing strategies—that is, a case of the Parrondo's paradox.

3.4. Is the Metastatic Behavior a Strategy within Parrondo's Paradox?

Metastasis remains the leading cause of mortality for cancer patients [69]. However, the proximate and ultimate causes of this phenomenon are not completely understood [70]. As proposed by many evolutionary ecologists and cancer biologists, metastasis could be considered through the lens of biological dispersal [71]. In the light of the previous literature on the colonist/nomad game (see Box 2), it is useful to explore the metastatic process within the framework of Parrondo's paradox. The fact that less than 0.1% of cells disseminating from the primary tumor form metastases (a phenomenon referred to as 'metastatic inefficiency' [72]) suggests that leaving the primary tumor to disperse is a costly, extremely frequent, losing strategy [73–75]. Can the non-metastatic behavior also be viewed as a losing strategy, i.e., equivalent to a colonist losing strategy? This possibility is consistent with events that unfold in certain parts of solid tumors, especially as resources and/or space inevitably become limiting. Angiogenesis is critical for tumor survival, and once solid tumors reach a substantial size, vascularization becomes irregular in the core regions yielding tumor necrosis [76,77]. Within the Parrondo's logic, there should be a positive correlation between the frequency of necrosis phenomena and the propensity of malignant cells to adopt a dispersing lifestyle. Accordingly, clinical observations clearly indicate that tumor necrosis of solid tumors is an accurate indication of metastatic tumors [78–84]. In the same context, a link has been established between acidification [85] or the lack of oxygen [86], and the propensity to metastasis, revealing that a deleterious environment for the colonists in the primary tumor can lead to another mainly inefficient and deadly strategy for malignant cells, with very rare successes. Thus, both loosing strategies could lead to a winning outcome when a few metastatic cells succeed in dispersing, invading new tissues and resume proliferation, because they have both the prerequisite malignant phenotypes and the available resource and space in the new environment.

4. Therapeutic Implications

Malignant cells can become resistant to many different types of drugs [87,88]. High and continuous doses of drugs typically allow resistant cancer cells to win twice. First, they are not killed; second, they are not outcompeted by sensitive cancer cells. Because of this combination of effects, treating cancers with such protocols seems to be a losing strategy because it ultimately results in fatal disseminated cancers. However, applying no treatment is often a losing strategy too because cancer progression is not prevented, yielding also to disseminated cancers that kill patients. Alternating these two strategies can produce a better option, as elegantly illustrated by the adaptive therapy approach [89]. Acquisition of chemo-resistance generally requires significant investment of resources, and because of the 'cost' of phenotypic resistance, cancer cells are subject to an evolutionary trade-off between resistance and proliferation. Adaptive therapy alternates treatment and treatment breaks in an adaptive fashion to enforce a stable tumor burden by permitting the persistence of a significant population of chemo-sensitive cells. In so doing, chemo-sensitive cells can compete with chemo-resistant subpopulations hence limiting their expansion. Instead of introducing treatment breaks, another strategy consists in providing fake drugs. Cells that are resistant to multiple drugs often have efflux pumps that remove the drugs. The cell pumps, however, require energy to run. Fake drugs proposed by Kam et al. [90], also called 'ersatzdroges', are non (or minimally) toxic substances that activate efflux pumps in resistant cells and cause them to expend energy, without actually giving these cells a survival benefit over non-resistant cells. Applying only these ersatzdroges is a losing strategy because it does not prevent the proliferation of cells that do not possess pumps and the tumor can continue to grow. However, alternating a fake drug with a real drug can keep the size of tumor constant, without resistance selection.

Inspired by the processes responsible for the extinction of species, Gatenby et al. [91] demonstrated that once an initial therapy reduced population size and diversity of a large tumor, several less aggressive therapies, each unable to eradicate large tumors, can successfully eliminate small and spatially fragmented malignancies, without selecting for resistance (e.g., neo-adjuvant chemotherapy). Again, while each category of treatment is unable to prevent cancer progression, alternating them in the order proposed above yields a relevant therapeutic option, illustrating the use of Parrondo's paradox in a therapeutical framework.

5. Concluding Remarks

Although tumor progression has stimulated a large number of clinical and theoretical studies, its underlying mechanisms still remain elusive. The extent to which oncogenic selection promotes malignant strategies relying on Parrondo's paradox principles is a legitimate question, at least because the instability and the multidimensionality of cancer cells' phenotypes do not systematically correspond to alterations that effectively contribute to increase their fitness at any time. Studies that focus on occasional snapshots of the tumor state cannot capture these Parrondo's dynamics. Further work is also necessary to explore the extent to which malignant cell switching rates match optimally or not the rate of micro environmental instability (i.e., as for the random phase variation with bacteria discussed above). Another relevant direction would be to test the hypothesis that selection for Parrondo dynamics generates syndromes in malignant cells, that is the simultaneous alteration of multiple phenotypic traits, that could appear not optimal individually but that collectively produce a winning outcome. These syndromes would result from some major physiological disruptions reflecting the need to combine many losing strategies to acquire full malignancy. The complexity and the sub-optimality of malignant phenotypes could, thus, at least partially, result from few major physiological effects selected under Parrondo's logic, followed by a cascade of phenotypic effects forming a syndrome. In conclusion, future exploration of the Parrondo's paradox and its relevance to the phenotypic diversity of tumor cells could provide novel insight into the complexity of cancer dynamics and progression.

Author Contributions: All authors (J.-P.C., A.M.N., A.M.D., B.R., F.C., B.U., C.A.-P. and F.T.) have participated to the conceptualization and the writing of the manuscript. All authors have read and agreed to the published version of the manuscript.

Funding: This work is supported by an ANR TRANSCAN (ANR-18-CE35-0009), the MAVA Foundation, a CNRS International Associated Laboratory Grant, the National Institute of Cancer (INCa, http://www.e-cancer.fr), the SIRIC Montpellier Cancer Grant INCa_Inserm_DGOS_12553, as well as by the European Union Horizon 2020 Research and Innovation program under the Marie Skłodowska-Curie grant agreement No 765492.

Institutional Review Board Statement: Not applicable.

Informed Consent Statement: Not applicable.

Data Availability Statement: Not applicable.

Conflicts of Interest: Catherine Alix-Panabières received honorarium from Menarini.

References

1. Boumahdi, S.; de Sauvage, F.J. The great escape: Tumour cell plasticity in resistance to targeted therapy. *Nat. Rev. Drug Discov.* **2020**, *19*, 39–56. [CrossRef]
2. Bell, C.C.; Gilan, O. Principles and mechanisms of non-genetic resistance in cancer. *Br. J. Cancer* **2020**, *122*, 465–472. [CrossRef]
3. Salgia, R.; Kulkarni, P. The Genetic/Non-genetic Duality of Drug 'Resistance' in Cancer. *Trends Cancer* **2018**, *4*, 110–118. [CrossRef]
4. LaFave, L.M.; Kartha, V.K.; Ma, S.; Meli, K.; Del Priore, I.; Lareau, C.; Naranjo, S.; Westcott, P.M.K.; Duarte, F.M.; Sankar, V.; et al. Epigenomic State Transitions Characterize Tumor Progression in Mouse Lung Adenocarcinoma. *Cancer Cell* **2020**, *38*, 212–228.e213. [CrossRef]
5. Marjanovic, N.D.; Hofree, M.; Chan, J.E.; Canner, D.; Wu, K.; Trakala, M.; Hartmann, G.G.; Smith, O.C.; Kim, J.Y.; Evans, K.V.; et al. Emergence of a High-Plasticity Cell State during Lung Cancer Evolution. *Cancer Cell* **2020**, *38*, 229–246.e213. [CrossRef] [PubMed]
6. Pastore, A.; Gaiti, F.; Lu, S.X.; Brand, R.M.; Kulm, S.; Chaligne, R.; Gu, H.; Huang, K.Y.; Stamenova, E.K.; Beguelin, W.; et al. Corrupted coordination of epigenetic modifications leads to diverging chromatin states and transcriptional heterogeneity in CLL. *Nat. Commun.* **2019**, *10*, 1874. [CrossRef]
7. Hinohara, K.; Wu, H.J.; Vigneau, S.; McDonald, T.O.; Igarashi, K.J.; Yamamoto, K.N.; Madsen, T.; Fassl, A.; Egri, S.B.; Papanastasiou, M.; et al. KDM5 Histone Demethylase Activity Links Cellular Transcriptomic Heterogeneity to Therapeutic Resistance. *Cancer Cell* **2018**, *34*, 939–953.e939. [CrossRef] [PubMed]
8. Ackermann, M. A functional perspective on phenotypic heterogeneity in microorganisms. *Nat. Rev. Microbiol.* **2015**, *13*, 497–508. [CrossRef] [PubMed]
9. Capp, J.P.; Thomas, F. A Similar Speciation Process Relying on Cellular Stochasticity in Microbial and Cancer Cell Populations. *iScience* **2020**, *23*, 101531. [CrossRef]
10. Fraser, D.; Kaern, M. A chance at survival: Gene expression noise and phenotypic diversification strategies. *Mol. Microbiol.* **2009**, *71*, 1333–1340. [CrossRef]
11. Gupta, P.B.; Fillmore, C.M.; Jiang, G.; Shapira, S.D.; Tao, K.; Kuperwasser, C.; Lander, E.S. Stochastic state transitions give rise to phenotypic equilibrium in populations of cancer cells. *Cell* **2011**, *146*, 633–644. [CrossRef]
12. Huang, S.; Ernberg, I.; Kauffman, S. Cancer attractors: A systems view of tumors from a gene network dynamics and developmental perspective. *Semin. Cell Dev. Biol.* **2009**, *20*, 869–876. [CrossRef] [PubMed]
13. Li, Q.; Wennborg, A.; Aurell, E.; Dekel, E.; Zou, J.Z.; Xu, Y.; Huang, S.; Ernberg, I. Dynamics inside the cancer cell attractor reveal cell heterogeneity, limits of stability, and escape. *Proc. Natl. Acad. Sci. USA* **2016**, *113*, 2672–2677. [CrossRef]
14. Waddington, C.H. *The Strategy of the Genes*; A Discussion of Some Aspects of Theoretical Biology; Allen & Unwin: London, UK, 1957; 262p.
15. Goldberg, A.D.; Allis, C.D.; Bernstein, E. Epigenetics: A landscape takes shape. *Cell* **2007**, *128*, 635–638. [CrossRef]
16. Stergachis, A.B.; Neph, S.; Reynolds, A.; Humbert, R.; Miller, B.; Paige, S.L.; Vernot, B.; Cheng, J.B.; Thurman, R.E.; Sandstrom, R.; et al. Developmental fate and cellular maturity encoded in human regulatory DNA landscapes. *Cell* **2013**, *154*, 888–903. [CrossRef] [PubMed]
17. Jenkinson, G.; Pujadas, E.; Goutsias, J.; Feinberg, A.P. Potential energy landscapes identify the information-theoretic nature of the epigenome. *Nat. Genet.* **2017**, *49*, 719–729. [CrossRef] [PubMed]
18. Amir el, A.D.; Davis, K.L.; Tadmor, M.D.; Simonds, E.F.; Levine, J.H.; Bendall, S.C.; Shenfeld, D.K.; Krishnaswamy, S.; Nolan, G.P.; Pe'er, D. viSNE enables visualization of high dimensional single-cell data and reveals phenotypic heterogeneity of leukemia. *Nat. Biotechnol.* **2013**, *31*, 545–552. [CrossRef]
19. Lawson, D.A.; Kessenbrock, K.; Davis, R.T.; Pervolarakis, N.; Werb, Z. Tumour heterogeneity and metastasis at single-cell resolution. *Nat. Cell Biol.* **2018**, *20*, 1349–1360. [CrossRef]
20. Xiong, S.; Feng, Y.; Cheng, L. Cellular Reprogramming as a Therapeutic Target in Cancer. *Trends Cell Biol.* **2019**, *29*, 623–634. [CrossRef]
21. Udyavar, A.R.; Wooten, D.J.; Hoeksema, M.; Bansal, M.; Califano, A.; Estrada, L.; Schnell, S.; Irish, J.M.; Massion, P.P.; Quaranta, V. Novel Hybrid Phenotype Revealed in Small Cell Lung Cancer by a Transcription Factor Network Model That Can Explain Tumor Heterogeneity. *Cancer Res.* **2017**, *77*, 1063–1074. [CrossRef]
22. Jolly, M.K.; Huang, B.; Lu, M.; Mani, S.A.; Levine, H.; Ben-Jacob, E. Towards elucidating the connection between epithelial-mesenchymal transitions and stemness. *J. R. Soc. Interface* **2014**, *11*, 20140962. [CrossRef]
23. Dongre, A.; Weinberg, R.A. New insights into the mechanisms of epithelial-mesenchymal transition and implications for cancer. *Nat. Rev. Mol. Cell Biol.* **2019**, *20*, 69–84. [CrossRef] [PubMed]
24. Rios, A.C.; Capaldo, B.D.; Vaillant, F.; Pal, B.; van Ineveld, R.; Dawson, C.A.; Chen, Y.; Nolan, E.; Fu, N.Y.; Group, D.; et al. Intraclonal Plasticity in Mammary Tumors Revealed through Large-Scale Single-Cell Resolution 3D Imaging. *Cancer Cell* **2019**, *35*, 618–632.e616. [CrossRef] [PubMed]

25. Capp, J.P.; Laforge, B. A Darwinian and Physical Look at Stem Cell Biology Helps Understanding the Role of Stochasticity in Development. *Front. Cell Dev. Biol.* **2020**, *8*, 659. [CrossRef]
26. Capp, J.P. Tissue disruption increases stochastic gene expression thus producing tumors: Cancer initiation without driver mutation. *Int. J. Cancer* **2017**, *140*, 2408–2413. [CrossRef] [PubMed]
27. Capp, J.P.; Bataille, R. Multiple Myeloma as a Bone Disease? The Tissue Disruption-Induced Cell Stochasticity (TiDiS) Theory. *Cancers* **2020**, *12*, 2158. [CrossRef]
28. Capp, J.P.; Thomas, F. Tissue-disruption-induced cellular stochasticity and epigenetic drift: Common origins of aging and cancer? *Bioessays* **2021**, *43*, e2000140. [CrossRef]
29. Shaffer, S.M.; Dunagin, M.C.; Torborg, S.R.; Torre, E.A.; Emert, B.; Krepler, C.; Beqiri, M.; Sproesser, K.; Brafford, P.A.; Xiao, M.; et al. Rare cell variability and drug-induced reprogramming as a mode of cancer drug resistance. *Nature* **2017**, *546*, 431–435. [CrossRef]
30. Shaffer, S.M.; Emert, B.L.; Reyes Hueros, R.A.; Cote, C.; Harmange, G.; Schaff, D.L.; Sizemore, A.E.; Gupte, R.; Torre, E.; Singh, A.; et al. Memory Sequencing Reveals Heritable Single-Cell Gene Expression Programs Associated with Distinct Cellular Behaviors. *Cell* **2020**, *182*, 947–959.e917. [CrossRef]
31. Magee, J.A.; Piskounova, E.; Morrison, S.J. Cancer stem cells: Impact, heterogeneity, and uncertainty. *Cancer Cell* **2012**, *21*, 283–296. [CrossRef]
32. Harmer, G.P.; Abbott, D. Losing strategies can win by Parrondo's paradox. *Nature* **1999**, *402*, 864. [CrossRef]
33. Harmer, G.P.; Abbott, D.; Taylor, P.G.; Parrondo, J.M. Brownian ratchets and Parrondo's games. *Chaos* **2001**, *11*, 705–714. [CrossRef] [PubMed]
34. Shu, J.J.; Wang, Q.W. Beyond Parrondo's paradox. *Sci. Rep.* **2014**, *4*, 4244. [CrossRef] [PubMed]
35. Lai, J.W.; Cheong, K.H. Parrondo's paradox from classical to quantum: A review. *Nonlinear Dyn.* **2020**, *100*, 849–861. [CrossRef]
36. Parrondo, J.M.; Harmer, G.P.; Abbott, D. New paradoxical games based on brownian ratchets. *Phys. Rev. Lett.* **2000**, *85*, 5226–5229. [CrossRef] [PubMed]
37. Arena, P.; Fazzino, S.; Fortuna, L.; Maniscalco, P. Game theory and non-linear dynamics: The Parrondo Paradox case study. *Chaos Solit. Fract.* **2003**, *17*, 545–555. [CrossRef]
38. Ye, Y.; Wang, L.; Xie, N. Parrondo's games based on complex networks and the paradoxical effect. *PLoS ONE* **2013**, *8*, e67924. [CrossRef]
39. Harmer, G.P.; Abbott, D.; Taylor, P.G. The paradox of Parrondo's games. *Proc. R. Soc. Lond. Ser. A Math. Phys. Eng. Sci.* **2000**, *456*, 247–259. [CrossRef]
40. Amengual, P.; Allison, A.; Toral, R.; Abbott, D. Discrete–time ratchets, the Fokker–Planck equation and Parrondo's paradox. *Proc. R. Soc. Lond. Ser. A Math. Phys. Eng. Sci.* **2004**, *460*, 2269–2284. [CrossRef]
41. Abbott, D. Asymmetry and disorder: A decade of Parrondo's paradox. *Fluct. Noise Lett.* **2010**, *09*, 129–156. [CrossRef]
42. Allison, A.; Abbott, D. Control systems with stochastic feedback. *Chaos* **2001**, *11*, 715–724. [CrossRef]
43. Almeida, J.; Peralta-Salas, D.; Romera, M. Can two chaotic systems give rise to order? *Physica D* **2005**, *200*, 124–132. [CrossRef]
44. Boyarsky, A.; Góra, P.; Islam, M.S. Randomly chosen chaotic maps can give rise to nearly ordered behavior. *Physica D* **2005**, *210*, 284–294. [CrossRef]
45. Fulai, W. Improvement and empirical research on chaos control by theory of "chaos + chaos = order". *Chaos* **2012**, *22*, 043145. [CrossRef]
46. Danca, M.F.; Lai, D. Parrondo's game model to find numerically stable attraactors of a tumor growth model. *Int. J. Bifurcat Chaos* **2012**, *22*, 1250258. [CrossRef]
47. Danca, M.F.; Romera, M.; Pastor, G.; Montoya, F. Finding attractors of continuous-time systems by parameter switching. *Nonlinear Dyn.* **2012**, *67*, 2317–2342. [CrossRef]
48. Cheong, K.H.; Koh, J.M.; Jones, M.C. Paradoxical Survival: Examining the Parrondo Effect across Biology. *Bioessays* **2019**, *41*, e1900027. [CrossRef] [PubMed]
49. Reed, F.A. Two-locus epistasis with sexually antagonistic selection: A genetic Parrondo's paradox. *Genetics* **2007**, *176*, 1923–1929. [CrossRef]
50. Wolf, D.M.; Vazirani, V.V.; Arkin, A.P. Diversity in times of adversity: Probabilistic strategies in microbial survival games. *J. Theor. Biol.* **2005**, *234*, 227–253. [CrossRef]
51. Kussell, E.; Leibler, S. Phenotypic diversity, population growth, and information in fluctuating environments. *Science* **2005**, *309*, 2075–2078. [CrossRef]
52. Acar, M.; Mettetal, J.T.; van Oudenaarden, A. Stochastic switching as a survival strategy in fluctuating environments. *Nat. Genet.* **2008**, *40*, 471–475. [CrossRef] [PubMed]
53. Cheong, K.H.; Tan, Z.X.; Xie, N.G.; Jones, M.C. A Paradoxical Evolutionary Mechanism in Stochastically Switching Environments. *Sci. Rep.* **2016**, *6*, 34889. [CrossRef]
54. Jansen, V.A.; Yoshimura, J. Populations can persist in an environment consisting of sink habitats only. *Proc. Natl. Acad. Sci. USA* **1998**, *95*, 3696–3698. [CrossRef] [PubMed]
55. Tan, Z.X.; Cheong, K.H. Nomadic-colonial life strategies enable paradoxical survival and growth despite habitat destruction. *Elife* **2017**, *6*. [CrossRef]

56. Cheong, K.H.; Tan, Z.X.; Ling, Y.H. A time-based switching scheme for nomadic-colonial alternation under noisy conditions. *Commun. Nonlinear Sci. Numer. Simul.* **2018**, *60*, 107–114. [CrossRef]
57. Koh, J.M.; Xie, N.; Cheong, K.H. Nomadic-colonial switching with stochastic noise: Subsidence-recovery cycles and long-term growth. *Nonlinear Dyn.* **2018**, *94*, 1467–1477. [CrossRef]
58. Tarnita, C.E.; Washburne, A.; Martinez-Garcia, R.; Sgro, A.E.; Levin, S.A. Fitness tradeoffs between spores and nonaggregating cells can explain the coexistence of diverse genotypes in cellular slime molds. *Proc. Natl. Acad. Sci. USA* **2015**, *112*, 2776–2781. [CrossRef] [PubMed]
59. Nelson, P.; Masel, J. Intercellular competition and the inevitability of multicellular aging. *Proc. Natl. Acad. Sci. USA* **2017**, *114*, 12982–12987. [CrossRef]
60. Cheong, K.H.; Koh, J.M.; Jones, M.C. Multicellular survival as a consequence of Parrondo's paradox. *Proc. Natl. Acad. Sci. USA* **2018**, *115*, E5258–E5259. [CrossRef]
61. Bindra, R.S.; Glazer, P.M. Genetic instability and the tumor microenvironment: Towards the concept of microenvironment-induced mutagenesis. *Mutat. Res.* **2005**, *569*, 75–85. [CrossRef]
62. Sonugur, F.G.; Akbulut, H. The Role of Tumor Microenvironment in Genomic Instability of Malignant Tumors. *Front. Genet.* **2019**, *10*, 1063. [CrossRef]
63. Ammerpohl, O.; Hattermann, K.; Held-Feindt, J.; Röcken, C.; Schäfer, H.; Schem, C.; Schewe, D.; Schulenburg, H.; Sebens, S.; Synowitz, M.; et al. Chapter 20-Dormancy: An Evolutionary Key Phenomenon in Cancer Developmenta. In *Ecology and Evolution of Cancer*; Ujvari, B., Roche, B., Thomas, F., Eds.; Academic Press: Cambridge, MA, USA, 2017; pp. 235–242. [CrossRef]
64. Almog, N. Molecular mechanisms underlying tumor dormancy. *Cancer Lett.* **2010**, *294*, 139–146. [CrossRef] [PubMed]
65. Paez, D.; Labonte, M.J.; Bohanes, P.; Zhang, W.; Benhanim, L.; Ning, Y.; Wakatsuki, T.; Loupakis, F.; Lenz, H.J. Cancer dormancy: A model of early dissemination and late cancer recurrence. *Clin. Cancer Res.* **2012**, *18*, 645–653. [CrossRef]
66. West-Eberhard, M.J. Phenotypic Plasticity and the Origins of Diversity. *Annu. Rev. Ecol. Syst.* **1989**, *20*, 249–278. [CrossRef]
67. Arnal, A.; Ujvari, B.; Crespi, B.; Gatenby, R.A.; Tissot, T.; Vittecoq, M.; Ewald, P.W.; Casali, A.; Ducasse, H.; Jacqueline, C.; et al. Evolutionary perspective of cancer: Myth, metaphors, and reality. *Evol. Appl.* **2015**, *8*, 541–544. [CrossRef] [PubMed]
68. Childs, D.Z.; Metcalf, C.J.; Rees, M. Evolutionary bet-hedging in the real world: Empirical evidence and challenges revealed by plants. *Proc. Biol. Sci.* **2010**, *277*, 3055–3064. [CrossRef]
69. Dillekas, H.; Rogers, M.S.; Straume, O. Are 90% of deaths from cancer caused by metastases? *Cancer Med.* **2019**, *8*, 5574–5576. [CrossRef]
70. Lloyd, M.C.; Gatenby, R.A.; Brown, J.S. Chapter 11-Ecology of the Metastatic Process. In *Ecology and Evolution of Cancer*; Ujvari, B., Roche, B., Thomas, F., Eds.; Academic Press: Cambridge, MA, USA, 2017; pp. 153–165. [CrossRef]
71. Tissot, T.; Massol, F.; Ujvari, B.; Alix-Panabieres, C.; Loeuille, N.; Thomas, F. Metastasis and the evolution of dispersal. *Proc. Biol. Sci.* **2019**, *286*, 20192186. [CrossRef]
72. Luzzi, K.J.; MacDonald, I.C.; Schmidt, E.E.; Kerkvliet, N.; Morris, V.L.; Chambers, A.F.; Groom, A.C. Multistep nature of metastatic inefficiency: Dormancy of solitary cells after successful extravasation and limited survival of early micrometastases. *Am. J. Pathol.* **1998**, *153*, 865–873. [CrossRef]
73. Chambers, A.F.; Groom, A.C.; MacDonald, I.C. Dissemination and growth of cancer cells in metastatic sites. *Nat. Rev. Cancer* **2002**, *2*, 563–572. [CrossRef]
74. Mehlen, P.; Puisieux, A. Metastasis: A question of life or death. *Nat. Rev. Cancer* **2006**, *6*, 449–458. [CrossRef]
75. Ganesh, K.; Massague, J. Targeting metastatic cancer. *Nat. Med.* **2021**, *27*, 34–44. [CrossRef] [PubMed]
76. Brown, J.M.; Wilson, W.R. Exploiting tumour hypoxia in cancer treatment. *Nat. Rev. Cancer* **2004**, *4*, 437–447. [CrossRef] [PubMed]
77. Gatenby, R.A.; Gillies, R.J. Why do cancers have high aerobic glycolysis? *Nat. Rev. Cancer* **2004**, *4*, 891–899. [CrossRef]
78. Richards, C.H.; Mohammed, Z.; Qayyum, T.; Horgan, P.G.; McMillan, D.C. The prognostic value of histological tumor necrosis in solid organ malignant disease: A systematic review. *Future Oncol.* **2011**, *7*, 1223–1235. [CrossRef]
79. Caruso, R.; Parisi, A.; Bonanno, A.; Paparo, D.; Quattrocchi, E.; Branca, G.; Scardigno, M.; Fedele, F. Histologic coagulative tumour necrosis as a prognostic indicator of aggressiveness in renal, lung, thyroid and colorectal carcinomas: A brief review. *Oncol. Lett.* **2012**, *3*, 16–18. [CrossRef]
80. Makki, J. Diversity of Breast Carcinoma: Histological Subtypes and Clinical Relevance. *Clin. Med. Insights Pathol.* **2015**, *8*, 23–31. [CrossRef] [PubMed]
81. Wang, T.; Jin, Y.; Yang, W.; Zhang, L.; Jin, X.; Liu, X.; He, Y.; Li, X. Necroptosis in cancer: An angel or a demon? *Tumour Biol.* **2017**, *39*, 1010428317711539. [CrossRef] [PubMed]
82. Jiao, D.; Cai, Z.; Choksi, S.; Ma, D.; Choe, M.; Kwon, H.J.; Baik, J.Y.; Rowan, B.G.; Liu, C.; Liu, Z.G. Necroptosis of tumor cells leads to tumor necrosis and promotes tumor metastasis. *Cell Res.* **2018**, *28*, 868–870. [CrossRef]
83. Liu, Z.G.; Jiao, D. Necroptosis, tumor necrosis and tumorigenesis. *Cell Stress* **2020**, *4*, 1–8. [CrossRef]
84. Della Torre, L.; Nebbioso, A.; Stunnenberg, H.G.; Martens, J.H.A.; Carafa, V.; Altucci, L. The Role of Necroptosis: Biological Relevance and Its Involvement in Cancer. *Cancers* **2021**, *13*, 684. [CrossRef]
85. Gatenby, R.A.; Gawlinski, E.T.; Gmitro, A.F.; Kaylor, B.; Gillies, R.J. Acid-mediated tumor invasion: A multidisciplinary study. *Cancer Res.* **2006**, *66*, 5216–5223. [CrossRef] [PubMed]

86. Donato, C.; Kunz, L.; Castro-Giner, F.; Paasinen-Sohns, A.; Strittmatter, K.; Szczerba, B.M.; Scherrer, R.; Di Maggio, N.; Heusermann, W.; Biehlmaier, O.; et al. Hypoxia Triggers the Intravasation of Clustered Circulating Tumor Cells. *Cell Rep.* **2020**, *32*, 108105. [CrossRef] [PubMed]
87. Ujvari, B.; Roche, B.; Thomas, F. *Ecology and Evolution of Cancer*; Academic Press: Cambridge, MA, USA, 2017.
88. Gatenby, R.; Brown, J. The Evolution and Ecology of Resistance in Cancer Therapy. *Cold Spring Harb Perspect Med.* **2018**, *8*. [CrossRef]
89. Gatenby, R.A.; Silva, A.S.; Gillies, R.J.; Frieden, B.R. Adaptive therapy. *Cancer Res.* **2009**, *69*, 4894–4903. [CrossRef] [PubMed]
90. Kam, Y.; Das, T.; Tian, H.; Foroutan, P.; Ruiz, E.; Martinez, G.; Minton, S.; Gillies, R.J.; Gatenby, R.A. Sweat but no gain: Inhibiting proliferation of multidrug resistant cancer cells with "ersatzdroges". *Int. J. Cancer* **2015**, *136*, E188–E196. [CrossRef] [PubMed]
91. Gatenby, R.A.; Artzy-Randrup, Y.; Epstein, T.; Reed, D.R.; Brown, J.S. Eradicating Metastatic Cancer and the Eco-Evolutionary Dynamics of Anthropocene Extinctions. *Cancer Res.* **2020**, *80*, 613–623. [CrossRef]

cancers

Review

PPARs and Tumor Microenvironment: The Emerging Roles of the Metabolic Master Regulators in Tumor Stromal–Epithelial Crosstalk and Carcinogenesis

Hong Sheng Cheng [1,*], Yun Sheng Yip [1], Eldeen Kai Yi Lim [2], Walter Wahli [1,3,4] and Nguan Soon Tan [1,2,*]

1 Lee Kong Chian School of Medicine, Nanyang Technological University Singapore, 11 Mandalay Road, Singapore 308232, Singapore; ysyip@ntu.edu.sg (Y.S.Y.); walter.wahli@ntu.edu.sg (W.W.)
2 School of Biological Sciences, Nanyang Technological University Singapore, 60 Nanyang Drive, Singapore 637551, Singapore; LI0027YI@e.ntu.edu.sg
3 Toxalim (Research Center in Food Toxicology), INRAE, ENVT, INP-PURPAN, UMR 1331, UPS, Université de Toulouse, 31300 Toulouse, France
4 Center for Integrative Genomics, Université de Lausanne, Le Génopode, CH-1015 Lausanne, Switzerland
* Correspondence: hscheng@ntu.edu.sg (H.S.C.); nstan@ntu.edu.sg (N.S.T.); Tel.: +65-6904-1295 (N.S.T.)

Simple Summary: The roles of peroxisome proliferator-activated receptors (PPARs) in carcinogenesis are increasingly appreciated. With the growing interest in tumor stromal-epithelial crosstalk, we aim to provide an up-to-date overview of the implications of PPARs in the tumor microenvironment. In the tumor stromal cells, the nuclear receptors exhibit critical, but functionally diverse activities, rendering it hard to ascribe either an exclusive pro- or anti-tumorigenic role for different PPAR isotypes. Based on the existing evidence, we also highlight the knowledge gaps and future prospects of targeting PPARs in the tumor microenvironment. Essentially, a PPAR-based anticancer approach holds a great deal of untapped potential, but its success relies on innovative strategies for cell-specific or tumor microenvironment-triggered drug delivery systems.

Abstract: Peroxisome proliferator-activated receptors (PPARs) have been extensively studied for more than three decades. Consisting of three isotypes, PPARα, γ, and β/δ, these nuclear receptors are regarded as the master metabolic regulators which govern many aspects of the body energy homeostasis and cell fate. Their roles in malignancy are also increasingly recognized. With the growing interest in crosstalk between tumor stroma and epithelium, this review aims to highlight the current knowledge on the implications of PPARs in the tumor microenvironment. PPARγ plays a crucial role in the metabolic reprogramming of cancer-associated fibroblasts and adipocytes, coercing the two stromal cells to become substrate donors for cancer growth. Fibroblast PPARβ/δ can modify the risk of tumor initiation and cancer susceptibility. In endothelial cells, PPARβ/δ and PPARα are pro- and anti-angiogenic, respectively. Although the angiogenic role of PPARγ remains ambiguous, it is a crucial regulator in autocrine and paracrine signaling of cancer-associated fibroblasts and tumor-associated macrophages/immune cells. Of note, angiopoietin-like 4 (ANGPTL4), a secretory protein encoded by a target gene of PPARs, triggers critical oncogenic processes such as inflammatory signaling, extracellular matrix derangement, anoikis resistance and metastasis, making it a potential drug target for cancer treatment. To conclude, PPARs in the tumor microenvironment exhibit oncogenic activities which are highly controversial and dependent on many factors such as stromal cell types, cancer types, and oncogenesis stages. Thus, the success of PPAR-based anticancer treatment potentially relies on innovative strategies to modulate PPAR activity in a cell type-specific manner.

Keywords: peroxisome proliferation-activated receptor; metabolic reprogramming; cancer-associated fibroblast; cancer-associated adipocyte; tumor-associated macrophage

Citation: Cheng, H.S.; Yip, Y.S.; Lim, E.K.Y.; Wahli, W.; Tan, N.S. PPARs and Tumor Microenvironment: The Emerging Roles of the Metabolic Master Regulators in Tumor Stromal–Epithelial Crosstalk and Carcinogenesis. *Cancers* **2021**, *13*, 2153. https://doi.org/10.3390/cancers13092153

Academic Editors: José I. López and Ildefonso M. de la Fuente

Received: 31 March 2021
Accepted: 26 April 2021
Published: 29 April 2021

1. Introduction

The year 2020 marks the 30-year discovery of nuclear hormone receptor, peroxisome proliferator-activated receptors (PPARs). In 1990, the first isotype of PPAR, now called PPARα, was successfully cloned from the mouse liver and identified as a novel nuclear receptor that is essential for triglyceride and cholesterol homeostasis [1]. Two years later, all three PPAR isotypes, namely PPARα, PPARβ/δ, and PPARγ, were isolated from the *Xenopus laevis* ovary and liver [2]. The research on PPARs has expanded exponentially ever since. Compelling evidence supports their roles as master regulators in metabolism and body energy homeostasis [3]. The clinical significance of PPARs is underscored by their synthetic ligands which are used to treat different facets of metabolic syndrome. Even before the discovery of PPARs, fibrates, which are PPARα agonists, have been used as lipid-lowering drugs and continue to be a mainstream therapy for atherogenic dyslipidemia and atherosclerosis [4]. Major synthetic PPARγ agonists, the thiazolidinediones (TZDs), are potent glucose-lowering agents that improve insulin sensitivity in adipose tissues and skeletal muscles [5]. To date, no PPARβ/δ ligand has been approved for clinical use. The clinical successes of TZDs and fibrates have spurred extensive development of next-generation PPAR ligands (i.e., antagonist, dual- and pan-PPAR agonists) for various metabolic complications, ranging from pre-morbid conditions such as obesity to chronic morbidities such as non-alcoholic fatty liver disease and chronic kidney disease [6]. Clearly, the discovery of PPARs underscores an important milestone in medicine, given the profound and pervasive impacts of PPARs in the way we tackle modern metabolic diseases.

The clinical impact of PPARs extends beyond metabolic disorders. To date, PPAR agonists have been trialed in many human diseases, including neurodegenerative disorders, psychiatric disorders, autoimmune and inflammatory diseases, as well as malignancies, with varying degrees of success [6,7]. PPAR-related metabolic dysregulations, such as obesity and type 2 diabetes, are independent risk factors of carcinogenesis and cancer prognosis predictors [8,9]. Thus, there is intense research spotlight on exploiting PPARs for cancer therapy. Early investigations revealed that, in the majority of cases, the activation of PPARβ/δ is linked to tumor progression, whereas PPARα and PPARγ are associated with anti-tumorigenesis [10]. Nevertheless, existing cancer trials revealed a huge cancer-to-cancer discrepancy, undermining the potential of PPAR ligands in cancer therapy [6]. Such discordance between preclinical and clinical outcomes indicates unaccounted hidden players interacting with PPARs during carcinogenesis.

It is now well-recognized that cancer cells do not live in a rigid and homogenous mass, but rather in a highly dynamic and heterogeneous community comprising a wide variety of cell types such as fibroblasts, adipocytes, immune cells, endothelial cells, pericytes, and mesenchymal stem cells, collectively known as the tumor stromal cells [11]. The interplay between tumor stromal cells and the epithelium is crucial to every step of tumorigenesis, from initiation, progression, and metastasis, besides offering enhanced plasticity and resistance to various stressors and physiological cues in cancer cells [12]. Increasing evidence also implicates a profound role for PPARs in stromal cellular behaviors and eventual consequences in cancer hallmarks. Our review aims to consolidate the current understanding of PPAR-mediated activities in carcinogenesis and tumor stromal–epithelial communication.

2. The Roles of PPARs in Tumor Epithelium

2.1. Functional Diversity of PPARs in Tumorigenesis

2.1.1. PPARα

The three PPAR isotypes have diverse physiological functions and expression patterns in different tissues. Likewise, they also possess vastly different roles in cancer cells (Figure 1). Marked species differences are apparent in response to peroxisome proliferation induced by activated PPARα. Rats and mice are extremely sensitive, while humans appear to be relatively insensitive or non-responsive at dose levels that produce a marked carcinogenic response in rodents. Experimental evidence suggests a probable link between peroxisome-proliferator-elicited liver growth and the subsequent development of liver tumors in rats and mice. In rodents, the activation of PPARα induces miRNA-mediated neoplastic changes in the liver [13]. However, these oncogenic events are not recapitulated in PPARα humanized mice and human hepatocyte cell lines [14].

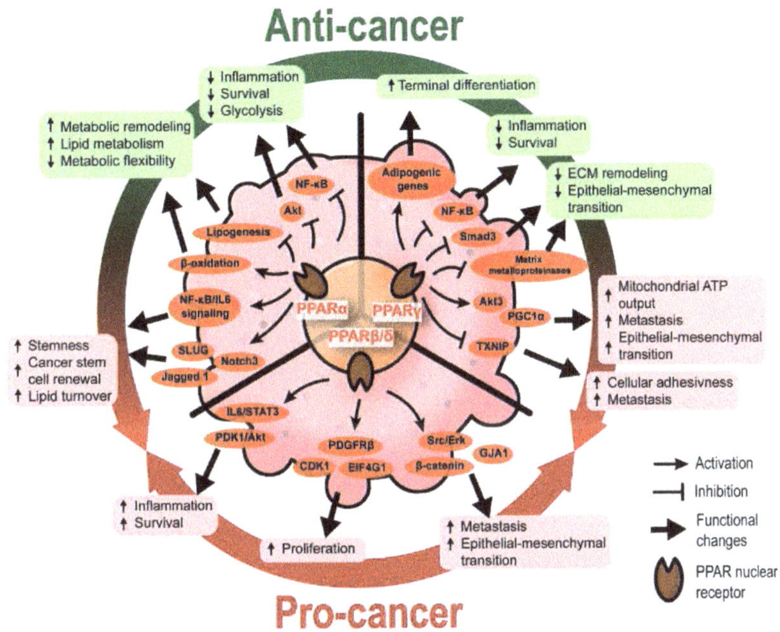

Figure 1. The cellular activities regulated by PPARs in tumor epithelium. In a tumor cell, PPARα and PPARγ exhibit controversial roles. They are generally linked to anticancer effects (green text boxes) by impairing the pro-inflammatory, pro-metastatic, and pro-survival responses, as well as reducing metabolic flexibility. However, their pro-cancer activities (red text boxes), including the maintenance of cancer stemness, meeting high energy demands of cancers and promoting metastasis, have been reported. On the other hand, PPARβ/δ activates signaling pathways and key mediators implicated in pro-cancer activities such as enhanced survival, proliferation, and epithelial–mesenchymal transition. ECM, extracellular matrix.

A few studies also reported a pro-carcinogenic role of PPARα. In a small-scale cross sectional study (n = 100 patients), the overexpression of PPARα in the tumor microenvironment (TME) of colorectal cancer has been linked to poorer prognosis [15]. In breast cancer stem cells, GW6471 (a PPARα antagonist) is anti-proliferative and pro-apoptotic, while Wy14643 (a PPARα agonist) induces the clonal expansion of breast cancer mammospheres by promoting the signaling activities of the nuclear receptor κB (NF-κB)/ interleukin-6 (IL-6) axis, SLUG, Notch3, and Jagged 1 [16,17]. PPARα signaling also ensures a high lipid turnover rate, sustaining the high energy demand to maintain stemness and self-renewal in pancreatic and colorectal cancer stem cells [18].

Based on a meta-analysis, the clinical use of fibrates, which can be traced back to the mid-1970s, does not significantly increase cancer incidence [19]. In fact, PPARα activities are primarily thought to be anticancer in humans. The nuclear receptor can repress the oncogenic roles of NF-κB and Akt, besides forcing the tumor cells to adopt a lipo-centric metabolism [20,21]. Consequently, the tumor cells which adapt poorly to the PPARα-mediated anti-inflammatory response and enhanced fatty acid oxidation may become less proliferative and undergo apoptotic, necrotic, or autophagic cell death.

2.1.2. PPARγ

Most studies support an anti-carcinogenic role for PPARγ, as summarized in a recent review [22]. A high expression of PPARγ is associated with a favorable prognosis in colorectal cancer patients [15]. The activation of PPARγ in cancer cells stimulates adipogenesis and disrupts the Hippo-YAP signaling pathway to force terminal differentiation and suppress proliferation [23–25]. Many cancer stem cells are also sensitive to the terminal differentiation directed by PPARγ [24,26,27]. PPARγ-mediated *PTEN* upregulation inhibits PI3K signaling to diminish the self-renewal and aggressiveness of cancer stem cells [28,29]. Furthermore, PPARγ agonists trigger NF-κB transrepression and modulate various BCL-2 family proteins such as BAX, BAD, Bcl-XL, Bcl2, and PI3K/Akt c-Jun to exert anti-inflammatory and pro-apoptotic properties [21]. PPARγ agonists, ciglitazone and 15-deoxy-Δ12,14-prostaglandin J2 (15d-PGJ2), inhibited cell viability and proliferation of brain tumor stem cells at least via the inhibition of Sox2 while enhancing Nanog expression [30]. The differential regulation of Sox2 and Nanog by PPARγ agonists suggests a critical role for these stemness factors in modulating the growth and differentiation of stem cells in glioma. However, the mechanism by which PPARγ agonists regulate differentiation and self-renewal remains unclear. Separately, by suppressing matrix metalloproteinases and antagonizing Smad3-dependent transcriptional activity, PPARγ also attenuates extracellular matrix (ECM) remodeling and epithelial–mesenchymal transition (EMT), which, in turn, leads to reduced tumor metastasis [31,32].

Reports on the pro-cancer effect of PPARγ are not uncommon. Yang et al. (2005) [33] and Pino et al. (2004) [34] concluded that the use of PPARγ agonists is associated with increased cancer incidence in genetic mice models of colorectal cancer. A few studies have also described the increased risk of PPARγ agonists for renal and bladder cancers [35,36]. Several molecular mechanisms for the pro-cancer effect of PPARγ have been proposed. For instance, Galbraith et al. (2021) [37] demonstrated that in prostate cancer, PPARγ overexpression promoted the activity of Akt3, which subsequently inhibited a nuclear export protein, CRM1, and enhanced the nuclear retention of PPARγ co-activator 1α (PGC1α). Such activity ramps up the mitochondrial ATP output in cancer cells to meet the exorbitant energy demand for EMT and metastasis. In human melanoma, the activation of PPARγ remodels the expression and localization of surface integrins, particularly integrin β-3 and integrin α-5, to increase cellular adhesiveness and distal metastatic seeding [38]. These metastatic phenotypes are linked to the suppression of thioredoxin-interacting protein (*TXNIP*), whose expression is negatively regulated by PPARγ [38]. Moreover, using a liver-specific *Pten* knockout mouse model, it was found that Akt2 promotes the activation and pro-tumorigenic signaling of PPARγ by repressing hepatocyte nuclear factor 1α (HNF1α) [39,40]. Although multiple pro-tumorigenic mechanisms of PPARγ have been found, to date, there is no consensus if these PPARγ-mediated pathways are ubiquitous in different cancer cell types. Notably, the genetic background could act as a strong modifier of the pro-tumor effect of PPARγ, as examplified by the predisposition of certain PPARγ polymorphisms (i.e., Pro12Ala and C161T) to breast cancer [41]. The genetic predisposition would also explain why certain ethnic groups may be more susceptible to the cancer onset with prolonged usage of PPARγ agonists even though TZDs are generally associated with protective effects against several common cancers [42,43].

2.1.3. PPARβ/δ

The dual role of PPARβ/δ in cancer has been thoroughly reviewed [44,45]. In summary, most of the studies are in favor of a pro-tumorigenic profile of PPARβ/δ. Extensive investigations were focused on colon cancers [15]. Pro-tumorigenic activities of PPARβ/δ have been demonstrated in many colon cancer mouse models, including $Apc^{Min/+}$ mice [46], azoxymethane-induced colon tumors [47], colitis-associated colon cancer [48], high-fat diet or PPARβ/δ agonist-treated mice [49]. PPARβ/δ overexpression exacerbates the activation of β-catenin and several pro-invasive pathways, namely connexin 43, PDGFRβ, Akt1, EIF4G1, and CDK1, to promote colorectal cancer progression [46]. PPARβ/δ also positively regulates IL-6/STAT3-mediated inflammation and many pro-metastatic genes [48,50]. PPARβ/δ is a key mediator of PDK1-mediated mammary carcinogenesis [51]. In a nonmelanoma skin cancer mouse model, PPARβ/δ activates the oncogene Src and the EGFR/Erk1/2 signaling pathways upon UV exposure, resulting in increased tumor burden and EMT [52]. Enhanced response of Erk to transforming growth factor β1 (TGF-β1) is also seen in prostate cancer cells, in response to PPARβ/δ-mediated activation of ABCA1 and caveolin-1, which results in TGF-β1-induced tumor growth, migration, and invasion [53]. In terms of cancer stem cells, the current understanding of the role of PPARβ/δ is somewhat lacking. A recent study revealed that PPARβ/δ upregulates Nanog expression in colorectal cancer cells, promoting metastasis when exposed to a fat-enriched environment [54]; yet, another study showed its suppressive effect on SOX2 expression, thus inhibiting neuroblastoma tumorigenesis [55]. While the pro-tumorigenic role of PPARβ/δ in tumor epithelium is well-supported, opposite findings have also been reported [56–58]. The conflicting results suggest other still hidden mechanisms that can fine-tune the cellular activity of PPARβ/δ towards pro- or anticancer effects.

2.2. Clinical Development of PPAR Modulators as Cancer Therapeutics

The tight entanglement of PPAR signaling and tumorigenesis leads to the repurposing of PPAR-targeting drugs for cancer treatment. Many early phase clinical trials have been conducted to examine the clinical feasibility of PPAR agonists, particularly PPARα and PPARγ agonists, against a wide range of cancers [6]. However, existing evidence does not support using any PPAR modulators to treat cancers because of underpowered study design, marginal effect size, and underwhelming outcomes. The discrepancy between preclinical and clinical results highlights a knowledge gap in our understanding of PPARs in carcinogenesis. In fact, PPAR activities may vary across different cancer types and stages. On top of that, the TME adds an extra layer of complexity to the regulatory roles of PPARs in oncogenic processes, which existing PPAR cancer research often fails to take into consideration. As PPARs may have vastly distinct roles in tumor stromal cells compared to epithelial cells during tumorigenesis, in the next section, we will provide an overview of the current understanding of PPARs in the TME and the interplay between tumor stroma and epithelium.

3. The Roles of PPARs in Stromal Cells in the Tumor Microenvironment

Most anticancer therapies target malignant cancer cells while largely ignoring the surrounding noncancer cell components of the tumor or TME. The TME or tumor stroma comprises nonmalignant host cellular and acellular components, including, but not limited to, fibroblasts, immune cells, endothelial cells, fat cells, and noncellular components of the tumor niche such as the basement membrane and ECM. Although most normal host cells in the stroma possess certain tumor-suppressing abilities, the stroma will change during malignancy, causing the tumor stromal cells to confer pro- or anti-tumor properties in a context- and cell type-dependent manner. Over the past decades, the role of the TME in determining every aspect of cancer progression and the efficacy of treatment has become evident. The functions of PPARs in these stromal cells are increasingly appreciated and have direct or indirect impacts on cancer progression.

3.1. PPARγ: A Master Regulator of Stromal Metabolic Reprogramming

3.1.1. Cancer-Associated Fibroblasts

Cancer metabolism and bioenergetics are vastly different from those of normal epithelial cells. A high basal metabolic rate, coupled with abnormal vasculatures in the TME, poses a tremendous challenge for cancer cells to fulfill their energy demand. While the cancer cells possess remarkable plasticity and versatility to utilize various substrates to meet their demand for cellular energy, the surrounding stromal cells also play an indispensable role during cancer progression.

Under the paracrine influences of cancer cells, stromal cells such as cancer-associated fibroblasts (CAFs) and cancer-associated adipocytes (CAAs) can transform into substrate donors to provide fuels and building blocks, namely glutamine, L-lactate, fatty acids, and ketone bodies. These metabolites are readily channeled into the Krebs cycle and oxidative phosphorylation of the cancer cells for ATP generation [59,60]. PPARγ governs many processes involved in the metabolic remodeling of stromal cells. Clinically, the expression of PPARγ is significantly upregulated in CAFs of cutaneous skin squamous cell carcinoma and colon adenocarcinoma [61,62]. In one study, immortalized human fibroblasts overexpressing PPARγ were more glycolytic, autophagic, and displayed a senescent phenotype [63]. L-lactate secretion also increased by 70% in PPARγ-overexpressing fibroblasts compared to wild-type counterparts [63]. These PPARγ-induced metabolic features are typical in a tumor-supporting stroma, as evidenced by accelerated tumor xenograft growth of MDA-MD-231 breast cancer cells when co-implanted with transgenic fibroblasts overexpressing PPARγ, but not with wild-type fibroblasts [63].

The hypoxic TME further aggravates the autophagic phenotype in tumor stromal cells, suggesting a modifying role of hypoxia-inducible factor 1α (HIF-1α) in PPARγ-dependent autophagy [63,64]. Furthermore, a study on a genetic defect (*MTO1* deficiency) in mitochondria reported that AMP-activated protein kinase (AMPK) and uncoupling protein 2 (UCP2) interacted closely with PPARγ and HIF-1α, generating a HIF1α-PPARγ-UCP2-AMPK axis, to influence mitochondrial bioenergetics and key metabolic processes such as glycolysis, fatty acid oxidation, and oxidative phosphorylation, leading to extensive metabolic reprogramming in fibroblasts [65]. AMPK ensures the maturation of autophagosome and lysosomal fusion during autophagy [66], besides modulating the genes responsible for mitochondrial integrity (*UCP2* and *PGC-1α*), autophagy (*BECN-1, LC3B, ATG5, ATG7*, and *SQSTM1*), and mitophagy (*PINK1, FUNDC1, BNIP3*, and *PRKN*) [67]. The expression of AMPK target genes is considerably disrupted in fibroblasts overexpressing PPARγ under normoxia and hypoxia [63]. As such, the interplay among PPARγ, HIF1α, and AMPK is pivotal in modulating CAF autophagy, but the exact mode of interaction remains largely elusive.

Following autophagy, glycolysis occurs to recycle cellular organelles and debris into basic building blocks reusable by cancer cells [68,69]. Many glycolytic genes are subject to PPARγ regulation [70,71]. Several studies also pointed to NF-κB as a key transcription factor of stromal autophagy and glycolysis [63,72], but its interaction with PPARγ remains elusive. In short, PPARγ regulates key genes and cellular events in CAFs to accomplish the metabolic coupling of tumor stroma and epithelium, essentially transforming CAFs into a powerhouse that constantly generates energetic biomolecules to support tumor growth.

In contrast to the tumor-supporting properties of CAFs overexpressing PPARγ, pharmacologic PPARγ activation in tumor epithelium confers anticancer effects by reducing tumor proliferation and neovascularization [63]. Thus, the activation of PPARγ metabolically reprograms CAFs to favor autophagic and glycolytic behaviors, allowing cancer cells to use nutrients from non-autonomous sources to sustain their uncontrolled proliferation and other activities.

3.1.2. Cancer-Associated Adipocytes

Like CAFs, CAAs also serve as storage sites and nutrient donors in the TME [73]. Fibroblasts and mesenchymal stromal cells readily undergo adipogenesis and differentiate into adipocytes upon exposure to adipogenic stimuli, especially the activation and upregulation of PPARγ [74,75]. Cancer exosomes loaded with miRNA-144 and miRNA-155 facilitate the beige/brown differentiation of CAAs by modulating the MAP3K8-Erk1/2-PPARγ axis, whereas those carrying miRNA-126 can disrupt IRS-GLUT4 signaling and promote AMPK- and HIF1α-mediated autophagy [76,77]. Cancer cells can also initiate the dedifferentiation of adjacent adipocytes, a process that is consistently observed when adipocytes are cocultured with cancer cells [78,79]. The process is characterized by the progressive loss of mature adipocyte markers such as leptin, adiponectin, HSL, and

PPARγ, increased expression of fibroblast markers such as matrix metalloproteinase 11 (MMP11), collagen I, and α-SMA, as well as the adoption of a fibroblast-like morphology in the cocultured adipocytes [78,79]. These dedifferentiated adipocytes exhibit transcriptional suppression of *GLUT4* and *IRS1* and inhibit insulin-induced Akt phosphorylation [78]. These aberrations occur alongside the downregulation of MAP3K8-Erk1/2-PPARγ, effectively escalating the catabolic capacity of CAAs to secrete pyruvate, L-lactate, and ketone bodies [76].

Moreover, diminished ligand activation of PPARγ through the constitutive expression of Notch1 induces adipocyte de-differentiation and tumor-like manifestations [80]. Treatment with rosiglitazone, a PPARγ agonist, effectively promoted adipocyte redifferentiation and attenuated the transformation of the adipocytes [80]. Consistent with these observations, the adipocyte-specific deletion of *PPARγ* in a chemically induced breast cancer model impaired *BRCA1* expression in CAAs and subsequently accelerated tumor formation and progression [81]. Undoubtedly, PPARγ is a critical mediator in the cellular fate and metabolic reprogramming of CAAs. Although the actual functionality of adipocyte dedifferentiation in tumor stroma remains unclear, it is generally associated with pro-tumorigenic activities [76,78]. Furthermore, dedifferentiated adipocytes can be redifferentiated into other cell lineages, including beige/brown adipocytes that readily release bioenergetic molecules into the TME [82]. Such plasticity of adipocytes entails the possibility for tumor cells to coerce the CAAs into other tumor supportive cells.

Taken together, CAFs and CAAs are two key stromal cells that undergo extensive metabolic reprogramming to act as energy reserves for cancer epithelium, as illustrated in Figure 2. PPARγ signaling is implicated in the remodeling of both stromal cells, but the activity is vastly different. Autophagic CAFs are triggered by PPARγ activation, while PPARγ is suppressed in dedifferentiated CAAs. This cell type-dependent disparity highlights a need for strategies to target PPARγ in a cell-specific manner so that the treatment is not counter-productive.

3.2. PPARβ/δ in CAFs Governs Redox Homeostasis and Affects Tumor Initiation

The differentiation of normal fibroblasts into CAFs is one of the cornerstones of early tumor initiation in many cancer types [83,84]. CAFs can disrupt the local ECM and deliver proliferative paracrine signals to support tumorigenic events. Interestingly, mice with fibroblast-selective *PPARβ/δ* deletion developed fewer and smaller skin tumors than wild-type mice exposed to topical carcinogens [85]. Similar results were recapitulated using chemically and genetically induced intestinal carcinogenesis in these mutant mice [86], indicating that PPARβ/δ activity in stromal fibroblasts promotes tumor initiation. The delayed tumor emergence in the mutant mice was due to an enhanced antioxidant response in the epithelium. Mechanistically, PPARβ/δ-knockout fibroblasts markedly increase the Nox4-derived H_2O_2 production in the adjacent epidermis, subsequently triggering an RAF/MEK-mediated NRF2 activation that elicits a strong antioxidant and cytoprotective response [85]. By reducing the phosphorylation of many tumor suppressors and oncogenes, NRF2 also increases the tumor suppressor activity of PTEN and reduces the oncogenic activity of Src and Akt, leading to delayed tumor growth [85]. Hence, reducing the expression and activity of *PPARβ/δ* in CAFs may provide a new therapeutic option to disrupt cancer susceptibility in the neighboring tumor epidermis.

Leucine-rich-alpha-2-glycoprotein 1 (LRG1) and TGFβ1 underpin a crucial process in the PPARβ/δ-mediated stromal–epithelial crosstalk. PPARβ/δ in fibroblasts upregulates the expression of LRG1, which blunts the epidermal response to TGFβ1 [87]. Furthermore, exogenous LRG1 can also ablate the influence of TGFβ1 on ROS generation and NRF2 activity [85]. In colorectal carcinoma and pancreatic ductal adenocarcinoma patients, the level of LRG1 in the TME and bloodstream is significantly higher than in healthy individuals and correlates positively with a more advanced cancer stage and poorer prognosis [88–90]. This observation suggests a pro-tumorigenic role of LRG1. Surprisingly, the *LRG1* promoter has two putative PPAR response elements [91]. The expression of *LRG1* is increased by a PPARβ/δ agonist, GW501516, which strongly suggests that LRG1 is a direct target of PPARβ/δ [91]. Therefore, during the early stage of tumorigenesis, CAF PPARβ/δ may stimulate LRG1 expression, which interferes with TGFβ1-dependent redox homeostasis, to support a sustained oncogenic transformation in the surrounding tumor epithelium.

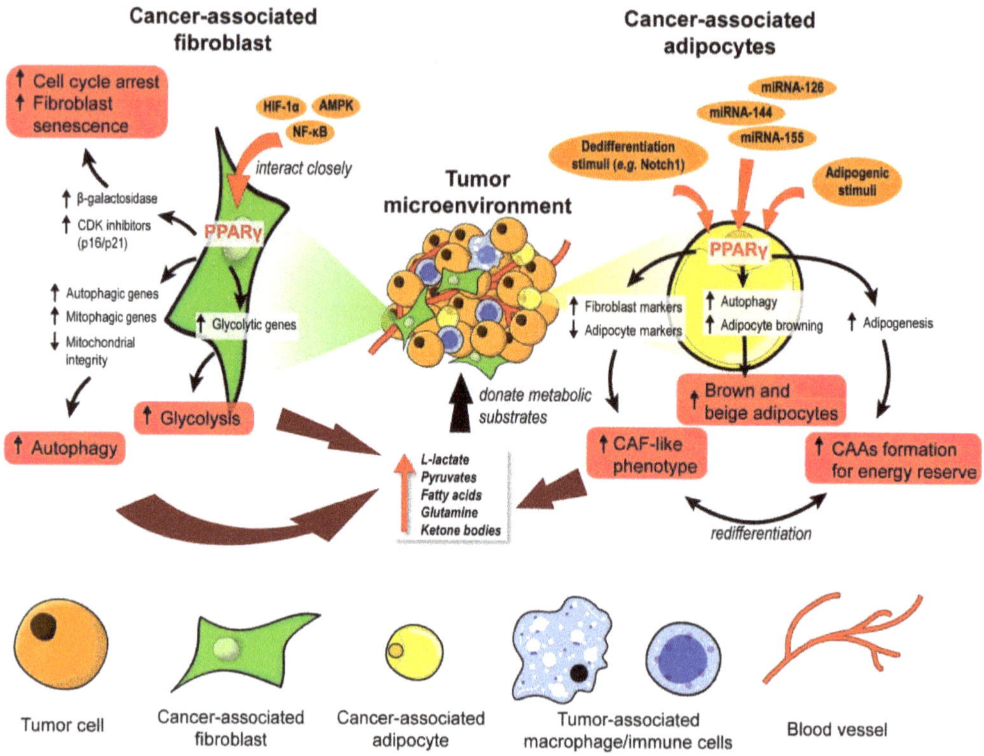

Figure 2. PPARγ orchestrates the metabolic reprogramming of cancer-associated fibroblasts and adipocytes. In cancer-associated fibroblasts (CAFs), PPARγ interacts closely with HIF-1α, AMPK, and NF-κB to promote cell cycle arrest, senescence, autophagy, and glycolysis. These functional changes unleash many metabolic substrates into the tumor microenvironment for the neighboring tumor cells. Similarly, PPARγ governs the fate and function of cancer-associated adipocytes (CAAs). Upon exposure to adipogenic stimuli, PPARγ mediates adipogenesis and formation of CAAs to act as an energy reserve. In contrast, exposure to dedifferentiation stimuli drives CAAs to adopt a CAF-like phenotype and act as a substrate doner in the tumor microenvironment. Certain miRNAs can suppress PPARγ to induce brown and beige differentiation of CAAs which are also energy donors for cancer progression.

Collectively, these findings uncover a major role for stromal PPARβ/δ in the epithelial–mesenchymal communication and cellular oxidative response in tumor development (Figure 3). Notably, this novel role of PPARβ/δ was primarily documented, so far, in nonmelanoma skin carcinoma and colorectal cancer models. Thus, further validation in other cancer models is necessary.

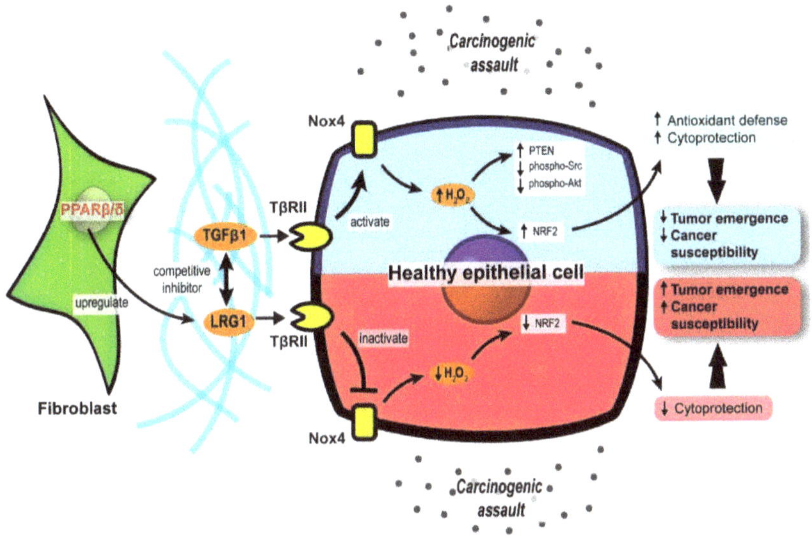

Figure 3. Stromal PPARβ/δ regulates epithelial redox homeostasis and oncogenesis. In carcinogenic assaults, TGFβ signaling in epithelial cells is activated to promote H_2O_2 synthesis, which subsequently activates NRF2 and reinforces the cytoprotection against carcinogens (blue upper compartment of the epithelial cell). However, fibroblast PPARβ/δ disrupts the protective mechanism by upregulating LRG1, which acts as a competitive inhibitor of TGFβ1 and dampens TGFβ signaling, resulting in increased cancer susceptibility and oncogenesis (red lower compartment of the epithelial cell).

3.3. Endothelial PPARs Affect Angiogenesis in the Tumor Microenvironment

Hypoxic regions often arise because of rapid tumor growth, which outgrows the oxygen perfusion and nutrient supply from existing vasculature [92]. Cancer cells mitigate the predicament by releasing pro-angiogenic factors that stimulate angiogenesis, which is affected by all three PPAR isotypes.

In terms of PPARα, synthetic PPARα agonists such as fenofibrate and Wy-14643 have demonstrated suppressive effects on endothelial cell proliferation, neovascularization, and tumor xenograft growth [93,94]. Such anti-angiogenic effects of PPARα agonists were lost in PPARα-deficient mice transplanted with PPARα-intact tumor cells, implying that PPARα activation in surrounding stromal cells, but not the tumor cells, attenuated tumor angiogenesis [93,94]. The underlying mechanism is associated with increased anti-angiogenic factors (i.e., thrombospondin-1 and endostatin) and the interference of pro-angiogenic factor biosynthesis (i.e., VEGF-A, angiopoietin-1, and angiopoietin-2), affecting VEGF- and FGF2-mediated endothelial proliferation and migration [93,95]. Furthermore, by transcriptionally suppressing the expression of endothelial P450 CYP2C epoxygenase, whose function is to catalyze arachidonic acid epoxidation, PPARα also diminishes the epoxygenase products, epoxyeicosatrienoic acids, which are pro-angiogenic [96]. Thus, PPARα activation in stromal endothelial cells inhibited the biosynthesis of pro-angiogenic factors while promoting the secretion of anti-angiogenic factors, thereby abrogating angiogenesis and limiting nutrient supply to attenuate tumor progression.

In contrast to PPARα, PPARβ/δ is a pro-angiogenic nuclear receptor in line with its wound healing properties [97–99]. The activation of PPARβ/δ in endothelial cells by synthetic ligands or genetic manipulation consistently results in aberrant biosynthesis of VEGF, PDGFR, and c-KI, as well as accelerated endothelial cell proliferation and vascular formation [100,101]. In the TME, these pro-angiogenic changes stimulate the formation of a tumor with a higher vessel density, enhancing tumor feeding, oxygen provision, and metastasis capacity of the cancer cells [101]. Interestingly, in PPARβ/δ knockout mice harboring experimental wild-type tumors, the endothelial cells forming the microvessels in the tumors appear immature, hyperplastic, and less well-organized, leading to abnormal microvasculature and restricted blood flow into the tumors [102,103]. Apart from conventional growth factors, other potential PPARβ/δ-dependent angiogenic mediators include CDKN1C [102], IL-8 [104], CLIC4, and CRBP1 [105]. Considering its regulatory effects on many angiogenic genes and the strong linkages with advanced cancer stages, tumor recurrence, and distant metastasis, PPARβ/δ is identified as one of the pro-angiogenic signaling hubs in cancers [103]. Thus, the pro-tumorigenic and pro-angiogenic activities of PPARβ/δ warrant the development of efficacious PPARβ/δ antagonists to be tested in cancer models.

Existing evidence on the role of PPARγ in angiogenesis remains ambiguous. Like PPARα, PPARγ activities in the TME are associated with the dysregulated production of angiogenic factors, especially platelet-derived endothelial cell growth factor (PD-ECGF) and fibroblast growth factor (FGF) [106,107]. Early studies generally concluded on an inhibitory effect of PPARγ ligands on endothelial cell proliferation in response to pro-angiogenic factors and endothelial tube formation [108,109], whereas subsequent investigations suggested otherwise [110,111]. Such conflicting findings may be attributable to the dosages of PPARγ ligands and endothelial cell types [112]. Regardless of the pro- or anti-angiogenic properties, VEGF/VEGFR signaling is coherently implicated in the PPARγ-mediated effect [108–110]. A recent study using endothelial-specific PPARγ knockout models shed new light on the role of this nuclear receptor in angiogenesis. In mature endothelial cells, PPARγ knockdown impaired proliferation, migratory properties, and tubule formation capacity [111]. These impairments translated into the loss of circulating endothelial progenitor cells and angiogenic capacity in endothelial-specific PPARγ-deficient mice, which was reversed by the transplantation of wild-type bone marrow [111]. Mechanistically, abolishing PPARγ in the endothelial cells disrupts E2F1-mediated Wnt signaling and GSK3B interacting protein activity, resulting in suppressed endothelial proliferation [111]. Conceivably, the genetic models reinforce the pro-angiogenic activity of PPARγ in endothelial cells.

In short, PPARα and PPARβ/δ exert anti- and pro-angiogenic activities in the endothelial cells of TME, respectively. On the other hand, opposing roles have been reported for PPARγ in angiogenesis. The roles of each PPAR isotype in angiogenesis are summarized in Figure 4. Notably, most findings on PPARγ are not established using oncogenic models. As the physiological cues in a TME are different from a normal condition, the true nature of PPARγ in cancer angiogenesis and tumor epithelium-endothelium crosstalk requires further investigation.

3.4. *PPAR-Dependent Autocrine and Paracrine Signaling*

Autocrine signaling facilities self-stimulation, while paracrine signaling allows local cell–cell communication. In the TME, both forms of cell signaling are imperative to coordinate every stage of oncogenesis, alerting the tumor cells how and when to proliferate, evade immune surveillance, escape from the existing microenvironment, and settle at a distal site. The transmission of complex messages in response to cellular stimuli is made possible by a plethora of secretory mediators, including cytokines, chemokines, growth factors, catalytic proteins, miRNAs, extracellular vesicles, and lipid compounds [113]. Many of these messengers are directly or indirectly regulated by PPARs (Figure 5). For instance, a new PPARγ agonist, CB13, remodels the exosomal contents from radio-resistant non-small cell lung cancer to promote endoplasmic reticulum stress and cell death via a PERK-eIF2α-ATF4-CHOP axis [114].

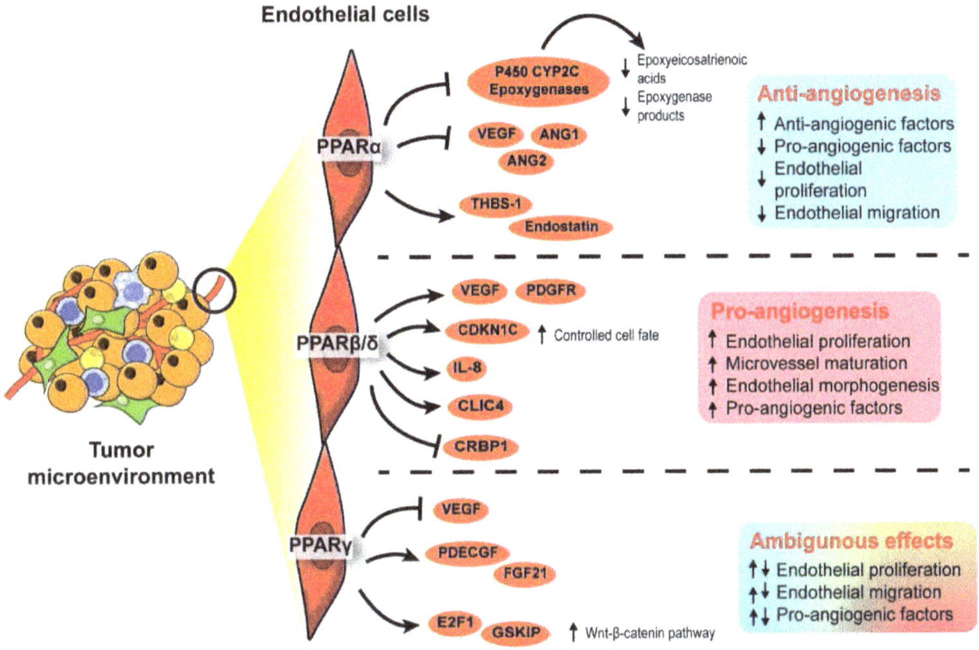

Figure 4. Angiogenic role of PPARs in endothelial cells. In the endothelial cells, PPARα exhibits an anti-angiogenic effect by inhibiting endothelial proliferation, whereas PPARβ/δ appears pro-angiogenic by ensuring proper endothelial morphogenesis and vascular maturation. The role of PPARγ in angiogenesis is conflicting and warrants further investigation.

3.4.1. Disruption of Pro-Tumor Signaling by PPARγ in CAFs

Eicosanoids, which are lipid signaling molecules and cognate ligands of PPARs, are the main drivers of PPAR activation in the TME. Major eicosanoid subfamilies include prostaglandins, thromboxanes, leukotrienes, and epoxygenated fatty acids, among which the prostaglandins are the most well-investigated. In colon cancers, cyclooxygenase-2 (COX-2), an enzyme that catalyzes the conversion of arachidonic acid to prostaglandin H2 (PGH2), is overexpressed in CAFs surrounding colon adenocarcinomas, leading to a buildup of intratumoral PGE2 [61,115]. However, the resultant activity of PPARs varies across different stromal cells. For instance, 15d-PGJ2 activates PPARγ and suppresses the proliferation of CAFs and expression of the ECM remodeling enzyme, MMP2 [116]. By inhibiting NF-κB, TZD-activated PPARγ substantially lowers the expression of pro-inflammatory, pro-angiogenic, and pro-metastatic signaling molecules in CAFs, including IL-6, IL-8, CXCR4, MMP2, and MMP9, which further dampens pro-tumor crosstalk in the TME [117,118]. The repression of PPARγ activity also disturbs the quiescent state of hepatic and pancreatic stellate cells, compelling their differentiation into CAFs with highly aggressive phenotypes and inducing desmoplasia in the TME [119–122]. Despite some conflicting results [123], PPARγ in CAFs can disrupt pro-tumorigenic paracrine signaling by suppressing the liberation of cytokines and chemokines.

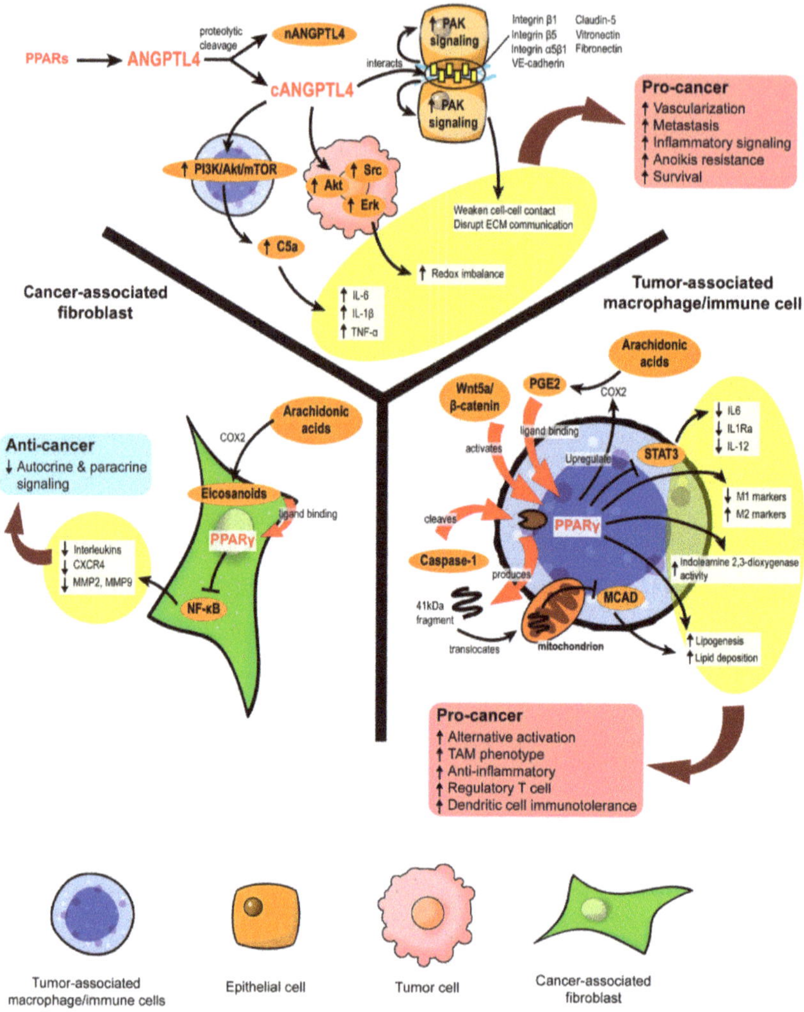

Figure 5. PPARs modulate stromal–epithelial crosstalk in the tumor microenvironment. PPARs affect autocrine and paracrine signaling in different stromal cells. In cancer-associated fibroblasts, PPARγ activation upon ligand binding represses NF-κB, alleviating the secretion of many autocrine and paracrine signals. However, in macrophages and immune cells, PPARγ activation is primarily linked to pro-cancer activities, such as the formation of tumor-associated macrophages (TAMs), increased regulatory T cells, and immunotolerance. ANGPTL4 is a target gene product of PPARs. Proteolytic cleavage of full-length ANGPTL4 yields nANGPTL4 and cANGPTL4 domains, of which the latter is a potent paracrine signal and key mediator of inflammatory signals, anoikis resistance, and metastasis.

3.4.2. PPARγ Propels the Formation of Tumor-Associated Macrophages

The role of PPARs in innate and adaptive immune cells has been extensively studied. Unlike CAFs, the activation of PPARα and PPARγ in macrophages favors an anti-inflammatory tumor-associated macrophage (TAM) phenotype [124,125]. Classical PPARγ ligands, namely rosiglitazone, *N*-docosahexaenoyl ethanolamide, and *N*-docosahexaenoyl serotonin, effectively block paracrine signals from cancer cells to sway the fate of macrophages to adopt alternative activation and reduce their STAT3-mediated pro-inflammatory response [125]. In macrophages challenged with pathogens, WY14643

(PPARα agonist) and 15d-PGJ2 (PPARγ agonist) tip the balance towards the M2 phenotype by enhancing the expression of arginase I, Ym1 (chitinase 3-like 3), mannose receptor, TGF-β and increasing phagocytic capacity while diminishing M1 macrophage biomarkers [126]. PPARγ antagonists and macrophage-specific PPARγ ablation attenuate these effects, clearly outlining the dependency of TAM differentiation on PPARγ [127,128].

Mechanistically, PPARγ agonism promotes lipid retention, lipogenesis, and PGE2 secretion in macrophages. The lipid metabolic changes are partly mediated by the Akt/mTOR pathway [129]. On top of its role as a nuclear receptor and transcription factor, PPARγ is subject to cleavage by caspase-1 to yield a 41 kDa fragment that translocates to mitochondria and inhibits medium-chain acyl-CoA dehydrogenase (MCAD). Such a non-canonical peptide–protein interaction can inhibit fatty acid oxidation, further aggravating lipid droplet accumulation and TAM formation [130]. Likewise, in dendritic cells residing in the TME, PPARγ activation directed by Wnt5a/β-catenin paracrine signaling disrupts fatty acid oxidation and indoleamine 2,3-dioxygenase-1 activity, subsequently leading to the generation of regulatory T cells, immunotolerance, and weakened immunotherapy response [131]. These PPARγ activities create a "friendly" TME for cancer survival, which also coincides with the functional trajectory of macrophage PPARβ/δ [132,133].

Nonetheless, some findings support counterarguments. For example, Cheng et al. (2016) [134] identified macrophage PPARγ as a key tumor suppressor and TAM modulator by abolishing Gpr132 expression. Van Ginderachter et al. (2006) [135] agreed that PPARγ was highly expressed in TAMs, but further stimulation with synthetic and natural ligands could sabotage TAM-induced cytotoxic T lymphocyte suppression to confer an anti-tumor effect. The overexpression of PPARγ in macrophages promotes the upregulation of *PTEN*, which is encapsulated in exosomes. The uptake of these macrophage-derived exosomes by adjacent cancer cells inhibits Akt, p38 MAPK, and migratory properties [136]. Many eicosanoids are also packaged in these exosomes to achieve paracrine stimulation of PPARγ and augment the inhibitory effect on tumor EMT [136].

Taken together, PPARγ acts as a master immuno-metabolic switch in immune cells that govern their fate and tumor-supporting role. Current consensus depicts that PPARγ exhibits a pro-tumorigenic effect in immune cells by promoting alternative activation, which contradicts its anticancer properties in tumor epithelium and CAFs. On the other hand, the related information on other PPAR isotypes in this aspect is somewhat limited. Interestingly, a recent study unveiled that fatty acid-enriched cancer exosomes markedly activate PPARα in tumor-infiltrating dendritic cells, resulting in mitochondrial overdrive and impaired dendritic cell-mediated $CD8^+$ cytotoxic T-cell priming [137]. These exciting findings strongly suggest an immuno-metabolic regulatory role of PPARα in the TME similar to PPARγ. Such a novel activity of PPARα warrants further investigation.

3.4.3. Role of ANGPTL4 in Stromal–Epithelial Crosstalk

Growing evidence suggests a role of angiopoietin-like 4 (ANGPTL4) in cancer and stromal-epithelial communication. ANGPTL4 is a secretory protein that belongs to a family of ANGPTL proteins that share high amino acid sequence similarity with the angiopoietin (ANG) family [138,139]. Its expression is regulated by all three PPAR isotypes and PGE2, especially during major metabolic challenges such as starvation and hypoxia [139–141]. The native full-length ANGPTL4 can undergo proteolytic cleavage to yield C-terminal (cANGPTL4) and N-terminal (nANGPTL4) chains, each with distinct biological activities [142]. The nANGPTL4 domain is primarily responsible for lipid and glucose metabolism, while the cANGPTL4 domain is closely linked to tumorigenic activities, notably angiogenesis, anoikis resistance, and metastasis [143]. Thus, we will be focusing more on the cANGPTL4 fragment.

High expression of ANGPTL4 has been reported in ovarian, urothelial, and breast tumor biopsies, particularly in the CAAs [144–146]. The ANGPTL4 overexpression in CAAs is directed by IL-1β from neighboring TAMs with activated NLRC4 inflammasome and can be exacerbated by tumor hypoxia [147], resulting in cANGPTL4 aggregation in the TME. The cANGPTL4 interacts with integrins β1, β5, α5β1, VE-cadherin, and claudin-5 to induce PAK signaling and weaken cell–cell contacts [148,149]. Moreover, it also disrupts cell–ECM communication through its interaction with vitronectin and fibronectin [150]. The destabilization of cell junctions is then translated to greater intratumoral vascularization and migratory capacity of the malignant cells [151–153].

By manipulating redox homeostasis and activating several pro-survival mechanisms such as FAK/Src, PI3K/Akt, Erk signaling, ANGPTL4 markedly sharpens the resilience of tumor cells and confers anoikis resistance [154–156]. Our latest report showed that exogenous ANGPTL4 activates macrophages and induces hypercytokinemia via PI3K/Akt-mediated complement component 5a (C5a) activation [157]. This finding indicates a modifying role of ANGPTL4 in TAM functionality and paracrine signaling in the TME. Thus, ANGPTL4 may act as a powerful autocrine and paracrine signaling effector of PPARs that can shape a supportive environment for cancer progression. Further investigations on the therapeutic feasibility of targeting ANGPTL4 are warranted.

3.5. Stromal PPARγ Modulates Tumor Metastasis

Only a handful of studies have investigated stromal PPAR activities on metastasis, and the results are conflicting. In myeloid-derived suppressor cells (MDSCs), deficiency of lysosomal acid lipase ($lal^{-/-}$) impaired the production of PPARγ ligands, which led to reduced PPARγ activity, ROS accumulation, and mTOR-mediated tumor metastasis [158]. Following intravenous injection of B16 melanoma cells, increased lung metastases were observed in mice with myeloid-specific PPARγ knockout, further reinforcing the role of MDSCs' PPARγ in metastasis. Contradictorily, a PPARγ agonist, pioglitazone, has been shown to promote alternative activation of macrophages in the TME [159]. These pro-tumorigenic myeloid cells can synthesize TGFβ1 to promote EMT of surrounding tumor cells [160]. Although the true role of stromal PPARγ in metastasis remains debatable, a recent study showed that astrocytes liberate polyunsaturated fatty acids, which are PPARγ agonists, to promote the extravasation of circulating cancer cells into the brain while PPARγ antagonists can reduce brain metastatic burden in vivo [161]. Astrocyte–cancer cell communication is also mediated by TGF-β2 and ANGPTL4, the latter of which is an effector of PPARs [162]. Hence, PPARγ may serve as a nutritional cue to provoke the invasion of metastatic cells into a nutrient-rich environment. The results also argue for the potential use of PPARγ blockade to treat brain metastasis.

4. Knowledge Gaps and Prospects of Targeting PPARs in Tumor Stroma

4.1. Pressing Questions in Current PPAR Cancer Research Paradigm

Our understanding of the role of PPARs in cancer and the TME has expanded exponentially in the past decade. As the master switch of metabolism, PPARs and their actions are deeply rooted in key tumor-supporting cells in the TME, namely CAFs, CAAs, endothelial cells, and immune cells. However, the outcomes of PPAR manipulation are not always consistent. Disagreements and even conflicting experimental results between different stromal cells are not unusual [63,120,125]. The high context dependency remains a puzzle and, to date, no hypothesis can substantially address the variations.

To explain the disparate findings, Youssef and Badr (2011) [163] put forward three postulations: (i) off-target effects of PPAR ligands, (ii) diverse pharmacokinetic properties of PPAR agonists, and (iii) cancer stage-dependent effect, of which the first two focus on the intrinsic characteristics of the synthetic PPAR ligands while the last one is linked to the biological context of the TME. Undeniably, synthetic ligands that are supposed to target the same PPAR isotype do not always have comparable efficacy, off-targets, turnover rate and toxicities [164]. Hence, PPAR-independent activities on the carcinogenesis caused by the non-specificity of the PPAR modulators cannot be eliminated. However, many functional studies of PPAR in the TME were also reinforced by results from genetically knockout models [85,111,127,128]. Therefore, the pharmacodynamic and pharmacokinetic variations of synthetic PPAR ligands may not fully account for the observed discrepancies.

We believe that the controversial roles of PPARs in carcinogenesis should also have underlying biological rationales. One overlooked aspect is the crosstalk between PPARs and other nuclear receptors in different cancer types and stages. Classically, all PPAR isotypes form heterodimers with RXR to coordinately modulate their target genes [165]. Nonetheless, PPARs can cooperate with other nuclear receptors such as glucocorticoid receptors, estrogen-related receptors, and photoreceptor-specific nuclear receptors to form atypical heterodimers transiently [166,167]. These atypical heterodimers may regulate the expression of different sets of genes from those of the classical heterodimers, leading to diverse cell fate and behaviors [167]. The fact that the atypical heterodimers are not commonly detected suggests that the protein–protein interaction is labile and can only be stabilized with a unique combination of physiological cues, microenvironment, bioavailability of the co-factors and cognate ligands. The striking intra- and inter-heterogeneity of the TME, coupled with numerous unorthodox cellular activities, may be adequate to accomplish all sorts of stringent biological environments necessary for the stabilization of different PPAR-dependent atypical heterodimers. Such a flexible and highly amendable transcriptional regulatory mechanism mediated by PPAR–nuclear receptor collaboration may answer some of the disparities observed in PPAR cancer research. Nonetheless, the concept remains highly speculative. While it may explain the context-dependency of the PPAR-related carcinogenic roles, the real challenge is to experimentally capture the transient heterodimers and dissect their endogenous biological activities [167]. Nevertheless, the ability to rewire the non-canonical nuclear receptor crosstalk in the TME may offer a new therapeutic strategy in oncology considering the marked druggability of most nuclear receptors.

Another pitfall in PPAR cancer research is that current drug development and research attention highly skew towards PPARα and PPARγ. Our knowledge on PPARβ/δ and choices of PPARβ/δ-targeting drugs is comparatively limited. Yet, unlike PPARα and PPARγ, PPARβ/δ, which is ubiquitously expressed in almost all tissues, displays an apparent pro-tumor activity. Hence, potent PPARβ/δ antagonists may offer some fruitful outcomes in cancer treatment.

4.2. Future Prospects and Strategies to Target Stromal PPARs for Precision Oncology

Owing to the controversial roles of PPAR in the TME, the success of PPAR-based anticancer treatment potentially relies on innovative strategies for cell-type-specific drug delivery or TME-triggered drug release systems [168] (Figure 6). In this context, exosomes are excellent candidates to be developed into precise drug carriers. They are naturally occurring, hence exhibiting remarkable biocompatibility and bioavailability with limited immunogenicity [169]. Furthermore, by modifying the membrane protein compositions, exosomes have shown excellent specificity to recognize a selected protein [170] or cell type [171]. They also possess high drug loading and unloading capacity [172]. The phospholipid bilayer effectively contains the cargo from systemic drug release [169]. These striking features of exosomes allow them to be used as a targeted drug delivery system for pharmacotherapy. In fact, exosomes loaded with natural PPAR ligands such as fatty acids and eicosanoids are easily internalized, leading to high intracellular retention of the biomolecules [136,137,173]. Therefore, by carefully selecting the membrane protein targets of the exosomes, it may become possible to achieve stromal-specific administration of PPAR agonists or antagonists.

Recent advancements in TME-responsive drug release with nanoparticles are remarkable [174]. Unlike exosomes, which depend on membrane proteins to promote targeted delivery, the TME-sensing moieties of nanoparticles are usually based on physico-chemical alterations of the TME such as acidic pH, redox imbalance, high ATP and the enrichment of extracellular enzymes (MMPs and β-galactosidases) or paracrine signals (PDL-1) [174]. Nanoparticles of about 100 nm in diameter demonstrate desirable cellular uptake, and for deep tumor penetration, nanoparticles smaller than 30 nm should be used [175]. Superparamagnetic iron oxide nanoparticles functionalized with conjugated linoleic acid have been shown to increase PPARγ activity, subsequently triggering necrotic cell death in cancer cells [176]. Clearly, the nanoparticle-mediated delivery of PPAR ligands is a viable anticancer strategy. By incorporating different combinations of TME-sensing moieties within a single carrier, we can fabricate multi-sensing nanocarriers which execute drug release only when a specific cell type or set of physiological conditions is met [174]. However, singularly targeting one stromal cell type is not sufficient. For example, fibroblast activation protein-α (FAP) is a transmembrane prolyl endopeptidase highly expressed in CAFs [177]. Sibrotuzumab, a FAP-neutralizing antibody, failed to achieve even one complete or partial remission in a phase II trial involving 25 patients with metastatic colon cancer [178]. Another phase II trial with talabostat, a small molecule inhibitor of FAP, also yielded disappointing patient outcomes [179].

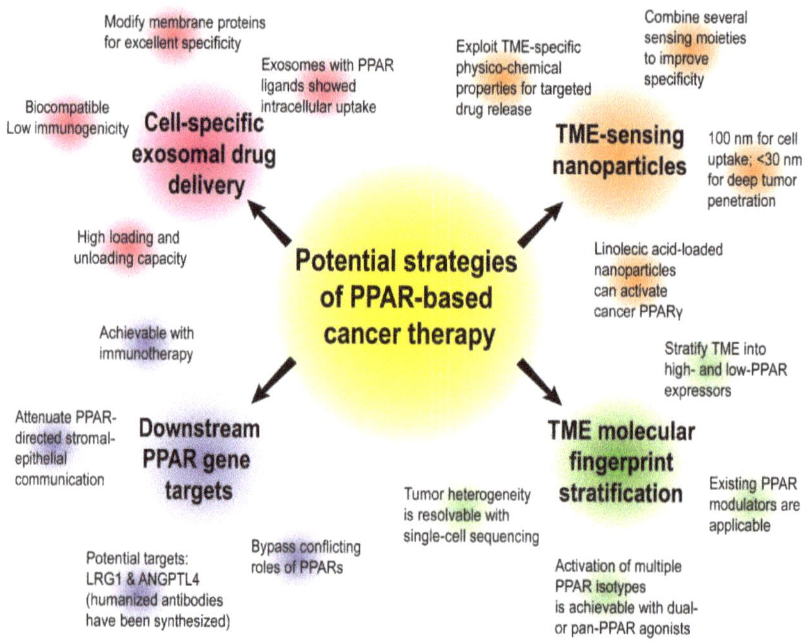

Figure 6. Possible strategies to target stromal PPARs for precision oncology. In this review, we propose four strategies to achieve PPAR-based precision oncology, including (i) cell-specific exosomes, (ii) TME-sensing nanoparticles, (iii) targeting pro-tumorigenic PPAR gene targets with immunotherapy, and (iv) stratification of PPAR-related TME molecular fingerprints. The features of each strategy are summarized in the figure.

We can further restrict stromal–epithelial crosstalk by targeting downstream paracrine signals with immunotherapy. With careful selection of the drug candidates, immunotherapy can effectively shut down critical communication conduits between cancer cells and stromal cells. We have previously examined the feasibility of a nuclear receptor-based partitional strategy by targeting CAFs of skin squamous carcinoma [62]. The treatment disrupted stromal–epithelial communication, reduced xenograft tumor growth, and prevented the recurrence of chemoresistant cancer. Mounting evidence also supports the exploitation of molecular targets downstream to PPARs. In this review, we highlight LRG1 and ANGPTL4, which are key mediators of metastasis and EMT. Immunotherapy targeting these two molecules may effectively shut down PPAR-directed communication between tumor epithelium and stroma. Importantly, humanized neutralizing antibodies targeting these proteins are readily available [154,180].

Another step towards effective PPAR-mediated therapy is by stratifying cancer patients and predicting their susceptibility to PPAR drugs based on tumor genetic and transcriptomic profiles. Cancer patients may be stratified into low- and high-expressors of a specific PPAR isotype either in the stromal cells or cancer cells. New generation dual PPARs agonists may be administered to maximize their anticancer effect on the stromal and cancer cells. The heterogeneity of tumors is a technical challenge, which can be addressed using single-cell sequencing. Identifying molecular fingerprints between stromal and tumor cells in the actual TME will also be critical for a highly precise stratification strategy that enables existing PPAR-targeting drugs to be put to clinical use immediately. Additionally, the emergence of next-generation PPAR modulators [6], such as the selective PPARα modulator, pemafibrate, and dual- and pan-PPAR agonists such as saroglitazar, elafibranor, lanifibranor, and chiglitazar, brings about new prospects to PPAR cancer research. We anticipate that the investigation of newer PPAR modulators and their anticancer effect in the TME will gain momentum in the years to come.

5. Conclusions

Despite the impacts of PPAR activities on different aspects of tumor stromal–epithelial communication and tumor progression, it is not possible to ascribe either an exclusive pro- or anti-tumorigenic role for different PPAR isotypes. This is due to controversies and/or PPAR dual activities on cancer types and different stromal cell types. Likewise, conventional agonists and antagonists which target PPARs systemically may be counterproductive, considering their differential role in cancer and stromal cells, as reflected by the outcome of existing clinical trials. Targeting PPARs in the TME still holds a great deal of untapped potential. However, there is an urgent need to devise highly specific and precise strategies to target the nuclear receptors in different stromal cells to accomplish precision medicine in cancer therapy.

Author Contributions: Conceptualization, H.S.C., W.W., N.S.T.; data curation, H.S.C., Y.S.Y., and E.K.Y.L.; writing—original draft preparation, H.S.C., Y.S.Y. and E.K.Y.L.; writing—review and editing, all authors; supervision, W.W., N.S.T. All authors have read and agreed to the published version of the manuscript.

Funding: This research received no external funding.

Acknowledgments: We would like to thank William Tan Wei Ren for his kind suggestions on the first draft and illustrations of the manuscript.

Conflicts of Interest: The authors declare no conflict of interest.

References

1. Issemann, I.; Green, S. Activation of a member of the steroid hormone receptor superfamily by peroxisome proliferators. *Nature* **1990**, *347*, 645–650. [CrossRef]
2. Dreyer, C.; Krey, G.; Keller, H.; Givel, F.; Helftenbein, G.; Wahli, W. Control of the peroxisomal beta-oxidation pathway by a novel family of nuclear hormone receptors. *Cell* **1992**, *68*, 879–887. [CrossRef]
3. Lamichane, S.; Dahal Lamichane, B.; Kwon, S.M. Pivotal Roles of Peroxisome Proliferator-Activated Receptors (PPARs) and Their Signal Cascade for Cellular and Whole-Body Energy Homeostasis. *Int. J. Mol. Sci.* **2018**, *19*, 949. [CrossRef] [PubMed]
4. Jellinger, P.S.; Handelsman, Y.; Rosenblit, P.D.; Bloomgarden, Z.T.; Fonseca, V.A.; Garber, A.J.; Grunberger, G.; Guerin, C.K.; Bell, D.S.H.; Mechanick, J.I.; et al. American Association of Clinical Endocrinologists and American College of Endocrinology Guidelines for Management of Dyslipidemia and Prevention of Cardiovascular Disease. *Endocr. Pract.* **2017**, *23*, 1–87. [CrossRef] [PubMed]
5. Davies, M.J.; D'Alessio, D.A.; Fradkin, J.; Kernan, W.N.; Mathieu, C.; Mingrone, G.; Rossing, P.; Tsapas, A.; Wexler, D.J.; Buse, J.B. Management of Hyperglycemia in Type 2 Diabetes, 2018. A Consensus Report by the American Diabetes Association (ADA) and the European Association for the Study of Diabetes (EASD). *Diabetes Care* **2018**, *41*, 2669–2701. [CrossRef] [PubMed]
6. Cheng, H.S.; Tan, W.R.; Low, Z.S.; Marvalim, C.; Lee, J.Y.H.; Tan, N.S. Exploration and Development of PPAR Modulators in Health and Disease: An Update of Clinical Evidence. *Int. J. Mol. Sci.* **2019**, *20*, 5055. [CrossRef]
7. Hong, F.; Xu, P.; Zhai, Y. The Opportunities and Challenges of Peroxisome Proliferator-Activated Receptors Ligands in Clinical Drug Discovery and Development. *Int. J. Mol. Sci.* **2018**, *19*, 2189. [CrossRef] [PubMed]

8. Hopkins, B.D.; Goncalves, M.D.; Cantley, L.C. Obesity and Cancer Mechanisms: Cancer Metabolism. *J. Clin. Oncol.* **2016**, *34*, 4277–4283. [CrossRef]
9. Johnson, J.A.; Carstensen, B.; Witte, D.; Bowker, S.L.; Lipscombe, L.; Renehan, A.G.; Diabetes; Cancer Research, C. Diabetes and cancer (1): Evaluating the temporal relationship between type 2 diabetes and cancer incidence. *Diabetologia* **2012**, *55*, 1607–1618. [CrossRef] [PubMed]
10. Gou, Q.; Gong, X.; Jin, J.; Shi, J.; Hou, Y. Peroxisome proliferator-activated receptors (PPARs) are potential drug targets for cancer therapy. *Oncotarget* **2017**, *8*, 60704–60709. [CrossRef] [PubMed]
11. Balkwill, F.R.; Capasso, M.; Hagemann, T. The tumor microenvironment at a glance. *J. Cell Sci.* **2012**, *125*, 5591–5596. [CrossRef] [PubMed]
12. Wang, M.; Zhao, J.; Zhang, L.; Wei, F.; Lian, Y.; Wu, Y.; Gong, Z.; Zhang, S.; Zhou, J.; Cao, K.; et al. Role of tumor microenvironment in tumorigenesis. *J. Cancer* **2017**, *8*, 761–773. [CrossRef]
13. Shah, Y.M.; Morimura, K.; Yang, Q.; Tanabe, T.; Takagi, M.; Gonzalez, F.J. Peroxisome proliferator-activated receptor alpha regulates a microRNA-mediated signaling cascade responsible for hepatocellular proliferation. *Mol. Cell. Biol.* **2007**, *27*, 4238–4247. [CrossRef]
14. Cheung, C.; Akiyama, T.E.; Ward, J.M.; Nicol, C.J.; Feigenbaum, L.; Vinson, C.; Gonzalez, F.J. Diminished hepatocellular proliferation in mice humanized for the nuclear receptor peroxisome proliferator-activated receptor alpha. *Cancer Res.* **2004**, *64*, 3849–3854. [CrossRef]
15. Yaghoubizadeh, M.; Pishkar, L.; Basati, G. Aberrant Expression of Peroxisome Proliferator-Activated Receptors in Colorectal Cancer and Their Association with Cancer Progression and Prognosis. *Gastrointest. Tumors* **2020**, *7*, 11–20. [CrossRef]
16. Castelli, V.; Catanesi, M.; Alfonsetti, M.; Laezza, C.; Lombardi, F.; Cinque, B.; Cifone, M.G.; Ippoliti, R.; Benedetti, E.; Cimini, A.; et al. PPARalpha-Selective Antagonist GW6471 Inhibits Cell Growth in Breast Cancer Stem Cells Inducing Energy Imbalance and Metabolic Stress. *Biomedicines* **2021**, *9*, 127. [CrossRef]
17. Papi, A.; Guarnieri, T.; Storci, G.; Santini, D.; Ceccarelli, C.; Taffurelli, M.; De Carolis, S.; Avenia, N.; Sanguinetti, A.; Sidoni, A.; et al. Nuclear receptors agonists exert opposing effects on the inflammation dependent survival of breast cancer stem cells. *Cell Death Differ.* **2012**, *19*, 1208–1219. [CrossRef]
18. Kuramoto, K.; Yamamoto, M.; Suzuki, S.; Togashi, K.; Sanomachi, T.; Kitanaka, C.; Okada, M. Inhibition of the Lipid Droplet-Peroxisome Proliferator-Activated Receptor alpha Axis Suppresses Cancer Stem Cell Properties. *Genes* **2021**, *12*, 99. [CrossRef]
19. Bonovas, S.; Nikolopoulos, G.K.; Bagos, P.G. Use of fibrates and cancer risk: A systematic review and meta-analysis of 17 long-term randomized placebo-controlled trials. *PLoS ONE* **2012**, *7*, e45259. [CrossRef] [PubMed]
20. Grabacka, M.; Reiss, K. Anticancer Properties of PPARalpha-Effects on Cellular Metabolism and Inflammation. *PPAR Res* **2008**, *2008*, 930705. [CrossRef] [PubMed]
21. Peters, J.M.; Shah, Y.M.; Gonzalez, F.J. The role of peroxisome proliferator-activated receptors in carcinogenesis and chemoprevention. *Nat. Rev. Cancer* **2012**, *12*, 181–195. [CrossRef]
22. Augimeri, G.; Gelsomino, L.; Plastina, P.; Giordano, C.; Barone, I.; Catalano, S.; Ando, S.; Bonofiglio, D. Natural and Synthetic PPARgamma Ligands in Tumor Microenvironment: A New Potential Strategy against Breast Cancer. *Int. J. Mol. Sci.* **2020**, *21*, 9721. [CrossRef]
23. Sarraf, P.; Mueller, E.; Jones, D.; King, F.J.; DeAngelo, D.J.; Partridge, J.B.; Holden, S.A.; Chen, L.B.; Singer, S.; Fletcher, C.; et al. Differentiation and reversal of malignant changes in colon cancer through PPARgamma. *Nat. Med.* **1998**, *4*, 1046–1052. [CrossRef] [PubMed]
24. Basu-Roy, U.; Han, E.; Rattanakorn, K.; Gadi, A.; Verma, N.; Maurizi, G.; Gunaratne, P.H.; Coarfa, C.; Kennedy, O.D.; Garabedian, M.J.; et al. PPARgamma agonists promote differentiation of cancer stem cells by restraining YAP transcriptional activity. *Oncotarget* **2016**, *7*, 60954–60970. [CrossRef] [PubMed]
25. Mueller, E.; Sarraf, P.; Tontonoz, P.; Evans, R.M.; Martin, K.J.; Zhang, M.; Fletcher, C.; Singer, S.; Spiegelman, B.M. Terminal differentiation of human breast cancer through PPAR gamma. *Mol. Cell* **1998**, *1*, 465–470. [CrossRef]
26. Wang, Y.; Tan, H.; Xu, D.; Ma, A.; Zhang, L.; Sun, J.; Yang, Z.; Liu, Y.; Shi, G. The combinatory effects of PPAR-gamma agonist and survivin inhibition on the cancer stem-like phenotype and cell proliferation in bladder cancer cells. *Int. J. Mol. Med.* **2014**, *34*, 262–268. [CrossRef] [PubMed]
27. Kramer, K.; Wu, J.; Crowe, D.L. Tumor suppressor control of the cancer stem cell niche. *Oncogene* **2016**, *35*, 4165–4178. [CrossRef]
28. Liu, L.; Yang, Z.; Xu, Y.; Li, J.; Xu, D.; Zhang, L.; Sun, J.; Xia, S.; Zou, F.; Liu, Y. Inhibition of oxidative stress-elicited AKT activation facilitates PPARgamma agonist-mediated inhibition of stem cell character and tumor growth of liver cancer cells. *PLoS ONE* **2013**, *8*, e73038.
29. Bigoni-Ordonez, G.D.; Ortiz-Sanchez, E.; Rosendo-Chalma, P.; Valencia-Gonzalez, H.A.; Aceves, C.; Garcia-Carranca, A. Molecular iodine inhibits the expression of stemness markers on cancer stem-like cells of established cell lines derived from cervical cancer. *BMC Cancer* **2018**, *18*, 928. [CrossRef]
30. Pestereva, E.; Kanakasabai, S.; Bright, J.J. PPARgamma agonists regulate the expression of stemness and differentiation genes in brain tumour stem cells. *Br. J. Cancer* **2012**, *106*, 1702–1712. [CrossRef]
31. Shen, B.; Chu, E.S.; Zhao, G.; Man, K.; Wu, C.W.; Cheng, J.T.; Li, G.; Nie, Y.; Lo, C.M.; Teoh, N.; et al. PPARgamma inhibits hepatocellular carcinoma metastases in vitro and in mice. *Br. J. Cancer* **2012**, *106*, 1486–1494. [CrossRef]

32. Reka, A.K.; Kurapati, H.; Narala, V.R.; Bommer, G.; Chen, J.; Standiford, T.J.; Keshamouni, V.G. Peroxisome proliferator-activated receptor-gamma activation inhibits tumor metastasis by antagonizing Smad3-mediated epithelial-mesenchymal transition. *Mol. Cancer Ther.* **2010**, *9*, 3221–3232. [CrossRef]
33. Yang, K.; Fan, K.H.; Lamprecht, S.A.; Edelmann, W.; Kopelovich, L.; Kucherlapati, R.; Lipkin, M. Peroxisome proliferator-activated receptor gamma agonist troglitazone induces colon tumors in normal C57BL/6J mice and enhances colonic carcinogenesis in Apc1638 N/+ Mlh1+/- double mutant mice. *Int. J. Cancer* **2005**, *116*, 495–499. [CrossRef] [PubMed]
34. Pino, M.V.; Kelley, M.F.; Jayyosi, Z. Promotion of colon tumors in C57BL/6J-APC(min)/+ mice by thiazolidinedione PPARgamma agonists and a structurally unrelated PPARgamma agonist. *Toxicol. Pathol.* **2004**, *32*, 58–63. [CrossRef] [PubMed]
35. Taub, M. Cancer drug troglitazone stimulates the growth and response of renal cells to hypoxia inducible factors. *Biochem. Biophys. Res. Commun.* **2016**, *471*, 342–347. [CrossRef] [PubMed]
36. Han, E.; Jang, S.Y.; Kim, G.; Lee, Y.H.; Choe, E.Y.; Nam, C.M.; Kang, E.S. Rosiglitazone Use and the Risk of Bladder Cancer in Patients With Type 2 Diabetes. *Medicine* **2016**, *95*, e2786. [CrossRef] [PubMed]
37. Galbraith, L.C.A.; Mui, E.; Nixon, C.; Hedley, A.; Strachan, D.; MacKay, G.; Sumpton, D.; Sansom, O.J.; Leung, H.Y.; Ahmad, I. PPAR-gamma induced AKT3 expression increases levels of mitochondrial biogenesis driving prostate cancer. *Oncogene* **2021**, *40*, 2355–2366. [CrossRef] [PubMed]
38. Meylan, P.; Pich, C.; Winkler, C.; Ginster, S.; Mury, L.; Sgandurra, M.; Dreos, R.; Frederick, D.T.; Hammond, M.; Boland, G.M.; et al. Low expression of the PPARgamma-regulated gene thioredoxin-interacting protein accompanies human melanoma progression and promotes experimental lung metastases. *Sci. Rep.* **2021**, *11*, 7847. [CrossRef]
39. Patitucci, C.; Couchy, G.; Bagattin, A.; Caneque, T.; de Reynies, A.; Scoazec, J.Y.; Rodriguez, R.; Pontoglio, M.; Zucman-Rossi, J.; Pende, M.; et al. Hepatocyte nuclear factor 1alpha suppresses steatosis-associated liver cancer by inhibiting PPARgamma transcription. *J. Clin. Investig.* **2017**, *127*, 1873–1888. [CrossRef] [PubMed]
40. Panasyuk, G.; Espeillac, C.; Chauvin, C.; Pradelli, L.A.; Horie, Y.; Suzuki, A.; Annicotte, J.S.; Fajas, L.; Foretz, M.; Verdeguer, F.; et al. PPARgamma contributes to PKM2 and HK2 expression in fatty liver. *Nat. Commun.* **2012**, *3*, 672. [CrossRef]
41. Unal, E.; Aslan, E.I.; Ozturk, T.; Kurnaz Gomleksiz, O.; Kucukhuseyin, O.; Tuzuner, M.B.; Seyhan, M.F.; Ozturk, O.; Yilmaz Aydogan, H. Peroxisome Proliferator-Activated Receptor Gamma Pro12Ala/C161T Genotypes and Risky Haplotype Altering Risk of Breast Cancer: A Turkish Case-Control Study. *Biochem. Genet.* **2021**. [CrossRef]
42. Qu, H.; Zheng, Y.; Wang, Y.; Zhang, R.; Ruan, X.; Yang, G.; Liu, Z.; Zheng, H. Global and Regional Effects of Bladder Cancer Risk Associated with Pioglitazone Therapy in Patients with Diabetes. *Sci. Rep.* **2017**, *7*, 15804. [CrossRef]
43. Soccio, R.E.; Chen, E.R.; Lazar, M.A. Thiazolidinediones and the promise of insulin sensitization in type 2 diabetes. *Cell Metab.* **2014**, *20*, 573–591. [CrossRef] [PubMed]
44. Wagner, N.; Wagner, K.D. PPAR Beta/Delta and the Hallmarks of Cancer. *Cells* **2020**, *9*, 1133. [CrossRef] [PubMed]
45. Tan, N.S.; Vazquez-Carrera, M.; Montagner, A.; Sng, M.K.; Guillou, H.; Wahli, W. Transcriptional control of physiological and pathological processes by the nuclear receptor PPARbeta/delta. *Prog. Lipid Res.* **2016**, *64*, 98–122. [CrossRef] [PubMed]
46. Liu, Y.; Deguchi, Y.; Tian, R.; Wei, D.; Wu, L.; Chen, W.; Xu, W.; Xu, M.; Liu, F.; Gao, S.; et al. Pleiotropic Effects of PPARD Accelerate Colorectal Tumorigenesis, Progression, and Invasion. *Cancer Res.* **2019**, *79*, 954–969. [CrossRef]
47. Zuo, X.; Peng, Z.; Moussalli, M.J.; Morris, J.S.; Broaddus, R.R.; Fischer, S.M.; Shureiqi, I. Targeted genetic disruption of peroxisome proliferator-activated receptor-delta and colonic tumorigenesis. *J. Natl. Cancer. Inst.* **2009**, *101*, 762–767. [CrossRef] [PubMed]
48. Mao, F.; Xu, M.; Zuo, X.; Yu, J.; Xu, W.; Moussalli, M.J.; Elias, E.; Li, H.S.; Watowich, S.S.; Shureiqi, I. 15-Lipoxygenase-1 suppression of colitis-associated colon cancer through inhibition of the IL-6/STAT3 signaling pathway. *FASEB J.* **2015**, *29*, 2359–2370. [CrossRef]
49. Beyaz, S.; Mana, M.D.; Roper, J.; Kedrin, D.; Saadatpour, A.; Hong, S.J.; Bauer-Rowe, K.E.; Xifaras, M.E.; Akkad, A.; Arias, E.; et al. High-fat diet enhances stemness and tumorigenicity of intestinal progenitors. *Nature* **2016**, *531*, 53–58. [CrossRef] [PubMed]
50. Zuo, X.; Xu, W.; Xu, M.; Tian, R.; Moussalli, M.J.; Mao, F.; Zheng, X.; Wang, J.; Morris, J.S.; Gagea, M.; et al. Metastasis regulation by PPARD expression in cancer cells. *JCI Insight* **2017**, *2*, e91419. [CrossRef] [PubMed]
51. Pollock, C.B.; Yin, Y.; Yuan, H.; Zeng, X.; King, S.; Li, X.; Kopelovich, L.; Albanese, C.; Glazer, R.I. PPARdelta activation acts cooperatively with 3-phosphoinositide-dependent protein kinase-1 to enhance mammary tumorigenesis. *PLoS ONE* **2011**, *6*, e16215. [CrossRef]
52. Montagner, A.; Delgado, M.B.; Tallichet-Blanc, C.; Chan, J.S.; Sng, M.K.; Mottaz, H.; Degueurce, G.; Lippi, Y.; Moret, C.; Baruchet, M.; et al. Src is activated by the nuclear receptor peroxisome proliferator-activated receptor beta/delta in ultraviolet radiation-induced skin cancer. *EMBO Mol. Med.* **2014**, *6*, 80–98. [CrossRef] [PubMed]
53. Her, N.G.; Jeong, S.I.; Cho, K.; Ha, T.K.; Han, J.; Ko, K.P.; Park, S.K.; Lee, J.H.; Lee, M.G.; Ryu, B.K.; et al. PPARdelta promotes oncogenic redirection of TGF-beta1 signaling through the activation of the ABCA1-Cav1 pathway. *Cell Cycle* **2013**, *12*, 1521–1535. [CrossRef]
54. Wang, D.; Fu, L.; Wei, J.; Xiong, Y.; DuBois, R.N. PPARdelta Mediates the Effect of Dietary Fat in Promoting Colorectal Cancer Metastasis. *Cancer Res.* **2019**, *79*, 4480–4490. [CrossRef]
55. Yao, P.L.; Chen, L.; Dobrzanski, T.P.; Zhu, B.; Kang, B.H.; Muller, R.; Gonzalez, F.J.; Peters, J.M. Peroxisome proliferator-activated receptor-beta/delta inhibits human neuroblastoma cell tumorigenesis by inducing p53- and SOX2-mediated cell differentiation. *Mol. Carcinog.* **2017**, *56*, 1472–1483. [CrossRef] [PubMed]

56. Harman, F.S.; Nicol, C.J.; Marin, H.E.; Ward, J.M.; Gonzalez, F.J.; Peters, J.M. Peroxisome proliferator-activated receptor-delta attenuates colon carcinogenesis. *Nat. Med.* **2004**, *10*, 481–483. [CrossRef] [PubMed]
57. Yang, L.; Zhou, J.; Ma, Q.; Wang, C.; Chen, K.; Meng, W.; Yu, Y.; Zhou, Z.; Sun, X. Knockdown of PPAR delta gene promotes the growth of colon cancer and reduces the sensitivity to bevacizumab in nude mice model. *PLoS ONE* **2013**, *8*, e60715.
58. Foreman, J.E.; Chang, W.C.; Palkar, P.S.; Zhu, B.; Borland, M.G.; Williams, J.L.; Kramer, L.R.; Clapper, M.L.; Gonzalez, F.J.; Peters, J.M. Functional characterization of peroxisome proliferator-activated receptor-beta/delta expression in colon cancer. *Mol. Carcinog.* **2011**, *50*, 884–900. [CrossRef] [PubMed]
59. Alam, M.M.; Lal, S.; FitzGerald, K.E.; Zhang, L. A holistic view of cancer bioenergetics: Mitochondrial function and respiration play fundamental roles in the development and progression of diverse tumors. *Clin. Transl. Med.* **2016**, *5*, 3. [CrossRef] [PubMed]
60. Martinez-Outschoorn, U.E.; Lisanti, M.P.; Sotgia, F. Catabolic cancer-associated fibroblasts transfer energy and biomass to anabolic cancer cells, fueling tumor growth. *Semin. Cancer Biol.* **2014**, *25*, 47–60. [CrossRef]
61. Vandoros, G.P.; Konstantinopoulos, P.A.; Sotiropoulou-Bonikou, G.; Kominea, A.; Papachristou, G.I.; Karamouzis, M.V.; Gkermpesi, M.; Varakis, I.; Papavassiliou, A.G. PPAR-gamma is expressed and NF-κB pathway is activated and correlates positively with COX-2 expression in stromal myofibroblasts surrounding colon adenocarcinomas. *J. Cancer Res. Clin. Oncol.* **2006**, *132*, 76–84. [CrossRef]
62. Chan, J.S.K.; Sng, M.K.; Teo, Z.Q.; Chong, H.C.; Twang, J.S.; Tan, N.S. Targeting nuclear receptors in cancer-associated fibroblasts as concurrent therapy to inhibit development of chemoresistant tumors. *Oncogene* **2018**, *37*, 160–173. [CrossRef]
63. Avena, P.; Anselmo, W.; Whitaker-Menezes, D.; Wang, C.; Pestell, R.G.; Lamb, R.S.; Hulit, J.; Casaburi, I.; Ando, S.; Martinez-Outschoorn, U.E.; et al. Compartment-specific activation of PPARgamma governs breast cancer tumor growth, via metabolic reprogramming and symbiosis. *Cell Cycle* **2013**, *12*, 1360–1370. [CrossRef] [PubMed]
64. Pavlides, S.; Vera, I.; Gandara, R.; Sneddon, S.; Pestell, R.G.; Mercier, I.; Martinez-Outschoorn, U.E.; Whitaker-Menezes, D.; Howell, A.; Sotgia, F.; et al. Warburg meets autophagy: Cancer-associated fibroblasts accelerate tumor growth and metastasis via oxidative stress, mitophagy, and aerobic glycolysis. *Antioxid. Redox Signal.* **2012**, *16*, 1264–1284. [CrossRef]
65. Boutoual, R.; Meseguer, S.; Villarroya, M.; Martin-Hernandez, E.; Errami, M.; Martin, M.A.; Casado, M.; Armengod, M.E. Defects in the mitochondrial-tRNA modification enzymes MTO1 and GTPBP3 promote different metabolic reprogramming through a HIF-PPARgamma-UCP2-AMPK axis. *Sci. Rep.* **2018**, *8*, 1163. [CrossRef] [PubMed]
66. Jang, M.; Park, R.; Kim, H.; Namkoong, S.; Jo, D.; Huh, Y.H.; Jang, I.S.; Lee, J.I.; Park, J. AMPK contributes to autophagosome maturation and lysosomal fusion. *Sci. Rep.* **2018**, *8*, 12637. [CrossRef]
67. Wang, S.; Kandadi, M.R.; Ren, J. Double knockout of Akt2 and AMPK predisposes cardiac aging without affecting lifespan: Role of autophagy and mitophagy. *Biochim. Biophys. Acta. Mol. Basis Dis.* **2019**, *1865*, 1865–1875. [CrossRef] [PubMed]
68. Jiao, L.; Zhang, H.L.; Li, D.D.; Yang, K.L.; Tang, J.; Li, X.; Ji, J.; Yu, Y.; Wu, R.Y.; Ravichandran, S.; et al. Regulation of glycolytic metabolism by autophagy in liver cancer involves selective autophagic degradation of HK2 (hexokinase 2). *Autophagy* **2018**, *14*, 671–684. [CrossRef] [PubMed]
69. Fan, Q.; Yang, L.; Zhang, X.; Ma, Y.; Li, Y.; Dong, L.; Zong, Z.; Hua, X.; Su, D.; Li, H.; et al. Autophagy promotes metastasis and glycolysis by upregulating MCT1 expression and Wnt/beta-catenin signaling pathway activation in hepatocellular carcinoma cells. *J. Exp. Clin. Cancer Res.* **2018**, *37*, 9. [CrossRef]
70. Shashni, B.; Sakharkar, K.R.; Nagasaki, Y.; Sakharkar, M.K. Glycolytic enzymes PGK1 and PKM2 as novel transcriptional targets of PPARgamma in breast cancer pathophysiology. *J. Drug Target.* **2013**, *21*, 161–174. [CrossRef]
71. Zhang, H.; Li, L.; Chen, Q.; Li, M.; Feng, J.; Sun, Y.; Zhao, R.; Zhu, Y.; Lv, Y.; Zhu, Z.; et al. PGC1beta regulates multiple myeloma tumor growth through LDHA-mediated glycolytic metabolism. *Mol. Oncol.* **2018**, *12*, 1579–1595. [CrossRef] [PubMed]
72. Londhe, P.; Yu, P.Y.; Ijiri, Y.; Ladner, K.J.; Fenger, J.M.; London, C.; Houghton, P.J.; Guttridge, D.C. Classical NF-kappaB Metabolically Reprograms Sarcoma Cells Through Regulation of Hexokinase 2. *Front. Oncol.* **2018**, *8*, 104. [CrossRef]
73. Wu, Q.; Li, B.; Li, Z.; Li, J.; Sun, S.; Sun, S. Cancer-associated adipocytes: Key players in breast cancer progression. *J. Hematol. Oncol.* **2019**, *12*, 95. [CrossRef] [PubMed]
74. Wolins, N.E.; Quaynor, B.K.; Skinner, J.R.; Tzekov, A.; Park, C.; Choi, K.; Bickel, P.E. OP9 mouse stromal cells rapidly differentiate into adipocytes: Characterization of a useful new model of adipogenesis. *J. Lipid Res.* **2006**, *47*, 450–460. [CrossRef]
75. Chen, J.H.; Goh, K.J.; Rocha, N.; Groeneveld, M.P.; Minic, M.; Barrett, T.G.; Savage, D.; Semple, R.K. Evaluation of human dermal fibroblasts directly reprogrammed to adipocyte-like cells as a metabolic disease model. *Dis. Model. Mech.* **2017**, *10*, 1411–1420. [CrossRef] [PubMed]
76. Wu, Q.; Li, J.; Li, Z.; Sun, S.; Zhu, S.; Wang, L.; Wu, J.; Yuan, J.; Zhang, Y.; Sun, S.; et al. Exosomes from the tumour-adipocyte interplay stimulate beige/brown differentiation and reprogram metabolism in stromal adipocytes to promote tumour progression. *J. Exp. Clin. Cancer Res.* **2019**, *38*, 223. [CrossRef] [PubMed]
77. Wu, Q.; Sun, S.; Li, Z.; Yang, Q.; Li, B.; Zhu, S.; Wang, L.; Wu, J.; Yuan, J.; Yang, C.; et al. Tumour-originated exosomal miR-155 triggers cancer-associated cachexia to promote tumour progression. *Mol. Cancer* **2018**, *17*, 155. [CrossRef]
78. Cai, Z.; Liang, Y.; Xing, C.; Wang, H.; Hu, P.; Li, J.; Huang, H.; Wang, W.; Jiang, C. Cancer associated adipocytes exhibit distinct phenotypes and facilitate tumor progression in pancreatic cancer. *Oncol. Rep.* **2019**, *42*, 2537–2549. [CrossRef] [PubMed]
79. Zoico, E.; Darra, E.; Rizzatti, V.; Budui, S.; Franceschetti, G.; Mazzali, G.; Rossi, A.P.; Fantin, F.; Menegazzi, M.; Cinti, S.; et al. Adipocytes WNT5a mediated dedifferentiation: A possible target in pancreatic cancer microenvironment. *Oncotarget* **2016**, *7*, 20223–20235. [CrossRef]

80. Bi, P.; Yue, F.; Karki, A.; Castro, B.; Wirbisky, S.E.; Wang, C.; Durkes, A.; Elzey, B.D.; Andrisani, O.M.; Bidwell, C.A.; et al. Notch activation drives adipocyte dedifferentiation and tumorigenic transformation in mice. *J. Exp. Med.* **2016**, *213*, 2019–2037. [CrossRef] [PubMed]
81. Skelhorne-Gross, G.; Reid, A.L.; Apostoli, A.J.; Di Lena, M.A.; Rubino, R.E.; Peterson, N.T.; Schneider, M.; SenGupta, S.K.; Gonzalez, F.J.; Nicol, C.J. Stromal adipocyte PPARgamma protects against breast tumorigenesis. *Carcinogenesis* **2012**, *33*, 1412–1420. [CrossRef] [PubMed]
82. Matsumoto, T.; Kano, K.; Kondo, D.; Fukuda, N.; Iribe, Y.; Tanaka, N.; Matsubara, Y.; Sakuma, T.; Satomi, A.; Otaki, M.; et al. Mature adipocyte-derived dedifferentiated fat cells exhibit multilineage potential. *J. Cell Physiol.* **2008**, *215*, 210–222. [CrossRef] [PubMed]
83. Pereira, B.A.; Vennin, C.; Papanicolaou, M.; Chambers, C.R.; Herrmann, D.; Morton, J.P.; Cox, T.R.; Timpson, P. CAF Subpopulations: A New Reservoir of Stromal Targets in Pancreatic Cancer. *Trends Cancer* **2019**, *5*, 724–741. [CrossRef]
84. Yoshida, G.J.; Azuma, A.; Miura, Y.; Orimo, A. Activated Fibroblast Program Orchestrates Tumor Initiation and Progression; Molecular Mechanisms and the Associated Therapeutic Strategies. *Int. J. Mol. Sci.* **2019**, *20*, 2256. [CrossRef] [PubMed]
85. Tan, M.W.Y.; Sng, M.K.; Cheng, H.S.; Low, Z.S.; Leong, B.J.J.; Chua, D.; Tan, E.H.P.; Chan, J.S.K.; Yip, Y.S.; Lee, Y.H.; et al. Deficiency in fibroblast PPARbeta/delta reduces nonmelanoma skin cancers in mice. *Cell Death Differ.* **2020**. [CrossRef]
86. Tan, E.H.P.; Sng, M.K.; How, I.S.B.; Chan, J.S.K.; Chen, J.; Tan, C.K.; Wahli, W.; Tan, N.S. ROS release by PPARbeta/delta-null fibroblasts reduces tumor load through epithelial antioxidant response. *Oncogene* **2018**, *37*, 2067–2078. [CrossRef]
87. Sng, M.K.; Chan, J.S.K.; Teo, Z.; Phua, T.; Tan, E.H.P.; Wee, J.W.K.; Koh, N.J.N.; Tan, C.K.; Chen, J.P.; Pal, M.; et al. Selective deletion of PPARbeta/delta in fibroblasts causes dermal fibrosis by attenuated LRG1 expression. *Cell Discov.* **2018**, *4*, 15. [CrossRef]
88. Zhang, Q.; Huang, R.; Tang, Q.; Yu, Y.; Huang, Q.; Chen, Y.; Wang, G.; Wang, X. Leucine-rich alpha-2-glycoprotein-1 is up-regulated in colorectal cancer and is a tumor promoter. *Onco Targets Ther.* **2018**, *11*, 2745–2752. [CrossRef]
89. Xie, Z.B.; Zhang, Y.F.; Jin, C.; Mao, Y.S.; Fu, D.L. LRG-1 promotes pancreatic cancer growth and metastasis via modulation of the EGFR/p38 signaling. *J. Exp. Clin. Cancer Res.* **2019**, *38*, 75. [CrossRef]
90. Zhou, Y.; Zhang, X.; Zhang, J.; Fang, J.; Ge, Z.; Li, X. LRG1 promotes proliferation and inhibits apoptosis in colorectal cancer cells via RUNX1 activation. *PLoS ONE* **2017**, *12*, e0175122. [CrossRef]
91. Liu, C.; Lim, S.T.; Teo, M.H.Y.; Tan, M.S.Y.; Kulkarni, M.D.; Qiu, B.; Li, A.; Lal, S.; Dos Remedios, C.G.; Tan, N.S.; et al. Collaborative Regulation of LRG1 by TGF-beta1 and PPAR-beta/delta Modulates Chronic Pressure Overload-Induced Cardiac Fibrosis. *Circ. Heart Fail.* **2019**, *12*, e005962. [CrossRef]
92. Eales, K.L.; Hollinshead, K.E.; Tennant, D.A. Hypoxia and metabolic adaptation of cancer cells. *Oncogenesis* **2016**, *5*, e190. [CrossRef]
93. Panigrahy, D.; Kaipainen, A.; Huang, S.; Butterfield, C.E.; Barnes, C.M.; Fannon, M.; Laforme, A.M.; Chaponis, D.M.; Folkman, J.; Kieran, M.W. PPARalpha agonist fenofibrate suppresses tumor growth through direct and indirect angiogenesis inhibition. *Proc. Natl. Acad. Sci. USA* **2008**, *105*, 985–990. [CrossRef] [PubMed]
94. Pozzi, A.; Ibanez, M.R.; Gatica, A.E.; Yang, S.; Wei, S.; Mei, S.; Falck, J.R.; Capdevila, J.H. Peroxisomal proliferator-activated receptor-alpha-dependent inhibition of endothelial cell proliferation and tumorigenesis. *J. Biol. Chem.* **2007**, *282*, 17685–17695. [CrossRef] [PubMed]
95. Arima, T.; Uchiyama, M.; Nakano, Y.; Nagasaka, S.; Kang, D.; Shimizu, A.; Takahashi, H. Peroxisome proliferator-activated receptor alpha agonist suppresses neovascularization by reducing both vascular endothelial growth factor and angiopoietin-2 in corneal alkali burn. *Sci. Rep.* **2017**, *7*, 17763. [CrossRef]
96. Pozzi, A.; Popescu, V.; Yang, S.; Mei, S.; Shi, M.; Puolitaival, S.M.; Caprioli, R.M.; Capdevila, J.H. The anti-tumorigenic properties of peroxisomal proliferator-activated receptor alpha are arachidonic acid epoxygenase-mediated. *J. Biol. Chem.* **2010**, *285*, 12840–12850. [CrossRef] [PubMed]
97. Leu, J.-G.; Chiang, M.-H.; Chen, C.-Y.; Lin, J.-T.; Chen, H.-M.; Chen, Y.-L.; Liang, Y.-J. Adenine accelerated the diabetic wound healing by PPAR delta and angiogenic regulation. *Eur. J. Pharmacol.* **2018**, *818*, 569–577. [CrossRef]
98. Montagner, A.; Wahli, W.; Tan, N.S. Nuclear receptor peroxisome proliferator activated receptor (PPAR) beta/delta in skin wound healing and cancer. *Eur. J. Dermatol.* **2015**, *25* (Suppl. 1), 4–11. [PubMed]
99. Tan, N.S.; Icre, G.; Montagner, A.; Bordier-ten-Heggeler, B.; Wahli, W.; Michalik, L. The nuclear hormone receptor peroxisome proliferator-activated receptor beta/delta potentiates cell chemotactism, polarization, and migration. *Mol. Cell Biol.* **2007**, *27*, 7161–7175. [CrossRef]
100. Piqueras, L.; Reynolds, A.R.; Hodivala-Dilke, K.M.; Alfranca, A.; Redondo, J.M.; Hatae, T.; Tanabe, T.; Warner, T.D.; Bishop-Bailey, D. Activation of PPARbeta/delta induces endothelial cell proliferation and angiogenesis. *Arterioscler. Thromb. Vasc. Biol.* **2007**, *27*, 63–69. [CrossRef]
101. Wagner, K.D.; Du, S.; Martin, L.; Leccia, N.; Michiels, J.F.; Wagner, N. Vascular PPARbeta/delta Promotes Tumor Angiogenesis and Progression. *Cells* **2019**, *8*, 1623. [CrossRef]
102. Muller-Brusselbach, S.; Komhoff, M.; Rieck, M.; Meissner, W.; Kaddatz, K.; Adamkiewicz, J.; Keil, B.; Klose, K.J.; Moll, R.; Burdick, A.D.; et al. Deregulation of tumor angiogenesis and blockade of tumor growth in PPARbeta-deficient mice. *EMBO J.* **2007**, *26*, 3686–3698. [CrossRef]

103. Abdollahi, A.; Schwager, C.; Kleeff, J.; Esposito, I.; Domhan, S.; Peschke, P.; Hauser, K.; Hahnfeldt, P.; Hlatky, L.; Debus, J.; et al. Transcriptional network governing the angiogenic switch in human pancreatic cancer. *Proc. Natl. Acad. Sci. USA* **2007**, *104*, 12890–12895. [CrossRef]
104. Meissner, M.; Hrgovic, I.; Doll, M.; Naidenow, J.; Reichenbach, G.; Hailemariam-Jahn, T.; Michailidou, D.; Gille, J.; Kaufmann, R. Peroxisome proliferator-activated receptor δ activators induce IL-8 expression in nonstimulated endothelial cells in a transcriptional and posttranscriptional manner. *J. Biol. Chem.* **2010**, *285*, 33797–33804. [CrossRef] [PubMed]
105. Adamkiewicz, J.; Kaddatz, K.; Rieck, M.; Wilke, B.; Muller-Brusselbach, S.; Muller, R. Proteomic profile of mouse fibroblasts with a targeted disruption of the peroxisome proliferator activated receptor-beta/delta gene. *Proteomics* **2007**, *7*, 1208–1216. [CrossRef]
106. Possati, L.; Rocchetti, R.; Talevi, S.; Beatrici, V.; Margiotta, C.; Ferrante, L.; Calza, R.; Sagrini, D.; Ferri, A. The role of peroxisome proliferator-activated receptor gamma in bladder cancer in relation to angiogenesis and progression. *Gen. Pharmacol.* **2000**, *35*, 269–275. [CrossRef]
107. Huang, W.; Shao, M.; Liu, H.; Chen, J.; Hu, J.; Zhu, L.; Liu, F.; Wang, D.; Zou, Y.; Xiong, Y.; et al. Fibroblast growth factor 21 enhances angiogenesis and wound healing of human brain microvascular endothelial cells by activating PPARgamma. *J. Pharmacol. Sci.* **2019**, *140*, 120–127. [CrossRef]
108. Xin, X.; Yang, S.; Kowalski, J.; Gerritsen, M.E. Peroxisome proliferator-activated receptor gamma ligands are potent inhibitors of angiogenesis in vitro and in vivo. *J. Biol. Chem.* **1999**, *274*, 9116–9121. [CrossRef] [PubMed]
109. Sarayba, M.A.; Li, L.; Tungsiripat, T.; Liu, N.H.; Sweet, P.M.; Patel, A.J.; Osann, K.E.; Chittiboyina, A.; Benson, S.C.; Pershadsingh, H.A.; et al. Inhibition of corneal neovascularization by a peroxisome proliferator-activated receptor-gamma ligand. *Exp. Eye Res.* **2005**, *80*, 435–442. [CrossRef] [PubMed]
110. Biscetti, F.; Gaetani, E.; Flex, A.; Aprahamian, T.; Hopkins, T.; Straface, G.; Pecorini, G.; Stigliano, E.; Smith, R.C.; Angelini, F.; et al. Selective activation of peroxisome proliferator-activated receptor (PPAR)alpha and PPAR gamma induces neoangiogenesis through a vascular endothelial growth factor-dependent mechanism. *Diabetes* **2008**, *57*, 1394–1404. [CrossRef] [PubMed]
111. Vattulainen-Collanus, S.; Akinrinade, O.; Li, M.; Koskenvuo, M.; Li, C.G.; Rao, S.P.; de Jesus Perez, V.; Yuan, K.; Sawada, H.; Koskenvuo, J.W.; et al. Loss of PPARgamma in endothelial cells leads to impaired angiogenesis. *J. Cell Sci.* **2016**, *129*, 693–705. [CrossRef]
112. Bishop-Bailey, D.; Swales, K.E. The Role of PPARs in the Endothelium: Implications for Cancer Therapy. *PPAR Res.* **2008**, *2008*, 904251. [CrossRef]
113. Kusuma, G.D.; Carthew, J.; Lim, R.; Frith, J.E. Effect of the Microenvironment on Mesenchymal Stem Cell Paracrine Signaling: Opportunities to Engineer the Therapeutic Effect. *Stem. Cells Dev.* **2017**, *26*, 617–631. [CrossRef]
114. Kim, T.W.; Hong, D.W.; Hong, S.H. CB13, a novel PPARgamma ligand, overcomes radio-resistance via ROS generation and ER stress in human non-small cell lung cancer. *Cell Death Dis.* **2020**, *11*, 848. [CrossRef]
115. Rigas, B.; Goldman, I.S.; Levine, L. Altered eicosanoid levels in human colon cancer. *J. Lab. Clin. Med.* **1993**, *122*, 518–523.
116. Paulitschke, V.; Gruber, S.; Hofstatter, E.; Haudek-Prinz, V.; Klepeisz, P.; Schicher, N.; Jonak, C.; Petzelbauer, P.; Pehamberger, H.; Gerner, C.; et al. Proteome analysis identified the PPARgamma ligand 15d-PGJ2 as a novel drug inhibiting melanoma progression and interfering with tumor-stroma interaction. *PLoS ONE* **2012**, *7*, e46103. [CrossRef]
117. Rovito, D.; Gionfriddo, G.; Barone, I.; Giordano, C.; Grande, F.; De Amicis, F.; Lanzino, M.; Catalano, S.; Ando, S.; Bonofiglio, D. Ligand-activated PPARgamma downregulates CXCR4 gene expression through a novel identified PPAR response element and inhibits breast cancer progression. *Oncotarget* **2016**, *7*, 65109–65124. [CrossRef] [PubMed]
118. Papi, A.; De Carolis, S.; Bertoni, S.; Storci, G.; Sceberras, V.; Santini, D.; Ceccarelli, C.; Taffurelli, M.; Orlandi, M.; Bonafe, M. PPARgamma and RXR ligands disrupt the inflammatory cross-talk in the hypoxic breast cancer stem cells niche. *J. Cell. Physiol.* **2014**, *229*, 1595–1606. [CrossRef] [PubMed]
119. Kim, N.; Choi, S.; Lim, C.; Lee, H.; Oh, J. Albumin mediates PPAR-gamma or C/EBP-alpha-induced phenotypic changes in pancreatic stellate cells. *Biochem. Biophys. Res. Commun.* **2010**, *391*, 640–644. [CrossRef] [PubMed]
120. Sharvit, E.; Abramovitch, S.; Reif, S.; Bruck, R. Amplified inhibition of stellate cell activation pathways by PPAR-gamma, RAR and RXR agonists. *PLoS ONE* **2013**, *8*, e76541. [CrossRef]
121. Zhang, Q.; Xiang, S.; Liu, Q.; Gu, T.; Yao, Y.; Lu, X. PPARgamma Antagonizes Hypoxia-Induced Activation of Hepatic Stellate Cell through Cross Mediating PI3K/AKT and cGMP/PKG Signaling. *PPAR Res.* **2018**, *2018*, 6970407. [CrossRef] [PubMed]
122. Shimizu, K.; Kobayashi, M.; Tahara, J.; Shiratori, K. Cytokines and peroxisome proliferator-activated receptor gamma ligand regulate phagocytosis by pancreatic stellate cells. *Gastroenterology* **2005**, *128*, 2105–2118. [CrossRef]
123. Pich, C.; Meylan, P.; Mastelic-Gavillet, B.; Nguyen, T.N.; Loyon, R.; Trang, B.K.; Moser, H.; Moret, C.; Goepfert, C.; Hafner, J.; et al. Induction of Paracrine Signaling in Metastatic Melanoma Cells by PPARgamma Agonist Rosiglitazone Activates Stromal Cells and Enhances Tumor Growth. *Cancer Res.* **2018**, *78*, 6447–6461. [CrossRef]
124. Christofides, A.; Konstantinidou, E.; Jani, C.; Boussiotis, V.A. The role of peroxisome proliferator-activated receptors (PPAR) in immune responses. *Metabolism* **2021**, *114*, 154338. [CrossRef] [PubMed]
125. Gionfriddo, G.; Plastina, P.; Augimeri, G.; Catalano, S.; Giordano, C.; Barone, I.; Morelli, C.; Giordano, F.; Gelsomino, L.; Sisci, D.; et al. Modulating Tumor-Associated Macrophage Polarization by Synthetic and Natural PPARgamma Ligands as a Potential Target in Breast Cancer. *Cells* **2020**, *9*, 174. [CrossRef]

126. Penas, F.; Mirkin, G.A.; Vera, M.; Cevey, A.; Gonzalez, C.D.; Gomez, M.I.; Sales, M.E.; Goren, N.B. Treatment in vitro with PPARalpha and PPARgamma ligands drives M1-to-M2 polarization of macrophages from *T. cruzi*-infected mice. *Biochim. Biophys. Acta* **2015**, *1852*, 893–904. [CrossRef] [PubMed]
127. Odegaard, J.I.; Ricardo-Gonzalez, R.R.; Goforth, M.H.; Morel, C.R.; Subramanian, V.; Mukundan, L.; Red Eagle, A.; Vats, D.; Brombacher, F.; Ferrante, A.W.; et al. Macrophage-specific PPARgamma controls alternative activation and improves insulin resistance. *Nature* **2007**, *447*, 1116–1120. [CrossRef] [PubMed]
128. Vats, D.; Mukundan, L.; Odegaard, J.I.; Zhang, L.; Smith, K.L.; Morel, C.R.; Wagner, R.A.; Greaves, D.R.; Murray, P.J.; Chawla, A. Oxidative metabolism and PGC-1beta attenuate macrophage-mediated inflammation. *Cell Metab.* **2006**, *4*, 13–24. [CrossRef]
129. Souza-Moreira, L.; Soares, V.C.; Dias, S.; Bozza, P.T. Adipose-derived Mesenchymal Stromal Cells Modulate Lipid Metabolism and Lipid Droplet Biogenesis via AKT/mTOR -PPARgamma Signalling in Macrophages. *Sci. Rep.* **2019**, *9*, 20304. [CrossRef]
130. Niu, Z.; Shi, Q.; Zhang, W.; Shu, Y.; Yang, N.; Chen, B.; Wang, Q.; Zhao, X.; Chen, J.; Cheng, N.; et al. Caspase-1 cleaves PPARgamma for potentiating the pro-tumor action of TAMs. *Nat. Commun.* **2017**, *8*, 766. [CrossRef] [PubMed]
131. Zhao, F.; Xiao, C.; Evans, K.S.; Theivanthiran, T.; DeVito, N.; Holtzhausen, A.; Liu, J.; Liu, X.; Boczkowski, D.; Nair, S.; et al. Paracrine Wnt5a-beta-Catenin Signaling Triggers a Metabolic Program that Drives Dendritic Cell Tolerization. *Immunity* **2018**, *48*, 147–160 e147. [CrossRef]
132. Schumann, T.; Adhikary, T.; Wortmann, A.; Finkernagel, F.; Lieber, S.; Schnitzer, E.; Legrand, N.; Schober, Y.; Nockher, W.A.; Toth, P.M.; et al. Deregulation of PPARbeta/delta target genes in tumor-associated macrophages by fatty acid ligands in the ovarian cancer microenvironment. *Oncotarget* **2015**, *6*, 13416–13433. [CrossRef]
133. Odegaard, J.I.; Ricardo-Gonzalez, R.R.; Red Eagle, A.; Vats, D.; Morel, C.R.; Goforth, M.H.; Subramanian, V.; Mukundan, L.; Ferrante, A.W.; Chawla, A. Alternative M2 activation of Kupffer cells by PPARdelta ameliorates obesity-induced insulin resistance. *Cell Metab.* **2008**, *7*, 496–507. [CrossRef] [PubMed]
134. Cheng, W.Y.; Huynh, H.; Chen, P.; Pena-Llopis, S.; Wan, Y. Macrophage PPARgamma inhibits Gpr132 to mediate the anti-tumor effects of rosiglitazone. *Elife* **2016**, *5*, e18502. [CrossRef]
135. Van Ginderachter, J.A.; Meerschaut, S.; Liu, Y.; Brys, L.; De Groeve, K.; Hassanzadeh Ghassabeh, G.; Raes, G.; De Baetselier, P. Peroxisome proliferator-activated receptor gamma (PPARgamma) ligands reverse CTL suppression by alternatively activated (M2) macrophages in cancer. *Blood* **2006**, *108*, 525–535. [CrossRef] [PubMed]
136. Kim, Y.B.; Ahn, Y.H.; Jung, J.H.; Lee, Y.J.; Lee, J.H.; Kang, J.L. Programming of macrophages by UV-irradiated apoptotic cancer cells inhibits cancer progression and lung metastasis. *Cell. Mol. Immunol.* **2019**, *16*, 851–867. [CrossRef]
137. Yin, X.; Zeng, W.; Wu, B.; Wang, L.; Wang, Z.; Tian, H.; Wang, L.; Jiang, Y.; Clay, R.; Wei, X.; et al. PPARalpha Inhibition Overcomes Tumor-Derived Exosomal Lipid-Induced Dendritic Cell Dysfunction. *Cell Rep.* **2020**, *33*, 108278. [CrossRef]
138. Tan, M.J.; Teo, Z.; Sng, M.K.; Zhu, P.; Tan, N.S. Emerging roles of angiopoietin-like 4 in human cancer. *Mol. Cancer Res.* **2012**, *10*, 677–688. [CrossRef] [PubMed]
139. Zhu, P.; Goh, Y.Y.; Chin, H.F.; Kersten, S.; Tan, N.S. Angiopoietin-like 4: A decade of research. *Biosci. Rep.* **2012**, *32*, 211–219. [CrossRef]
140. Kim, S.H.; Park, Y.Y.; Kim, S.W.; Lee, J.S.; Wang, D.; DuBois, R.N. ANGPTL4 induction by prostaglandin E2 under hypoxic conditions promotes colorectal cancer progression. *Cancer Res.* **2011**, *71*, 7010–7020. [CrossRef] [PubMed]
141. Kersten, S.; Mandard, S.; Tan, N.S.; Escher, P.; Metzger, D.; Chambon, P.; Gonzalez, F.J.; Desvergne, B.; Wahli, W. Characterization of the fasting-induced adipose factor FIAF, a novel peroxisome proliferator-activated receptor target gene. *J. Biol. Chem.* **2000**, *275*, 28488–28493. [CrossRef] [PubMed]
142. Ge, H.; Yang, G.; Huang, L.; Motola, D.L.; Pourbahrami, T.; Li, C. Oligomerization and regulated proteolytic processing of angiopoietin-like protein 4. *J. Biol. Chem.* **2004**, *279*, 2038–2045. [CrossRef] [PubMed]
143. La Paglia, L.; Listi, A.; Caruso, S.; Amodeo, V.; Passiglia, F.; Bazan, V.; Fanale, D. Potential Role of ANGPTL4 in the Cross Talk between Metabolism and Cancer through PPAR Signaling Pathway. *PPAR Res.* **2017**, *2017*, 8187235. [CrossRef]
144. Zhou, S.; Wang, R.; Xiao, H. Adipocytes induce the resistance of ovarian cancer to carboplatin through ANGPTL4. *Oncol. Rep.* **2020**, *44*, 927–938. [CrossRef] [PubMed]
145. Cai, Y.C.; Yang, H.; Wang, K.F.; Chen, T.H.; Jiang, W.Q.; Shi, Y.X. ANGPTL4 overexpression inhibits tumor cell adhesion and migration and predicts favorable prognosis of triple-negative breast cancer. *BMC Cancer* **2020**, *20*, 878. [CrossRef]
146. Hsieh, H.Y.; Jou, Y.C.; Tung, C.L.; Tsai, Y.S.; Wang, Y.H.; Chi, C.L.; Lin, R.I.; Hung, S.K.; Chuang, Y.M.; Wu, S.F.; et al. Epigenetic silencing of the dual-role signal mediator, ANGPTL4 in tumor tissues and its overexpression in the urothelial carcinoma microenvironment. *Oncogene* **2018**, *37*, 673–686. [CrossRef] [PubMed]
147. Kolb, R.; Kluz, P.; Tan, Z.W.; Borcherding, N.; Bormann, N.; Vishwakarma, A.; Balcziak, L.; Zhu, P.; Davies, B.S.; Gourronc, F.; et al. Obesity-associated inflammation promotes angiogenesis and breast cancer via angiopoietin-like 4. *Oncogene* **2019**, *38*, 2351–2363. [CrossRef]
148. Goh, Y.Y.; Pal, M.; Chong, H.C.; Zhu, P.; Tan, M.J.; Punugu, L.; Lam, C.R.; Yau, Y.H.; Tan, C.K.; Huang, R.L.; et al. Angiopoietin-like 4 interacts with integrins beta1 and beta5 to modulate keratinocyte migration. *Am. J. Pathol.* **2010**, *177*, 2791–2803. [CrossRef] [PubMed]
149. Huang, R.L.; Teo, Z.; Chong, H.C.; Zhu, P.; Tan, M.J.; Tan, C.K.; Lam, C.R.; Sng, M.K.; Leong, D.T.; Tan, S.M.; et al. ANGPTL4 modulates vascular junction integrity by integrin signaling and disruption of intercellular VE-cadherin and claudin-5 clusters. *Blood* **2011**, *118*, 3990–4002. [CrossRef] [PubMed]

150. Goh, Y.Y.; Pal, M.; Chong, H.C.; Zhu, P.; Tan, M.J.; Punugu, L.; Tan, C.K.; Huang, R.L.; Sze, S.K.; Tang, M.B.; et al. Angiopoietin-like 4 interacts with matrix proteins to modulate wound healing. *J. Biol. Chem.* **2010**, *285*, 32999–33009. [CrossRef] [PubMed]
151. Nakayama, T.; Hirakawa, H.; Shibata, K.; Nazneen, A.; Abe, K.; Nagayasu, T.; Taguchi, T. Expression of angiopoietin-like 4 (ANGPTL4) in human colorectal cancer: ANGPTL4 promotes venous invasion and distant metastasis. *Oncol. Rep.* **2011**, *25*, 929–935. [CrossRef] [PubMed]
152. Ma, T.; Jham, B.C.; Hu, J.; Friedman, E.R.; Basile, J.R.; Molinolo, A.; Sodhi, A.; Montaner, S. Viral G protein-coupled receptor up-regulates Angiopoietin-like 4 promoting angiogenesis and vascular permeability in Kaposi's sarcoma. *Proc. Natl. Acad. Sci. USA* **2010**, *107*, 14363–14368. [CrossRef] [PubMed]
153. Huang, X.F.; Han, J.; Hu, X.T.; He, C. Mechanisms involved in biological behavior changes associated with Angptl4 expression in colon cancer cell lines. *Oncol. Rep.* **2012**, *27*, 1541–1547.
154. Zhu, P.; Tan, M.J.; Huang, R.L.; Tan, C.K.; Chong, H.C.; Pal, M.; Lam, C.R.; Boukamp, P.; Pan, J.Y.; Tan, S.H.; et al. Angiopoietin-like 4 protein elevates the prosurvival intracellular O2(-):H2O2 ratio and confers anoikis resistance to tumors. *Cancer Cell* **2011**, *19*, 401–415. [CrossRef]
155. Baba, K.; Kitajima, Y.; Miyake, S.; Nakamura, J.; Wakiyama, K.; Sato, H.; Okuyama, K.; Kitagawa, H.; Tanaka, T.; Hiraki, M.; et al. Hypoxia-induced ANGPTL4 sustains tumour growth and anoikis resistance through different mechanisms in scirrhous gastric cancer cell lines. *Sci. Rep.* **2017**, *7*, 11127. [CrossRef] [PubMed]
156. Liao, Y.H.; Chiang, K.H.; Shieh, J.M.; Huang, C.R.; Shen, C.J.; Huang, W.C.; Chen, B.K. Epidermal growth factor-induced ANGPTL4 enhances anoikis resistance and tumour metastasis in head and neck squamous cell carcinoma. *Oncogene* **2017**, *36*, 2228–2242. [CrossRef]
157. Jung, K.H.; Son, M.K.; Yan, H.H.; Fang, Z.; Kim, J.; Kim, S.J.; Park, J.H.; Lee, J.E.; Yoon, Y.C.; Seo, M.S.; et al. ANGPTL4 exacerbates pancreatitis by augmenting acinar cell injury through upregulation of C5a. *EMBO Mol. Med.* **2020**, *12*, e11222. [CrossRef]
158. Zhao, T.; Du, H.; Blum, J.S.; Yan, C. Critical role of PPARgamma in myeloid-derived suppressor cell-stimulated cancer cell proliferation and metastasis. *Oncotarget* **2016**, *7*, 1529–1543. [CrossRef]
159. Li, H.; Sorenson, A.L.; Poczobutt, J.; Amin, J.; Joyal, T.; Sullivan, T.; Crossno, J.T., Jr.; Weiser-Evans, M.C.; Nemenoff, R.A. Activation of PPARgamma in myeloid cells promotes lung cancer progression and metastasis. *PLoS ONE* **2011**, *6*, e28133. [CrossRef] [PubMed]
160. Sippel, T.R.; Johnson, A.M.; Li, H.Y.; Hanson, D.; Nguyen, T.T.; Bullock, B.L.; Poczobutt, J.M.; Kwak, J.W.; Kleczko, E.K.; Weiser-Evans, M.C.; et al. Activation of PPARgamma in Myeloid Cells Promotes Progression of Epithelial Lung Tumors through TGFbeta1. *Mol. Cancer Res.* **2019**, *17*, 1748–1758. [CrossRef] [PubMed]
161. Zou, Y.; Watters, A.; Cheng, N.; Perry, C.E.; Xu, K.; Alicea, G.M.; Parris, J.L.D.; Baraban, E.; Ray, P.; Nayak, A.; et al. Polyunsaturated Fatty Acids from Astrocytes Activate PPARgamma Signaling in Cancer Cells to Promote Brain Metastasis. *Cancer Discov.* **2019**, *9*, 1720–1735. [CrossRef] [PubMed]
162. Gong, X.; Hou, Z.; Endsley, M.P.; Gronseth, E.I.; Rarick, K.R.; Jorns, J.M.; Yang, Q.; Du, Z.; Yan, K.; Bordas, M.L.; et al. Interaction of tumor cells and astrocytes promotes breast cancer brain metastases through TGF-β2/ANGPTL4 axes. *NPJ Precis. Oncol.* **2019**, *3*, 24. [CrossRef] [PubMed]
163. Youssef, J.; Badr, M. Peroxisome proliferator-activated receptors and cancer: Challenges and opportunities. *Br. J. Pharmacol.* **2011**, *164*, 68–82. [CrossRef]
164. Xi, Y.; Zhang, Y.; Zhu, S.; Luo, Y.; Xu, P.; Huang, Z. PPAR-Mediated Toxicology and Applied Pharmacology. *Cells* **2020**, *9*, 352. [CrossRef]
165. Plutzky, J. The PPAR-RXR transcriptional complex in the vasculature: Energy in the balance. *Circ. Res.* **2011**, *108*, 1002–1016. [CrossRef] [PubMed]
166. Fulton, J.; Mazumder, B.; Whitchurch, J.B.; Monteiro, C.J.; Collins, H.M.; Chan, C.M.; Clemente, M.P.; Hernandez-Quiles, M.; Stewart, E.A.; Amoaku, W.M.; et al. Heterodimers of photoreceptor-specific nuclear receptor (PNR/NR2E3) and peroxisome proliferator-activated receptor-gamma (PPARgamma) are disrupted by retinal disease-associated mutations. *Cell. Death. Dis.* **2017**, *8*, e2677. [CrossRef] [PubMed]
167. De Bosscher, K.; Desmet, S.J.; Clarisse, D.; Estebanez-Perpina, E.; Brunsveld, L. Nuclear receptor crosstalk—defining the mechanisms for therapeutic innovation. *Nat. Rev. Endocrinol.* **2020**, *16*, 363–377. [CrossRef] [PubMed]
168. Thomas, R.G.; Surendran, S.P.; Jeong, Y.Y. Tumor Microenvironment-Stimuli Responsive Nanoparticles for Anticancer Therapy. *Front. Mol. Biosci.* **2020**, *7*, 610533. [CrossRef] [PubMed]
169. Lutz, H.; Hu, S.; Dinh, P.-U.; Cheng, K. Cells and cell derivatives as drug carriers for targeted delivery. *Med. Drug Discov.* **2019**, *3*, 100014. [CrossRef]
170. Zhang, X.; Wang, C.; Wang, J.; Hu, Q.; Langworthy, B.; Ye, Y.; Sun, W.; Lin, J.; Wang, T.; Fine, J.; et al. PD-1 Blockade Cellular Vesicles for Cancer Immunotherapy. *Adv. Mater.* **2018**, *30*, e1707112. [CrossRef] [PubMed]
171. Tian, Y.; Li, S.; Song, J.; Ji, T.; Zhu, M.; Anderson, G.J.; Wei, J.; Nie, G. A doxorubicin delivery platform using engineered natural membrane vesicle exosomes for targeted tumor therapy. *Biomaterials* **2014**, *35*, 2383–2390. [CrossRef] [PubMed]
172. Kim, M.S.; Haney, M.J.; Zhao, Y.; Mahajan, V.; Deygen, I.; Klyachko, N.L.; Inskoe, E.; Piroyan, A.; Sokolsky, M.; Okolie, O.; et al. Development of exosome-encapsulated paclitaxel to overcome MDR in cancer cells. *Nanomedicine* **2016**, *12*, 655–664. [CrossRef]

173. Subra, C.; Grand, D.; Laulagnier, K.; Stella, A.; Lambeau, G.; Paillasse, M.; De Medina, P.; Monsarrat, B.; Perret, B.; Silvente-Poirot, S.; et al. Exosomes account for vesicle-mediated transcellular transport of activatable phospholipases and prostaglandins. *J. Lipid Res.* **2010**, *51*, 2105–2120. [CrossRef] [PubMed]
174. He, Q.; Chen, J.; Yan, J.; Cai, S.; Xiong, H.; Liu, Y.; Peng, D.; Mo, M.; Liu, Z. Tumor microenvironment responsive drug delivery systems. *Asian J. Pharm. Sci.* **2020**, *15*, 416–448. [CrossRef] [PubMed]
175. Sun, Q.; Zhou, Z.; Qiu, N.; Shen, Y. Rational Design of Cancer Nanomedicine: Nanoproperty Integration and Synchronization. *Adv. Mater.* **2017**, *29*, 1606628. [CrossRef] [PubMed]
176. Ricci, M.; Miola, M.; Multari, C.; Borroni, E.; Canuto, R.A.; Congiusta, N.; Verne, E.; Follenzi, A.; Muzio, G. PPARs are mediators of anti-cancer properties of superparamagnetic iron oxide nanoparticles (SPIONs) functionalized with conjugated linoleic acid. *Chem. Biol. Interact.* **2018**, *292*, 9–14. [CrossRef] [PubMed]
177. Koczorowska, M.M.; Tholen, S.; Bucher, F.; Lutz, L.; Kizhakkedathu, J.N.; De Wever, O.; Wellner, U.F.; Biniossek, M.L.; Stahl, A.; Lassmann, S.; et al. Fibroblast activation protein-alpha, a stromal cell surface protease, shapes key features of cancer associated fibroblasts through proteome and degradome alterations. *Mol. Oncol.* **2016**, *10*, 40–58. [CrossRef]
178. Hofheinz, R.D.; al-Batran, S.E.; Hartmann, F.; Hartung, G.; Jager, D.; Renner, C.; Tanswell, P.; Kunz, U.; Amelsberg, A.; Kuthan, H.; et al. Stromal antigen targeting by a humanised monoclonal antibody: An early phase II trial of sibrotuzumab in patients with metastatic colorectal cancer. *Onkologie* **2003**, *26*, 44–48. [CrossRef]
179. Narra, K.; Mullins, S.R.; Lee, H.O.; Strzemkowski-Brun, B.; Magalong, K.; Christiansen, V.J.; McKee, P.A.; Egleston, B.; Cohen, S.J.; Weiner, L.M.; et al. Phase II trial of single agent Val-boroPro (Talabostat) inhibiting Fibroblast Activation Protein in patients with metastatic colorectal cancer. *Cancer Biol. Ther.* **2007**, *6*, 1691–1699. [CrossRef]
180. Kallenberg, D.; Tripathi, V.; Javaid, F.; Pilotti, C.; George, J.; Davis, S.; Blackburn, J.W.; O'Connor, M.; Dowsett, L.; Bowers, C.E.; et al. A Humanized Antibody against LRG1 that Inhibits Angiogenesis and Reduces Retinal Vascular Leakage. *bioRxiv* 2020. [CrossRef]

cancers

MDPI

Review

Cytoskeletal Dynamics in Epithelial-Mesenchymal Transition: Insights into Therapeutic Targets for Cancer Metastasis

Arpita Datta [1], Shuo Deng [1], Vennila Gopal [1], Kenneth Chun-Hong Yap [1,2], Clarissa Esmeralda Halim [1], Mun Leng Lye [1], Mei Shan Ong [1], Tuan Zea Tan [3], Gautam Sethi [2,4], Shing Chuan Hooi [1,4], Alan Prem Kumar [2,3,4,5,*] and Celestial T. Yap [1,4,5,*]

1 Department of Physiology, Yong Loo Lin School of Medicine, National University of Singapore, Singapore 117593, Singapore; phsarpi@nus.edu.sg (A.D.); phsdes@nus.edu.sg (S.D.); e0512812@u.nus.edu (V.G.); e0176787@u.nus.edu (K.C.-H.Y.); phsceh@nus.edu.sg (C.E.H.); e0370773@u.nus.edu (M.L.L.); e0013225@u.nus.edu (M.S.O.); phshsc@nus.edu.sg (S.C.H.)
2 Department of Pharmacology, Yong Loo Lin School of Medicine, National University of Singapore, Singapore 117593, Singapore; phcgs@nus.edu.sg
3 Cancer Science Institute of Singapore, National University of Singapore, Singapore 117593, Singapore; csittz@nus.edu.sg
4 Cancer Translational Research Programme, Yong Loo Lin School of Medicine, National University of Singapore, Singapore 117593, Singapore
5 National University Cancer Institute, National University Health System, Singapore 119074, Singapore
* Correspondence: apkumar@nus.edu.sg (A.P.K.); phsyapc@nus.edu.sg (C.T.Y); Tel.: +65-6873-5456 (A.P.K.); +65-6516-3294 (C.T.Y.); Fax: +65-6873-9664 (A.P.K.); +65-6778-8161 (C.T.Y.)

Citation: Datta, A.; Deng, S.; Gopal, V.; Yap, K.C.-H.; Halim, C.E.; Lye, M.L.; Ong, M.S.; Tan, T.Z.; Sethi, G.; Hooi, S.C.; et al. Cytoskeletal Dynamics in Epithelial-Mesenchymal Transition: Insights into Therapeutic Targets for Cancer Metastasis. *Cancers* **2021**, *13*, 1882. https://doi.org/10.3390/cancers13081882

Academic Editors: José I. López and Ildefonso M. de la Fuente

Received: 19 March 2021
Accepted: 12 April 2021
Published: 14 April 2021

Simple Summary: The epithelial to mesenchymal transition (EMT) is a well-documented process in the study of cancer metastases. The cytoskeleton is an intricate network involved in various cellular activities and impacts cell shape, division, trafficking, and motility. However, several functions and activities of the cytoskeleton, which plays a pivotal role in EMT, are not fully understood. This review aims to provide significant insights into the cytoskeleton's physiological functions and the crucial role in the EMT process. Our review focuses on the participation of actin filaments, intermediate filaments, and microtubules in promoting EMT and their influence on cancer metastasis. We have also highlighted potential therapeutic targets associated with EMT activation for clinical intervention. A better understanding of multi-drug resistance (MDR) mechanisms in cancer cells with the cytoskeleton could accelerate the discovery of new therapies for aggressive cancer.

Abstract: In cancer cells, a vital cellular process during metastasis is the transformation of epithelial cells towards motile mesenchymal cells called the epithelial to mesenchymal transition (EMT). The cytoskeleton is an active network of three intracellular filaments: actin cytoskeleton, microtubules, and intermediate filaments. These filaments play a central role in the structural design and cell behavior and are necessary for EMT. During EMT, epithelial cells undergo a cellular transformation as manifested by cell elongation, migration, and invasion, coordinated by actin cytoskeleton reorganization. The actin cytoskeleton is an extremely dynamic structure, controlled by a balance of assembly and disassembly of actin filaments. Actin-binding proteins regulate the process of actin polymerization and depolymerization. Microtubule reorganization also plays an important role in cell migration and polarization. Intermediate filaments are rearranged, switching to a vimentin-rich network, and this protein is used as a marker for a mesenchymal cell. Hence, targeting EMT by regulating the activities of their key components may be a potential solution to metastasis. This review summarizes the research done on the physiological functions of the cytoskeleton, its role in the EMT process, and its effect on multidrug-resistant (MDR) cancer cells—highlight some future perspectives in cancer therapy by targeting cytoskeleton.

Keywords: actin cytoskeleton; epithelial to mesenchymal transition; metastasis; multidrug resistance

1. Introduction

Cancer metastases continue to be a significant clinical hurdle in cancer diagnosis and treatment. Hence, much focus has been given to the so-called epithelial-mesenchymal transition (EMT), a vital process by which epithelial cells change biochemically to achieve mesenchymal phenotypes and is a critical part of cancer metastases. EMT can be classified into three sub-processes. The first comprises EMT's role during developmental processes, such as gastrulation [1–5]. The second sub-process of EMT is involved in the wound repair and fibrosis processes, which are activated by inflammation [1]. The third sub-process of EMT occurs during the cancer progression, which aids the invasion and metastasis of tumor cells to distant sites and promotes the chemoresistance capabilities of these cells [1,6–8]. This review paper primarily focuses on the third sub-process of EMT.

Polarized, non-motile epithelial cancer cells undergo EMT, wherein they lose cell-cell adherence through the loss of cell junctions, such as the adherens and tight junctions [9,10]. The apical-basal polarity of the epithelial cells also changes to a front-rear polarity, with cytoskeletal reorganization, which induces changes in cell shape and a restructuring of the cells' attachment to the extracellular matrix (ECM) [10]. Once EMT is completed, the cells become mesenchymal and motile. Underlying these phenomena are the molecular changes occurring in the cells. TGF-β, IGF-II, FGF, and EGF signaling promote EMT by activating transcription factors such as Snail, Twist, and ZEB, and they mediate gene expression changes in the cells [10–12]. Epithelial cell markers such as E-cadherin, a significant component of the cell junctions, are downregulated [9,10]. On the other hand, N-cadherin, fibronectin, vitronectin, and MMPs are transcriptionally upregulated by these three transcription factors, contributing to the mesenchymal phenotypes of the cells [4,10,13,14]. Originally, EMT was defined as a morphological conversion, but recent advances in biochemical studies have revealed that EMT acts as a central mechanism for carcinoma progression and metastasis. The transcriptional program controlling trans-differentiation and morphological changes during EMT has been comprehensively studied and documented. In contrast, the dynamic remodeling of the cytoskeleton and how it is regulated still lacks comprehensive understanding; especially, the structural mechanism of how the cytoskeleton is remodeled is still being deciphered. Moreover, rearrangement of the actin cytoskeleton into F-actin stress fibers during EMT also aids the formation of membrane ruffles and lamellipodia, promoting the motility of the resultant mesenchymal cells [9–11]. This review explores the role of the actin cytoskeleton, microtubules, and intermediate filaments in EMT and how these cytoskeleton proteins can be used as a potential biomarker.

EMT/MET in the Metastatic Cascade

EMT is a reversible process—EMT followed by a MET is crucial for cancer metastasis [15]. This cancer metastasis process is well-documented and consists of the following key steps that highlight the importance of EMT in the metastasis process (Figure 1).

EMT in Primary Tumors

The first step in cancer metastasis is the local invasion of the tissues surrounding the primary tumor [16,17]. In order to do so, the epithelial cancer cells need first to lose their cell-cell junctions and become motile. This is followed by the degradation of the underlying basement membrane and ECM [18–20]. The ability to degrade the ECM and invade into the surrounding tissue parenchyma is obtained by activating the EMT program, where the epithelial cancer cells transform into a more invasive and motile mesenchymal state [21,22]. During EMT, adherens junctions' dissolution is stabilized by the cleavage and degradation of E-cadherin at the cell membrane [9]. Furthermore, the expression of E-cadherin is suppressed by core transcription factors (TFs) of the EMT program, namely Snail1, Snail2, Zeb1, and Zeb2 [23–25]. Apart from repressing E-cadherin expression, the EMT TFs are also responsible for downregulating the expression of other epithelial genes such as desmoplakin, plakophilin, and plakoglobin [10,26], which are crucial for the formation of desmosomes intercellular junctions that have been reported to work synergistically with adherens junctions to strengthen epithelial cell-cell contact [27]. Hence, the dissolution of these intercellular junctions during EMT allows the cancer cells to separate from each other, thus promoting migration [28].

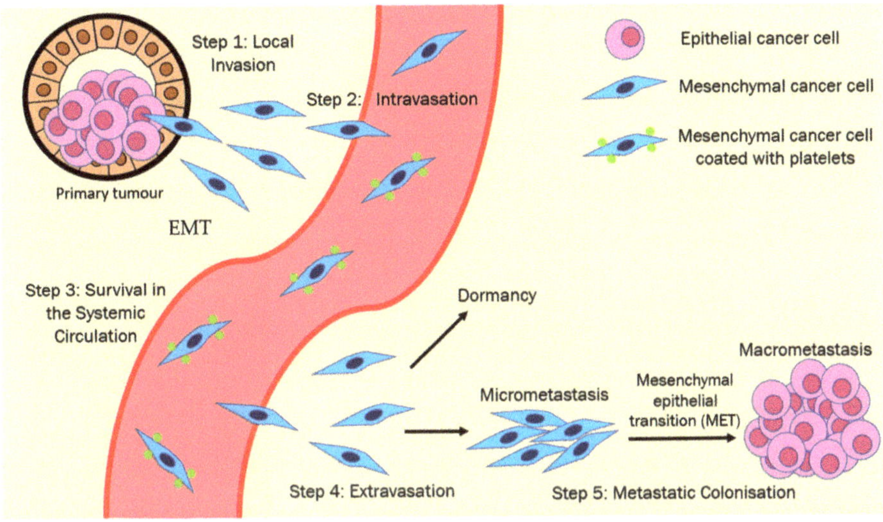

Figure 1. EMT-MET model for the metastatic cascade: Epithelial cancer cells undergo EMT, which causes them to lose their cell-cell junctions and gain the ability to invade the surrounding tissue parenchyma (Step 1). These EMT-induced cells may then intravasate into the systemic circulation (Step 2) and must survive in the circulation (Step 3) before reaching the target organ site. Upon reaching the target organ site, the cells must then extravasate into the tissue parenchyma (Step 4), following which they may either enter a state of dormancy or form micro metastases. Subsequent development into clinically detectable and potentially life-threatening macro metastases requires MET activation (Step 5).

EMT in Intravasation

Following the local invasion of the tissue parenchyma, the cancer cells must cross the endothelial barrier via an intravasation process to be disseminated in the systemic circulation. As the epithelial cells change to a mesenchymal phenotype, such a transformation promotes cell migration and the creation of actin-rich protrusions. Matrixins (MMPs) are actively involved in the invasion—studies have shown that MMP-2 and MMP-9 promote metastatic pathways, such as ECM degradation, cell proliferation, apoptosis, invasion, and morphological changes [29,30]. The increased activity of MMPs results in the enzymatic degradation of adherens junctions and ECM fibers, which increases cancer cell motility and helps break through the basement membrane and the invasion of neighboring tissue [31]. EMT also facilitates intravasation by promoting tumor angiogenesis that can promote tumor spread [32,33].

EMT in the Systemic Circulation

Following the entry into the systemic circulation, cancer cells must overcome several challenges before reaching the target organ site, the first of which is anoikis. Cancer cells lose integrin-mediated anchorage to the ECM upon entering the systemic circulation. As interactions between integrins and the ECM produce pro-survival signals, a lack of these interactions halts the production of these signals, eventually leading to the onset of anoikis, a form of programmed cell death [28].

EMT in Extravasation

Upon reaching the target organ site, the cancer cells can cross the endothelial barrier in an extravasation process to reach the parenchyma. Upon reaching the parenchyma, the tumor cells establish integrin-mediated adhesions with the ECM, enabled by the filopodium-like protrusions (FLPs), which are essential for metastasis. Experimental data has shown a direct correlation between tumor cells' ability to create FLPs and their mesenchymal states, and the FLPs formation can be prompted by the expression of Twist1 and Snail1 [34]. In 2018, podocalyxin (PODXL), a cell surface protein whose expression is upregulated during EMT, was found to be essential for mediating the extravasation process of human breast cancer cells by interacting with actin cytoskeletal linker protein ezrin, thereby causing cytoskeletal rearrangements that promote the transition of the cancer cell into a shape that is optimal for extravasation [35].

Reversion of EMT in Colonization and Establishment of Macrometastases

After reaching the target tissue parenchyma, the disseminated tumor cells (DTCs) can either go into a state of dormancy where they stop proliferation [36] or go on to form micrometastases that may eventually develop into much larger macrometastases. As histological analyses have revealed that macrometastases exhibit epithelial phenotypes rather than mesenchymal [37], the mesenchymal-to-epithelial transition (MET) theory was proposed to explain this phenomenon. This theory posits that DTCs undergo MET to revert back from a mesenchymal state to an epithelial state that allows them to proliferate at the metastatic site and develop into macrometastases. There is some experimental evidence to support MET in cancer metastasis, albeit scarce. Tsai et al. showed in a mouse skin tumor model that the reversion of EMT at the metastatic site via the withdrawal of a Twist1-inducing signal was required for the formation of distant metastases [38]. Additionally, Ocaña et al. found that the silencing of EMT inducer Prrx in BT-549 human breast cancer cells was required for metastatic colonization of the lungs. In fact, silencing both Prrx and Twist1 simultaneously resulted in increased metastatic foci [39]. Taken together, these two studies present strong evidence for the occurrence of MET at the metastatic site.

2. Structure and Functions of the Cytoskeleton

The cytoskeleton is a dynamic and adaptive network of filamentous and tubular protein polymers in the cytoplasm, which provides structural support for cells. The cytoskeleton consists of three major components, which are microtubules, microfilaments, and intermediate filaments. As a fundamental structure, the cytoskeleton serves multiple roles in cells. It compartmentalizes the organelles and other cellular contents, controls the cell's shape and movement, and enables the communication between the cell and extracellular environment. The component polymers that make up the cytoskeleton, together with their regulatory proteins, continuously reorganize themselves in response to a stimulus to support the biological processes and functions.

2.1. Microfilament

Microfilaments are made of actin monomers polymerized into asymmetric strands with barbed and pointed ends [40,41]. An abundance of actin-binding proteins crosslinks and rearrange the thin filaments into organized and stiff actin filament networks, such as bundled networks and branched networks. The actin cytoskeleton is a critical component in a broad diversity of cellular events, ranging from cell motility, cell differentiation, vesicular trafficking to cell proliferation and cell death regulator [42]. It affects the structure and motility of a cell [43], aids in muscle contraction [44], cytokinesis [45], and interacts with myosins to transport vesicles within the cell [46,47]. The actin cytoskeleton's ability to participate in various cellular processes is mainly dependent on its intrinsic dynamic reorganization, which continually happens under the regulation of actin-binding proteins (ABPs) in response to cellular changes [48]. More than a hundred ABPs fall into seven prominent families, namely actin-monomer-binding proteins, severing proteins, nucleation proteins, actin filament polymerases, capping proteins crosslinking proteins, and filament-binding proteins [40].

2.2. Microtubule (MT)

Microtubules are composed of numerous tubulin subunits made of homologous α- or β-dimers. Similar to microfilament, the microtubule is highly dynamic [49], whereby it alternates between the states of gradual extension and rapid shortening [50]. It also has a plus end and a minus end, where β-subunits and α-subunits are exposed, respectively. Microtubule assembly is regulated by microtubule-binding proteins (MTBPs), including stabilizers, destabilizers, capping proteins, bundlers/cross-linkers, molecular motors, cytoplasmic linker proteins (CLIPs), and cytoskeletal integrators [49].

Microtubules have various physiological functions like the microfilament counterpart, where it is critical for cell cycle, intracellular trafficking, cell growth, and death. The balance between assembly and disassembly of microtubules requires tight regulation to ensure proper function being executed inside cells. For instance, during mitosis, disassembly of pre-existing MT network coupled with the assembly of new MTs to form the mitotic spindles is the pre-requisite for the cell to proceed with mitotic phases. When cells are exiting mitosis, the reverse process, where mitotic spindles resolve and MT network reforms, would result in two functional daughter cells [49,51].

2.3. Intermediate Filament (IF)

The intermediate filament protein is characterized by a long, rod-like, α-helical, coiled-coil structure, with both ends flanked by additional residues [52]. Unlike microfilament and microtubule, mature intermediate filaments are not polarized and lack directionality. Compared to the other two cytoskeletal components, intermediate filaments are more stable with less fracturing in terms of biochemical properties, with no know motor proteins to travel along [41]. Intermediate filaments provide the structural support for the cell, where the extensive intermediate filament network in association with plasma membrane reinforces the shape and morphology of the cell [52–55]. They also respond to external mechanical stresses to transduce the signal into cells.

2.4. Cytoskeleton in Cancer Progression and Metastasis

It has been long known that altered cytoskeleton is crucial for the development of many pathological conditions, including cancer. Due to the essential roles of cytoskeleton in cells, many hallmarks of cancer require the participation of at least one cytoskeletal component. For instance, in the case of the actin cytoskeleton, extensive studies have shown that the microfilaments and ABPs are crucial for resisting cell death, sustaining growth and proliferation, promoting invasion and metastasis, inducing angiogenesis, and avoiding immune response [42,56–58].

Perhaps one of the most established roles of the actin cytoskeleton in cancer progression is its ability to influence the metastasis of cancer cells. Numerous studies have shown that the actin cytoskeleton is reorganized at distinct parts of cells like invadopodia, enabling the movement and migration of cells [59]. Moreover, such alternation of the cytoskeleton also aids the transformation of cells from a stationed epithelial type into a migratory mesenchymal type (discussed in detail later). A large group of ABPs belonging to different types and families plays a critical role in regulating many aspects of metastasis, highlighting the importance of actin dynamics and regulation during metastasis. These family members include the Rho GTPases [60], depolymerization factors the gelsolin family members and ADF family member cofilin [61], actin motor proteins like myosin family members [62], actin nucleation and branching factors Arp2/3 and its regulators [63], and capping proteins like CapG [64]. Though less understood, there is emerging evidence showing that microtubule can play a key role in metastasis. The role of microtubule in metastasis regulation might be subtype-specific, where α-Tubulin and βIII-tubulin have been linked to metastasis [65,66].

Moreover, microtubules can regulate metastasis via the crosstalk with actin. One evidence comes from a gastric adenocarcinoma study that microtubule alternation promotes cell motility via Rho GTPase [67]. Several classes of intermediate filaments have been shown to be mediators of metastasis. Vimentin has been shown to be promoting both cell motility and EMT [68]. In fact, one of the most widely used mesenchymal markers is vimentin, highlighting the critical roles of intermediate filaments in metastasis-related changes. Utilizing the highly interactive property among the cytoskeleton components, it is worth identifying new cooperative patterns and proteins which interconnect different cytoskeletal components to aid cancer metastasis.

3. Role of Cellular Cytoskeleton in EMT

The cytoskeleton is a pivotal contributor to the cell's structural framework and is responsible for the mechanical strength and integrity needed to establish cell shape and movement. During EMT, the epithelial cytoskeleton is restructured, such that loss of cell polarity, disruption of cell-cell junctions, and degradation of the underlying basement membrane, and reorganization of the extracellular matrix (ECM) occur (Figure 2). Then the cells become motile and acquire invasive capacity [4]. The following sections describe the essential role of the cytoskeleton in the EMT process.

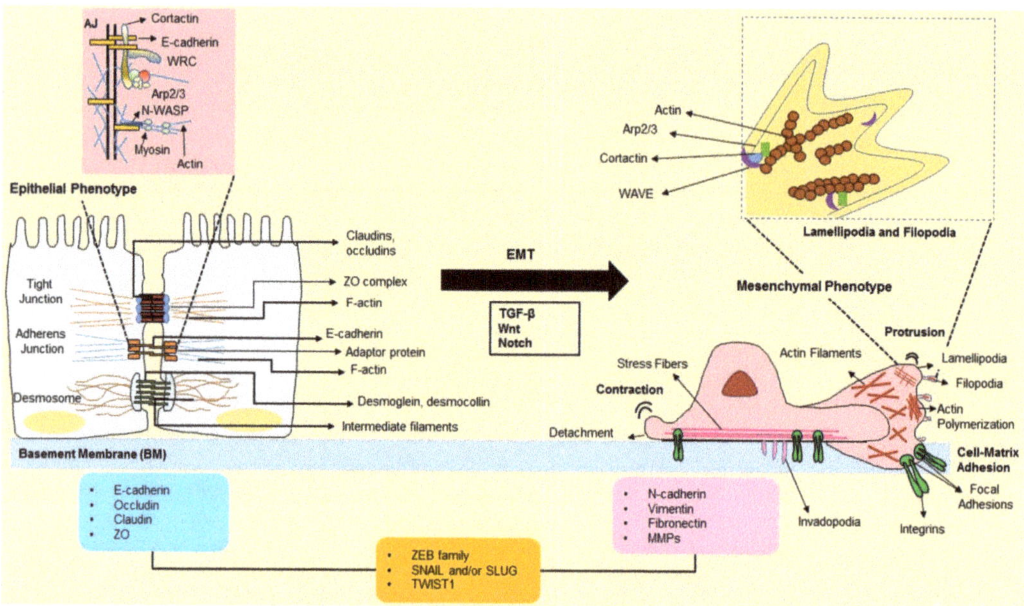

Figure 2. Layout of an epithelial-mesenchymal transition (EMT) transition process and forward migration of cells. Epithelial cells are bound together by tight junctions, adherens junctions, and desmosomes. The adherens junctions are cadherin-based and actin filament-associated cell-to-cell junctions that are composed of defined protein complexes. Epithelial cells are tightly secured to the basement membrane via highly specialized integrin-mediated attachment structures. Signaling pathways are said to trigger the EMT process—propagated by various EMT-TFs, such as ZEB, SNAIL, and TWIST that curb gene expression (listed in the blue box) related with the epithelial state and induce expression of genes associated with the mesenchymal state (listed in the pink box). Mesenchymal cells contain vimentin-based intermediate filaments and use integrin-containing focal adhesions to attach to the ECM. In contrast to epithelial cells, mesenchymal cell migration presents a leading and trailing edge and an extensively reorganized cytoskeleton. Lamellipodia is formed by polymerization of actin by the WAVE-Arp2/3 nucleation mechanism.

3.1. Actin Cytoskeleton

Actin is one of the essential components of the cytoskeleton, and the remodeling of actin filaments is closely related to EMT [69]. Actin exists in two forms: a monomeric unit G-actin (globular actin) and a polymeric filament, F-actin (fibrous actin). G-actin is evenly distributed between the cytoplasm and nucleus. G-actin readily polymerizes under certain physiological conditions to form F-actin with the concomitant hydrolysis of ATP. Furthermore, the distribution of F-actin filaments depends on the cell type as well as the cell cycle phases. Cell spreading and adhesion to ECM are mediated by actomyosin by forming prominent bundles of F-actin, which are known as stress fibers. Stress fibers connect to focal adhesions and hence play an important role in cell adhesion and morphogenesis. Actin filaments interact with actin-binding proteins and myosin II within the leading cell edge and deliver F-actin. This provides an important mechanism for cell movement. In the assembly and disassembly of the actin cytoskeleton, myosin II is considered to play a central role through its ATP- dependent motor function [70]. Actin organization is vital for different cellular processes like cell motility, organelle movement, maintenance of cell junctions, and cell shape [71,72].

The EMT process is regulated by gene expression, post-translational modification of proteins, and reorganization of the cytoskeleton [73]. Epithelial cells are held together by tight junctions (TJs), adherens junctions (AJs), and desmosomes and are also connected to the extracellular matrix (ECM) through integrins [9]. Maciej et al. show that endothelial cells maintain cell-cell junctions (adherens and tight junctions) through the stabilization of F-actin. F-actin filaments are stabilized by amplifying β-catenin and the ZO-1 proteins in the cells that overexpress tropomyosin1 [74]. They also show that α-catenin has a role in suppressing actin polymerization in the area of cell-cell junction [75]. Microscopy studies have shown that during early EMT, dynamic changes happen at the cell-cell boundaries, which weaken the AJ and cell-cell adhesion [76], resulting in the destabilization of the cell-cell junctions. Separation starts with the sequential loss of TJ, AJ, and desmosome integrity, commencing the transition of epithelial cells into a mesenchymal state [26]. This transition occurs through variable intermediate-hybrid states [77]. Recent work has revealed that cells in intermediate states of EMT acquire an augmented capacity for tumor-cell dissemination. Molecular markers have characterized these states, but the structural features and the cellular mechanisms underlying these invasive characteristics are yet to be researched.

E-cadherin-mediated cell-cell adhesion complexes are attached to the actin cytoskeleton via β-catenin and α-catenin. E-cadherin complex is destabilized by post-translation modifications, such as increased phosphorylation, internalization and degradation of E-cadherin, and β-catenin [78]. This triggers the destabilization and degradation of AJs. It has also been shown that E-cadherin complexes are attached to the dynamic actin framework via α-catenin and stabilized by suppressing the activity of Rho A and activating Rac and cdc42 [79,80]. On the other hand, cell-surface receptors, such as the integrins, bind to ECM components and play a key role in modifying the cell attachment required for motility and invasion. A multi-protein complex achieves integrin-mediated cell-matrix adhesion and links to the actin cytoskeleton. The majority of integrins, a cell-surface receptor, connect with the actin cytoskeleton in cell-matrix adhesions via cytoskeletal linker proteins, e.g., talin, paxillin, and vinculin [81,82]. FAK is a tyrosine kinase, and it has a role in actin remodeling dynamics during cell adhesion, and motility and its expression and activity correlate with increased metastatic phenotype [83]. These connections between integrins and the actin cytoskeleton are necessary for the activation of downstream pathways. Snail (transcription factor) induces the expression of αvβ3- integrin, which is localized in the invading front of cancer and enhances cell detachment [84]. It has been shown during the EMT process, engagement of integrins by collagen type I results in a loss of E-cadherin mediated cell-cell contact and activation of the β-catenin pathway in pancreatic cells [85]. Thus, integrins provide a link between the outside environment and cellular responses related to motility, such as immune cell trafficking, hemostasis, and migration of cancer cells.

3.1.1. Actin-Binding Proteins

The actin cytoskeleton is a collection of microfilaments (actin) and a vast array of actin-binding proteins (ABPs). Research studies have shown that a major proportion of ABPs travels between the cytoplasm and the nucleus. The ABPs control the process of assembly and disassembly of actin microfilaments. This plays a pivotal role in cell movement, division, membrane organization, and cancer progression, all of which require the coordinated turnover and remodeling of the actin filaments [18]. The polymerization is associated with the formation of flat, sheet-like membrane protrusions called lamellipodia, or finger-like extensions at the edge of lamellipodia called filopodia [19]. In lamellipodia, actin filaments are arranged in a loosely organized meshwork, often referred to as dendritic networks [20], whereas, in filopodia, actin filaments arrange into parallel bundles [21]. These two different types of organizations rely on the action of specific actin-organizing proteins. During migration, actin depolymerization and debranching occur that allow the dynamic remodeling of the actin network and the cyclic extension and retraction of lamellipodia—this generates the pushing force that provides the propulsion for the cell to move forward. Due to the actin filaments' contraction, the cell body follows the direction of the front lamellipodia. Filopodia are formed of tightly bundled parallel actin filaments with their tapered ends facing toward the plasma membrane. Filopodia filaments are primarily bundled by small crosslinking actin-binding proteins like fascin [22].

The extravasation step is also dependent on actin cytoskeleton dynamics. The entire blood vessel escape process involves attaching the cell to the endothelium, crossing the adhesive endothelium, and then finally establishing a secondary tumor site [33]. It is believed that cells can breach the tissue barrier due to the formation of F-actin protrusions, called invadopodia, that degrade the ECM, thereby enabling cell penetration [34]. We can conclude that lamellipodia and filopodia are involved in the process of forwarding movement of the cell. In contrast, invadopodia are actin-rich protrusions that are associated with the degradation of the ECM through the local deposition of proteases and are involved in cell penetration.

Extensive research has broadened our view on how ABPs affect the rate and extent of polymerization through their wide range of functions—maintaining the pool of monomeric actin (profilin), regulating the state of polymerization of actin filaments (ADF/cofilin), regulating actin filament dynamics and capping (gelsolin, villin), severing actin filaments (cofilin, gelsolin), actin filament nucleation (Arp2/3 complex, WASP), binding to the sides of actin filaments (gelsolin, Arp2/3) bundling and crosslinking (fascin, fimbrin) [86], they have been at the forefront of cancer research. Actin polymerization is a tightly regulated activity. The Arp2/3 (actin-related protein2/3) complex is a seven-subunit protein, controlled by its link with the WAVE and WASP family of WH2 domain-containing proteins (WAVE1, 2, & 3, WASP and N-WASP) that bind both the Arp2/3 complex and actin monomers [87,88]. This, in turn, brings the actin monomers close to the Arp2/3 complex, thereby increasing the rate of Arp2/3-mediated actin polymerization. Arp2/3 is a protein complex involved in the origination of actin filament polymerization. Arp2/3 is frequently overexpressed in malignant tumors, such as breast and liver carcinomas, suggesting a strong correlation between dynamic actin reorganization and cancer progression [89]. WAVE3 is essential for the EMT process to start through the involvement of DNA synthesis, the migration, and the formation of protrusions in breast cancer cells. In-vivo studies show that knockdown of WAVE3 decreases the number of lung metastasis of breast cancer in SCID mice [90–92]. Through their binding partner, WASP family proteins, poly-proline motif, also bind profilin, which further helps recruit actin monomers to the Arp2/3 complex. WASH protein is overexpressed in a breast cancer cell line SKBr3 and may be a potential biomarker for EMT [93]. The actin-binding protein cortactin also binds to Arp2/3, and this helps to locate active Arp2/3 complexes to the sides of existing actin filaments leading to branched arrays of F-actin. The overexpression of cortactin has been identified during metastasis [94,95].

ADF/cofilin is often referred to as a depolymerization factor because it is binding to slow-growing ends of actin filaments that accelerate depolymerization. During the EMT process, filopodia are stabilized through the LIMK/cofilin signaling pathway, suppressing actin fibers' cleavage [69]. Another actin-binding protein fascin is upregulated during migration by stimulating the formation of invadopodia and filopodia [96]. It is reported that gelsolin is required to form lamellipodia and podosomes, which are important protrusions for motile cells [97]. The expression and secretion of urokinase plasminogen activator (uPA), an important protein triggering a cascade to degrade extracellular matrix, is dependent on gelsolin, suggesting the role of gelsolin to enhance invasion [98]. Formins are the actin nucleating proteins that regulate cell movement and organization. It has been reported that formin expression is upregulated at the leading edge in mesenchymal-transformed cells upon EMT [99]. Cortactin, MENA, and Tks proteins form the core structure of invadopodia and play an essential role in actin polymerization, cell signaling, membrane penetration, ECM adhesion, and degradation. Research has also shown that Tks protein is primarily required in invadopodia formation and invasion activity in various human cancer cells [100]. Contactin induces EMT in different types of cancer [101,102]. Karamanou et al. showed that contactin-mediated cell movement induces EMT and participates in tumor migration and invasion [103].

Studies have shown that the gene coding for ABPs displays altered transcription or translation in different cancer types. Alterations in the actin cytoskeleton are a general feature of tumor cells since the ABP expressions are changed in various cancer types. For example, expression of gelsolin, cofilin, CapZ, and thymosin β10 are altered in ovarian cancer, whereas CapZ, CapG, profilin1, cortactin, gelsolin, N-WASp, and WAVE expression levels are changed in prostate cancer [104]. Similarly, data has shown that increased levels of Tβ4 are associated with chronic liver disease and are involved in liver fibrosis by regulating the proliferation and activation of HSC [105]. Other studies have highlighted that increased levels of N-cadherin and decreased levels of E-cadherin have a direct correlation with poor prognosis and cancer progression in prostate cancer patients [106]. Androgen receptor (AR) is associated with EMT and metastasis. In metastatic castration-sensitive prostate cancer, AR deprivation therapy is used as part of combination therapies [107]. Expression of ARPC2 (actin-related protein2/3 complex) is higher in breast cancer tissues, and higher expression of ARPC2 significantly contributes to EMT and metastasis [108]. Research suggests that EMT promotes actin remodeling, which in turn makes the breast cancer cells resistant to NK-cell–mediated killing—actin polymerization is impaired by knocking down N-WASP or CDC42 [109]. Filamin deficit is predominant among carcinomas, including colon, prostate, and breast cancer [110]. Studies have shown that FLNα acts as a promoter in metastasis and invasion in the cytoplasm but acts as a tumor suppressor in the nucleus. Both colorectal and breast cancer cells expressed a high level of TAGLN (22-kDa actin-binding protein). This, in turn, enhances migration and correlates with poor prognosis [111,112]. Higher expression of α-actinin (actin filament cross-linker) is associated with poor prognosis in breast cancer and significantly associated with the degree of clinical advancement and lymph node status [113,114].

3.1.2. Rho GTPases

During EMT, actin and MT dynamics are regulated by Rho GTPases. In different types of cancer, it has been shown that high expression of Rho GTPases genes correlates with a metastatic phenotype. The intracellular protein signaling cascades control the binding of the monomeric or polymeric form by actin-binding proteins (ABPs) when the GTPases of the Rho family are activated. The Rho family of GTPases, including Rac1, RhoA, and Cdc42, are well known for their regulation of actin cytoskeleton organization— such as cytoskeletal dynamics, directional sensing, cell-cell junction assembly/disassembly, and integrin-matrix adhesion. Controlling the activities of Rho GTPases is critical during the growth-factor-induced EMT. Rho activates actin stress fiber formation and regulates cytoskeleton changes, affecting cell-cell or cell-matrix adhesion. The role of Rho signaling in the regulation of actin-myosin contraction is vital and causes actin reorganization to induce stress fibers. Rho-associated kinase (ROCK) promotes myosin light chain phosphorylation, activates actin severing factor cofilin through LIM kinase. Rho signaling pathway affects the activity of various other proteins that control actin polymerization, including profilin and FH proteins [115]. Rac1 stimulates lamellipodia formation through regulating branched actin polymerization [116]. Both Rac1 and Cdc42 are present in the front edge towards the direction of migration. Cdc42 is also upregulated in different types of cancer [117,118]. Hepatocyte growth factor (HGF) activates CdC42 and Rac, which induces filopodia and lamellipodia formation and cell scattering through actin cytoskeleton rearrangement [119]. TGF-β induces activation of RhoA, results in disruption of cell-cell adhesion and formation of stress fibers [120]. RhoA is also involved in microtubules-mediated cell-matrix adhesion and basement membrane integrity [121]—multiple downstream effectors of Rac, Cdc42, and RhoA participating in the reorganization of the actin cytoskeleton. Recent studies have shown that many of these regulatory pathways become deregulated in cancer cells and most likely add to the invasive behavior during metastasis [122].

3.2. Microtubule (MT)

In EMT, the aspects of the actin cytoskeleton and intermediate filaments are well identified, but the functions of microtubules (MT) are still under discovery. MTs are significant parts of the cytoskeleton, play an essential role in motility, intracellular trafficking, and support the cell shape [123]. MTs are composed of α and β-tubulin dimers, and they mostly grow and shrink from the plus end and create dynamic instability [124]. The functions of MTs are dependent on their assembly and stability, which are regulated by post-translation modification and interaction with various stabilizing and destabilizing proteins [125].

The stability of the MT network has been involved in the control of reattachment and cell migration through α-tubulin acetylation at lysine40 (post-translation modification). The acetylation leads to the formation of cell protrusion and tumor cell reattachment, which promote breast cancer metastasis [65]. Calmodulin-regulated spectrin-associated protein (CAMSAP3) is an MT binding protein required to maintain MT organization. It has been shown that loss of CAMSAP3 promotes Akt-dependent EMT by tubulin acetylation [126]. In the EMT program, the transcription factors (TWIST and SNAIL) expressions are enhanced, promoting α-tubulin detyrosination. This further promotes the formation of tubulin-based micro tentacles. These then enhance the reattachment of circulating tumor cells to the endothelial cells [127].

Research has shown that microtubule-interacting protein EB1(end binding protein) co-localizes and interacts with microtubules. EB1 is a negative regulator of microtubule stability and promotes tumor cell migration. It modulates MT dynamics both in vitro and in vivo [128–130]. Furthermore, MT-associated protein ATIP3 is encoded by the tumor suppressor gene MTUS1. Breast cancer cell migration is enhanced by loss of ATIP3 and associated with altered MT dynamics [131]. Decreased expression of ATIP3 inhibits MT tips from reaching the cell cortex during migration, which is essential for cell polarity and migration [132]. ATIP3 is used as an important prognostic marker for breast cancer patients.

Stathmin is an MT regulator protein that depolymerizes MT and enhances and regulates MT dynamics. The destabilization of MT is linked to the phosphorylation of stathmin at its four serine residues [133]. In some human cancer, like sarcomas and Wilms tumors, stathmin has been upregulated and linked to more aggressive metastasis [134].

During EMT, MT has a significant role in cell migration. The anti-MT drugs work via inhibiting cell division on the one hand but also by inhibiting cell migration by stopping the forming of MT network-based membrane protrusions [135,136]. Several strands of research exist on the role of drugs in cell migration [137,138]. It has been shown that the subtoxic dosage of drugs reduces only cell migration without affecting cell division. However, a higher concentration of drugs inhibits cell division but exhibits loss of directionality. It has been shown that MT restrains cell movement as well as establishes directionality [139,140]. But despite the various bodies of research as stated before, the complete role of MT in EMT is still not fully understood, and further research needs to be done to analyze how MT dynamics are correlated with EMT.

During cell migration, one of the critical phenomena is the interaction of MT with the actin cytoskeleton [141]. In response to extrinsic signals, cells migrate due to the activity of RhoA, Rac1, and Cdc42 and their downstream targets. This further mediates a change in the actin cytoskeleton and maintains the stability of MT. Furthermore, variation in MT stability regulates cortical F-actin through activation or inhibition of different Rho GTPases [142].

Apart from their role in cell division and migration, MT also plays an essential role in cell polarization. Research studies have shown that MI s are nucleated at their minus ends, which confines mostly at the centrosome, and the plus ends are stabilized at the leading edge. It had been shown that selective stabilization of the plus ends of MTs enables the centrosome to reorient towards one particular leading edge, which results in a directed movement of the cell [132,143]. Cortical regulation of MT supposedly plays a vital role in creating a polarized MT needed for morphogenesis and cell migration. MT indirectly contributes to cell-cell adhesion by dynamic remodeling of actin network, but the role of MT to function with cell-cell adhesion to regulate migration or EMT is still under active research. Byrne et al. show that MT-interacting protein stathmin is involved in cell migration and metastasis through crosstalk between MT and actin cytoskeleton [144]. Utilizing this interaction where the actin cytoskeleton is targeted via MT, novel pharmacological strategies could be designed that could surpass the toxic effects associated with some actin-based therapies.

3.3. Intermediate Filament (IF)

Intermediate filaments (IFs) are essential cytoskeleton components that give structural support and mechanical strength. More than 50 different IF proteins are encoded by one of the largest families of genes in the human genome, which inscribe five different categories of IF. Types I–IV are localized in the cytoplasm, which includes vimentin, which is a classical marker for EMT, and its expression correlates with the invasive phenotype of epithelial cancers. In order to maintain cell shapes, IFs are associated with the plasma membrane and other elements of the cytoskeleton [145]. IFs also exhibit distinct tissue expression patterns as compared to the actin cytoskeleton and MT.

Type I IF keratins are epithelial-specific and essential for the mechanical stability of epithelial cells. These filaments are associated with desmosome and hemidesmosome through a complex network that extends from the periphery of the nucleus to the plasma membrane. During EMT, the reduction of keratin proteins is often considered a histological and biochemical feature of cancer cells [146,147]. Desmosomes are essential for epithelial integrity, and keratin stabilizes desmosome-mediated intercellular contacts [148]. In epithelial cells, the expression of vimentin activates the destabilization of desmosomes and increases focal adhesion dynamics to promote migration [149].

A type III IF, vimentin, is a canonical marker of the EMT. Vimentin expression is upregulated during EMT in epithelial cells, and increased vimentin expression has been reported in various cancer cell lines and tissues, including prostate cancer, breast cancer, malignant melanoma, and colorectal cancer. It is used as an indicator of poor prognosis [71,150,151]. In the reverse process, the mesenchymal to epithelial transition (MET), vimentin expression is downregulated as cell motility decreases, and cells get epithelial morphology [152]. During EMT, vimentin contributes to the determination and maintenance of cell shape. In breast cancer, vimentin plays a significant role in the EMT processes, and its knockdown results in a reduction in genes linked with breast cancer invasion and the basal-like phenotype [153]. Recent studies have revealed that vimentin expression is linked with motile prostate cancer cell lines, and its knockdown significantly decreases tumor cell motility and invasive activity [154]. Xuan et al. show that vimentin expression is significantly high in polyploidal giant cancer cells (PGCCs). Vimentin intermediate filament is responsible for enlarged morphology and increased migration [155]. Collectively, vimentin expression is preeminently characterized in the EMT process, including tumor cell migration and invasion.

A type VI IF, nestin, is known as a stem cell marker in embryonic and adult central nervous system (CNS) stem cells [156]. Furthermore, from recent research, the role of nestin has been amplified to show that it is also a CSC marker in different forms of cancer, like brain tumors, ovarian, glioblastoma, lung tumors, and head and neck cancers [157]. It has been shown that nestin interacts with IFs like vimentin and desmin to form heterodimers or polymers (which provide cellular support, maintain cellular membranes), and regulate apoptosis-related factors that support cytoskeleton reorganization during mitosis [158]. It has been shown that nestin is involved in the cellular migration and metastasis processes by modulating E-cadherin and Snail expression [159]. Nestin is essential for TGF-β1/Smad mediated EMT in pancreatic cancer. Overexpression of nestin is a positive feedback regulator of the TGF-β1 signaling pathway. This implies a significant role of nestin in the regulation of TGF- Induced EMT, thereby serving as a potential treatment for pancreatic cancers [160,161].

Furthermore, data from clinical samples (shown in Table 1) establishes the cytoskeleton's role in the EMT process [162]. Table 1 shows the correlation between cytoskeleton genes and EMT of various cancers in clinical samples. The data shows about 65% of cytoskeleton genes are positively correlated with EMT.

Table 1. Correlation (Rho) of cytoskeleton genes expression (FPKM) with EMT score in clinical samples (TCGA cohorts, n = 12,290.) A higher Rho indicates that a sample has more mesenchymal-like phenotype, whereas a lower Rho indicates that a sample has a more epithelial-like phenotype. The correlation significance was assessed using the Spearman correlation coefficient test.

Gene	Spearman Correlation Coefficient Rho	Spearman Correlation Coefficient *p*-Value
ACTB	−0.044	8.75E−07
ACTR2	−0.124	5.61E−44
GSN	+0.220	6.81E−135
LIMK1	+0.168	1.28E−78
CFL1	−0.207	3.13E−119
WASF1	+0.509	0.00E+00
FN1	+0.152	5.68E−65
CTTN	−0.392	0.00E+00
WAS	+0.413	0.00E+00
WASL	−0.027	2.11E−03
FSCN1	+0.329	0.00E+00
TUBA1A	+0.656	0.00E+00
VIM	+0.593	0.00E+00
NES	+0.580	0.00E+00
TPM1	−0.129	3.78E−47

Gene expression (FPKM) data from TCGA cohorts were downloaded from Broad firehose, version 2016_01_28 (Reference: Broad Institute TCGA Genome Data Analysis Center (2016): Firehose stddata__2016_01_28 run. Broad Institute of MIT and Harvard. doi:10.7908/C11G0KM9). The EMT score was computed using a previously defined EMT signature and the two-sample Kolmogorov-Smirnov-based method [162].

4. Clinical Evidence for the Actin Cytoskeleton in EMT and Therapeutic Implications

There are primarily two forms of drug resistance in cancer: intrinsic and developed resistance. Intrinsic, as the name suggests, exists before the start of any cancer therapy, and it results in the ability of cancer cells to survive any drug treatment [163,164]. Developed resistance, on the other hand, is when the patient shows an initial positive response to treatment, but, over a period of time, the cancer cell acquires protein alterations, which results in unresponsiveness to treatment [165,166]. Recent studies have shown that scientists focus on combination therapy targeting multiple molecules in the same signaling pathway, multiple pathways in the same tumor, or targeting both cancer cells and immune cells [167,168]. Combination therapies are still under investigation and will eventually better our understanding of drug resistance mechanisms. Different pharmacological strategies have been used to target EMT, such as extracellular inducers and transcription factors. Both have some advantages but also have some distinct drawbacks. For instance, direct inhibition of transcription factors (TF) has been chemically challenging, and successful studies regarding direct targeting of EMT -TFs have been few and far between. The efficacy is limited due to the presence of a large variety of TFs that can initiate EMT. The other disadvantage is that this needs to be initiated in the early stages of carcinoma [169,170]. Hence, new ideas have suggested that targeting EMT and cytoskeletal proteins would be novel in combating cancer drug resistance.

Actin is essential for normal cell physiology. Hence potential actin-specific chemotherapies, despite their promise in-vitro and in-vivo, have not been successful due to their non-specific targeting of normal tissues causing cardiotoxicity and renal problems [171,172]. Recently it has been shown that anti-tropomyosin compounds, which only target tropomyosin-containing filaments in cancer cells, can be used to treat a wide variety of cancer [173]. Studies have revealed that suppressing ROCK, LIMK, and cofilin inhibit cancer metastasis. Over the years, inhibitors of ROCK, LIMK, and cofilin have been investigated in preclinical and clinical models as anti-cancer agents. A few inhibitors, such as Y-276432, have been developed for ROCK1/ROCK2 and prevent MDA-MB-231 breast cancer cell metastasis. Fasudil is the only clinically approved ROCK inhibitor used in humans for systemic applications [174]. MRCK regulates actin-myosin contractility and has a role in cell invasion and metastasis. BDP5290, a potent inhibitor, strongly inhibits the invasion of human squamous cell carcinoma [175].

Similarly, JG-6, an oligosaccharide, is a cofilin-inhibitor that can induce actin depolymerization and suppression of migration and metastasis in MDA-MB-435 and an orthotopic xenograft model [176]. Increasing evidence suggests that the increase in the level of EMT-related actin-binding proteins (ABPs) associated with the actin cytoskeleton reorganization is due to the initiation of the EMT process and metastasis. Therefore, management of ABP expression can possibly help in suppressing migration and promote cancer cells' sensitivity towards drug treatments. Many studies have focused on Arp2/3, cortactin, formins, and fascin. However, the role of other ABPs, which can also be potential targets in carcinogenesis, is understudied. Another challenge for anti-cancer therapy is that the actin cytoskeleton and ABPs are difficult to target actin, and ABPs are involved in the formation of contractile structures in cardiac and skeletal muscles [104].

Intermediate filaments vimentin and nestin are associated with different types of cancer. Vimentin is a marker for mesenchymal cells while participating in EMT. Withaferin-A, a naturally derived anti-cancer drug, works by apoptosis induction in vimentin expressing cancer cells [177,178]. Another drug, moscatilin, is proven to inhibit EMT and sensitizes anoikis, causing programmed cell death [179,180]. Another recently discovered drug, FOXC3 inhibiting vimentin effector1(FiVe1), shows promising results, and this specifically targets vimentin [181].

Anti-tumor drugs are found to alter microtubule dynamics, which subsequently affect mitosis and apoptosis [182]. Taxol was known to be the first drug to initiate tubulin assembly and inhibit microtubules disassembly from halting mitosis [183]. Nanoparticle albumin-bound paclitaxel (Abraxane®®) is an intravenously administered microtubule inhibitor. Nab-paclitaxel plus gemcitabine have been shown to have a good outcome in metastatic pancreatic cancer. This combination of drugs leads to a break in the cells' reproduction activity [184]. In children with recurrent neuroblastoma, ABT-751, a type of orally active drug, works by inhibiting microtubule polymerization by binding to β-tubulin [185]. During EMT, microtubules have significant control in tumor migration and invasion. These anti-tumor drugs inhibit cell division and formation of the membrane protrusions formed by the network-based microtubules, which trigger cell migration and invasion. Eribulin is a MI depolymerization drug used to treat patients with metastatic breast cancer. This drug inhibits angiogenesis, vascular remodeling, and EMT in breast cancer [186,187]. The compound 2-hydroxy-4-methoxy-2′,3′-benzochalcone (HymnPro) disrupts microtubule assembly, which leads to mitotic arrest and progressive activation of the caspase pathway leads to the anti-tumor property, ensuing apoptosis [188]. BPR0C305 is an orally active drug that inhibits tubulin polymerization and disrupts cellular microtubule assembly [136]. The diaryloxazole PC-046 is an anti-tumor drug also with high oral bioavailability. It is a small molecule microtubule destabilizing agent that is synthetically derived. This drug is known to have the advantage of having fewer MDR cross-resistance compared to other prevailing microtubule destabilizing agents [135]. Table 2 summarizes the various drugs that target the cytoskeleton protein, including their specific mode of action.

Table 2. Examples of various drugs targeting cytoskeletal molecules and their action mechanism.

Cytoskeleton Target Protein		Drug Therapy	Mode of Action	References
Actin		Cytochalasins	Inhibits polymerization by binding F-actin	[189,190]
		Latrunculin	Inhibits polymerization Enhances depolymerization through interaction with G-actin.	[191]
		Jasplakinolide	Enhances polymerization by binding F-actin at multiple sites.	[192,193]
Actin-binding Protein	Tropomyosin	TR-100	Inhibits tropomyosin (TPM3.1) in the tropomyosin-dependent actin filament function to promote anti-cancer drug development.	[194]
	ROCK1/ROCK2	Y-276432	Inhibits the kinase activities of ROCK1/ROCK2.	[195]
	ROCK1	Fasudil	Inhibits ROCK in the vascular system and is a calcium channel blocker.	[196]
	Actin-Myosin	BDP5290	Blocks MLC phosphorylation on stress fibers and actin bundles.	[175]
	Cofilin	JG-6	Induces actin depolymerization and suppression of migration.	[176]
	LIMK1	4-Pyridocarbazolone	Inhibits cofilin and actin dynamics.	[197]
	LIMK1 and LIMK2	CRT0105950, CRT0105446	Inhibits cofilin phosphorylation.	[198]
Intermediate Filament		Withaferin-A	Binds and inhibits vimentin.	[177,178]
		Moscatilin	Suppresses AKT phosphorylation and also suppresses the expression of vimentin, SLUG, and SNAIL. Inhibits EMT and sensitizes anoikis.	[179,180]
		FOXC3(FiVe1)	Promotes vimentin disorganization, leading to mitotic catastrophe.	[181]
Microtubules		Taxol	Promotes tubulin assembly and inhibits MT disassembly from halting mitosis.	[184]
		ABT-751	Inhibits MT polymerization by binding to β-tubulin.	[185]
		Eribulin	Inhibits angiogenesis and vascular remodeling and is an MT depolymerization drug.	[187]
		BPR0C305	Inhibits tubulin polymerization and disrupts cellular microtubule assembly.	[136]

EMT Related Cytoskeleton Proteins- Associated Multi-Drug Resistance (MDR)

Multiple signaling pathways involved in EMT and cytoskeletal proteins play an important role in drug resistance in cancer cells [199]. EMT cells have an increase in anti-apoptotic effects and drug efflux pumps. Hence, new ideas have suggested that targeting EMT and cytoskeletal proteins would be novel in combating cancer drug resistance. Chemotherapy is widely used in cancer treatment as monotherapy or as a combination with radiotherapy or surgical intervention. In recent years, multiple discoveries have been made in cancer treatment as drug resistance, which has been one of the significant causes of cancer mortality, is on the rise [200–202]. Many targeted therapy drugs (e.g., Erlotinib, Gefitinib) have shown promising results during the initial trials. However, the majority of them develop drug resistance after long-term drug therapy [203].

EMT and drug resistance have been associated with one another for almost two decades [200]. Multiple findings have shown a significant link between metastatic cancer cells and EMT. One of the common reasons that have been a significant impediment to the success of cancer pharmacotherapies is the overexpression of ATP-binding cassette (ABC) efflux transporters in cancer cells. Recent studies have established that angiopoietin-like 4 (ANGPTL4) protein plays a pivotal role in the metastatic distribution of cancer cells and boosts MDR in the cancerous cell during the EMT process by transcriptionally upregulating the ABC transporters expression via the Myc and NF-κB signaling pathways [204]. It has also been observed that ANGPTL4 increases ABC transporter activity, which results in pushing anticancer drugs out of cells, which causes chemotherapy failure. However, the function of ABC transporters beyond their drug-efflux capacity remains mostly unexplored. There are various schools of thought, and one of the ways suggested to overcome MDR is to develop ABC efflux transporter inhibitors to alert cancer cells to chemotherapeutic drugs [55].

EMT cells are speculated to have selective growth ability in the drug-filled environment. Though some papers suggest that EMT does not entirely contribute to cancer metastasis, other papers show that drug resistance in cancer cells is highly associated with

EMT. This includes bladder cancer [205], pancreatic cancer [7], breast cancer [206], lung cancer [207] and ovarian cancer [208].

Adriamycin-resistant MCF-7 cell lines and vinblastine-resistant ZR-75-B cell lines are shown to have undergone EMT, whereby adriamycin-resistant MCF-7 cells show high vimentin levels and have suppressed the formation of desmosomes and tight junctions, which are specific phenotypes for EMT. [209]. Recent studies have focused on ACTN4, an actin-binding protein whose expression increases with cell motility and EMT. It has been shown that during EMT, ACTN4 interacts with Akt signaling, and this may lead to resistance to DNA damaging drugs, which are used in cancer therapy [210,211].

A vast body of research has described that modifications in the drug target, such as changed microtubule dynamics, tubulin mutations, modified tubulin isotype expression, and altered microtubule regulatory proteins, are the critical targets of anti-microtubule drug resistance. Research has also indicated that other cytoskeletal proteins that can regulate microtubule regulations through signaling or structural connections may be essential factors of anti-microtubule resistance [212]. This resistance to anti-microtubule agents can be either congenital or acquired over the years due to the mentioned factors. The following (Figure 3) is a schematic diagram that illustrates the resistance mechanisms associated with anti-microtubule drugs.

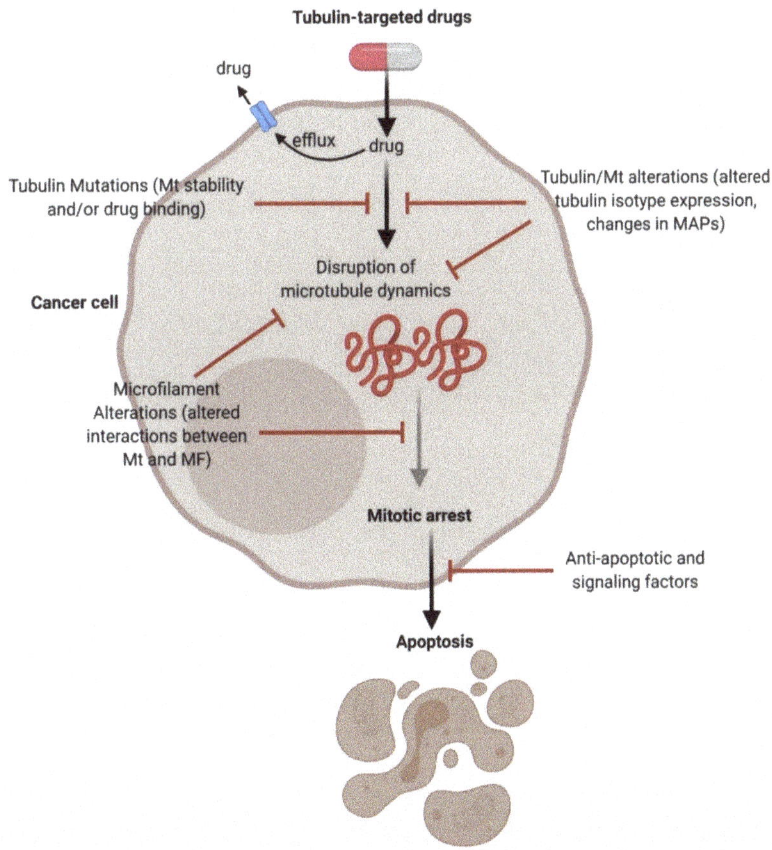

Figure 3. A schematic diagram illustrating the resistance mechanisms associated with anti-microtubule drugs. These drugs cross the lipid bilayer of the cell membrane and then bind themselves to β-tubulin to alter the microtubule dynamics, causing mitotic arrest and consequent apoptosis. Drugs can be effluxed

before reaching the cellular target with the aid of functional drug transport protein. Disruption in the tubulin/microtubule system can avert anti-microtubule drugs from disrupting the microtubules and leading to drug resistance. Before drug binding occurs, signaling and anti-apoptotic factors may also contribute to drug resistance. MF: microfilaments, Mt: microtubules. [212]. The figure was created with BioRender.com (Accessed on 1 February 2021) and was exported under a paid subscription.

Antibody-drug conjugates (ADCs) are a new class of targeted anticancer therapy found to be efficient in MDR cancer. A key mode in which these ADCs cause apoptosis in tumor cells is when high-affinity antibody (Ab) couples with the drug and drives a targeted drug delivery into the cell. This Ab-drug conjugate also blocks the cells' pro-survival receptor besides forming a cytotoxic load coupled by a selective tumor cell killing (Figure 4). Initially, two ADCs—Mylotarg and Adcetris—were approved by the US FDA for treating hematological malignancies. However, the significant discovery was when breast cancer-targeting ADC, Kadcyla was found. To improve the efficacy and attenuate the side effects, integrated 'drug: antibody ratio' (DAR) has been attained [213]. In a discovery by Endo et al. on ADC, cytoskeleton-associated protein 5 (CKAP5), which is a microtubule-associated protein, has been shown to serve as a cell surface target for T-DM1. The binding of these two molecules is mediated by payload (DM1). Upon forming this complex, cell membrane damage occurs which leads to calcium influx, disrupting microtubule network, and apoptosis [214]. The discovery of ADC can lead to other combination therapies, including immunotherapy. Extensive research is currently ongoing to develop strategies to enhance the efficacy and targetability of ADCs in treating tumors. We conclude this section with a table (Table 3) that shows anti-cancer drugs which target the cytoskeletal proteins to alter or inhibit EMT in cancer therapy.

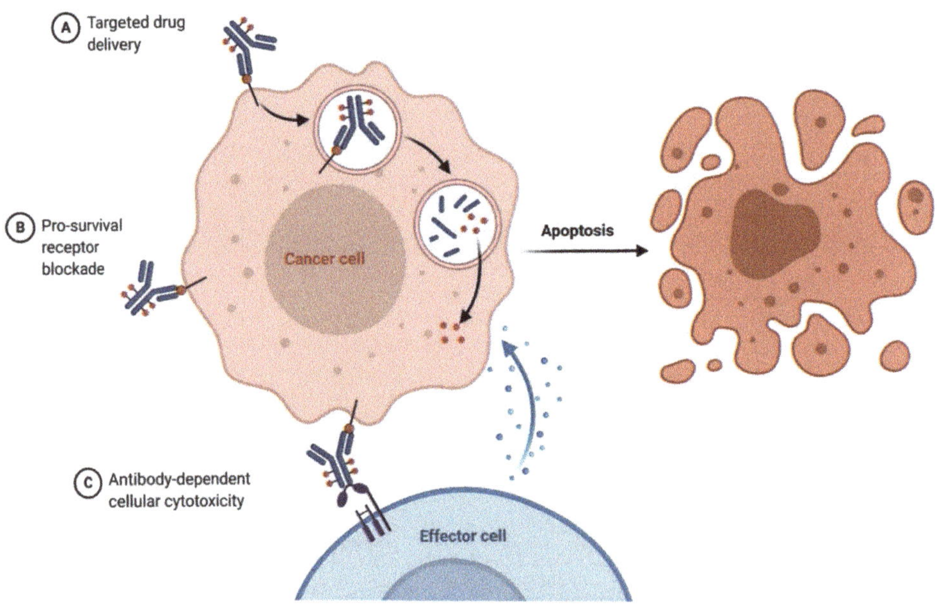

Figure 4. A diagram illustrating the mechanism of action of Antibody-Drug Conjugate (ADC) in a cancer cell. (**A**): High-affinity Antibody binds to the drug, forming ADC, entering the double lipid-membrane layer of the cell to cause cell death. (**B**): ADC binds to the pro-survival receptor of cancer cells, inhibiting its function, commencing apoptosis. (**C**): ADC binds to both the membrane-surface antigen of the cancer cell and an effector cell in the immune system, inducing cellular cytotoxicity lysing the cancer cells. The figure was created with BioRender.com (Accessed on 1 February 2021) and was exported under a paid subscription.

Table 3. Anti-cancer drugs which target the cytoskeletal proteins to alter or inhibit EMT in cancer therapy. The associated cancers and drugs which are resistant to these cancers are also laid out.

Target Cytoskeletal Proteins	Features	Functions	Anti-Cancer Drugs	Function of Anti-Cancer Drugs	Associated Cancers	Drugs Resistant to These Cancers
Vimentin	Central intermediate filament (IF) protein of mesenchymal cells	Organizer of several critical proteins involved in attachment, migration, and cell signaling	Moscatilin FiVe1	Inhibits EMT and sensitizes anoikis FiVe1 disrupts mitotic progression	Lung cancer [179,180] Brest cancer [181]	
α-Actinin	Cellular protrusions, stress fibers, lamellipodia, microvilli, invadopodia of multiple cell types	Crosslinks actin into parallel bundles by forming dimers head to tail	Not Available clinically		Expression in breast, ovary, pancreas, lung, astrocytoma cancers. [55,215–220]	Docetaxel, carboplatin, tamoxifen (Ovary and breast) [221,222]
γ-actin	Distributed along perinuclear and nearby cytoplasm, suggesting a distribution based on diffusion or restriction to nearby cytoplasm. [223]	Regulates cellular morphologies, extending processes, and ruffling edges that reflect cell movement [223]	Not Available clinically		Acute lymphoblastic leukemia	Vinblastine, Desoxyepothilone [224]
		Distributes β-actin and form actin-rich retraction fibers during mitosis	Paclitaxel	Targets the microtubule and causes mitotic arrest and apoptosis	Breast cancer [225]	
					Neuroblastoma	Paclitaxel, vinblastine, epothilone [224]
F-actin	Formed by the polymerization of G-actin under physiological conditions, with the concomitant hydrolysis of ATP.	Cell adhesion, migration, and division	Jasplakinolide (Jas)	Stimulates actin polymerization but disrupts F-actin fibers	Breast and prostate cancer [226,227]	
Eplin	Stress fibers of multiple cell types	Actin filament bundling and side-binding	Not Available clinically		Downregulation correlates with progression and metastasis in prostate cancer. Potential tumor suppressor in breast cancer [228–230]	
β-Tubulin	polymerize into microtubules, a significant component of the eukaryotic cytoskeleton	Involved in many essential cellular processes, including mitosis	Taxanes (paclitaxel), epothilones, and Vinca alkaloids	Binds to β-tubulin and disrupts microtubule dynamics by inducing a potent mitotic block and subsequent cell deathVinca alkaloids inhibit MT polymerization.	Breast, ovarian, lung cancer [212]	
					Acute lymphoblastic leukemia	Vincristine, vinblastine and desoxyepothilone B [231,232]
					Non-small cell lung cancer	Paclitaxel [233]
Cytoskeleton-associated protein 5 (CKAP5)	A microtubule-associated protein which is encoded by the CKAP5 gene	Regulates microtubule organization, nucleation, elongation, and microtubule dynamics by binding to the plus end of the microtubule. Serves as a cell surface target for T-DM1	T-DM1	Upon forming the T-DM1-CKAP5 complex, cell membrane damage occurs, which leads to calcium influx, disrupting microtubule dynamics causing apoptosis	Heptocellular carcinoma [214]	

5. Conclusions

EMT is a highly active process of conversion of epithelial cells to mesenchymal cells. The transformation in the phenotype of an epithelial cell to mesenchymal involves the cell gaining features such as invasiveness, motility, multi-drug resistance, immune-evasiveness, and immunosuppressive properties. In turn, the cell migration occurs because of the swift reorganization of the actin cytoskeleton consisting of polymerization and disintegration of actin filaments. Research findings suggest the vital role of actin-binding proteins in regulating the polymerization and depolymerization process of actin filaments [86]. Studies have also proved that the loss of E-cadherin, which localizes the adherens junctions, is one of the critical features in EMT transitions. The Rho family GTPases also play a vital role in controlling the dynamics of the actin cytoskeleton in both epithelial and mesenchymal cells. Numerous studies have also shown that actin-binding proteins (ABPs) perform many distinct functions that affect the rate and extent of polymerization—nucleating, capping, severing, sequestering, bundling, and crosslinking [86]. Over the years, studies have elucidated that microtubules and intermediate filaments also play a vital role in EMT. Microtubules play an essential role in motility, intracellular trafficking, supporting the cell shape, and produce pushing and pulling forces to support protrusion. The role of intermediate filaments has been a subject that has evoked much interest in recent years. There is clear evidence that points to the fact that EMT is associated with vimentin protein expression, which undergoes phosphorylation and reorientation in cells, regulating cell contraction and focal adhesion assembly/disassembly.

Furthermore, crosstalk between different components of the cytoskeleton is present during metastasis. tActin, IF, and MT cytoskeletons work together in cell migration and metastasis [234]. Recent evidence paints a strong relationship between cytoskeleton dynamics and EMT, which can be utilized to identify potential biomarkers.

6. Future Perspectives

In recent years, it is gradually becoming evident that targeting EMT in cancer treatment may lead to new targets for the development of anti-cancer therapies. In recent research, it has also emerged that several metastatic and invasive cancer have lacked signs of EMT [235] (loss of epithelial feature or increase of mesenchymal proteins). Hence, further research must be done to understand the mechanism of the underlying regulation of actin cytoskeleton and cancer cell EMT. For all these reasons, actin presents itself as a hypothetically attractive anti-cancer therapeutic target. However, in reality, results have proved that actin has been a poor target because of toxic side effects primarily due to the inability of therapeutics to distinguish between actin isoforms [236]. In the past few years, pharmaceutical research studies have pivoted their direction from actin to actin-binding proteins such as the Arp2/3 complex and tropomyosin, which are promising therapeutic targets in cancer drug discovery plans, as these proteins offer many isoforms for selective targeting and the prospect to avoid toxic side effects [237]. Recent studies have highlighted a key characteristic of the protein that makes it a lucrative candidate for further research as a therapeutic target, its specific modulation in activity levels and expression in cancer cell lines. Examining actin-binding proteins as novel therapeutic targets offer great potential for the development of specific cancer therapies—researchers also need to consider a lot of procedural considerations when using phenotype screening to obtain positive outcomes. These new findings and analysis are an active area of interest since it can lead to breakthrough results—combining conventional cancer therapy with EMT-related mechanisms in our fight against cancer and drug-resistant cancer cells.

Funding: C.T.Y was supported by grants from the National Medical Research Council (NMRC r-185-000-353-213) of Singapore. The work was supported by a grant from the Singapore Ministry of Education (MOE-T2EP30120-0016) to A.P.K., A.P.K. and T.Z.T. are also supported by the National Medical Research Council of Singapore and the Singapore Ministry of Education under its Research Centers of Excellence initiative to Cancer Science Institute of Singapore, National University of Singapore.

Conflicts of Interest: The authors declare no conflict of interest.

References

1. Kalluri, R.; Weinberg, R.A. The basics of epithelial-mesenchymal transition. *J. Clin. Invest.* **2009**, *119*, 1420–1428. [CrossRef]
2. Dai, X.; Ahn, K.S.; Wang, L.Z.; Kim, C.; Deivasigamni, A.; Arfuso, F.; Um, J.Y.; Kumar, A.P.; Chang, Y.C.; Kumar, D.; et al. Ascochlorin Enhances the Sensitivity of Doxorubicin Leading to the Reversal of Epithelial-to-Mesenchymal Transition in Hepatocellular Carcinoma. *Mol. Cancer* **2016**, *15*, 2966–2976. [CrossRef]
3. Loo, S.Y.; Hirpara, J.L.; Pandey, V.; Tan, T.Z.; Yap, C.T.; Lobie, P.E.; Thiery, J.P.; Goh, B.C.; Pervaiz, S.; Clement, M.V.; et al. Manganese Superoxide Dismutase Expression Regulates the Switch Between an Epithelial and a Mesenchymal-Like Phenotype in Breast Carcinoma. *Antioxid Redox Signal.* **2016**, *25*, 283–299. [CrossRef]
4. Dongre, A.; Weinberg, R.A. New insights into the mechanisms of epithelial-mesenchymal transition and implications for cancer. *Nat. Rev. Mol. Cell Biol.* **2019**, *20*, 69–84. [CrossRef]
5. Friedman, S.L. Molecular regulation of hepatic fibrosis, an integrated cellular response to tissue injury. *J. Biol. Chem.* **2000**, *275*, 2247–2250. [CrossRef] [PubMed]
6. Zheng, X.; Carstens, J.L.; Kim, J.; Scheible, M.; Kaye, J.; Sugimoto, H.; Wu, C.C.; LeBleu, V.S.; Kalluri, R. Epithelial-to-mesenchymal transition is dispensable for metastasis but induces chemoresistance in pancreatic cancer. *Nature* **2015**, *527*, 525–530. [CrossRef] [PubMed]
7. Arumugam, T.; Ramachandran, V.; Fournier, K.F.; Wang, H.; Marquis, L.; Abbruzzese, J.L.; Gallick, G.E.; Logsdon, C.D.; McConkey, D.J.; Choi, W. Epithelial to mesenchymal transition contributes to drug resistance in pancreatic cancer. *Cancer Res.* **2009**, *69*, 5820–5828. [CrossRef] [PubMed]
8. Fischer, K.R.; Durrans, A.; Lee, S.; Sheng, J.; Li, F.; Wong, S.T.; Choi, H.; El Rayes, T.; Ryu, S.; Troeger, J.; et al. Epithelial-to-mesenchymal transition is not required for lung metastasis but contributes to chemoresistance. *Nature* **2015**, *527*, 472–476. [CrossRef] [PubMed]
9. Yilmaz, M.; Christofori, G. EMT, the cytoskeleton, and cancer cell invasion. *Cancer Metastasis Rev.* **2009**, *28*, 15–33. [CrossRef]
10. Lamouille, S.; Xu, J.; Derynck, R. Molecular mechanisms of epithelial-mesenchymal transition. *Nat. Rev. Mol. Cell Biol.* **2014**, *15*, 178–196. [CrossRef]
11. Kalluri, R.; Neilson, E.G. Epithelial-mesenchymal transition and its implications for fibrosis. *J. Clin. Invest.* **2003**, *112*, 1776–1784. [CrossRef] [PubMed]
12. Ashrafizadeh, M.; Zarrabi, A.; Hushmandi, K.; Kalantari, M.; Mohammadinejad, R.; Javaheri, T.; Sethi, G. Association of the Epithelial-Mesenchymal Transition (EMT) with Cisplatin Resistance. *Int. J. Mol. Sci.* **2020**, *21*. [CrossRef]
13. Cheng, J.T.; Wang, L.; Wang, H.; Tang, F.R.; Cai, W.Q.; Sethi, G.; Xin, H.W.; Ma, Z. Insights into Biological Role of LncRNAs in Epithelial-Mesenchymal Transition. *Cells* **2019**, *8*. [CrossRef]
14. Loh, C.Y.; Chai, J.Y.; Tang, T.F.; Wong, W.F.; Sethi, G.; Shanmugam, M.K.; Chong, P.P.; Looi, C.Y. The E-Cadherin and N-Cadherin Switch in Epithelial-to-Mesenchymal Transition: Signaling, Therapeutic Implications, and Challenges. *Cells* **2019**, *8*. [CrossRef]
15. Yang, J.; Antin, P.; Berx, G.; Blanpain, C.; Brabletz, T.; Bronner, M.; Campbell, K.; Cano, A.; Casanova, J.; Christofori, G.; et al. Guidelines and definitions for research on epithelial-mesenchymal transition. *Nat. Rev. Mol. Cell Biol.* **2020**, *21*, 341–352. [CrossRef]
16. Shanmugam, M.K.; Ahn, K.S.; Hsu, A.; Woo, C.C.; Yuan, Y.; Tan, K.H.B.; Chinnathambi, A.; Alahmadi, T.A.; Alharbi, S.A.; Koh, A.P.F.; et al. Thymoquinone Inhibits Bone Metastasis of Breast Cancer Cells Through Abrogation of the CXCR4 Signaling Axis. *Front. Pharm.* **2018**, *9*, 1294. [CrossRef]
17. Wang, C.; Kar, S.; Lai, X.; Cai, W.; Arfuso, F.; Sethi, G.; Lobie, P.E.; Goh, B.C.; Lim, L.H.K.; Hartman, M.; et al. Triple negative breast cancer in Asia: An insider's view. *Cancer Treat. Rev.* **2018**, *62*, 29–38. [CrossRef]
18. Ashrafizadeh, M.; Hushmandi, K.; Hashemi, M.; Akbari, M.E.; Kubatka, P.; Raei, M.; Koklesova, L.; Shahinozzaman, M.; Mohammadinejad, R.; Najafi, M.; et al. Role of microRNA/Epithelial-to-Mesenchymal Transition Axis in the Metastasis of Bladder Cancer. *Biomolecules* **2020**, *10*. [CrossRef] [PubMed]
19. Ashrafizadeh, M.; Najafi, M.; Ang, H.L.; Moghadam, E.R.; Mahabady, M.K.; Zabolian, A.; Jafaripour, L.; Bejandi, A.K.; Hushmandi, K.; Saleki, H.; et al. PTEN, a Barrier for Proliferation and Metastasis of Gastric Cancer Cells: From Molecular Pathways to Targeting and Regulation. *Biomedicines* **2020**, *8*. [CrossRef]
20. Derynck, R.; Weinberg, R.A. EMT and Cancer: More Than Meets the Eye. *Dev. Cell* **2019**, *49*, 313–316. [CrossRef] [PubMed]
21. Hwang, S.T.; Yang, M.H.; Kumar, A.P.; Sethi, G.; Ahn, K.S. Corilagin Represses Epithelial to Mesenchymal Transition Process Through Modulating Wnt/beta-Catenin Signaling Cascade. *Biomolecules* **2020**, *10*. [CrossRef] [PubMed]
22. Shin, E.M.; Hay, H.S.; Lee, M.H.; Goh, J.N.; Tan, T.Z.; Sen, Y.P.; Lim, S.W.; Yousef, E.M.; Ong, H.T.; Thike, A.A.; et al. DEAD-box helicase DP103 defines metastatic potential of human breast cancers. *J. Clin. Invest.* **2014**, *124*, 3807–3824. [CrossRef]
23. Tsai, J.H.; Yang, J. Epithelial-mesenchymal plasticity in carcinoma metastasis. *Genes Dev.* **2013**, *27*, 2192–2206. [CrossRef] [PubMed]
24. Ashrafizadeh, M.; Ang, H.L.; Moghadam, E.R.; Mohammadi, S.; Zarrin, V.; Hushmandi, K.; Samarghandian, S.; Zarrabi, A.; Najafi, M.; Mohammadinejad, R.; et al. MicroRNAs and Their Influence on the ZEB Family: Mechanistic Aspects and Therapeutic Applications in Cancer Therapy. *Biomolecules* **2020**, *10*. [CrossRef] [PubMed]

25. Lee, J.H.; Chinnathambi, A.; Alharbi, S.A.; Shair, O.H.M.; Sethi, G.; Ahn, K.S. Farnesol abrogates epithelial to mesenchymal transition process through regulating Akt/mTOR pathway. *Pharm. Res.* **2019**, *150*, 104504. [CrossRef] [PubMed]
26. Huang, R.Y.; Guilford, P.; Thiery, J.P. Early events in cell adhesion and polarity during epithelial-mesenchymal transition. *J. Cell Sci.* **2012**, *125*, 4417–4422. [CrossRef] [PubMed]
27. Huen, A.C.; Park, J.K.; Godsel, L.M.; Chen, X.; Bannon, L.J.; Amargo, E.V.; Hudson, T.Y.; Mongiu, A.K.; Leigh, I.M.; Kelsell, D.P.; et al. Intermediate filament–membrane attachments function synergistically with actin-dependent contacts to regulate intercellular adhesive strength. *J. Cell Biol.* **2002**, *159*, 1005–1017. [CrossRef]
28. Heerboth, S.; Housman, G.; Leary, M.; Longacre, M.; Byler, S.; Lapinska, K.; Willbanks, A.; Sarkar, S. EMT and tumor metastasis. *Clin. Transl. Med.* **2015**, *4*, 6. [CrossRef]
29. Gonzalez-Avila, G.; Sommer, B.; Mendoza-Posada, D.A.; Ramos, C.; Garcia-Hernandez, A.A.; Falfan-Valencia, R. Matrix metalloproteinases participation in the metastatic process and their diagnostic and therapeutic applications in cancer. *Crit. Rev. Oncol. Hematol.* **2019**, *137*, 57–83. [CrossRef]
30. Velinov, N.; Poptodorov, G.; Gabrovski, N.; Gabrovski, S. The role of matrixmetalloproteinases in the tumor growth and metastasis. *Khirurgiia* **2010**, *1*, 44–49.
31. Skovierova, H.; Okajcekova, T.; Strnadel, J.; Vidomanova, E.; Halasova, E. Molecular regulation of epithelial-to-mesenchymal transition in tumorigenesis (Review). *Int. J. Mol. Med.* **2018**, *41*, 1187–1200. [CrossRef] [PubMed]
32. Shanmugam, M.K.; Warrier, S.; Kumar, A.P.; Sethi, G.; Arfuso, F. Potential Role of Natural Compounds as Anti-Angiogenic Agents in Cancer. *Curr. Vasc. Pharm.* **2017**, *15*, 503–519. [CrossRef] [PubMed]
33. Sim, W.J.; Iyengar, P.V.; Lama, D.; Lui, S.K.L.; Ng, H.C.; Haviv-Shapira, L.; Domany, E.; Kappei, D.; Tan, T.Z.; Saei, A.; et al. c-Met activation leads to the establishment of a TGFbeta-receptor regulatory network in bladder cancer progression. *Nat. Commun.* **2019**, *10*, 4349. [CrossRef]
34. Sibony-Benyamini, H.; Gil-Henn, H. Invadopodia: The leading force. *Eur. J. Cell Biol.* **2012**, *91*, 896–901. [CrossRef] [PubMed]
35. Fröse, J.; Chen, M.B.; Hebron, K.E.; Reinhardt, F.; Hajal, C.; Zijlstra, A.; Kamm, R.D.; Weinberg, R.A. Epithelial-Mesenchymal Transition Induces Podocalyxin to Promote Extravasation via Ezrin Signaling. *Cell Rep.* **2018**, *24*, 962–972. [CrossRef]
36. Neophytou, C.M.; Kyriakou, T.-C.; Papageorgis, P. Mechanisms of Metastatic Tumor Dormancy and Implications for Cancer Therapy. *Int. J. Mol. Sci.* **2019**, *20*, 6158. [CrossRef]
37. Banyard, J.; Bielenberg, D.R. The role of EMT and MET in cancer dissemination. *Connect. Tissue Res.* **2015**, *56*, 403–413. [CrossRef]
38. Tsai, J.H.; Donaher, J.L.; Murphy, D.A.; Chau, S.; Yang, J. Spatiotemporal regulation of epithelial-mesenchymal transition is essential for squamous cell carcinoma metastasis. *Cancer Cell* **2012**, *22*, 725–736. [CrossRef]
39. Ocaña, O.H.; Córcoles, R.; Fabra, A.; Moreno-Bueno, G.; Acloque, H.; Vega, S.; Barrallo-Gimeno, A.; Cano, A.; Nieto, M.A. Metastatic colonization requires the repression of the epithelial-mesenchymal transition inducer Prrx1. *Cancer Cell* **2012**, *22*, 709–724. [CrossRef]
40. Pollard, T.D. Actin and actin-binding proteins. *Cold Spring Harb. Perspect. Biol.* **2016**, *8*, a018226. [CrossRef]
41. Pollard, T.D.; Goldman, R.D. Overview of the cytoskeleton from an evolutionary perspective. *Cold Spring Harb. Perspect. Biol.* **2018**, *10*, a030288. [CrossRef]
42. Desouza, M.; Gunning, P.W.; Stehn, J.R. The actin cytoskeleton as a sensor and mediator of apoptosis. *Bioarchitecture* **2012**, *2*, 75–87. [CrossRef]
43. Svitkina, T. The actin cytoskeleton and actin-based motility. *Cold Spring Harb. Perspect. Biol.* **2018**, *10*, a018267. [CrossRef]
44. Sweeney, H.L.; Hammers, D.W. Muscle contraction. *Cold Spring Harb. Perspect. Biol.* **2018**, *10*, a023200. [CrossRef] [PubMed]
45. Glotzer, M. Cytokinesis in metazoa and fungi. *Cold Spring Harb. Perspect. Biol.* **2017**, *9*, a022343. [CrossRef] [PubMed]
46. Sweeney, H.L.; Holzbaur, E.L. Motor proteins. *Cold Spring Harb. Perspect. Biol.* **2018**, *10*, a021931. [CrossRef] [PubMed]
47. Titus, M.A. Myosin-driven intracellular transport. *Cold Spring Harb. Perspect. Biol.* **2018**, *10*, a021972. [CrossRef] [PubMed]
48. Gunning, P.; O'Neill, G.; Hardeman, E. Tropomyosin-based regulation of the actin cytoskeleton in time and space. *Physiol. Rev.* **2008**, *88*, 1–35. [CrossRef]
49. Goodson, H.V.; Jonasson, E.M. Microtubules and microtubule-associated proteins. *Cold Spring Harb. Perspect. Biol.* **2018**, *10*, a022608. [CrossRef]
50. Pasquier, E.; Kavallaris, M. Microtubules: A dynamic target in cancer therapy. *IUBMB Life* **2008**, *60*, 165–170. [CrossRef]
51. Jordan, M.A.; Wilson, L. Microtubules as a target for anticancer drugs. *Nat. Rev. Cancer* **2004**, *4*, 253–265. [CrossRef] [PubMed]
52. Herrmann, H.; Aebi, U. Intermediate filaments: Structure and assembly. *Cold Spring Harb. Perspect. Biol.* **2016**, *8*, a018242. [CrossRef]
53. Cheng, F.; Eriksson, J.E. Intermediate filaments and the regulation of cell motility during regeneration and wound healing. *Cold Spring Harb. Perspect. Biol.* **2017**, *9*, a022046. [CrossRef] [PubMed]
54. Pegoraro, A.F.; Janmey, P.; Weitz, D.A. Mechanical Properties of the Cytoskeleton and Cells. *Cold Spring Harb. Perspect. Biol.* **2017**, *9*, a022038. [CrossRef]
55. Yamamoto, S.; Tsuda, H.; Honda, K.; Onozato, K.; Takano, M.; Tamai, S.; Imoto, I.; Inazawa, J.; Yamada, T.; Matsubara, O. Actinin-4 gene amplification in ovarian cancer: A candidate oncogene associated with poor patient prognosis and tumor chemoresistance. *Mod. Pathol.* **2009**, *22*, 499–507. [CrossRef] [PubMed]
56. Stevenson, R.P.; Veltman, D.; Machesky, L.M. Actin-bundling proteins in cancer progression at a glance. *J. Cell Sci.* **2012**, *125*, 1073–1079. [CrossRef]

57. Zankov, D.P.; Ogita, H. Actin-tethered junctional complexes in angiogenesis and lymphangiogenesis in association with vascular endothelial growth factor. *Biomed. Res. Int.* **2015**, *2015*, 314178. [CrossRef]
58. Wurzer, H.; Hoffmann, C.; Al Absi, A.; Thomas, C. Actin Cytoskeleton Straddling the Immunological Synapse between Cytotoxic Lymphocytes and Cancer Cells. *Cells* **2019**, *8*. [CrossRef]
59. Yamazaki, D.; Kurisu, S.; Takenawa, T. Regulation of cancer cell motility through actin reorganization. *Cancer Sci.* **2005**, *96*, 379–386. [CrossRef]
60. Karlsson, R.; Pedersen, E.D.; Wang, Z.; Brakebusch, C. Rho GTPase function in tumorigenesis. *Biochim. Biophys. Acta* **2009**, *1796*, 91–98. [CrossRef]
61. Wang, W.; Eddy, R.; Condeelis, J. The cofilin pathway in breast cancer invasion and metastasis. *Nat. Rev. Cancer* **2007**, *7*, 429–440. [CrossRef] [PubMed]
62. Ouderkirk, J.L.; Krendel, M. Non-muscle myosins in tumor progression, cancer cell invasion, and metastasis. *Cytoskelet. (Hoboken)* **2014**, *71*, 447–463. [CrossRef] [PubMed]
63. Molinie, N.; Gautreau, A. The Arp2/3 Regulatory System and Its Deregulation in Cancer. *Physiol. Rev.* **2018**, *98*, 215–238. [CrossRef] [PubMed]
64. Westbrook, J.A.; Cairns, D.A.; Peng, J.; Speirs, V.; Hanby, A.M.; Holen, I.; Wood, S.L.; Ottewell, P.D.; Marshall, H.; Banks, R.E.; et al. CAPG and GIPC1: Breast Cancer Biomarkers for Bone Metastasis Development and Treatment. *J. Natl. Cancer Inst.* **2016**, *108*. [CrossRef]
65. Boggs, A.E.; Vitolo, M.I.; Whipple, R.A.; Charpentier, M.S.; Goloubeva, O.G.; Ioffe, O.B.; Tuttle, K.C.; Slovic, J.; Lu, Y.; Mills, G.B.; et al. α-Tubulin acetylation elevated in metastatic and basal-like breast cancer cells promotes microtentacle formation, adhesion, and invasive migration. *Cancer Res.* **2015**, *75*, 203–215. [CrossRef]
66. Kanojia, D.; Morshed, R.A.; Zhang, L.; Miska, J.M.; Qiao, J.; Kim, J.W.; Pytel, P.; Balyasnikova, I.V.; Lesniak, M.S.; Ahmed, A.U. βIII-Tubulin Regulates Breast Cancer Metastases to the Brain. *Mol. Cancer* **2015**, *14*, 1152–1161. [CrossRef]
67. Eitaki, M.; Yamamori, T.; Meike, S.; Yasui, H.; Inanami, O. Vincristine enhances amoeboid-like motility via GEF-H1/RhoA/ROCK/Myosin light chain signaling in MKN45 cells. *BMC Cancer* **2012**, *12*, 469. [CrossRef] [PubMed]
68. Satelli, A.; Li, S. Vimentin in cancer and its potential as a molecular target for cancer therapy. *Cell Mol. Life Sci.* **2011**, *68*, 3033–3046. [CrossRef] [PubMed]
69. Shankar, J.; Messenberg, A.; Chan, J.; Underhill, T.M.; Foster, L.J.; Nabi, I.R. Pseudopodial actin dynamics control epithelial-mesenchymal transition in metastatic cancer cells. *Cancer Res.* **2010**, *70*, 3780–3790. [CrossRef] [PubMed]
70. Anderson, T.W.; Vaughan, A.N.; Cramer, L.P. Retrograde flow and myosin II activity within the leading cell edge deliver F-actin to the lamella to seed the formation of graded polarity actomyosin II filament bundles in migrating fibroblasts. *Mol. Biol. Cell* **2008**, *19*, 5006–5018. [CrossRef]
71. Bohnert, K.A.; Willet, A.H.; Kovar, D.R.; Gould, K.L. Formin-based control of the actin cytoskeleton during cytokinesis. *Biochem. Soc. Trans.* **2013**, *41*, 1750–1754. [CrossRef] [PubMed]
72. Grzanka, D.; Gagat, M.; Izdebska, M. Involvement of the SATB1/F-actin complex in chromatin reorganization during active cell death. *Int. J. Mol. Med.* **2014**, *33*, 1441–1450. [CrossRef]
73. Morris, H.T.; Machesky, L.M. Actin cytoskeletal control during epithelial to mesenchymal transition: Focus on the pancreas and intestinal tract. *Br. J. Cancer* **2015**, *112*, 613–620. [CrossRef] [PubMed]
74. Gagat, M.; Grzanka, D.; Izdebska, M.; Grzanka, A. Effect of L-homocysteine on endothelial cell-cell junctions following F-actin stabilization through tropomyosin-1 overexpression. *Int. J. Mol. Med.* **2013**, *32*, 115–129. [CrossRef] [PubMed]
75. Drees, F.; Pokutta, S.; Yamada, S.; Nelson, W.J.; Weis, W.I. Alpha-catenin is a molecular switch that binds E-cadherin-beta-catenin and regulates actin-filament assembly. *Cell* **2005**, *123*, 903–915. [CrossRef] [PubMed]
76. Zhitnyak, I.Y.; Rubtsova, S.N.; Litovka, N.I.; Gloushankova, N.A. Early Events in Actin Cytoskeleton Dynamics and E-Cadherin-Mediated Cell-Cell Adhesion during Epithelial-Mesenchymal Transition. *Cells* **2020**, *9*. [CrossRef] [PubMed]
77. Jordan, N.V.; Johnson, G.L.; Abell, A.N. Tracking the intermediate stages of epithelial-mesenchymal transition in epithelial stem cells and cancer. *Cell Cycle* **2011**, *10*, 2865–2873. [CrossRef]
78. Harris, T.J.; Tepass, U. Adherens junctions: From molecules to morphogenesis. *Nat. Rev. Mol. Cell Biol.* **2010**, *11*, 502–514. [CrossRef]
79. Noren, N.K.; Liu, B.P.; Burridge, K.; Kreft, B. p120 catenin regulates the actin cytoskeleton via Rho family GTPases. *J. Cell Biol.* **2000**, *150*, 567–580. [CrossRef]
80. Noren, N.K.; Niessen, C.M.; Gumbiner, B.M.; Burridge, K. Cadherin engagement regulates Rho family GTPases. *J. Biol. Chem.* **2001**, *276*, 33305–33308. [CrossRef]
81. Humphries, J.D.; Wang, P.; Streuli, C.; Geiger, B.; Humphries, M.J.; Ballestrem, C. Vinculin controls focal adhesion formation by direct interactions with talin and actin. *J. Cell Biol.* **2007**, *179*, 1043–1057. [CrossRef]
82. Zamir, E.; Katz, B.Z.; Aota, S.; Yamada, K.M.; Geiger, B.; Kam, Z. Molecular diversity of cell-matrix adhesions. *J. Cell Sci.* **1999**, *112*, 1655–1669.
83. De Vicente, J.C.; Rosado, P.; Lequerica-Fernandez, P.; Allonca, E.; Villallain, L.; Hernandez-Vallejo, G. Focal adhesion kinase overexpression: Correlation with lymph node metastasis and shorter survival in oral squamous cell carcinoma. *Head Neck* **2013**, *35*, 826–830. [CrossRef]

84. Haraguchi, M.; Okubo, T.; Miyashita, Y.; Miyamoto, Y.; Hayashi, M.; Crotti, T.N.; McHugh, K.P.; Ozawa, M. Snail regulates cell-matrix adhesion by regulation of the expression of integrins and basement membrane proteins. *J. Biol. Chem.* **2008**, *283*, 23514–23523. [CrossRef]
85. Koenig, A.; Mueller, C.; Hasel, C.; Adler, G.; Menke, A. Collagen type I induces disruption of E-cadherin-mediated cell-cell contacts and promotes proliferation of pancreatic carcinoma cells. *Cancer Res.* **2006**, *66*, 4662–4671. [CrossRef] [PubMed]
86. Izdebska, M.; Zielinska, W.; Grzanka, D.; Gagat, M. The Role of Actin Dynamics and Actin-Binding Proteins Expression in Epithelial-to-Mesenchymal Transition and Its Association with Cancer Progression and Evaluation of Possible Therapeutic Targets. *Biomed. Res. Int.* **2018**, *2018*, 4578373. [CrossRef]
87. Pollard, T.D. Regulation of actin filament assembly by Arp2/3 complex and formins. *Annu. Rev. Biophys. Biomol. Struct.* **2007**, *36*, 451–477. [CrossRef] [PubMed]
88. Takenawa, T.; Suetsugu, S. The WASP-WAVE protein network: Connecting the membrane to the cytoskeleton. *Nat. Rev. Mol. Cell Biol.* **2007**, *8*, 37–48. [CrossRef]
89. Sun, B.O.; Fang, Y.; Li, Z.; Chen, Z.; Xiang, J. Role of cellular cytoskeleton in epithelial-mesenchymal transition process during cancer progression. *Biomed. Rep.* **2015**, *3*, 603–610. [CrossRef] [PubMed]
90. Sossey-Alaoui, K.; Li, X.; Ranalli, T.A.; Cowell, J.K. WAVE3-mediated cell migration and lamellipodia formation are regulated downstream of phosphatidylinositol 3-kinase. *J. Biol. Chem.* **2005**, *280*, 21748–21755. [CrossRef] [PubMed]
91. Sossey-Alaoui, K.; Ranalli, T.A.; Li, X.; Bakin, A.V.; Cowell, J.K. WAVE3 promotes cell motility and invasion through the regulation of MMP-1, MMP-3, and MMP-9 expression. *Exp. Cell Res.* **2005**, *308*, 135–145. [CrossRef] [PubMed]
92. Sossey-Alaoui, K.; Safina, A.; Li, X.; Vaughan, M.M.; Hicks, D.G.; Bakin, A.V.; Cowell, J.K. Down-regulation of WAVE3, a metastasis promoter gene, inhibits invasion and metastasis of breast cancer cells. *Am. J. Pathol.* **2007**, *170*, 2112–2121. [CrossRef]
93. Leirdal, M.; Shadidy, M.; Rosok, O.; Sioud, M. Identification of genes differentially expressed in breast cancer cell line SKBR3: Potential identification of new prognostic biomarkers. *Int. J. Mol. Med.* **2004**, *14*, 217–222. [CrossRef]
94. Han, S.P.; Gambin, Y.; Gomez, G.A.; Verma, S.; Giles, N.; Michael, M.; Wu, S.K.; Guo, Z.; Johnston, W.; Sierecki, E.; et al. Cortactin scaffolds Arp2/3 and WAVE2 at the epithelial zonula adherens. *J. Biol. Chem.* **2014**, *289*, 7764–7775. [CrossRef]
95. Helgeson, L.A.; Prendergast, J.G.; Wagner, A.R.; Rodnick-Smith, M.; Nolen, B.J. Interactions with actin monomers, actin filaments, and Arp2/3 complex define the roles of WASP family proteins and cortactin in coordinately regulating branched actin networks. *J. Biol. Chem.* **2014**, *289*, 28856–28869. [CrossRef] [PubMed]
96. Adams, J.C. Fascin-1 as a biomarker and prospective therapeutic target in colorectal cancer. *Expert Rev. Mol. Diagn* **2015**, *15*, 41–48. [CrossRef] [PubMed]
97. Chellaiah, M.; Kizer, N.; Silva, M.; Alvarez, U.; Kwiatkowski, D.; Hruska, K.A. Gelsolin deficiency blocks podosome assembly and produces increased bone mass and strength. *J. Cell Biol.* **2000**, *148*, 665–678. [CrossRef]
98. Zhuo, J.; Tan, E.H.; Yan, B.; Tochhawng, L.; Jayapal, M.; Koh, S.; Tay, H.K.; Maciver, S.K.; Hooi, S.C.; Salto-Tellez, M.; et al. Gelsolin induces colorectal tumor cell invasion via modulation of the urokinase-type plasminogen activator cascade. *PLoS ONE* **2012**, *7*, e43594. [CrossRef]
99. Jaiswal, R.; Breitsprecher, D.; Collins, A.; Correa, I.R., Jr.; Xu, M.Q.; Goode, B.L. The formin Daam1 and fascin directly collaborate to promote filopodia formation. *Curr. Biol.* **2013**, *23*, 1373–1379. [CrossRef]
100. Courtneidge, S.A.; Azucena, E.F.; Pass, I.; Seals, D.F.; Tesfay, L. The SRC substrate Tks5, podosomes (invadopodia), and cancer cell invasion. *Cold Spring Harb Symp. Quant. Biol.* **2005**, *70*, 167–171. [CrossRef]
101. Huang, D.; Cao, L.; Xiao, L.; Song, J.X.; Zhang, Y.J.; Zheng, P.; Zheng, S.G. Hypoxia induces actin cytoskeleton remodeling by regulating the binding of CAPZA1 to F-actin via PIP2 to drive EMT in hepatocellular carcinoma. *Cancer Lett.* **2019**, *448*, 117–127. [CrossRef]
102. Ji, R.; Zhu, X.J.; Wang, Z.R.; Huang, L.Q. Cortactin in Epithelial-Mesenchymal Transition. *Front. Cell Dev. Biol.* **2020**, *8*, 585619. [CrossRef]
103. Karamanou, K.; Franchi, M.; Vynios, D.; Brezillon, S. Epithelial-to-mesenchymal transition and invadopodia markers in breast cancer: Lumican a key regulator. *Semin Cancer Biol.* **2020**, *62*, 125–133. [CrossRef] [PubMed]
104. Izdebska, M.; Zielinska, W.; Halas-Wisniewska, M.; Grzanka, A. Involvement of Actin and Actin-Binding Proteins in Carcinogenesis. *Cells* **2020**, *9*. [CrossRef] [PubMed]
105. Kim, J.; Jung, Y. Potential role of thymosin Beta 4 in liver fibrosis. *Int. J. Mol. Sci.* **2015**, *16*, 10624–10635. [CrossRef]
106. Pisolato, R.; Lombardi, A.P.; Vicente, C.M.; Lucas, T.F.; Lazari, M.F.; Porto, C.S. Expression and regulation of the estrogen receptors in PC-3 human prostate cancer cells. *Steroids* **2016**, *107*, 74–86. [CrossRef] [PubMed]
107. Ferro, M.; Lucarelli, G.; Crocetto, F.; Dolce, P.; Verde, A.; La Civita, E.; Zappavigna, S.; de Cobelli, O.; Di Lorenzo, G.; Facchini, B.A.; et al. First-line systemic therapy for metastatic castration-sensitive prostate cancer: An updated systematic review with novel findings. *Crit. Rev. Oncol. Hematol.* **2021**, *157*, 103198. [CrossRef]
108. Cheng, Z.; Wei, W.; Wu, Z.; Wang, J.; Ding, X.; Sheng, Y.; Han, Y.; Wu, Q. ARPC2 promotes breast cancer proliferation and metastasis. *Oncol. Rep.* **2019**, *41*, 3189–3200. [CrossRef]
109. Al Absi, A.; Wurzer, H.; Guerin, C.; Hoffmann, C.; Moreau, F.; Mao, X.; Brown-Clay, J.; Petrolli, R.; Casellas, C.P.; Dieterle, M.; et al. Actin Cytoskeleton Remodeling Drives Breast Cancer Cell Escape from Natural Killer-Mediated Cytotoxicity. *Cancer Res.* **2018**, *78*, 5631–5643. [CrossRef]

110. Savoy, R.M.; Ghosh, P.M. The dual role of filamin A in cancer: Can't live with (too much of) it, can't live without it. *Endocr. Relat. Cancer* **2013**, *20*, R341–R356. [CrossRef]
111. Hao, R.; Liu, Y.; Du, Q.; Liu, L.; Chen, S.; You, H.; Dong, Y. Transgelin-2 expression in breast cancer and its relationships with clinicopathological features and patient outcome. *Breast Cancer* **2019**, *26*, 776–783. [CrossRef] [PubMed]
112. Zhou, H.; Zhang, Y.; Wu, L.; Xie, W.; Li, L.; Yuan, Y.; Chen, Y.; Lin, Y.; He, X. Elevated transgelin/TNS1 expression is a potential biomarker in human colorectal cancer. *Oncotarget* **2018**, *9*, 1107–1113. [CrossRef] [PubMed]
113. Fang, C.; Li, J.J.; Deng, T.; Li, B.H.; Geng, P.L.; Zeng, X.T. Actinin-4 as a Diagnostic Biomarker in Serum of Breast Cancer Patients. *Med. Sci. Monit.* **2019**, *25*, 3298–3302. [CrossRef] [PubMed]
114. Kovac, B.; Makela, T.P.; Vallenius, T. Increased alpha-actinin-1 destabilizes E-cadherin-based adhesions and associates with poor prognosis in basal-like breast cancer. *PLoS ONE* **2018**, *13*, e0196986. [CrossRef] [PubMed]
115. Scott, R.W.; Olson, M.F. LIM kinases: Function, regulation and association with human disease. *J. Mol. Med.* **2007**, *85*, 555–568. [CrossRef] [PubMed]
116. El-Sibai, M.; Pertz, O.; Pang, H.; Yip, S.C.; Lorenz, M.; Symons, M.; Condeelis, J.S.; Hahn, K.M.; Backer, J.M. RhoA/ROCK-mediated switching between Cdc42- and Rac1-dependent protrusion in MTLn3 carcinoma cells. *Exp. Cell Res.* **2008**, *314*, 1540–1552. [CrossRef] [PubMed]
117. Chen, Q.Y.; Jiao, D.M.; Yao, Q.H.; Yan, J.; Song, J.; Chen, F.Y.; Lu, G.H.; Zhou, J.Y. Expression analysis of Cdc42 in lung cancer and modulation of its expression by curcumin in lung cancer cell lines. *Int. J. Oncol.* **2012**, *40*, 1561–1568. [CrossRef]
118. Kamai, T.; Yamanishi, T.; Shirataki, H.; Takagi, K.; Asami, H.; Ito, Y.; Yoshida, K. Overexpression of RhoA, Rac1, and Cdc42 GTPases is associated with progression in testicular cancer. *Clin. Cancer Res.* **2004**, *10*, 4799–4805. [CrossRef]
119. Royal, I.; Lamarche-Vane, N.; Lamorte, L.; Kaibuchi, K.; Park, M. Activation of cdc42, rac, PAK, and rho-kinase in response to hepatocyte growth factor differentially regulates epithelial cell colony spreading and dissociation. *Mol. Biol. Cell* **2000**, *11*, 1709–1725. [CrossRef]
120. Bhowmick, N.A.; Ghiassi, M.; Bakin, A.; Aakre, M.; Lundquist, C.A.; Engel, M.E.; Arteaga, C.L.; Moses, H.L. Transforming growth factor-beta1 mediates epithelial to mesenchymal transdifferentiation through a RhoA-dependent mechanism. *Mol. Biol. Cell* **2001**, *12*, 27–36. [CrossRef]
121. Nakaya, Y.; Sukowati, E.W.; Wu, Y.; Sheng, G. RhoA and microtubule dynamics control cell-basement membrane interaction in EMT during gastrulation. *Nat. Cell Biol.* **2008**, *10*, 765–775. [CrossRef] [PubMed]
122. Olson, M.F.; Sahai, E. The actin cytoskeleton in cancer cell motility. *Clin. Exp. Metastasis* **2009**, *26*, 273–287. [CrossRef] [PubMed]
123. Etienne-Manneville, S. Microtubules in cell migration. *Annu. Rev. Cell Dev. Biol.* **2013**, *29*, 471–499. [CrossRef] [PubMed]
124. Toya, M.; Takeichi, M. Organization of Non-centrosomal Microtubules in Epithelial Cells. *Cell Struct. Funct.* **2016**, *41*, 127–135. [CrossRef]
125. Luduena, R.F. A hypothesis on the origin and evolution of tubulin. *Int. Rev. Cell Mol. Biol.* **2013**, *302*, 41–185. [CrossRef]
126. Pongrakhananon, V.; Wattanathamsan, O.; Takeichi, M.; Chetprayoon, P.; Chanvorachote, P. Loss of CAMSAP3 promotes EMT via the modification of microtubule-Akt machinery. *J. Cell Sci.* **2018**, 131. [CrossRef] [PubMed]
127. Whipple, R.A.; Matrone, M.A.; Cho, E.H.; Balzer, E.M.; Vitolo, M.I.; Yoon, J.R.; Ioffe, O.B.; Tuttle, K.C.; Yang, J.; Martin, S.S. Epithelial-to-mesenchymal transition promotes tubulin detyrosination and microtentacles that enhance endothelial engagement. *Cancer Res.* **2010**, *70*, 8127–8137. [CrossRef]
128. Coquelle, F.M.; Vitre, B.; Arnal, I. Structural basis of EB1 effects on microtubule dynamics. *Biochem. Soc. Trans.* **2009**, *37*, 997–1001. [CrossRef] [PubMed]
129. Li, W.; Miki, T.; Watanabe, T.; Kakeno, M.; Sugiyama, I.; Kaibuchi, K.; Goshima, G. EB1 promotes microtubule dynamics by recruiting Sentin in Drosophila cells. *J. Cell Biol.* **2011**, *193*, 973–983. [CrossRef] [PubMed]
130. Zhang, T.; Zaal, K.J.; Sheridan, J.; Mehta, A.; Gundersen, G.G.; Ralston, E. Microtubule plus-end binding protein EB1 is necessary for muscle cell differentiation, elongation and fusion. *J. Cell Sci.* **2009**, *122*, 1401–1409. [CrossRef]
131. Molina, A.; Velot, L.; Ghouinem, L.; Abdelkarim, M.; Bouchet, B.P.; Luissint, A.C.; Bouhlel, I.; Morel, M.; Sapharikas, E.; Di Tommaso, A.; et al. ATIP3, a novel prognostic marker of breast cancer patient survival, limits cancer cell migration and slows metastatic progression by regulating microtubule dynamics. *Cancer Res.* **2013**, *73*, 2905–2915. [CrossRef]
132. Kaverina, I.; Straube, A. Regulation of cell migration by dynamic microtubules. *Semin Cell Dev. Biol.* **2011**, *22*, 968–974. [CrossRef] [PubMed]
133. Belmont, L.D.; Mitchison, T.J. Identification of a protein that interacts with tubulin dimers and increases the catastrophe rate of microtubules. *Cell* **1996**, *84*, 623–631. [CrossRef]
134. Baldassarre, G.; Belletti, B.; Nicoloso, M.S.; Schiappacassi, M.; Vecchione, A.; Spessotto, P.; Morrione, A.; Canzonieri, V.; Colombatti, A. p27(Kip1)-stathmin interaction influences sarcoma cell migration and invasion. *Cancer Cell* **2005**, *7*, 51–63. [CrossRef] [PubMed]
135. Landowski, T.H.; Samulitis, B.K.; Dorr, R.T. The diaryl oxazole PC-046 is a tubulin-binding agent with experimental anti-tumor efficacy in hematologic cancers. *Investig. New Drugs* **2013**, *31*, 1616–1625. [CrossRef]
136. Li, W.T.; Yeh, T.K.; Song, J.S.; Yang, Y.N.; Chen, T.W.; Lin, C.H.; Chen, C.P.; Shen, C.C.; Hsieh, C.C.; Lin, H.L.; et al. BPR0C305, an orally active microtubule-disrupting anticancer agent. *Anticancer Drugs* **2013**, *24*, 1047–1057. [CrossRef]
137. Puar, Y.R.; Shanmugam, M.K.; Fan, L.; Arfuso, F.; Sethi, G.; Tergaonkar, V. Evidence for the Involvement of the Master Transcription Factor NF-kappaB in Cancer Initiation and Progression. *Biomedicines* **2018**, *6*. [CrossRef]

138. Shanmugam, M.K.; Ahn, K.S.; Lee, J.H.; Kannaiyan, R.; Mustafa, N.; Manu, K.A.; Siveen, K.S.; Sethi, G.; Chng, W.J.; Kumar, A.P. Celastrol Attenuates the Invasion and Migration and Augments the Anticancer Effects of Bortezomib in a Xenograft Mouse Model of Multiple Myeloma. *Front. Pharm.* **2018**, *9*, 365. [CrossRef]
139. Ganguly, A.; Yang, H.; Sharma, R.; Patel, K.D.; Cabral, F. The role of microtubules and their dynamics in cell migration. *J. Biol. Chem.* **2012**, *287*, 43359–43369. [CrossRef]
140. Yang, H.; Ganguly, A.; Cabral, F. Inhibition of cell migration and cell division correlates with distinct effects of microtubule inhibiting drugs. *J. Biol. Chem.* **2010**, *285*, 32242–32250. [CrossRef]
141. Akhshi, T.K.; Wernike, D.; Piekny, A. Microtubules and actin crosstalk in cell migration and division. *Cytoskelet. (Hoboken)* **2014**, *71*, 1–23. [CrossRef]
142. Etienne-Manneville, S.; Hall, A. Integrin-mediated activation of Cdc42 controls cell polarity in migrating astrocytes through PKCzeta. *Cell* **2001**, *106*, 489–498. [CrossRef]
143. Watanabe, T.; Noritake, J.; Kaibuchi, K. Regulation of microtubules in cell migration. *Trends Cell Biol.* **2005**, *15*, 76–83. [CrossRef]
144. Byrne, F.L.; Yang, L.; Phillips, P.A.; Hansford, L.M.; Fletcher, J.I.; Ormandy, C.J.; McCarroll, J.A.; Kavallaris, M. RNAi-mediated stathmin suppression reduces lung metastasis in an orthotopic neuroblastoma mouse model. *Oncogene* **2014**, *33*, 882–890. [CrossRef]
145. Gan, Z.; Ding, L.; Burckhardt, C.J.; Lowery, J.; Zaritsky, A.; Sitterley, K.; Mota, A.; Costigliola, N.; Starker, C.G.; Voytas, D.F.; et al. Vimentin Intermediate Filaments Template Microtubule Networks to Enhance Persistence in Cell Polarity and Directed Migration. *Cell Syst.* **2016**, *3*, 252–263.e258. [CrossRef]
146. Kim, S.; Coulombe, P.A. Intermediate filament scaffolds fulfill mechanical, organizational, and signaling functions in the cytoplasm. *Genes Dev.* **2007**, *21*, 1581–1597. [CrossRef]
147. Kim, S.; Kellner, J.; Lee, C.H.; Coulombe, P.A. Interaction between the keratin cytoskeleton and eEF1Bgamma affects protein synthesis in epithelial cells. *Nat. Struct. Mol. Biol.* **2007**, *14*, 982–983. [CrossRef] [PubMed]
148. Kroger, C.; Loschke, F.; Schwarz, N.; Windoffer, R.; Leube, R.E.; Magin, T.M. Keratins control intercellular adhesion involving PKC-alpha-mediated desmoplakin phosphorylation. *J. Cell Biol.* **2013**, *201*, 681–692. [CrossRef] [PubMed]
149. Mendez, M.G.; Kojima, S.; Goldman, R.D. Vimentin induces changes in cell shape, motility, and adhesion during the epithelial to mesenchymal transition. *FASEB J.* **2010**, *24*, 1838–1851. [CrossRef] [PubMed]
150. Hugo, H.; Ackland, M.L.; Blick, T.; Lawrence, M.G.; Clements, J.A.; Williams, E.D.; Thompson, E.W. Epithelial–mesenchymal and mesenchymal–epithelial transitions in carcinoma progression. *J. Cell Physiol.* **2007**, *213*, 374–383. [CrossRef]
151. Lang, S.H.; Hyde, C.; Reid, I.N.; Hitchcock, I.S.; Hart, C.A.; Bryden, A.A.; Villette, J.M.; Stower, M.J.; Maitland, N.J. Enhanced expression of vimentin in motile prostate cell lines and in poorly differentiated and metastatic prostate carcinoma. *Prostate* **2002**, *52*, 253–263. [CrossRef] [PubMed]
152. Chaffer, C.L.; Brennan, J.P.; Slavin, J.L.; Blick, T.; Thompson, E.W.; Williams, E.D. Mesenchymal-to-epithelial transition facilitates bladder cancer metastasis: Role of fibroblast growth factor receptor-2. *Cancer Res.* **2006**, *66*, 11271–11278. [CrossRef] [PubMed]
153. Vuoriluoto, K.; Haugen, H.; Kiviluoto, S.; Mpindi, J.P.; Nevo, J.; Gjerdrum, C.; Tiron, C.; Lorens, J.B.; Ivaska, J. Vimentin regulates EMT induction by Slug and oncogenic H-Ras and migration by governing Axl expression in breast cancer. *Oncogene* **2011**, *30*, 1436–1448. [CrossRef]
154. Zhao, Y.; Yan, Q.; Long, X.; Chen, X.; Wang, Y. Vimentin affects the mobility and invasiveness of prostate cancer cells. *Cell Biochem. Funct.* **2008**, *26*, 571–577. [CrossRef]
155. Xuan, B.; Ghosh, D.; Jiang, J.; Shao, R.; Dawson, M.R. Vimentin filaments drive migratory persistence in polyploidal cancer cells. *Proc. Natl. Acad. Sci. USA* **2020**, *117*, 26756–26765. [CrossRef] [PubMed]
156. Lendahl, U.; Zimmerman, L.B.; McKay, R.D. CNS stem cells express a new class of intermediate filament protein. *Cell* **1990**, *60*, 585–595. [CrossRef]
157. Narita, K.; Matsuda, Y.; Seike, M.; Naito, Z.; Gemma, A.; Ishiwata, T. Nestin regulates proliferation, migration, invasion and stemness of lung adenocarcinoma. *Int. J. Oncol.* **2014**, *44*, 1118–1130. [CrossRef]
158. Traub, P.; Kuhn, S.; Grub, S. Separation and characterization of homo and hetero-oligomers of the intermediate filament proteins desmin and vimentin. *J. Mol. Biol.* **1993**, *230*, 837–856. [CrossRef]
159. Hagio, M.; Matsuda, Y.; Suzuki, T.; Ishiwata, T. Nestin regulates epithelial-mesenchymal transition marker expression in pancreatic ductal adenocarcinoma cell lines. *Mol. Clin. Oncol.* **2013**, *1*, 83–87. [CrossRef]
160. Matsuda, Y.; Naito, Z.; Kawahara, K.; Nakazawa, N.; Korc, M.; Ishiwata, T. Nestin is a novel target for suppressing pancreatic cancer cell migration, invasion and metastasis. *Cancer Biol.* **2011**, *11*, 512–523. [CrossRef]
161. Su, H.T.; Weng, C.C.; Hsiao, P.J.; Chen, L.H.; Kuo, T.L.; Chen, Y.W.; Kuo, K.K.; Cheng, K.H. Stem cell marker nestin is critical for TGF-beta1-mediated tumor progression in pancreatic cancer. *Mol. Cancer Res.* **2013**, *11*, 768–779. [CrossRef]
162. Tan, T.Z.; Miow, Q.H.; Miki, Y.; Noda, T.; Mori, S.; Huang, R.Y.; Thiery, J.P. Epithelial-mesenchymal transition spectrum quantification and its efficacy in deciphering survival and drug responses of cancer patients. *EMBO Mol. Med.* **2014**, *6*, 1279–1293. [CrossRef] [PubMed]
163. Chen, L.; Yuan, Y.; Kar, S.; Kanchi, M.M.; Arora, S.; Kim, J.E.; Koh, P.F.; Yousef, E.; Samy, R.P.; Shanmugam, M.K.; et al. PPARgamma Ligand-induced Annexin A1 Expression Determines Chemotherapy Response via Deubiquitination of Death Domain Kinase RIP in Triple-negative Breast Cancers. *Mol. Cancer* **2017**, *16*, 2528–2542. [CrossRef] [PubMed]

164. Garg, M.; Shanmugam, M.K.; Bhardwaj, V.; Goel, A.; Gupta, R.; Sharma, A.; Baligar, P.; Kumar, A.P.; Goh, B.C.; Wang, L.; et al. The pleiotropic role of transcription factor STAT3 in oncogenesis and its targeting through natural products for cancer prevention and therapy. *Med. Res. Rev.* **2020**. [CrossRef]
165. Chatterjee, N.; Bivona, T.G. Polytherapy and Targeted Cancer Drug Resistance. *Trends Cancer* **2019**, *5*, 170–182. [CrossRef]
166. Sethi, G.; Shanmugam, M.K.; Warrier, S.; Merarchi, M.; Arfuso, F.; Kumar, A.P.; Bishayee, A. Pro-Apoptotic and Anti-Cancer Properties of Diosgenin: A Comprehensive and Critical Review. *Nutrients* **2018**, *10*. [CrossRef]
167. Li, F.; Shanmugam, M.K.; Chen, L.; Chatterjee, S.; Basha, J.; Kumar, A.P.; Kundu, T.K.; Sethi, G. Garcinol, a polyisoprenylated benzophenone modulates multiple proinflammatory signaling cascades leading to the suppression of growth and survival of head and neck carcinoma. *Cancer Prev. Res.* **2013**, *6*, 843–854. [CrossRef]
168. Shanmugam, M.K.; Ong, T.H.; Kumar, A.P.; Lun, C.K.; Ho, P.C.; Wong, P.T.; Hui, K.M.; Sethi, G. Ursolic acid inhibits the initiation, progression of prostate cancer and prolongs the survival of TRAMP mice by modulating pro-inflammatory pathways. *PLoS ONE* **2012**, *7*, e32476. [CrossRef]
169. Davis, F.M.; Stewart, T.A.; Thompson, E.W.; Monteith, G.R. Targeting EMT in cancer: Opportunities for pharmacological intervention. *Trends Pharm. Sci.* **2014**, *35*, 479–488. [CrossRef]
170. Singh, M.; Yelle, N.; Venugopal, C.; Singh, S.K. EMT: Mechanisms and therapeutic implications. *Pharm. Ther.* **2018**, *182*, 80–94. [CrossRef]
171. Newman, D.J.; Cragg, G.M. Marine natural products and related compounds in clinical and advanced preclinical trials. *J. Nat. Prod.* **2004**, *67*, 1216–1238. [CrossRef] [PubMed]
172. Senderowicz, A.M.; Kaur, G.; Sainz, E.; Laing, C.; Inman, W.D.; Rodriguez, J.; Crews, P.; Malspeis, L.; Grever, M.R.; Sausville, E.A.; et al. Jasplakinolide's inhibition of the growth of prostate carcinoma cells in vitro with disruption of the actin cytoskeleton. *J. Natl. Cancer Inst.* **1995**, *87*, 46–51. [CrossRef] [PubMed]
173. Stehn, J.R.; Haass, N.K.; Bonello, T.; Desouza, M.; Kottyan, G.; Treutlein, H.; Zeng, J.; Nascimento, P.R.; Sequeira, V.B.; Butler, T.L.; et al. A novel class of anticancer compounds targets the actin cytoskeleton in tumor cells. *Cancer Res.* **2013**, *73*, 5169–5182. [CrossRef]
174. Feng, Y.; LoGrasso, P.V.; Defert, O.; Li, R. Rho Kinase (ROCK) Inhibitors and Their Therapeutic Potential. *J. Med. Chem.* **2016**, *59*, 2269–2300. [CrossRef]
175. Unbekandt, M.; Croft, D.R.; Crighton, D.; Mezna, M.; McArthur, D.; McConnell, P.; Schuttelkopf, A.W.; Belshaw, S.; Pannifer, A.; Sime, M.; et al. A novel small-molecule MRCK inhibitor blocks cancer cell invasion. *Cell Commun. Signal.* **2014**, *12*, 54. [CrossRef] [PubMed]
176. Lee, M.H.; Kundu, J.K.; Chae, J.I.; Shim, J.H. Targeting ROCK/LIMK/cofilin signaling pathway in cancer. *Arch. Pharm. Res.* **2019**, *42*, 481–491. [CrossRef] [PubMed]
177. Hahm, E.R.; Singh, S.V. Withaferin A-induced apoptosis in human breast cancer cells is associated with suppression of inhibitor of apoptosis family protein expression. *Cancer Lett.* **2013**, *334*, 101–108. [CrossRef]
178. Lahat, G.; Zhu, Q.S.; Huang, K.L.; Wang, S.; Bolshakov, S.; Liu, J.; Torres, K.; Langley, R.R.; Lazar, A.J.; Hung, M.C.; et al. Vimentin is a novel anti-cancer therapeutic target; insights from in vitro and in vivo mice xenograft studies. *PLoS ONE* **2010**, *5*, e10105. [CrossRef] [PubMed]
179. Busaranon, K.; Plaimee, P.; Sritularak, B.; Chanvorachote, P. Moscatilin inhibits epithelial-to-mesenchymal transition and sensitizes anoikis in human lung cancer H460 cells. *J. Nat. Med.* **2016**, *70*, 18–27. [CrossRef]
180. Ivaska, J.; Pallari, H.-M.; Nevo, J.; Eriksson, J.E. Novel functions of vimentin in cell adhesion, migration, and signaling. *Exp. Cell Res.* **2007**, *313*, 2050–2062. [CrossRef]
181. Bollong, M.J.; Pietila, M.; Pearson, A.D.; Sarkar, T.R.; Ahmad, I.; Soundararajan, R.; Lyssiotis, C.A.; Mani, S.A.; Schultz, P.G.; Lairson, L.L. A vimentin binding small molecule leads to mitotic disruption in mesenchymal cancers. *Proc. Natl. Acad. Sci. USA* **2017**, *114*, E9903–E9912. [CrossRef]
182. Manu, K.A.; Shanmugam, M.K.; Li, F.; Chen, L.; Siveen, K.S.; Ahn, K.S.; Kumar, A.P.; Sethi, G. Simvastatin sensitizes human gastric cancer xenograft in nude mice to capecitabine by suppressing nuclear factor-kappa B-regulated gene products. *J. Mol. Med.* **2014**, *92*, 267–276. [CrossRef] [PubMed]
183. Lee, J.H.; Kim, C.; Sethi, G.; Ahn, K.S. Brassinin inhibits STAT3 signaling pathway through modulation of PIAS-3 and SOCS-3 expression and sensitizes human lung cancer xenograft in nude mice to paclitaxel. *Oncotarget* **2015**, *6*, 6386–6405. [CrossRef]
184. Hoy, S.M. Albumin-bound paclitaxel: A review of its use for the first-line combination treatment of metastatic pancreatic cancer. *Drugs* **2014**, *74*, 1757–1768. [CrossRef]
185. Meany, H.J.; Sackett, D.L.; Maris, J.M.; Ward, Y.; Krivoshik, A.; Cohn, S.L.; Steinberg, S.M.; Balis, F.M.; Fox, E. Clinical outcome in children with recurrent neuroblastoma treated with ABT-751 and effect of ABT-751 on proliferation of neuroblastoma cell lines and on tubulin polymerization in vitro. *Pediatric. Blood Cancer* **2010**, *54*, 47–54. [CrossRef]
186. Pedersini, R.; Vassalli, L.; Claps, M.; Tulla, A.; Rodella, F.; Grisanti, S.; Amoroso, V.; Roca, E.; Simoncini, E.L.; Berruti, A. Eribulin in Heavily Pretreated Metastatic Breast Cancer Patients in the Real World: A Retrospective Study. *Oncology* **2018**, *94* (Suppl. S1), 10–15. [CrossRef]
187. Pizzuti, L.; Krasniqi, E.; Barchiesi, G.; Mazzotta, M.; Barba, M.; Amodio, A.; Massimiani, G.; Pelle, F.; Kayal, R.; Vizza, E.; et al. Eribulin in Triple Negative Metastatic Breast Cancer: Critic Interpretation of Current Evidence and Projection for Future Scenarios. *J. Cancer* **2019**, *10*, 5903–5914. [CrossRef]

188. Shin, S.Y.; Kim, J.-H.; Yoon, H.; Choi, Y.-K.; Koh, D.; Lim, Y.; Lee, Y.H. Novel antimitotic activity of 2-hydroxy-4-methoxy-2′, 3′-benzochalcone (HymnPro) through the inhibition of tubulin polymerization. *J. Agric. Food Chem.* **2013**, *61*, 12588–12597. [CrossRef] [PubMed]
189. Calaghan, S.C.; White, E.; Bedut, S.; Le Guennec, J.Y. Cytochalasin D reduces Ca2+ sensitivity and maximum tension via interactions with myofilaments in skinned rat cardiac myocytes. *J. Physiol.* **2000**, *529*, 405–411. [CrossRef] [PubMed]
190. Van Goietsenoven, G.; Mathieu, V.; Andolfi, A.; Cimmino, A.; Lefranc, F.; Kiss, R.; Evidente, A. In vitro growth inhibitory effects of cytochalasins and derivatives in cancer cells. *Planta Med.* **2011**, *77*, 711–717. [CrossRef] [PubMed]
191. Morton, W.M.; Ayscough, K.R.; McLaughlin, P.J. Latrunculin alters the actin-monomer subunit interface to prevent polymerization. *Nat. Cell Biol.* **2000**, *2*, 376–378. [CrossRef]
192. Odaka, C.; Sanders, M.L.; Crews, P. Jasplakinolide induces apoptosis in various transformed cell lines by a caspase-3-like protease-dependent pathway. *Clin. Diagn. Lab. Immunol.* **2000**, *7*, 947–952. [CrossRef]
193. Schweikart, K.; Guo, L.; Shuler, Z.; Abrams, R.; Chiao, E.T.; Kolaja, K.L.; Davis, M. The effects of jaspamide on human cardiomyocyte function and cardiac ion channel activity. *Toxicol. Vitr.* **2013**, *27*, 745–751. [CrossRef]
194. Currier, M.A.; Stehn, J.R.; Swain, A.; Chen, D.; Hook, J.; Eiffe, E.; Heaton, A.; Brown, D.; Nartker, B.A.; Eaves, D.W.; et al. Identification of Cancer-Targeted Tropomyosin Inhibitors and Their Synergy with Microtubule Drugs. *Mol. Cancer* **2017**, *16*, 1555–1565. [CrossRef]
195. Ishizaki, T.; Uehata, M.; Tamechika, I.; Keel, J.; Nonomura, K.; Maekawa, M.; Narumiya, S. Pharmacological properties of Y-27632, a specific inhibitor of rho-associated kinases. *Mol. Pharm.* **2000**, *57*, 976–983.
196. So, S.; Lee, Y.; Choi, J.; Kang, S.; Lee, J.Y.; Hwang, J.; Shin, J.; Dutton, J.R.; Seo, E.J.; Lee, B.H.; et al. The Rho-associated kinase inhibitor fasudil can replace Y-27632 for use in human pluripotent stem cell research. *PLoS ONE* **2020**, *15*, e0233057. [CrossRef]
197. Prunier, C.; Prudent, R.; Kapur, R.; Sadoul, K.; Lafanechere, L. LIM kinases: Cofilin and beyond. *Oncotarget* **2017**, *8*, 41749–41763. [CrossRef] [PubMed]
198. Mardilovich, K.; Baugh, M.; Crighton, D.; Kowalczyk, D.; Gabrielsen, M.; Munro, J.; Croft, D.R.; Lourenco, F.; James, D.; Kalna, G.; et al. LIM kinase inhibitors disrupt mitotic microtubule organization and impair tumor cell proliferation. *Oncotarget* **2015**, *6*, 38469–38486. [CrossRef] [PubMed]
199. Monisha, J.; Roy, N.K.; Padmavathi, G.; Banik, K.; Bordoloi, D.; Khwairakpam, A.D.; Arfuso, F.; Chinnathambi, A.; Alahmadi, T.A.; Alharbi, S.A.; et al. NGAL is Downregulated in Oral Squamous Cell Carcinoma and Leads to Increased Survival, Proliferation, Migration and Chemoresistance. *Cancers* **2018**, *10*. [CrossRef] [PubMed]
200. Manu, K.A.; Shanmugam, M.K.; Ramachandran, L.; Li, F.; Siveen, K.S.; Chinnathambi, A.; Zayed, M.E.; Alharbi, S.A.; Arfuso, F.; Kumar, A.P.; et al. Isorhamnetin augments the anti-tumor effect of capecitabine through the negative regulation of NF-kappaB signaling cascade in gastric cancer. *Cancer Lett.* **2015**, *363*, 28–36. [CrossRef] [PubMed]
201. Deng, S.; Shanmugam, M.K.; Kumar, A.P.; Yap, C.T.; Sethi, G.; Bishayee, A. Targeting autophagy using natural compounds for cancer prevention and therapy. *Cancer* **2019**, *125*, 1228–1246. [CrossRef]
202. Mishra, S.; Verma, S.S.; Rai, V.; Awasthee, N.; Chava, S.; Hui, K.M.; Kumar, A.P.; Challagundla, K.B.; Sethi, G.; Gupta, S.C. Long non-coding RNAs are emerging targets of phytochemicals for cancer and other chronic diseases. *Cell Mol. Life Sci.* **2019**, *76*, 1947–1966. [CrossRef] [PubMed]
203. Juchum, M.; Günther, M.; Laufer, S.A. Fighting cancer drug resistance: Opportunities and challenges for mutation-specific EGFR inhibitors. *Drug Resist. Updates* **2015**, *20*, 12–28. [CrossRef]
204. Lim, M.M.K.; Wee, J.W.K.; Soong, J.C.; Chua, D.; Tan, W.R.; Lizwan, M.; Li, Y.; Teo, Z.; Goh, W.W.B.; Zhu, P.; et al. Targeting metabolic flexibility via angiopoietin-like 4 protein sensitizes metastatic cancer cells to chemotherapy drugs. *Mol. Cancer* **2018**, *17*, 152. [CrossRef]
205. McConkey, D.J.; Choi, W.; Marquis, L.; Martin, F.; Williams, M.B.; Shah, J.; Svatek, R.; Das, A.; Adam, L.; Kamat, A. Role of epithelial-to-mesenchymal transition (EMT) in drug sensitivity and metastasis in bladder cancer. *Cancer Metastasis Rev.* **2009**, *28*, 335–344. [CrossRef] [PubMed]
206. Huang, J.; Li, H.; Ren, G. Epithelial-mesenchymal transition and drug resistance in breast cancer. *Int. J. Oncol.* **2015**, *47*, 840–848. [CrossRef]
207. Nurwidya, F.; Takahashi, F.; Murakami, A.; Takahashi, K. Epithelial mesenchymal transition in drug resistance and metastasis of lung cancer. *Cancer Res. Treat. Off. J. Korean Cancer Assoc.* **2012**, *44*, 151. [CrossRef]
208. Chen, X.; Zhang, J.; Zhang, Z.; Li, H.; Cheng, W.; Liu, J. Cancer stem cells, epithelial-mesenchymal transition, and drug resistance in high-grade ovarian serous carcinoma. *Hum. Pathol.* **2013**, *44*, 2373–2384. [CrossRef] [PubMed]
209. Sommers, C.L.; Heckford, S.E.; Skerker, J.M.; Worland, P.; Torri, J.A.; Thompson, E.W.; Byers, S.W.; Gelmann, E.P. Loss of epithelial markers and acquisition of vimentin expression in adriamycin-and vinblastine-resistant human breast cancer cell lines. *Cancer Res.* **1992**, *52*, 5190–5197.
210. Jiang, G.B.; Fang, H.Y.; Tao, D.Y.; Chen, X.P.; Cao, F.L. COX-2 potentiates cisplatin resistance of non-small cell lung cancer cells by promoting EMT in an AKT signaling pathway-dependent manner. *Eur. Rev. Med. Pharm. Sci.* **2019**, *23*, 3838–3846. [CrossRef]
211. Tentler, D.; Lomert, E.; Novitskaya, K.; Barlev, N.A. Role of ACTN4 in Tumorigenesis, Metastasis, and EMT. *Cells* **2019**, *8*. [CrossRef]
212. Verrills, N.; Kavallaris, M. Improving the targeting of tubulin-binding agents: Lessons from drug resistance studies. *Curr. Pharm. Des.* **2005**, *11*, 1719–1733. [CrossRef]

213. Sau, S.; Alsaab, H.O.; Kashaw, S.K.; Tatiparti, K.; Iyer, A.K. Advances in antibody–drug conjugates: A new era of targeted cancer therapy. *Drug Discov. Today* **2017**, *22*, 1547–1556. [CrossRef]
214. Endo, Y.; Takeda, K.; Mohan, N.; Shen, Y.; Jiang, J.; Rotstein, D.; Wu, W.J. Payload of T-DM1 binds to cell surface cytoskeleton-associated protein 5 to mediate cytotoxicity of hepatocytes. *Oncotarget* **2018**, *9*, 37200. [CrossRef]
215. Fu, L.; Qin, Y.R.; Xie, D.; Chow, H.Y.; Ngai, S.M.; Kwong, D.L.; Li, Y.; Guan, X.Y. Identification of alpha-actinin 4 and 67 kDa laminin receptor as stage-specific markers in esophageal cancer via proteomic approaches. *Cancer: Interdiscip. Int. J. Am. Cancer Soc.* **2007**, *110*, 2672–2681. [CrossRef]
216. Hirooka, S.; Akashi, T.; Ando, N.; Suzuki, Y.; Ishida, N.; Kurata, M.; Takizawa, T.; Kayamori, K.; Sakamoto, K.; Fujiwara, N.; et al. Localization of the Invadopodia-Related Proteins Actinin-1 and Cortactin to Matrix-Contact-Side Cytoplasm of Cancer Cells in Surgically Resected Lung Adenocarcinomas. *Pathobiology* **2011**, *78*, 10–23. [CrossRef]
217. Honda, K.; Yamada, T.; Hayashida, Y.; Idogawa, M.; Sato, S.; Hasegawa, F.; Ino, Y.; Ono, M.; Hirohashi, S. Actinin-4 increases cell motility and promotes lymph node metastasis of colorectal cancer. *Gastroenterology* **2005**, *128*, 51–62. [CrossRef]
218. Honda, K.; Yamada, T.; Seike, M.; Hayashida, Y.; Idogawa, M.; Kondo, T.; Ino, Y.; Hirohashi, S. Alternative splice variant of actinin-4 in small cell lung cancer. *Oncogene* **2004**, *23*, 5257–5262. [CrossRef]
219. Kikuchi, S.; Honda, K.; Tsuda, H.; Hiraoka, N.; Imoto, I.; Kosuge, T.; Umaki, T.; Onozato, K.; Shitashige, M.; Yamaguchi, U. Expression and gene amplification of actinin-4 in invasive ductal carcinoma of the pancreas. *Clin. Cancer Res.* **2008**, *14*, 5348–5356. [CrossRef] [PubMed]
220. Menez, J.; Chansac, B.L.M.; Dorothée, G.; Vergnon, I.; Jalil, A.; Carlier, M.-F.; Chouaib, S.; Mami-Chouaib, F. Mutant α-actinin-4 promotes tumorigenicity and regulates cell motility of a human lung carcinoma. *Oncogene* **2004**, *23*, 2630–2639. [CrossRef]
221. Zhou, Y.; Zhang, X.; Klibanski, A. MEG3 noncoding RNA: A tumor suppressor. *J. Mol. Endocrinol.* **2012**, *48*, R45. [CrossRef]
222. He, J.; Whelan, S.A.; Lu, M.; Shen, D.; Chung, D.U.; Saxton, R.E.; Faull, K.F.; Whitelegge, J.P.; Chang, H.R. Proteomic-Based Biosignatures in Breast Cancer Classification and Prediction of Therapeutic Response. *Int. J. Proteom.* **2011**, *2011*, 896476. [CrossRef]
223. Hill, M.A.; Gunning, P. Beta and gamma actin mRNAs are differentially located within myoblasts. *J. Cell Biol.* **1993**, *122*, 825–832. [CrossRef]
224. Verrills, N.M.; Po'uha, S.T.; Liu, M.L.M.; Liaw, T.Y.E.; Larsen, M.R.; Ivery, M.T.; Marshall, G.M.; Gunning, P.W.; Kavallaris, M. Alterations in γ-Actin and Tubulin-Targeted Drug Resistance in Childhood Leukemia. *JNCI J. Natl. Cancer Inst.* **2006**, *98*, 1363–1374. [CrossRef]
225. Po'uha, S.T.; Kavallaris, M. Gamma-actin is involved in regulating centrosome function and mitotic progression in cancer cells. *Cell Cycle* **2015**, *14*, 3908–3919. [CrossRef]
226. Duncan, M.D.; Harmon, J.W.; Duncan, K. Actin disruption inhibits bombesin stimulation of focal adhesion kinase (pp125FAK) in prostate carcinoma. *J. Surg. Res.* **1996**, *63*, 359–363. [CrossRef]
227. Stingl, J.; Andersen, R.J.; Emerman, J.T. In vitro screening of crude extracts and pure metabolites obtained from marine invertebrates for the treatment of breast cancer. *Cancer Chemother. Pharmacol.* **1992**, *30*, 401–406. [CrossRef]
228. Jiang, W.G.; Martin, T.A.; Lewis-Russell, J.M.; Douglas-Jones, A.; Ye, L.; Mansel, R.E. Eplin-alpha expression in human breast cancer, the impact on cellular migration and clinical outcome. *Mol. Cancer* **2008**, *7*, 71. [CrossRef]
229. Sanders, A.J.; Martin, T.A.; Ye, L.; Mason, M.D.; Jiang, W.G. EPLIN is a negative regulator of prostate cancer growth and invasion. *J. Urol.* **2011**, *186*, 295–301. [CrossRef]
230. Sanders, A.J.; Ye, L.; Mason, M.D.; Jiang, W.G. The impact of EPLINα (Epithelial protein lost in neoplasm) on endothelial cells, angiogenesis and tumorigenesis. *Angiogenesis* **2010**, *13*, 317–326. [CrossRef]
231. Verrills, N.M.; Walsh, B.J.; Cobon, G.S.; Hains, P.G.; Kavallaris, M. Proteome analysis of vinca alkaloid response and resistance in acute lymphoblastic leukemia reveals novel cytoskeletal alterations. *J. Biol. Chem.* **2003**, *278*, 45082–45093. [CrossRef]
232. Verrills, N.M.; Flemming, C.L.; Liu, M.; Ivery, M.T.; Cobon, G.S.; Norris, M.D.; Haber, M.; Kavallaris, M. Microtubule alterations and mutations induced by desoxyepothilone B: Implications for drug-target interactions. *Chem. Biol.* **2003**, *10*, 597–607. [CrossRef]
233. Monzó, M.; Rosell, R.; Sánchez, J.J.; Lee, J.S.; O'Brate, A.; González-Larriba, J.L.; Alberola, V.; Lorenzo, J.C.; Núñez, L.; Ro, J.Y.; et al. Paclitaxel Resistance in Non–Small-Cell Lung Cancer Associated With Beta-Tubulin Gene Mutations. *J. Clin. Oncol.* **1999**, *17*, 1786–1793. [CrossRef]
234. Fife, C.M.; McCarroll, J.A.; Kavallaris, M. Movers and shakers: Cell cytoskeleton in cancer metastasis. *Br. J. Pharm.* **2014**, *171*, 5507–5523. [CrossRef] [PubMed]
235. Christiansen, J.J.; Rajasekaran, A.K. Reassessing epithelial to mesenchymal transition as a prerequisite for carcinoma invasion and metastasis. *Cancer Res.* **2006**, *66*, 8319–8326. [CrossRef]
236. Bonello, T.T.; Stehn, J.R.; Gunning, P.W. New approaches to targeting the actin cytoskeleton for chemotherapy. *Future Med. Chem.* **2009**, *1*, 1311–1331. [CrossRef] [PubMed]
237. Creed, S.J.; Bryce, N.; Naumanen, P.; Weinberger, R.; Lappalainen, P.; Stehn, J.; Gunning, P. Tropomyosin isoforms define distinct microfilament populations with different drug susceptibility. *Eur. J. Cell Biol.* **2008**, *87*, 709–720. [CrossRef] [PubMed]

cancers MDPI

Article

The Plasma Membrane Ca^{2+} Pump PMCA4b Regulates Melanoma Cell Migration through Remodeling of the Actin Cytoskeleton

Randa Naffa [1], Rita Padányi [2], Attila Ignácz [3], Zoltán Hegyi [4], Bálint Jezsó [4], Sarolta Tóth [1,5], Karolina Varga [6], László Homolya [4], Luca Hegedűs [7], Katalin Schlett [3] and Agnes Enyedi [1,*]

1 Department of Transfusiology, Semmelweis University, H-1089 Budapest, Hungary; naffaranda@student.elte.hu (R.N.); toth.sarolta@med.semmelweis-univ.hu (S.T.)
2 Department of Biophysics and Radiation Biology, Semmelweis University, H-1094 Budapest, Hungary; padanyi.rita@med.semmelweis-univ.hu
3 Department of Physiology and Neurobiology, Eötvös Loránd University, H-1117 Budapest, Hungary; ignacz.attila@ttk.elte.hu (A.I.); schlett.katalin@ttk.elte.hu (K.S.)
4 Institute of Enzymology, Research Centre for Natural Sciences, Magyar Tudosok krt.2, H-1117 Budapest, Hungary; hegyi.zoltan@bio-science.hu (Z.H.); jezso.balint@ttk.elte.hu (B.J.); homolya.laszlo@ttk.hu (L.H.)
5 Department of Anatomy, Cell and Developmental Biology, Eötvös Loránd University, H-1117 Budapest, Hungary
6 Versys Clinics, H-1138 Budapest, Hungary; research@versysclinics.com
7 Department of Thoracic Surgery, Ruhrlandklinik, University Clinic Essen, 45239 Essen, Germany; luca.hegedues@rlk.uk-essen.de
* Correspondence: enyedi.agnes@med.semmelweis-univ.hu

Citation: Naffa, R.; Padányi, R.; Ignácz, A.; Hegyi, Z.; Jezsó, B.; Tóth, S.; Varga, K.; Homolya, L.; Hegedűs, L.; Schlett, K.; et al. The Plasma Membrane Ca^{2+} Pump PMCA4b Regulates Melanoma Cell Migration through Remodeling of the Actin Cytoskeleton. *Cancers* **2021**, *13*, 1354. https://doi.org/10.3390/cancers13061354

Academic Editors: José I. López and Ildefonso M. de la Fuente

Received: 29 January 2021
Accepted: 14 March 2021
Published: 17 March 2021

Simple Summary: Earlier we demonstrated that the plasma membrane Ca^{2+} pump PMCA4b inhibits migration and metastatic activity of BRAF mutant melanoma cells, however, the exact mechanism has not been fully understood. Here we demonstrate that PMCA4b acted through actin cytoskeleton remodeling in generating a low migratory melanoma cell phenotype resulting in increased cell–cell connections, lamellipodia and stress fiber formation. Both proper trafficking and calcium transporting activity of the pump were essential to complete these tasks indicating that controlling Ca^{2+} concentration levels at specific plasma membrane locations such as the cell front played a role. Our findings suggest that PMCA4b downregulation is likely one of the mechanisms that leads to the perturbed cancer cell cytoskeleton organization resulting in enhanced melanoma cell migration and metastasis.

Abstract: We demonstrated that the plasma membrane Ca^{2+} ATPase PMCA4b inhibits migration and metastatic activity of BRAF mutant melanoma cells. Actin dynamics are essential for cells to move, invade and metastasize, therefore, we hypothesized that PMCA4b affected cell migration through remodeling of the actin cytoskeleton. We found that expression of PMCA4b in A375 BRAF mutant melanoma cells induced a profound change in cell shape, cell culture morphology, and displayed a polarized migratory character. Along with these changes the cells became more rounded with increased cell–cell connections, lamellipodia and stress fiber formation. Silencing PMCA4b in MCF-7 breast cancer cells had a similar effect, resulting in a dramatic loss of stress fibers. In addition, the PMCA4b expressing A375 cells maintained front-to-rear Ca^{2+} concentration gradient with the actin severing protein cofilin localizing to the lamellipodia, and preserved the integrity of the actin cytoskeleton from a destructive Ca^{2+} overload. We showed that both PMCA4b activity and trafficking were essential for the observed morphology and motility changes. In conclusion, our data suggest that PMCA4b plays a critical role in adopting front-to-rear polarity in a normally spindle-shaped cell type through F-actin rearrangement resulting in a less aggressive melanoma cell phenotype.

Keywords: plasma membrane Ca^{2+} ATPase 4b; BRAF mutant melanoma; actin cytoskeleton; cofilin; cell motility; metastasis

1. Introduction

Melanoma is a form of skin cancer that originates from the neural crest derived melanocytes that produce melanin. It is a highly invasive type of cancer that tends to metastasize and causes death. Metastasis is a multistep process that requires migration of cells from the primary tumor to other sites in the body [1]. It has been widely acknowledged that changes in the actin cytoskeleton arrangement are essential for cells to migrate and metastasize [2]. Many proteins are involved in the regulation of actin dynamics and any alteration in their expression, activity, or localization may contribute to cellular transformation and tumorigenesis [3,4].

Several studies suggested that cytoplasmic free Ca^{2+} concentration plays a role in actin-based changes of cell polarity, chemotaxis, and motility [5–7]. In addition, Ca^{2+} is involved in actin rearrangement, focal adhesion turnover, invadopodia, and lamellipodia formation during cell migration [8,9]. Cells maintain Ca^{2+} homeostasis using a "molecular toolkit" that includes Ca^{2+} channels to allow Ca^{2+} to enter the cell and pumps or exchangers to remove excess Ca^{2+} for proper cell functioning. Changes in the expression of any of these tools can result in altered Ca^{2+} homeostasis leading to uncontrolled cell proliferation and metastasis [10].

Overexpression of Ca^{2+} channels has been implicated in the progression of several cancer types [11]. A recent study [12] reported elevation of T-type calcium channels (TTCCs) Cav3.1 in BRAFV600E mutant melanomas that promoted melanoma cell proliferation and migration. In bladder cancer, an increased level of transient receptor potential melastatin 8 (TRPM8) promoted cell proliferation, motility, and migration [13]. In metastatic prostate cancer, increased expression of the transient receptor potential cation channel subfamily V member 2 (TRPV2) resulted in enhanced cell motility through the invasion associated enzymes MMP9 and cathepsin B [14].

Cell motility and migration require filamentous actin (F-actin) rearrangements, and free Ca^{2+} concentration was found to play a role in this process [5,15]. The involvement of the transient receptor potential melastatin 4 (TRPM4) has been demonstrated in Ca^{2+} dependent actin cytoskeleton reorganization and migration of bone marrow-derived mast cells (BMMCs) [16]. Another study showed that transient receptor potential cation channel subfamily V member 4 (TRPV4) increased the migration of breast cancer cells via remodeling of the actin cytoskeleton through the Ca^{2+}-dependent activation of AKT [17].

Several actin binding-proteins are affected by changes in cytoplasmic Ca^{2+} concentration [18]. An important regulator of actin dynamics is the actin severing protein cofilin [19]. Inhibition of cofilin activity may disrupt cell polarity, protrusion formation, and chemotaxis. Cofilin is a 19 kDa protein that catalyzes actin depolymerization but also acts as a polymerization factor through producing monomer actin for the generation of locomotory protrusions at the leading edge of migrating cells. Cofilin activity is regulated by phosphorylation at Ser3, and by binding to PIP_2 or cortactin [20]. A study has reported that high concentration of free intracellular Ca^{2+} can induce cofilin dephosphorylation and activation through the calcium-dependent phosphatase calcineurin [21].

The plasma membrane Ca^{2+}-ATPase (PMCA) is a key regulator of cytosolic Ca^{2+} concentration that expels Ca^{2+} from the cell using the energy of ATP. Four different genes (*ATP2B1-4*) encode PMCAs in mammals (PMCA1-4) from which more than twenty variants are transcribed as a result of alternative mRNA splicing [22]. PMCAs have been implicated in a variety of cancer types. While PMCA2 (*ATP2B2*) was upregulated in HER+ breast cancer tumors [23], PMCA4 (*ATP2B4*) was downregulated in colon and prostate cancers, and lymph node metastases in contrast to the relatively high PMCA4 protein level in normal tissues [24,25].

Our laboratory identified the PMCA4b variant as a putative metastasis suppressor using BRAF mutant melanoma cell models [26]. We found that in selected melanoma cells with RAS or BRAF mutations PMCA4b was expressed at low levels. Inhibition of BRAF, MEK, or p38 MAPK, however, increased the expression of PMCA4b without affecting the expression of other Ca^{2+} pumps or Ca^{2+} channels [26,27]. The enhanced PMCA4b expression caused a dramatic change in cell motility without having a significant effect on proliferation, characteristic features of metastasis suppressors. We also identified additional PMCA4b expression inducers, such as the histone deacetylase inhibitors some of which are already in clinical use in a variety of melanoma, breast, and colon cancer cells [28–30].

Since the free Ca^{2+} ion level is an important modulator of actin cytoskeletal dynamics, the goal of the present study was to investigate whether PMCA4b, by controlling intracellular Ca^{2+} levels, acts through the actin cytoskeleton in reducing motility of BRAF mutant melanoma cells. We investigated the role of PMCA4b activity and trafficking in maintaining the integrity of the actin cytoskeleton and we studied the formation of cell–cell connections, lamellipodia and stress fibers in cells expressing wild type and mutant PMCA4b proteins. We show that polarized distribution of PMCA4b maintains a gradient of cytosolic free Ca^{2+} levels and induces cofilin redistribution to the leading edge that ultimately leads to a low motility melanoma cell phenotype.

2. Materials and Methods

2.1. Cell Culture

BRAF (V600E) mutant melanoma (A375) and breast cancer (MCF-7) cells were purchased from ATCC. Cells were cultured in Dulbecco's modified Eagle's medium (DMEM) (Lonza, Walkersville, MD, USA) supplemented with 10% fetal bovine serum (FBS) (Thermo Fisher Scientific, Waltham, MA, USA), 1% penicillin–streptomycin (Lonza), and 2 mM L-glutamine (Lonza) in humidified 5% CO_2 incubator at 37 °C.

2.2. Chemicals and Reagents

Cytochalasin D (cytD), calcium ionophore (A23187), and Phalloidin-TRITC were dissolved in DMSO and stored at −20 °C as a stock solution. The final DMSO concentration in the experiment did not exceed 0.01%. All reagents were purchased from Sigma-Aldrich (St. Louis, MO, USA).

2.3. DNA Constructs

The DNA plasmid of pmCherry-C1 and pEGFP-actin was purchased from Clontech Laboratories Inc., Mountain View, CA, USA. The mCherry-PMCA4b plasmid was generated as described previously [31]. The trafficking mutant pEGFP-PMCA4b-$L^{1167-1169}$A construct was prepared previously [31]. The SB-CAG-GFP-PMCA4b-CAG-Puromycin and SB-CAG-GFP-PMCA4b-LA-Puromycin constructs were generated, as described [26]. The non-functional mutant mCherry-PMCA4b-DE was created by introducing the D672E point mutation into the mCherry-PMCA4b and GFP-PMCA4b plasmids using QuikChange II Site-Directed Mutagenesis Kit (Stratagene) as described previously [32]. pCAGGS-GCaMP2-actin was a gift from Karel Svoboda (Addgene plasmid # 18928; http://n2t.net/addgene:18928; Accessed date: January 2021; RRID: Addgene_18928) [33]. The Cofilin-pmCherryC1 was a gift from Christien Merrifield (Addgene plasmid # 27687; http://n2t.net/addgene:27687; Accessed date: January 2021; RRID: Addgene_27687) [34] and CMV-R-GECO1 was a gift from Robert Campbell (Addgene plasmid # 32444; http://n2t.net/addgene:32444; Accessed date: January 2021; RRID: Addgene_32444) [35].

2.4. Generation of Stable Cell Lines

To generate A375-GFP-PMCA4b, A375-GFP-PMCA4b-LA, and MCF-7-GFP-PMCA4b cell lines, A375 or MCF-7 cells were stably transfected with the SB-CAG-GFP-PMCA4b-CAG-puromycin or SB-CAG-GFP-PMCA4b-LA-CAG-puromycin using the protocols described previously [26]. To generate the MCF-7-Sh-PMCA4b cell line, MCF-7 cells were transfected with the PMCA4b shRNA plasmid (sc-42602-SH, Santa Cruz Biotechnology, Santa Cruz, CA, USA) using FuGENE HD transfection reagent (Promega, Madison, WI, USA) according to the manufacturer's instructions. After 48 h, the culture medium was changed to the fresh medium containing puromycin dihydrochloride (1 µg/mL) (sc-108071, Santa Cruz Biotechnology) for selection. The medium with puromycin was changed every

2–3 days for two weeks. To confirm PMCA4b silencing the PMCA4b protein level was analyzed by Western blot.

2.5. Transient Transfection

A375 cell lines were seeded in 8-well Lab-Tek II chambered coverglass (Nalge Nunc International, Rochester, NY, USA). Next day, the cells were transiently transfected with one of the following plasmid DNA constructs (or in combination with one another): pmCherry-C1, mCherry-PMCA4b, mCherry-PMCA4b-DE, cofilin-pmCherryC1, GFP-PMCA4b-DE, pEGFP-actin, pCAGGS-GCAMP2-actin, and CMV-R-GECO1, as indicated in the experiment, using the FuGENE HD transfection reagent (Promega, Madison, WI, USA) according to the manufacturer's recommendations. The next day, the culture medium was changed and cells were incubated for a further 48 h.

2.6. siRNA Transfection

PMCA4b was knocked down by siRNA treatment as described previously [27]. Briefly, A375-GFP-PMCA4b melanoma cells were seeded onto 8-well Lab-Tek II chambered coverglass (Nunc). Next day, cells were transfected with ON-Target plus SMARTpool PMCA4b (ATP2B4) siRNA (50 nM, cat. # L-006118-00-005, Dharmacon Research Inc.) or SignalSilence® control siRNA (50 nM, Cell Signaling Technology, Danvers, MA, USA, cat. #6568S) using the DharmaFECT 1 transfection reagent (Dharmacon Research Inc., Cambridge, UK) according to the manufacturer's protocol. After 24-h transfection, the medium was changed, and the cells were incubated for an additional 48 h.

2.7. Cell Morphology Analysis

A375, A375-GFP-PMCA4b, and A375-GFP-PMCA4b-LA cells were cultured in a 6-well plate. Phase-contrast microscope (Olympus, Japan) images were taken either after overnight cell culture (low confluence) at 10× magnification or after a 48-h cell culture period (high confluence) at 4× magnification, as stated in the figure legends. The general morphology of the cells, including area and circularity, was analyzed by applying a black mask to display the contour of the cells using ImageJ software, v1.42q (National Institutes of Health, Bethesda, MD, USA).

2.8. Nearest Neighbour Distance Analysis

A375-GFP, A375-GFP-PMCA4b, A375-GFP-PMCA4b-LA, MCF-7, MCF-7-GFP-PMCA4b, and MCF-7-Sh-PMCA4b cells were cultured in a 6-well plate for 48 h. Phase-contrast microscope images were taken and cell centers were obtained by using the "Particle analysis" function of the ImageJ software. Nearest neighbor distance analysis was performed on binary images of cell centers using the graph plugin of ImageJ.

2.9. Non-Directional Cell Motility Assay

A375, A375-GFP-PMCA4b, and A375-GFP-PMCA4b-LA cells were cultured in 96-well plates. Next day, the nuclei of the cells were stained with 0.1 μM Hoechst 33342 for 1 h. The fluorescence signals for Hoechst and GFP were acquired automatically every 30 min for 24-h at 37 °C and 5% CO_2 using the ImageXpress Micro XL (Molecular Devices, Sunnyvale, CA, USA) high content screening system using a Nikon CFI Super Plan Fluor ELWD ADM 10× objective. For motility analysis, we used the multidimensional motion analysis module of MetaXpress High Content Image Acquisition and Analysis Software Version 5.3., Molecular Devices, Sunnyvale, CA, USA as described previously [28]. The time-lapse video demonstrating single cell motility was created from the images taken from the measurements.

2.10. Directional Cell Migration Assay

The Boyden chamber assay was performed as described previously [27]. Briefly, A375-GFP, A375-GFP-PMCA4b, and A375-GFP-PMCA4b-LA cells were seeded on the upper chamber of a 48-well Boyden Chamber device (Neuro probe, Gaithersburg, MD, USA) and incubated for 3 h at 37 °C. During this period, cells migrate through the 10 μm thick membrane (Whatman, Merck KGaA, Darmstadt, Germany) toward the fibronectin attractant (100 μg/mL, Merck KGaA, Darmstadt, Germany) at the lower chamber. At the end of the incubation period the membrane was removed and cells were scrapped from the upper side of the filter while at its lower side cells were fixed and stained with Toluidine blue. Light microscope images (6 images) were captured using a 10× objective lens and the number of cells migrated through the filter were counted.

2.11. Immunofluorescence Microscopy

A375, A375-GFP-PMCA4b, A375-GFP-PMCA4b-LA, MCF-7, MCF-7-GFP-PMCA4b, and MCF7-Sh-PMCA4b cells were seeded in 8-well Lab-Tek II chambered coverglass (Nunc) and cultured for 48 h. In some experiments, A375 cells and A375-GFP-PMCA4b cells were transiently transfected with Cofilin-pmCherryC1 or cotransfected with GFP-PMCA4b-DE in A375 cells. 48 h after transfection, cells were fixed with 4% paraformaldehyde (PFA), washed two times with PBS and then the nuclei were stained with DAPI.

For F-actin staining the cells were fixed and permeabilized with 0.1% Triton X-100 for 10 min and washed with PBS followed by incubation with Phalloidin-TRITC (Sigma-Aldrich) (1:1000) for 20 min. After that, cells were washed and kept in PBS.

For vinculin immunostaining, the cells were fixed and permeabilized as above and incubated in the blocking buffer (PBS containing 2 mg/mL bovine serum albumin, 1% fish gelatin, 0.1% Triton-X 100, 5% goat serum) for 1 h at room temperature (RT) followed by incubation with the rabbit monoclonal antivinculin antibody (1:100, ThermoFisher scientific, cat. # 700062) for 1 h at RT. After three washes with PBS, cells were incubated with AlexaFlour-647 conjugated anti-rabbit IgG (Invitrogen, Waltham, MA, USA) as a secondary antibody for 1 h at RT. Then cells were washed three times with PBS followed by staining with Phalloidin-TRITC for 30 min at RT. For nuclear staining, cells were incubated with DAPI for 10 min, washed, and kept in PBS.

To study the effect of Ca^{2+} influx on the actin cytoskeleton, A375 and A375-GFP-PMCA4b cells were seeded in 8-well Lab-Tek II chambered coverglass (Nunc). Next day, A375 cells were transiently transfected with GFP-PMCA4b-DE plasmid and cultured for 48 h. Ca^{2+} influx was initiated by the addition of 2 μM A23187 in HBSS buffer containing 2 mM Ca^{2+} (20 mM HEPES, pH 7.4, 2 mM $CaCl_2$, 0.9 mM $MgCl_2$) for 10 min at 37 °C. For positive and negative controls, cells were treated with 2.5 μM cytochalasin D (cytD) or with 2 μM A23187 in HBSS buffer without Ca^{2+} (in the presence of 100 μM EGTA), respectively. Cells were fixed with 4% PFA for 10 min and washed three times with PBS. Labeling with Phalloidin-TRITC is as described previously. Morphological parameters of the cells including area and circularity were analyzed using the ImageJ software.

In all experiments, confocal microscopy images were taken by Zeiss LSM710 or LSM800 confocal laser scanning microscopes using a Plan-Apochromat 40× (N.A. = 1.4) oil immersion objective (Zeiss, Oberkochen, Germany). Blue, green, red, and far red fluorescence images were sequentially acquired at 405, 488, 543, and 633 nm excitations, respectively.

2.12. Live-Cell Imaging

A375 cells were cultured overnight in 8-well Lab-Tek II chambered coverglass (Nunc). The next day, the cells were transfected with a GFP-actin plasmid in combination of one of the following plasmids: pmCherry-C1, mCherry-PMCA4b, or mcherry-PMCA4b-DE. After a 48-h incubation, the cells were washed twice with HBSS buffer before live-cell imaging. Experiments were initiated by the addition of (1) 2 mM A23187, (2) 2.5 μM cytD (positive control), or (3) 2 μM A23187 in HBSS buffer without Ca^{2+} (in the presence of 100 μM EGTA) (negative control). Treatments lasted for 10 min at 37 °C. Live-cell imaging was performed by acquiring Z-stack images every 15 s in both green and red channels at 488 and 561 nm excitation, respectively, using 100× 1.4 N.A. oil immersion objective of a Carl Zeiss Cell Observer SD microscope equipped with a Yokogawa CSU-X1 spinning-disk confocal module (Zeiss, Germany). For data analysis, 3D images were generated at time zero, 5 and 10 min and a video with one Z-stack was created for each experiment using ZEN 2.3 (blue edition) software, Carl Zeiss Microscopy GmbH, Jena, Germany. The Kymograph (space versus time), circulatory and area calculations relative to the zero time for each cell was analyzed using the ImageJ software.

2.13. *Fluorescence Recovery after Photobleaching (FRAP)*

A375 cells were cultured in 8-well Lab-Tek II chambered coverglass (Nunc). After overnight culture, cells were transiently transfected with plasmid constructs as described previously. For FRAP experiment, culture medium was changed with phenol free complete DMEM media supplemented with 25 mM HEPES (Gibco) and cells were kept at 37 °C. For simultaneous dual detection of GFP and mCherry fluorescence signals, 488 nm and 546 nm solid state lasers of a Carl Zeiss Cell Observer SD microscope equipped with a Yokogawa CSU-X1 spinning-disk confocal module were used. Differential interference contrast (DIC) images were taken at the end of the experiment. A defined region of interest (ROI) was drawn at three different cell parts: cell connections, cell free edge, and ruffles (lamellipodia). GFP signal was photobleached using a 488 nm bleaching laser at 20–40% intensity (RAPP UGA-42 Firefly 2L system) with a 40× 1.4 N.A. oil objective. Live cell imaging was carried out with images acquired every 0.2 s over 120 s time interval. Mean fluorescence intensity of a ROI, non-bleached region and background were analyzed using the ImageJ software. Data were imported into Microsoft Excel software 2016 (Microsoft Corporation by Impressa systems, Santa Rosa, CA, USA) and relative GFP-actin fluorescence intensity was calculated as follows: background intensity was subtracted from every ROI and at every time point, then the resulting ROI intensities were divided by a reference area intensity taken from a surrounding non-bleached cell. Post-bleach intensities were normalized to the mean of the first 10 prebleach time points. For FRAP analysis, the first 90 s of the post-bleach data was inserted into GraphPad Prism software v5.01 (GraphPad Software Inc., La Jolla, CA, USA) and non-linear regression analysis was used on the post-bleach sections to calculate the mobile fraction and half time of FRAP recovery curve ($t_{1/2}$).

2.14. *Ca^{2+} Signal Measurements*

To detect near the actin Ca^{2+} signal, A375 cells were cultured overnight in 8-well Lab-Tek II chambered coverglass (Nunc). The next day, cells were transfected with pCAGGS-GCAMP2-actin together with one of the following plasmids: mCherry-PMCA4b, or mCherry-PMCA4b-DE and cultured for 48 h. Then cells were washed and media were replaced with HBSS buffer. Calcium influx was initiated by the addition of 2 μM A23187 for 10 min at 37 °C. Live cell imaging was performed by detecting both GCAMP2 and mCherry signals in every 15 s using the spinning-disk confocal microscope specified above and 100× 1.4 N.A. oil immersion objective. For data analysis, images were taken before, during (peak) and at 7-min after the addition of A23187. Videos were created using ZEN 2.3 (blue edition) software. Cells were analyzed for near actin Ca^{2+} signal using the ImageJ software. Relative fluorescence intensities (F/F0) were calculated with the GraphPad Prism software v5.01.

To test the distribution of basal cytosolic Ca^{2+} concentration, A375, A375-GFP-PMCA4b, and A375-GFP-PMCA4b-LA cells were seeded onto 8-well Lab Tek II chambered coverglass (Nunc). The next day, the cells were transiently transfected with CMV-R-GECO1 plasmid and cultured for 48 h. Confocal microscope images were taken using confocal laser scanning microscope, Zeiss LSM710 with 63× oil immersion objective (Zeiss, Germany). Line plot analysis of the fluorescence signal was performed using the ImageJ software.

2.15. *Western Blot Analysis*

A375, A375-GFP-PMCA4b, A375-GFP-PMCA4b-LA, MCF-7, MCF-7-GFP-4b, and MCF-7-Sh-PMCA4b cells were cultured in a 6-well plate for 48 h. The total protein content of the cells was precipitated with 6% TCA. Samples were separated by using 10% or 15% acrylamide gels, as appropriate, and electroblotted onto PVDF membranes (Biorad, Hercules, CA, USA), as described previously [28].

Blots were immunostained with the following rabbit monoclonal primary antibodies: antivinculin (1:100, ThermoFisher scientific, cat. # 700062), anti-P-cofilin (Ser3) (1:1000, Cell Signaling Technology, Danvers, MA, USA, cat. # 77G2), rabbit polyclonal antibody: anti-β-tubulin (1:1000, Abcam, cat. # ab6046), anti-PMCA1 (1:1000, Affinity BioReagents, cat. # PA1-914), mouse monoclonal antibodies: anti-PMCA4 (JA9, 1:1000, Sigma-Aldrich, cat. # P1494), anti-NA^+/K^+ ATPase (1:2000, Enzo Life Sciences, cat. # BML-SA247), and chicken polyclonal antibody: anti-GFP (1:5000, Aves, GFP-1020). Horseradish peroxidase-conjugated anti-rabbit, anti-mouse, or anti-chicken secondary antibodies were used for detection (Jackson ImmunoResearch, dilution 1: 10,000) and were visualized with Pierce ECL Western Blotting Substrate (Thermo Scientific). The ImageJ software was used for densitometry analysis.

2.16. Statistical Analysis

For the stress fiber and lamellipodia formation analysis, a Chi square test while for Western blots, an unpaired t-test was used. For FRAP analysis, area and circulatory calculations, the differences between the control and the experimental groups were determined by a one-way analysis of variance (ANOVA) followed by Dunnett's multiple comparison post-hoc test, or unpaired t-test for two groups comparison. For circularity and area data versus time for the three groups, two-way ANOVA followed with the Bonferroni post-hoc test was used. The difference was considered significant at $p < 0.05$. The asterisks *, **, and *** denote values <0.05, <0.01, and <0.001, respectively.

3. Results

3.1. Proper Trafficking of PMCA4b Is Required to Change A375 Melanoma Cell Morphology, and Migration.

Previously, we demonstrated that overexpression of PMCA4b induced a profound change in the shape and motility of A375 melanoma cells [26]. Endocytic trafficking has been suggested to regulate both cell shape and motility in a variety of cell models [36–38]. Our laboratory has identified a di-leucine-like ^{1167}LLL internalization signal at the C-tail of PMCA4b. Mutation of these leucines to alanines resulted in a trafficking mutant (PMCA4b-LA), which has been characterized by having impaired endocytosis and hence high cell surface expression [31]. To test if endocytic trafficking of PMCA4b was essential for the distinct migratory and cell shape character of the melanoma cells, we compared shape and migration of GFP or GFP-PMCA4b expressing cells to those of the trafficking mutant GFP-PMCA4b-LA (Figure 1, Figures S1A and S8). At the single cell level, GFP-PMCA4b expression resulted in transition from a spindle-shaped character with three to four protrusions per cell to a polarized mesenchymal appearance with a typical asymmetric lamellipodial architecture, similarly to that shown previously [26]. In contrast, the A375-GFP-PMCA4b-LA cells retained the spindle-shaped character of the control A375-GFP cells (Figure 1A), and no significant change in area and circularity parameters could be detected (Figure S1B). In subconfluent cell cultures, the A375-GFP-PMCA4b cells formed clusters, whereas the A375-GFP-PMCA4b-LA cells showed scattered distribution similarly to that seen in the A375-GFP cells (Figure 1B). The nearest neighbor distribution histogram of A375-GFP-PMCA4b cells was shifted to the left as compared to the control or to the A375-GFP-PMCA4b-LA cells suggesting closer contact between the PMCA4b expressing cells in subconfluent culture (Figure 1C).

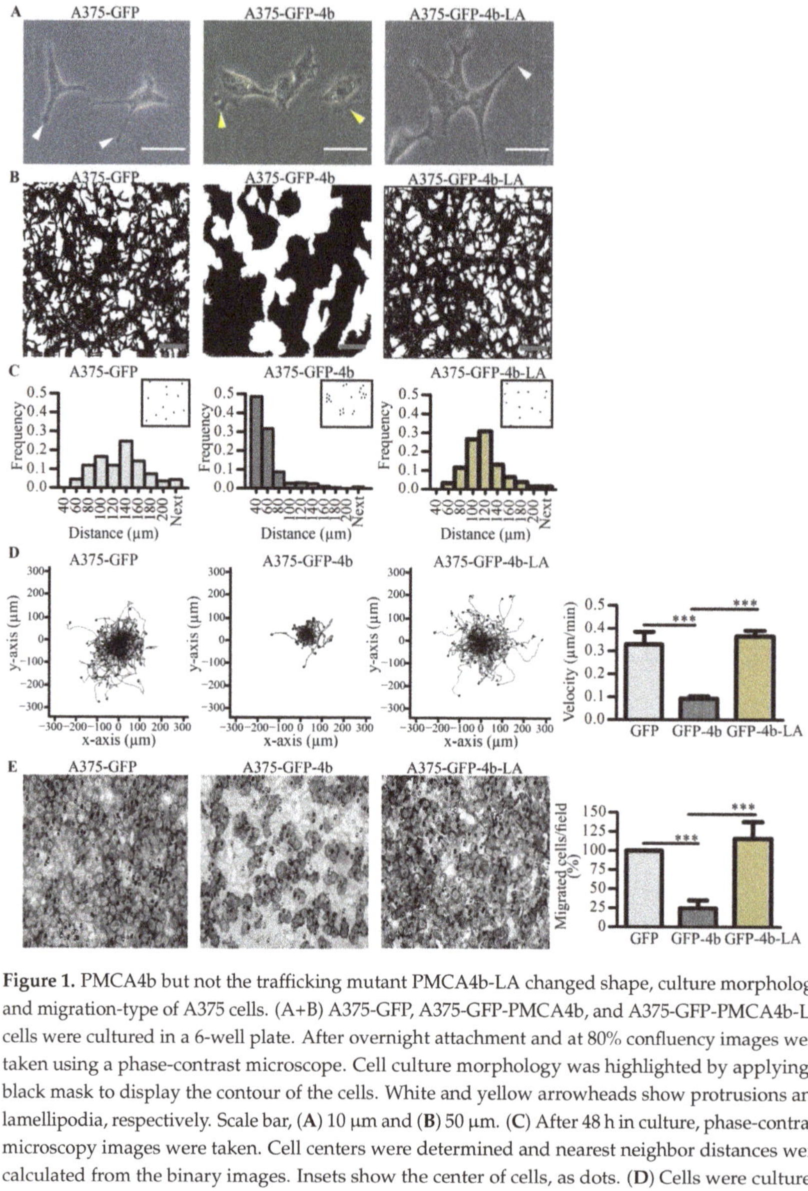

Figure 1. PMCA4b but not the trafficking mutant PMCA4b-LA changed shape, culture morphology, and migration-type of A375 cells. (A+B) A375-GFP, A375-GFP-PMCA4b, and A375-GFP-PMCA4b-LA cells were cultured in a 6-well plate. After overnight attachment and at 80% confluency images were taken using a phase-contrast microscope. Cell culture morphology was highlighted by applying a black mask to display the contour of the cells. White and yellow arrowheads show protrusions and lamellipodia, respectively. Scale bar, (**A**) 10 μm and (**B**) 50 μm. (**C**) After 48 h in culture, phase-contrast microscopy images were taken. Cell centers were determined and nearest neighbor distances were calculated from the binary images. Insets show the center of cells, as dots. (**D**) Cells were cultured in a 96-well plate and stained with Hoechst 33342. Migratory activity of the cells was followed by recording Hoechst and GFP signals by automated fluorescence microscopy for 24 h. Single cell trajectories of A375-GFP (n = 130), A375-GFP-PMCA4b (n = 77), and A375-GFP-PMCA4b-LA (n = 101) with the starting position of each trajectory translated to the origin of the plot are shown. Mean velocity ± S.D was determined from single cell trajectories (A375-GFP (n = 645), A375-PMCA4b (n = 941), and A375-PMCA4b-LA (n = 990) of 4–5 independent measurements. (**E**) For directional cell migration Boyden chamber assay was performed. Cells were seeded into the upper chamber and left to migrate for 3 h through the filter membrane towards the fibronectin at the lower chamber. Cells at the bottom of the filter membrane were fixed and stained with Toluidine blue. The number of migrated cells from six field of view was counted. Data show the means (% relative to the control) ± S.E.M from three independent experiments (*** $p < 0.001$).

A375 cells are a highly motile type of melanoma cells as they show dynamic protrusion and retraction activities with constantly changing direction of displacement. PMCA4b expression changed dramatically this character to a slow-moving mesenchymal-type with intense lamellipodia membrane ruffling and shorter displacement over time while the cells expressing the trafficking mutant A375-GFP-PMCA4b-LA, remained highly motile (Figure 1D and Figure S2, Videos S1 and S2). Similar results were obtained using a directional migration assay, in which the cells moved through the filter of a Boyden Chamber towards fibronectin as an attractant. Again, PMCA4b expression inhibited migration of the cells while expression of the mutant PMCA4b-LA had no effect (Figure 1E). Taken together, our data suggest that proper trafficking of PMCA4b was crucial for determining the shape and migratory behavior of these BRAF mutant melanoma cells.

3.2. PMCA4b Induces Actin Cytoskeleton Remodeling in A375 Melanoma Cells

Remodeling of the actin cytoskeleton plays a role in determining cell shape and migration [39]. One of the key regulators of actin dynamics is Ca^{2+} that acts through a variety of Ca^{2+}-dependent regulatory mechanisms [8,9]. Since PMCAs are considered as important regulators of intracellular Ca^{2+} concentration, we labeled A375 cells with Phalloidin-TRITC and studied the role of PMCA4b in F-actin organization with confocal laser scanning microscopy. We found that the A375-GFP-PMCA4b cells displayed an increased number of intercellular connections compared to the A375-GFP-PMCA4b-LA or A375-GFP cells where only few (one or two) connections could be detected (Figure 2A). Live-cell imaging experiments on mCherry-PMCA4b and GFP-actin coexpressing cells show that the mCherry-PMCA4b signal was followed by the thickening GFP-actin-based protrusions during the formation of cell–cell connections (Figure 2B and Video S3).

Figure 3A shows that A375-GFP-PMCA4b cells formed lamellipodia and actin stress fiber bundles in contrast to the A375-GFP-PMCA4b-LA and A375-GFP cells, which have more actin-rich membrane protrusions with significantly less stress fibers at the cell bottom. The graph indicates that GFP-PMCA4b expression increased the number of cells with stress fibers by more than 80% when compared to the GFP or the GFP-PMCA4b-LA mutant expressing cells. Treatment of the A375-GFP-PMCA4b cells with PMCA4b siRNA significantly reduced the number of cells with stress fibers confirming the role of PMCA4b in stress fiber formation (Figure S3A).

It has been demonstrated that free cytosolic Ca^{2+} concentration has a role in regulating focal adhesion turnover [40,41]. Therefore, we stained A375 and A375-GFP-PMCA4b cells for vinculin and F-actin, and found that vinculin dots clustered near the cell periphery towards the protrusions of the control A375-GFP cells, which was different from the punctate pattern at the focal adhesion sites of the stress fibers in the A375-GFP-PMCA4b cells (Figure 3B). This change was accompanied by decreased expression of vinculin in the A375-GFP-PMCA4b cells (Figure 3C). These data suggest stronger adhesion of the PMCA4b overexpressing cells to the substrate than the cells without PMCA4b.

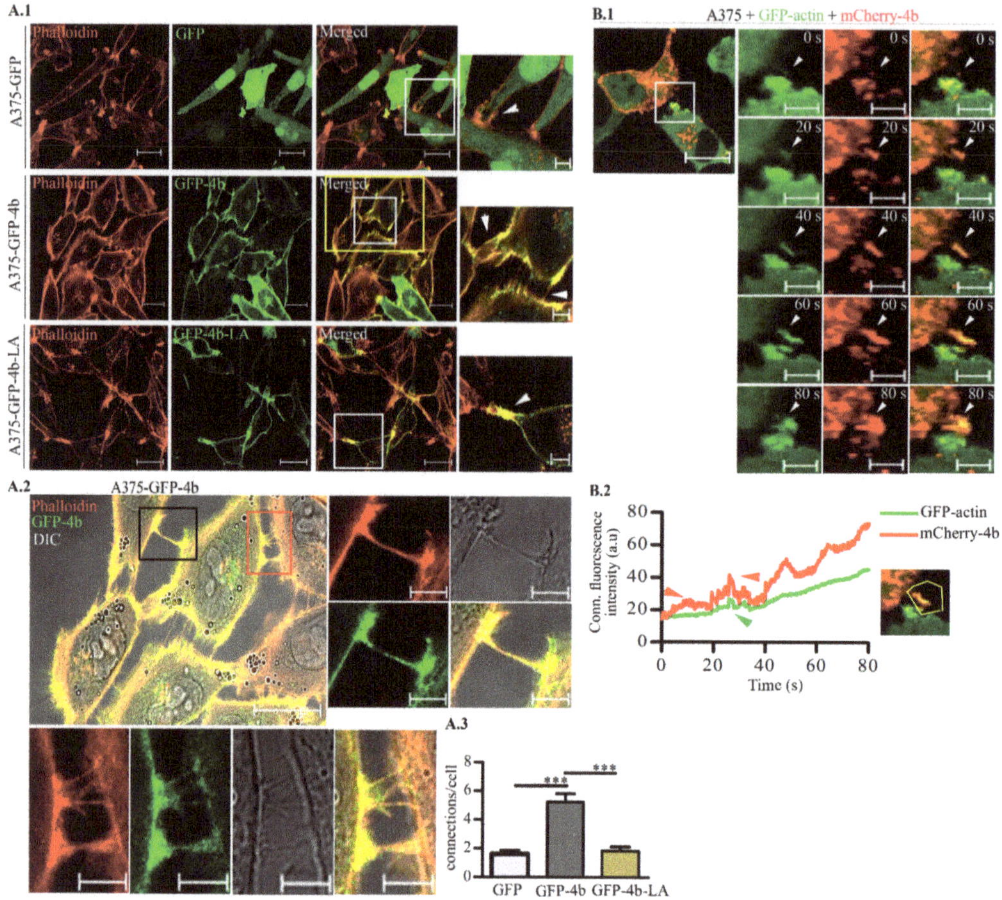

Figure 2. PMCA4b but not the trafficking mutant PMCA4b-LA increased cell–cell connections between A375 cells. (**A.1**) Confocal microscopy images of A375-GFP, A375-GFP-PMCA4b, and A375-GFP-PMCA4b-LA cells labeled with Phalloidin-TRITC. Scale bar, 20 µm. Insets show images with higher magnification of field marked with white rectangles; arrowhead indicate cell–cell connection. Scale bar, 5 µm. (**A.2**) High magnification DIC and fluorescence images of A375-GFP-PMCA4b cells are taken from the field marked with the yellow rectangle in (**A.1**). Scale bar, 20 µm. The black and red insets show images for the intercellular connections formed between cells. Scale bar, 5 µm (**A.3**) The graph represents the mean number of connections/cell for 12–13 cell. (**B.1**) Live cell imaging of A375 cells transiently expressing GFP-actin and mCherry-PMCA4b recorded by spinning-disc confocal microscopy. GFP and mCherry signals were recorded every 0.2 s for 180 s at 37 °C. Scale bar, 20 µm. Insets show the formation of connection between two cells with higher magnification at different times. Scale bar, 5 µm. (**B.2**) The graph depicts the time courses of GFP-actin and mCherry-PMCA4b signals determined in the region of interest (ROI) drawn around a newly forming connection (yellow polygon). Arrowheads indicate the increased signal (*** $p < 0.001$).

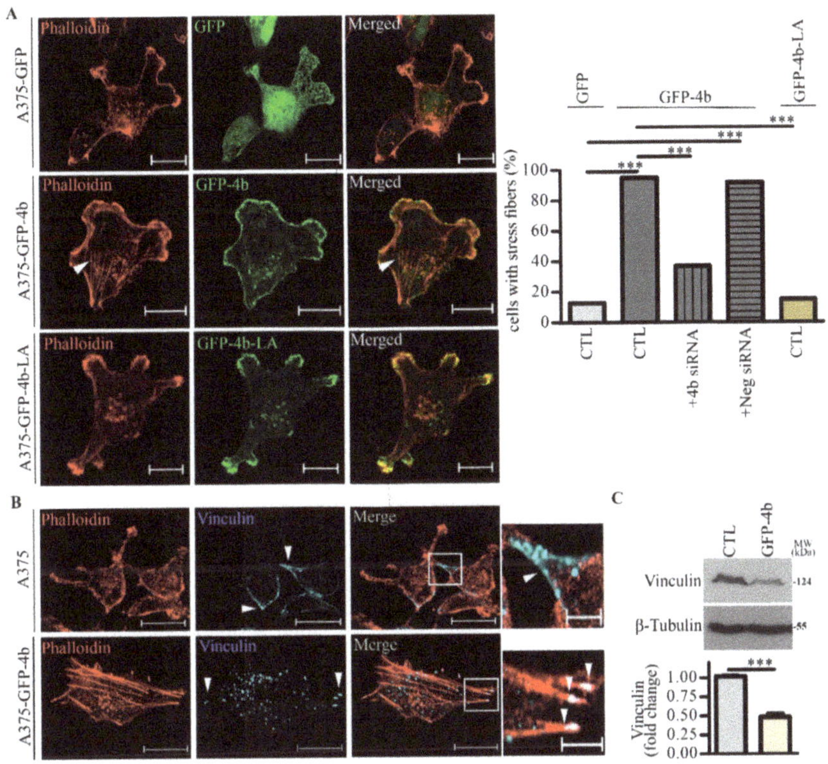

Figure 3. Lamellipodia and stress fiber formation are increased in A375 cells expressing GFP-PMCA4b but not in those expressing GFP-PMCA4b-LA. (**A**) Confocal microscopy images of A375-GFP, A375-GFP-PMCA4b, and A375-GFP-PMCA4b-LA cells labeled with Phalloidin-TRITC. Arrowheads show stress fibers. Scale bar, 20 μm. The fractions of stress fiber-positive cells in A375-GFP ($n = 55$), A375-GFP-PMCA4b-LA ($n = 72$), A375-GFP-PMCA4b ($n = 60$) ± PMCA4b siRNA ($n = 68$), and negative siRNA ($n = 49$) cultures were determined. siRNA confocal microscopy images are presented in Figure S3A. Relative abundance of cells with stress fibers is indicated by the bar graph. Confocal sections were taken from the bottom of the cells to show the stress fibers. (**B**) Confocal microscopy images of A375 and A375-GFP-PMCA4b cells immunostained with vinculin and labeled with Phalloidin-TRITC. Scale bar, 20 μm. Insets show part of the cells with higher magnification. Arrowheads show the differential localization of vinculin. Scale bar, 5 μm. (**C**) A375 and A375-GFP-PMCA4b cells were cultured in a 6-well plate for 48 h. Vinculin protein expression from total cell lysate was analyzed by Western blotting. β-tubulin was used as a loading control. Data represent mean ± S.E.M from three independent experiments (*** $p < 0.001$).

3.3. A Functional PMCA4b Pump Is Needed for Actin Cytoskeleton Remodeling

To test if PMCA4b function is required for actin cytoskeleton remodeling we used the non-functional mutant pump mCherry-PMCA4b-DE, in which we introduced an aspartate-to-glutamate substitution at position 672 [32]. Since this mutant cannot transport Ca^{2+}, it helped to dissect further the functional role of PMCA4b in actin cytoskeleton remodeling. In order to study this, we transiently expressed wild-type or mutant mCherry-tagged PMCA4b pump together with GFP-actin in A375 cells. It is worth noting that neither GFP nor mCherry tags affected PMCA4b activity, as described previously [32,42]. In the mCherry-PMCA4b expressing cells, GFP-actin concentrated to the lamellipodia at the cell front (Figure 4A) and formed stress fibers at the cell bottom (Figure S3B) as expected. In the non-functional mCherry-PMCA4b-DE expressing cells, however, the GFP-actin signal distributed evenly throughout the cell, and a significant reduction in stress fiber and lamellipodia formation was detected (Figure 4A,B and Figure S3B). A kymograph (Figure 4C) along the lines drawn across the lamellipodia of a mCherry-PMCA4b expressing cell shows highly polymerized actin, and high frequency of lamellipodial ruffling and retraction activity. We found mCherry-PMCA4b positive vesicles moving toward the edge of the lamellipodia and backwards with the GFP-actin retrograde flow (Figure 4D and Video S4). Together, these findings suggest that PMCA4b activity was essential for both stress fiber and lamellipodia formation, and that intense PMCA4b trafficking accompanied membrane ruffling at the lamellipodia.

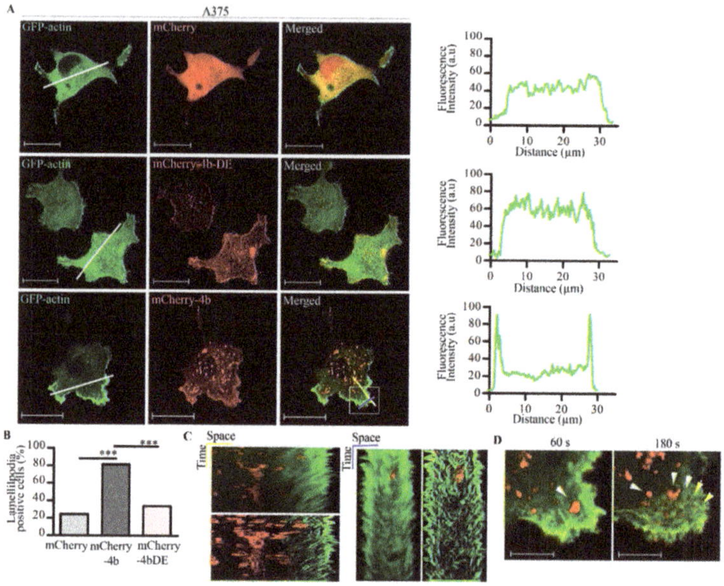

Figure 4. PMCA4b activity is necessary for lamellipodia formation in A375 melanoma cells. (**A**) A375 cells were transfected with GFP-actin together with one of the following plasmids: pmCherry-C1, mCherry-PMCA4b, mCherry-PMCA4b-DE, and cultured for 48 h. Confocal microscopy images of lamellipodia are shown. Scale bar, 20 μm. Graphs represent the GFP-actin intensity profile (green) for the line (white) drawn along the cell as indicated in the confocal image. Confocal sections were taken in the middle to visualize lamellipodia formation. (**B**) The graph shows the fraction of lamellipodia-positive cells in cultures transfected with GFP-actin and mCherry, mCherry-PMCA4b, or mCherry-PMCA4b-DE. (**C**,**D**) Live-cell imaging of cells expressing GFP-actin and mCherry-PMCA4b. Z-stacks of images were taken every 5 s for 5 min at 37 °C using spinning-disc confocal microscopy. (**C**) Kymographs were generated along the lines (yellow and blue) of the lamellipodia of the mCherry-PMCA4b cell shown in the image in (**A**), using the ImageJ software. Fine edges of ruffles are shown. (**D**) Magnified part of the lamellipodia from the same cell shown in (**A**). White arrowheads show mCherry-PMCA4b positive vesicles and the yellow arrowheads show GFP-actin retrograde flow. Scale bar, 5 μm (*** $p < 0.001$).

3.4. Silencing PMCA4b Expression Decreases the Number of Cells with Stress Fibers and Changes Cell Culture Morphology of MCF-7 Breast Cancer Cells

To confirm that the effect of PMCA4b on actin rearrangement is not cell type-specific; we used the estrogen receptor positive (ER+) luminal type of breast cancer cell line MCF-7. We showed that PMCA4b expression is relatively low in these cells that can be greatly upregulated with histone deacetylase (HDAC) inhibitors. Since we surmised that HDAC inhibitors could interfere with our studies, we either stably expressed or silenced PMCA4b in MCF-7 breast cancer cells. The Western blots in Figures S4 and S8 show that MCF-7 cells express PMCA4b endogenously and that silencing diminished its expression almost completely. In accordance with the relative PMCA4b abundance, changes in cell culture morphology were observed. Nearest neighbor distance histogram analysis shown in Figure 5A indicated that MCF-7-GFP-PMCA4b cells were located closer to each other than the parental or the PMCA4b silenced cells. In addition, PMCA4b silencing induced a dramatic—more than 60%—loss of stress fibers (Figure 5B,C) and increased significantly the area of individual cells (1.75×) (Figure 5A,D) underlying the importance of PMCA4b in actin cytoskeleton remodeling, and consequently changes in cell shape and cell culture morphology.

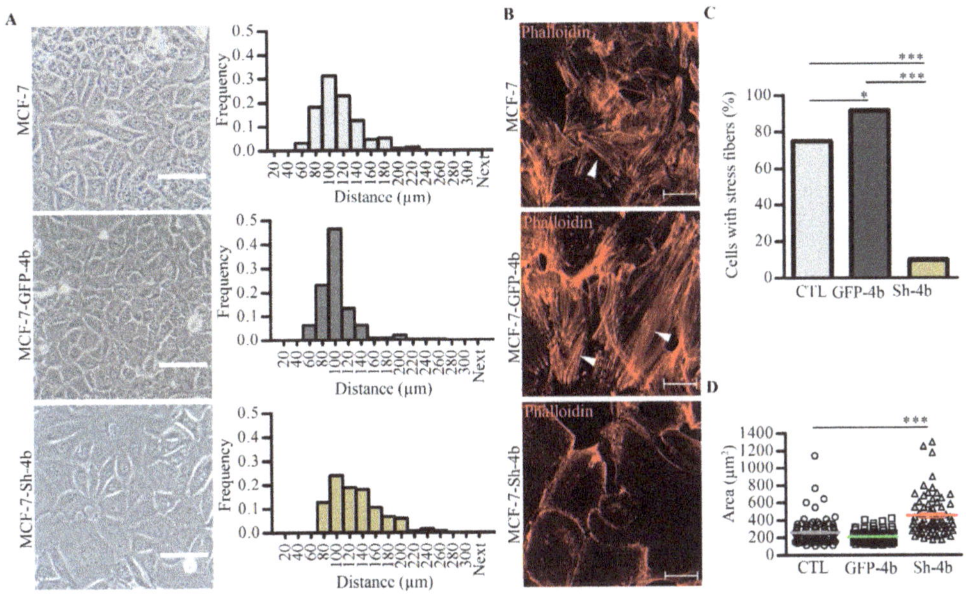

Figure 5. PMCA4b silencing in MCF-7 cells changed cell culture morphology and resulted in a dramatic loss of stress fibers. (**A**) Phase contrast images of MCF-7 cells, GFP-PMCA4b (MCF-7-GFP-4b), or shRNA of PMCA4b (MCF-7-Sh-4b). Scale bar, 50 μm. The distance between nearest neighbors was determined using the binary images of cell centers. (**B**) F-actin in the three cell types was labeled with Phalloidin-TRITC. Arrowheads show stress fibers. Scale bar, 20 μm. The fractions of stress fiber-positive cells were determined (n = 48, 83, and 100, respectively) (**C**) and the area of individual cells of the phase contrast images was calculated (n = 100, 88, and 70, respectively) (**D**). Bars represent the mean values for each cell type (* $p < 0.05$, *** $p < 0.001$).

3.5. PMCA4b Does Not Affect F-Actin Recovery after Photobleaching

Fluorescence recovery after photobleaching (FRAP) is often used to study actin cytoskeleton dynamics [43]. From FRAP analysis both the mobile fraction and turnover rate ($t_{1/2}$) of F-actin can be determined. To reveal whether changes in cell morphology was a result of F-actin dynamics, we performed FRAP analysis in cells transfected with GFP-actin and mCherry, mCherry-PMCA4b, or mCherry-PMCA4b-DE. FRAP was performed at three different parts of the cells: at the cell-free edge, at the cell–cell contacts (Figure S5), and at the lamellipodia (Figure 6). We found that PMCA4b expression did not affect either the mobile fraction or the recovery rate of F-actin at any of these locations. Although the mCherry-PMCA4b expressing cells showed slightly faster recovery rate (lower $t_{1/2}$) at the lamellipodia than the parental cells, the difference was not significant (Figure 6B and Figure S5B) suggesting that F-actin assembly was not affected by PMCA4b.

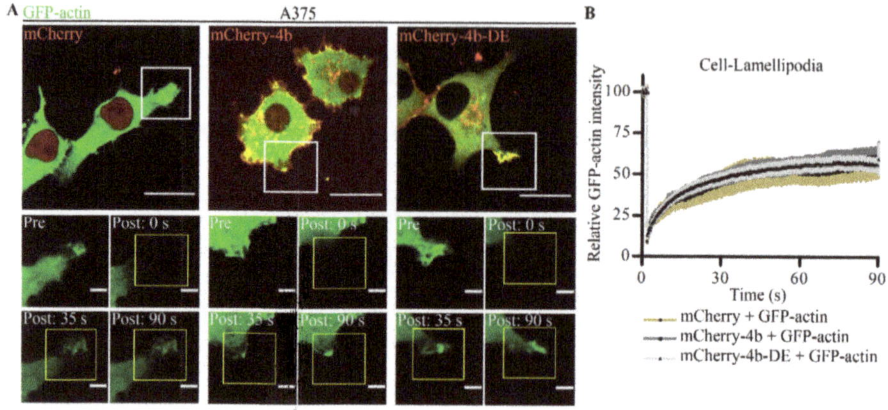

Figure 6. PMCA4b expression does not affect F-actin turnover in A375 cells. (**A**) A375 cells were transfected with GFP-actin together with one of the following plasmids, pmCherry-C1, mCherry-PMCA4b, mCherry-PMCA4b-DE, and cultured for 48 h. Media were changed to phenol free DMEM and cells were kept at 37 °C. GFP-actin was photobleached at the ruffles (mCherry: n = 8, mCherry-4b: n = 13, mCherry-4b-DE: n = 23) using spinning-disc confocal microscopy. GFP signal was recorded every 0.2 s for 90 s. Scale bar, 20 µm. Insets show magnified lamellipodia before (pre) and after (post) photobleaching at different time points. Scale bar, 2 µm. (**B**) Relative GFP-actin fluorescence intensity changes over time. Data represent mean ± S.E.M for three independent experiments.

3.6. PMCA4b Inhibits Ca^{2+} Induced F-Actin Depolymerization

Several studies have demonstrated that high cytosolic Ca^{2+} concentration can induce F-actin depolymerization [6,44–46]. To investigate if PMCA4b is able to protect the actin cytoskeleton from Ca^{2+} overload, we used A375 cells expressing GCaMP2- or GFP-actin together with mCherry, mCherry-PMCA4b, or the non-functional mutant mCherry-PMCA4b-DE. Cells were treated with the Ca^{2+} ionophore A23187 to allow Ca^{2+} influx. As expected, A23187 induced a sustained increase in GCaMP2-actin fluorescence in cells expressing the non-functional mutant PMCA4b-DE, while the fluorescence returned relatively quickly to the basal level in cells expressing the wild-type pump protein (Figure 7A). In correlation with the sustained increase in near-actin Ca^{2+} concentration, a dramatic loss of cell protrusions, intense membrane blebbing, and cell shrinkage was observed in the mCherry or the mCherry-PMCA4b-DE expressing cells (Figure 7B, see also Video S5). In correlation with these changes, a significant increase in circularity (1.64 and 1.83 fold) and a decrease in area (28.4% and 42.7%) were detected in the control and mutant pump expressing cells, respectively. In contrast, no change in these parameters was observed in the wild-type PMCA4b expressing cells displaying intact stress fibers at the cell bottom after treatment (Figure 7B). A kymograph in Figure S6A shows GFP-actin collapse at the cell periphery both in the mCherry and the mCherry-PMCA4b-DE expressing cells, while the mCherry-PMCA4b cells retained their original shape.

Similar results were obtained when the endogenous F-actin was labeled with Phalloidin-TRITC. Treatment of the parental and GFP-PMCA4b-DE expressing cells with A23187 led to cell shrinkage, rounding, and intense membrane blebbing, while the A375-GFP-PMCA4b cells did not show any of these changes even after a relatively long exposure to the ionophore (Figure 8 and Figure S7A). When ionophore was added in the absence of external Ca^{2+} (in the presence of EGTA), none of the cells showed change in morphology confirming the role of excess Ca^{2+} entry. As a control, the actin depolymerizing agent cytochalasin D destroyed the actin cytoskeleton independent of Ca^{2+} entry (Figures S6B and S7B, and Video S6). These results underlie the importance of the Ca^{2+} extrusion capacity of PMCA4b that can reduce near-actin Ca^{2+} concentration levels protecting the actin cytoskeleton from Ca^{2+} overload.

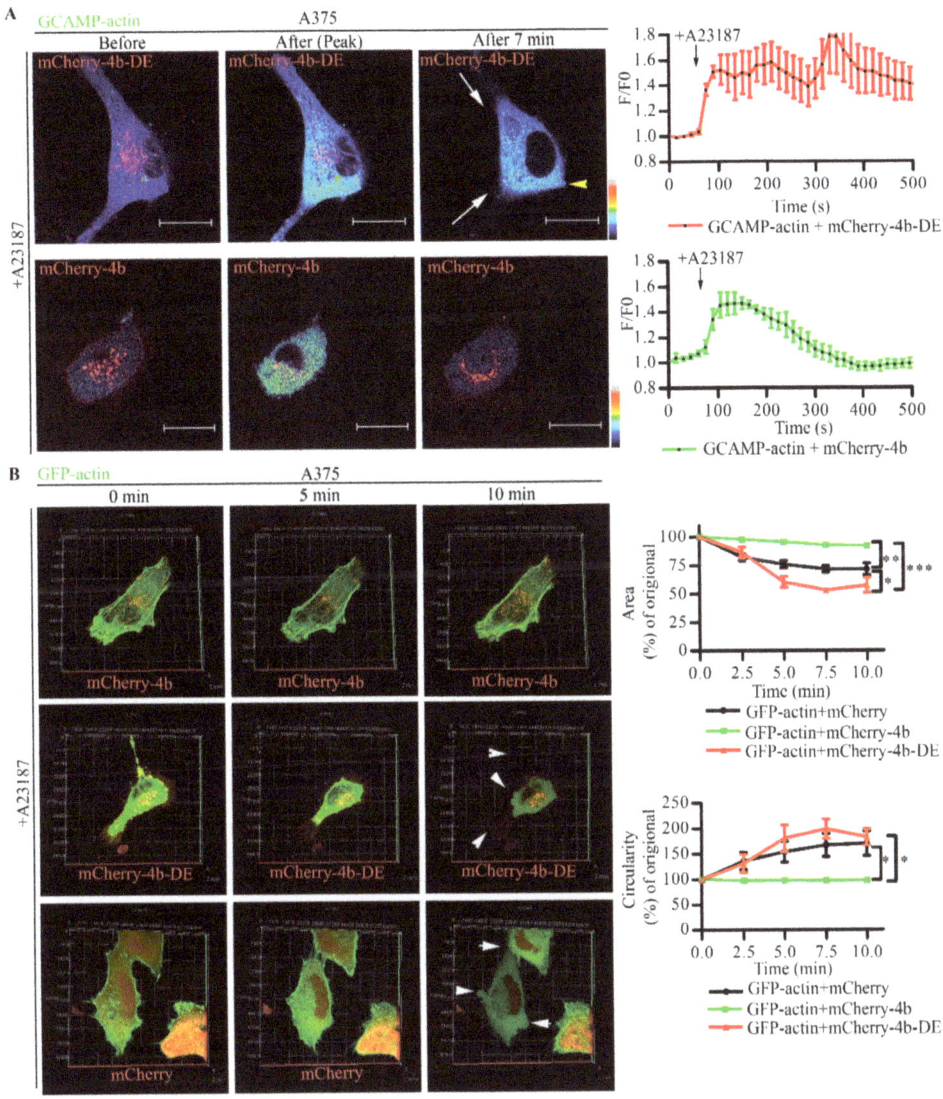

Figure 7. PMCA4b protects F-actin from Ca^{2+}-induced degradation. (**A**) Near actin Ca^{2+} signal was initiated in A375 cells transiently expressing GCAMP2-actin together with mCherry-PMCA4b or mCherry-PMCA4b-DE by the addition of 2 μM A23187. GCAMP2 and mCherry signals were recorded every 15 s for 10 min using a spinning-disc confocal microscopy. Confocal microscopy images show cells before, at the peak and 7 min after the addition of A23187, as indicated. Arrowheads show GCAMP2-actin signal retraction. Scale bar, 20 μm. Graphs represent fluorescence intensity values (F/F0) of the cell shown in (**A**). Error bars represent S.E.M obtained from two independent experiments (two cells analyzed in each, two ROIs per cell). Arrows indicate the addition of A23187. (**B**) 2 μM A23187 was added to A375 cells expressing GFP-actin together with pmCherry-C1, mCherry-PMCA4b, or mCherry-PMCA4b-DE, and Z-stack images of GFP and mCherry signals were recorded every 15 s for 10 min using a spinning-disc confocal microscope. Three-dimensional confocal images were created by the ZEN 2.3 (blue edition) software, and presented at 0, 5, and 10 min. Arrowheads indicate changes in cell shape. Area and circularity parameters for A375 cells transfected with the same plasmid combinations (n = 2–3 each) were analyzed by the ImageJ software. Data represent mean ± S.E.M. Significance is calculated for the 10-min time points (* $p < 0.05$, ** $p < 0.01$, *** $p < 0.001$).

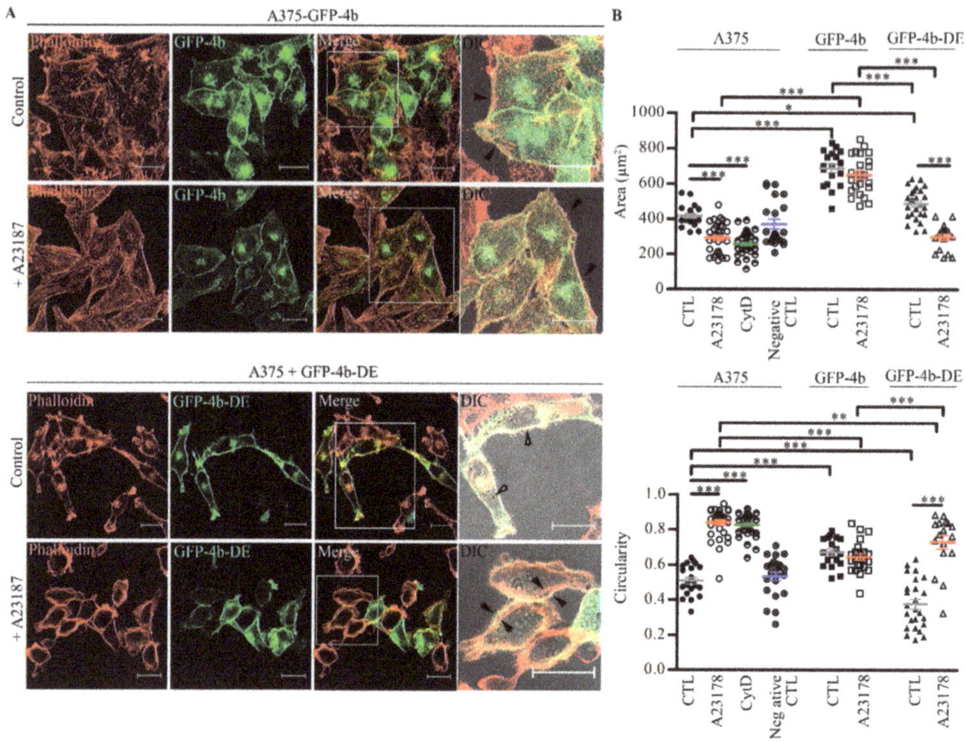

Figure 8. A375-GFP-PMCA4b but not the parental or the non-functional mutant expressing PMCA4b-DE cells maintain shape after Ca^{2+} overload. (**A**) A375-GFP-PMCA4b cells or A375 cells transiently expressing GFP-PMCA4b-DE were treated with 2 µM A23187 in HBSS buffer containing 2 µM Ca^{2+} for 10 min at 37 °C. Confocal and DIC microscopy images were taken after labeling with Phalloidin-TRITC. Scale bar, 20 µm. Right images show cells with higher magnification, scale bar, 20 µm. Arrowheads show the position of actin in relation to the cell periphery. (**B**) Area and circularity parameters were analyzed for each cell type (n = 17–33) by ImageJ software. The mean ± S.E.M values of data are presented as a scatter plot. Confocal microscopy images of cells incubated in the presence or absence of cytochalasin D (CytD) or A23187 are shown in Figure S7 (* $p < 0.05$, ** $p < 0.01$, *** $p < 0.001$).

3.7. PMCA4b Induces F-Actin Rearrangement through Cofilin Relocation.

Several studies described the role of cofilin in actin dynamics [19,47]. Cytosolic Ca^{2+} is an important regulator of cofilin activity [21], therefore, we investigated if PMCA4b expression altered the distribution of mCherry-cofilin in A375 cells. Figure 9A shows that mCherry-cofilin localized to the protrusions of the parental and the non-functional GFP-PMCA4bDE expressing cells, while in the A375 cells expressing active GFP-PMCA4b, it localized mostly to the leading edge of the lamellipodia. One of the mechanisms that regulate cofilin activity is its phosphorylation at serine 3. Therefore, we tested P-cofilin at the protein level, but no differences were detected between the parental and A375-GFP-PMCA4b cells (Figure 9B). These results suggest that cofilin relocalization to the leading edge rather than changes in its overall activity contributed to the actin cytoskeleton rearrangement of PMCA4b expressing cells.

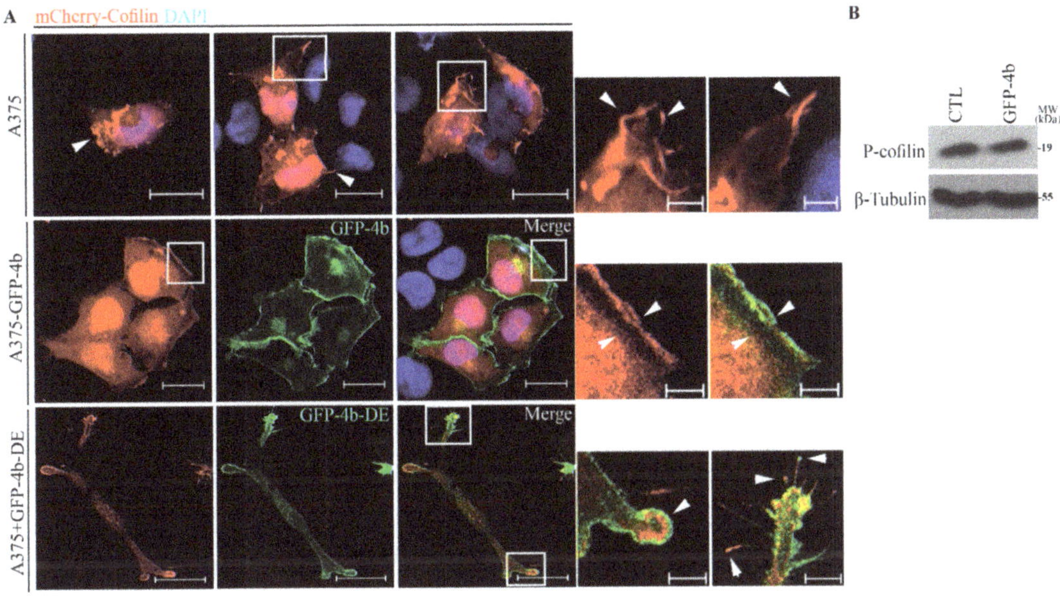

Figure 9. PMCA4b expression results in cofilin reorganization to the leading edge. (**A**) mCherry-Cofilin was transfected in A375 and A375-GFP-PMCA4b cells or cotransfected with GFP-PMCA4bDE in A375 cells. Confocal microscopy images were taken after nuclear staining with DAPI. Scale bar, 20 µm. Insets show a part of a cell with higher magnification. Arrowheads show the location of expressed cofilin at the leading edge in A375-GFP-PMCA4b cells and filopodia or protrusion in A375 cells and A375 expressing GFP-PMCA4bDE. Scale bar, 5 µm. (**B**) A375 and A375-GFP-PMCA4b cells were cultured in a 6-well plate for 48 h. Endogenous P-cofilin protein expression from total cell lysate was analyzed by Western blotting. β-tubulin was used as a loading control.

3.8. Proper PMCA4b Trafficking Is Essential in Managing Front-to-Rear Increasing Ca^{2+} Concentration Gradient in A375 Cells

A previous study suggested that polarized distribution of PMCA4b contributes to the front-to-rear Ca^{2+} concentration gradient during migration of Human Umbilical Vein Endothelial Cells (HUVECs) [48]. To investigate if PMCA4b localization to the leading edge resulted in a similar front-to-rear Ca^{2+} concentration gradient in A375 cells, we expressed the Ca^{2+} indicator R-GECO in the A375-GFP-PMCA4b or A375-GFP-PMCA4b-LA cells. Fluorescence intensities of R-GECO and GFP signals were recorded and analyzed across the lines shown on the confocal microscope images in Figure 10. The A375-GFP-PMCA4b cells displayed polarized distribution of the GFP-PMCA4b signal that was accompanied by an inverse distribution of the R-GECO signal. In contrast, the control and the trafficking mutant expressing cells displayed an even distribution of Ca^{2+} concentration across the entire cytoplasm (Figure 10). Our data suggest that PMCA4b is essentially involved in maintaining a Ca^{2+} concentration gradient increasing front-to-rear that may contribute to the actin cytoskeleton reorganization and the formation of the low-motility mesenchymal cell phenotype.

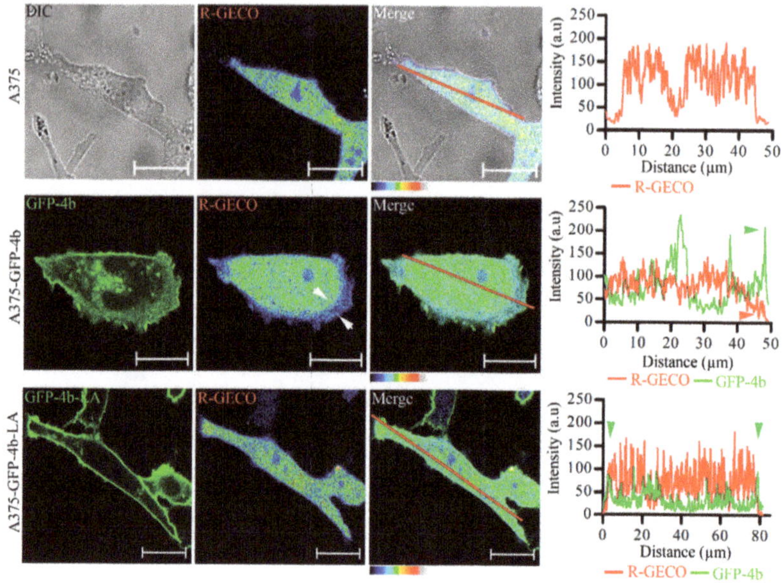

Figure 10. Proper trafficking of PMCA4b is needed to maintain a front-to-rear Ca^{2+} concentration gradient in A375 cells. A375, A375-GFP-PMCA4b, and A375-GFP-PMCA4b-LA cells were transfected with the CMV-R-GECO1 plasmid. Cells were fixed and confocal microscopy images were taken. Low Ca^{2+} levels are indicated by the arrowheads at the leading edge of A375-GFP-PMCA4b cells. Scale bar, 20 μm. Graphs represent GFP-PMCA4b (green) and R-GECO intensity (red) profiles for the line (red) drawn along the cells on the confocal images. Arrowheads show the GFP signal at the cell periphery. Line plots were drawn and analyzed using the ImageJ software.

4. Discussion

The calcium ion is considered as a modulator of actin dynamics, and a higher intracellular Ca^{2+} concentration is found to promote cell migration [8,11]. Previously, we identified the PMCA4b Ca^{2+} pump as a putative metastatic suppressor in BRAF mutant melanoma cells [26]. Since PMCAs are considered as key regulators of cellular Ca^{2+} homeostasis, we hypothesized that the reduced migration and morphology changes observed along with the reduced metastatic activity of these melanoma cells are correlated with changes in actin dynamics.

Melanocytes in the skin through their dendrites are in close contact with the surrounding keratinocytes and while melanocytes produce melanin to protect keratinocytes from UV radiation keratinocytes tightly control the proliferation capacity of melanocytes [49]. Across the epidermis, there is an increasing Ca^{2+} gradient that decreases proliferation and induces differentiation of keratinocytes up to the stratum corneum where Ca^{2+} concentration sharply decreases. This refers to both the extracellular Ca^{2+} concentration and the amount of Ca^{2+} in the intracellular Ca^{2+} stores. During differentiation expression of Ca^{2+} regulatory molecules such as Ca^{2+} channels and the Ca^{2+} sensing receptor are gradually changing that results in altered expression of differentiation markers and desmosome formation [50]. Interestingly, melanoma cells influence the differentiation pattern of the keratinocytes in the vicinity of the tumor through the production of growth factors and cytokines that induces the hyperplasia of the epidermis [51,52]. However, the role PMCA proteins in these processes is not known.

Melanoma cells tend to have great plasticity in shifting between mesenchymal and amoeboid motility style to allow cells to invade and metastasize [53,54]. Here we demonstrate that spontaneous movement of A375-GFP-PMCA4b cells has slower polar type motility in contrast to the fast-moving parental cells with dynamic actin-rich filopodia formation. Since a previous study from our laboratory did not find any significant changes in the expression of EMT/MET markers (E-cadherin, ZEB1, snail, and vimentin) [26], we surmise that PMCA4b expression induced a switch from a fast- to a slow-moving mesenchymal cell type rather than a transition between mesenchymal/epithelial phenotypes [55].

Several studies have reported on the role of cytoplasmic free Ca^{2+} in changing cell morphology through the induction of actin cytoskeleton reorganization in a variety of cell types [56,57]. A study on pulmonary endothelial cells reported that the activation of store-operated Ca^{2+} channels (SOCs) resulted in cell shape changes and this was dependent on site-specific reorganization of the actin cytoskeleton [58]. In agreement with these findings, we show that along with the switch between motility types, PMCA4b expression induced a dramatic change in cell morphology both at the single cell and cell culture levels including cell roundness, increased formation of cell–cell connections, lamellipodia formation, and stress fibers with localized distribution of focal adhesion sites. Interestingly, similar morphology changes were seen in the MCF-7 breast cancer cells suggesting a general role of PMCA4b in cell shape determination.

Stress fibers and their associated focal adhesions are important for cells to adhere [59]. It was found that highly motile cells have few, thin, and dynamic stress fibers compared to the thick and stable stress fibers of non-motile cells [60]. It was also suggested that cell motility inhibition could be a result of slow rearrangement of stress fiber actin bundles and focal adhesion [59]. This is in good correlation with the slow motility of A375-GFP-PMCA4b cells, which have thick stable stress fiber bundles in contrast to the fast-moving parental cells, which lack stress fibers almost completely. Again, the effect of PMCA4b on stress fiber formation was not restricted to the melanoma cell type; PMCA4b silencing caused a nearly complete loss of stress fibers in the breast cancer MCF-7 cell line, as well.

High free intracellular Ca^{2+} can increase the focal adhesion turnover rate and cell motility [8]. In the current study, we found punctate localization of vinculin at focal adhesion sites in the A375-GFP-PMCA4b cells as compared to the clustered vinculin dots near the cell periphery facing towards the protrusions in the parental cells. A study [61] showed that blocking store-operated Ca^{2+} influx (SOCE) in MDA-MB-231 cells resulted in different localization of vinculin that caused slow focal adhesion turnover rate and strong cell adherence. Similarly, another study on mesenchymal-like chemoresistant IGROV1 ovarian cancer cells showed enhanced cell migration as a result of enhanced focal adhesion turnover mediated by SOCE [62]. It has been suggested that PMCA4b has the ability to decrease near-membrane Ca^{2+} concentration in response to SOCE [63] that may explain, at least partly, the enhanced focal adhesion and reduced motility of the PMCA4b expressing melanoma cells.

High expression of vinculin has been found in cancerous cells and was used as a biomarker in pancreatic cancer [64]. A study reported that electromagnetic fields enhanced cell migration in bone marrow-derived mesenchymal cells in a Ca^{2+}-dependent manner by increasing the expression of several focal adhesion proteins including vinculin [65]. Using a Ca^{2+} channel blocker, vinculin expression decreased to its baseline level that reduced cell migration. Interestingly, PMCA4b expression resulted in decreased vinculin expression that might contribute to the reduced migratory activity of the PMCA4b expressing A375 melanoma cells.

Changes in the localization and trafficking of membrane proteins have been correlated to cell shape and motility [36–38]. In line with these findings we found that the trafficking mutant PMCA4b-LA did not affect shape, migration, and F-actin distribution of A375 cells. We showed that the characteristics of the PMCA4b-LA expressing cells resembled more that of the control cells than those of the wild-type PMCA4b expressing cells suggesting that proper localization was essential for the antimigratory behavior of the pump.

Further, we showed that expression of the non-functional PMCA4b-DE did not induce rearrangement of the actin cytoskeleton confirming that PMCA4b activity and hence, intracellular Ca^{2+} concentration played a role. Similarly, a study reported that activation of transient receptor potential melastatin 2 (TRPM2) Ca^{2+} channel by H_2O_2 in Hela and prostate cancer (PC)-3 cells resulted in filopodia formation, loss of stress fibers, and disassembly of focal adhesion that eventually caused an increase in cell migration [66]. Another study on highly metastatic osteosarcoma cell line U2OS reported that Ca^{2+} channel ORAI1 translocation to the leading edge was essential for formation of lamellipodia and cell directionality [67]. Our data are in good agreement with these findings and points to the importance of PMCA4b in mediating actin cytoskeleton rearrangement and cell motility through controlling cytosolic Ca^{2+} levels.

In polarized cells, actin reorganization at the lamellipodia of the leading edge directs cell migration [68]. We noticed polymerized actin at the lamellipodia occupying most of the cell front when a functional PMCA4b was expressed in contrast to the cells expressing the non-functional PMCA4b-DE mutant or the control cells where F-actin was more abundant at the pointed end of the cell protrusions. In addition, colocalization between F-actin and PMCA4b was observed at the lamellipodia and cell–cell contact sites that may indicate a direct or indirect interaction between these proteins, as suggested previously [69,70].

Using recovery after photobleaching (FRAP) measurements, no significant change in F-actin level or actin recovery rate was detected in response to PMCA4b expression suggesting that actin polymerization was not affected. Several studies have reported that an increase in intracellular Ca^{2+} concentration can destroy the cortical actin cytoskeleton with changes in cell shape in different cell types [5,44–46]. In this study we found that persistent increase in near-actin Ca^{2+} concentration—tested by the genetically encoded Ca^{2+} sensor fused to actin (GCaMP2-actin)—in response to Ca^{2+} ionophore treatment resulted in actin cytoskeleton collapse and cell rounding in the parental cells or in cells expressing the non-functional PMCA4b-DE. In contrast, the functional PMCA4b was able to protect the actin cytoskeleton from Ca^{2+} overload suggesting that PMCA4b can act as a negative modulator of Ca^{2+} induced F-actin depolymerization.

It has been suggested that an increased level of intracellular calcium can trigger cell motility by regulating proteins that interact with the actin cytoskeleton. Cofilin is an actin severing protein that catalyzes actin depolymerization and mediates actin polymerization by the formation of new barbed ends and supplying G-actin monomers. The activity of cofilin is regulated by Ca^{2+}, and was found to control lamellipodia and invadopodia formation [19]. A previous study reported the role of CRAC channels in lamellipodia formation through the regulation of cofilin activity by Ca^{2+} [5]. In our study, we observed close localization of GFP-PMCA4b and mCherry-cofilin in A375-GFP-PMCA4b cells at the lamellipodia compared to cofilin localization at the protrusions of the parental and GFP-PMCA4b-DE transfected A375 cells. This may indicate the importance of PMCA4b in localizing cofilin to the leading edge where it may inhibit cofilin activity by reducing nearby Ca^{2+} levels resulting in a less motile cell type. While PMCA4b expression at the leading edge induced cofilin relocalization, several studies indicate that inhibition or knocking down cofilin can reduce cell polarity [20,71]. Interestingly, cofilin can regulate store operated Ca^{2+} entry in platelets through dynamic F-actin remodeling [72]. Further studies are needed to test if cofilin affects PMCA4b localization, and consequently the establishment of the Ca^{2+} gradient in migrating cells.

An interesting point is that PMCA4b expression could affect matrix metalloproteinase (MMP) production because these proteins are calcium dependent zinc-containing endopeptidases, which are essential for the degradation of the extracellular matrix and hence they affect cell migration and metastasis. In correlation with this assumption many previous studies showed the involvement of Ca^{2+} channels in the production of MMPs. For example, an increase in the expression of the Ca^{2+} channels TRPV2 in prostate cancer, and TRPM8 in squamous cell carcinoma resulted in induction of MMP-2 and MMP-9, respectively [73].

It has been reported that localization of PMCA4b at the leading edge was responsible for maintaining the Ca^{2+} gradient and directional movement of HUVEC cells. The PMCA4b mediated high Ca^{2+} pumping rate resulted in a low basal Ca^{2+} level at the cell front that enabled effective local Ca^{2+} signaling by the STIM1/ORAI Ca^{2+} entry channels [48]. In line with these findings, we found that high PMCA4b levels at the cell front also resulted in a typical front-to-rear Ca^{2+} concentration gradient in A375 cells. This change in Ca^{2+} distribution can contribute to the actin-based shape change and PMCA4b induced switch in motility style of BRAF mutant melanoma cells.

Several lines of evidence suggested that non-cancerous epithelial cells express PMCA4b at a relatively high level that is lost or downregulated during tumorigenesis. Using two different cell types, we demonstrate here that the loss of PMCA4b can have a profound effect on cell shape and cell culture morphology through F-actin rearrangement. On the one hand, we show how a highly aggressive melanoma cell type with neural crest origin changes cell shape through actin cytoskeleton remodeling in response to PMCA4b expression. On the other hand, we demonstrate how the ER+ luminal breast cancer epithelial cell-type MCF-7 responds to overexpression or silencing of PMCA4b. Our findings for the role of PMCA4b in actin cytoskeleton remodeling using these two distinct types of cells may give a greater perspective for future studies.

Our data suggest that PMCA4b plays a critical role in regulating cell polarity through F-actin rearrangement that could associate with less aggressive cancer cell phenotype. This is in good agreement with our previous findings that identified PMCA4b as a putative metastasis suppressor in A375 melanoma cells [26]. In general, metastasis suppressors are hardly druggable since they usually disappear from the system during tumor progression. Therefore, our aims have been finding drugs that could counteract with the downregulation of PMCA4b that may help finding ways to prevent metastasis. Besides the conventional drugs targeting BRAF, vemurafenib and dabrafenib, we identified the stress response kinase p38 MAPK as a potential target. We found that PMCA4b is degraded in BRAF mutant cells in a p38 MAPK dependent manner and that specific inhibition of this kinase prevented its degradation [27]. Importantly, p38 inhibitors enhanced PMCA4b expression without affecting cell growth that could make them eligible to fulfill the requirements for the recently proposed group of drugs, "migrastatics" [74]. Another possible option could be the use of the epigenetic drugs, the HDAC inhibitors vorinostat (Zolinza) and/or valproic acid alone or in combination that has been shown to increase PMCA4b expression in a variety of melanoma and breast cancer tumor cells including A375 and MCF-7 [28,29,75].

5. Conclusions

Our findings indicate that both the expression and proper trafficking are essential for the antimigratory activity of the PMCA4b pump in BRAF mutant melanoma cells. We suggest that polarized distribution of a fully functional PMCA4b can generate and maintain a front-to-rear increasing Ca^{2+} concentration gradient, and induce redistribution of polymerized F-actin and cofilin from the dynamic protrusions to the leading edge, formation of stable stress fibers, increased cell–cell connections, and decreased vinculin expression resulting in a slow motility melanoma cell type. Manipulating PMCA4b abundance also induced characteristic redistribution of actin filaments in the MCF-7 breast cancer cells suggesting that downregulation of PMCA4b expression during carcinogenesis may contribute to aberrant cancer cell migration and tumor metastasis in different cancer types.

Supplementary Materials: The following are available online at https://www.mdpi.com/2072-6694/13/6/1354/s1, Figure S1: Complementary to Figure 1, Figure S2: Complementary to Figure 1D, Figure S3: PMCA4b activity is necessary for the stress fiber formation in A375 melanoma cells, Figure S4: Complementary to Figure 5, Figure S5: Complementary to Figure 6, Figure S6: Complementary to Figure 7, Figure S7: Complementary to Figure 8, Figure S8: Uncropped Western Blot Figures, Videos S1: Complementary to Figure 1D, Videos S2: Complementary to Figure 1D, Video S3: Complementary to Figure 2B, Video S4: Complementary to Figure 4D, Video S5: Complementary to Figure 7B, Video S6 Complementary to Figure 7.

Author Contributions: Conceptualization, R.N., R.P. and A.E.; methodology, R.N., R.P., A.I., Z.H., B.J., S.T., K.V. and L.H. (Luca Hegedűs); formal analysis, R.N., R.P. and A.I.; investigation, R.N., R.P., K.S. and A.E.; resources, L.H. (László Homolya), K.S.; writing—original draft preparation, R.N. and A.E.; writing—review and editing, R.P., L.H. (László Homolya), L.H. (Luca Hegedűs), K.S. and A.E.; Supervision, A.E.; funding acquisition, A.E, L.H. (László Homolya) and K.S. All authors have read and agreed to the published version of the manuscript.

Funding: Hungarian Scientific Research Development and Innovation Office (grant numbers: NKFIH K119223 and K135811 to A.E., K128123 to L.H. (László Homolya)), the Higher Education Institutional Excellence Program of the Ministry of Human Capacities in Hungary, within the framework of the Molecular Biology thematic program of the Semmelweis University (FIKP to A.E.), Stipendium Hungaricum Fellowships (awarded to R.N.), and VEKOP-2.3.3-15-2016-00007 (to K.S.).

Institutional Review Board Statement: Not applicable.

Informed Consent Statement: Not applicable.

Data Availability Statement: Data is contained within the article or Supplementary Material.

Acknowledgments: The authors thank John T. Penniston for his continuous support. We thank members of our laboratory for their help and encouragement.

Conflicts of Interest: The authors declare no conflict of interest.

References

1. Adler, N.R.; Haydon, A.; McLean, C.A.; Kelly, J.W.; Mar, V.J. Metastatic pathways in patients with cutaneous melanoma. *Pigment. Cell Melanoma Res.* **2017**, *30*, 13–27. [CrossRef]
2. Yamaguchi, H.; Condeelis, J. Regulation of the actin cytoskeleton in cancer cell migration and invasion. *Biochim. Biophys. Acta* **2007**, *1773*, 642–652. [CrossRef]
3. Xu, X.Z.; Garcia, M.V.; Li, T.Y.; Khor, L.Y.; Gajapathy, R.S.; Spittle, C.; Weed, S.; Lessin, S.R.; Wu, H. Cytoskeleton alterations in melanoma: Aberrant expression of cortactin, an actin-binding adapter protein, correlates with melanocytic tumor progression. *Mod. Pathol.* **2010**, *23*, 187–196. [CrossRef]
4. Pawlak, G.; Helfman, D.M. Cytoskeletal changes in cell transformation and tumorigenesis. *Curr. Opin. Genet. Dev.* **2001**, *11*, 41–47. [CrossRef]
5. Maus, M.; Medgyesi, D.; Kiss, E.; Schneider, A.E.; Enyedi, A.; Szilágyi, N.; Matkó, J.; Sármay, G. B cell receptor-induced Ca^{2+} mobilization mediates F-actin rearrangements and is indispensable for adhesion and spreading of B lymphocytes. *J. Leukoc. Biol.* **2013**, *93*, 537–547. [CrossRef]
6. Lee, J.; Ishihara, A.; Oxford, G.; Johnson, B.; Jacobson, K. Regulation of cell movement is mediated by stretch-activated calcium channels. *Nature* **1999**, *400*, 382–386. [CrossRef]
7. Brundage, R.A.; Fogarty, K.E.; Tuft, R.A.; Fay, F.S. Calcium gradients underlying polarization and chemotaxis of eosinophils. *Science* **1991**, *254*, 703–706. [CrossRef]
8. Martin-Romero, F.J.; Lopez-Guerrero, A.M.; Pascual-Caro, C.; Pozo-Guisado, E. The Interplay between Cytoskeleton and Calcium Dynamics. In *Cytoskeleton—Structure, Dynamics, Function and Disease*; Jimenez-Lopez, J.C., Ed.; IntechOpen: Rijeka, Croatia, 2017; Chapter 4; pp. 73–88.
9. Tsai, F.C.; Kuo, G.H.; Chang, S.W.; Tsai, P.J. Ca^{2+} signaling in cytoskeletal reorganization, cell migration, and cancer metastasis. *Biomed. Res. Int.* **2015**, *2015*, 409245. [CrossRef]
10. Marchi, S.; Pinton, P. Alterations of calcium homeostasis in cancer cells. *Curr. Opin. Pharmacol.* **2016**, *29*, 1–6. [CrossRef]
11. Déliot, N.; Constantin, B. Plasma membrane calcium channels in cancer: Alterations and consequences for cell proliferation and migration. *Biochim. Biophys. Acta* **2015**, *1848*, 2512–2522. [CrossRef]
12. Maiques, O.; Barceló, C.; Panosa, A.; Pijuan, J.; Orgaz, J.L.; Rodriguez-Hernandez, I.; Matas-Nadal, C.; Tell, G.; Vilella, R.; Fabra, A.; et al. T-type calcium channels drive migration/invasion in BRAFV600E melanoma cells through Snail1. *Pigment. Cell Melanoma Res.* **2018**, *31*, 484–495. [CrossRef]
13. Wang, G.; Cao, R.; Qian, K.; Peng, T.; Yuan, L.; Chen, L.; Cheng, S.; Xiong, Y.; Ju, L.; Wang, X.; et al. TRPM8 inhibition regulates the proliferation, migration and ROS metabolism of bladder cancer cells. *OncoTargets Ther.* **2020**, *13*, 8825–8835. [CrossRef]
14. Monet, M.; Lehen'kyi, V.; Gackiere, F.; Firlej, V.; Vandenberghe, M.; Roudbaraki, M.; Gkika, D.; Pourtier, A.; Bidaux, G.; Slomianny, C.; et al. Role of cationic channel TRPV2 in promoting prostate cancer migration and progression to androgen resistance. *Cancer Res.* **2010**, *70*, 1225–1235. [CrossRef]
15. Li, X.; Wang, J. Mechanical tumor microenvironment and transduction: Cytoskeleton mediates cancer cell invasion and metastasis. *Int. J. Biol. Sci.* **2020**, *16*, 2014–2028. [CrossRef]
16. Shimizu, T.; Owsianik, G.; Freichel, M.; Flockerzi, V.; Nilius, B.; Vennekens, R. TRPM4 regulates migration of mast cells in mice. *Cell Calcium* **2009**, *45*, 226–232. [CrossRef]

17. Lee, W.H.; Choong, L.Y.; Mon, N.N.; Lu, S.; Lin, Q.; Pang, B.; Yan, B.; Krishna, V.S.; Singh, H.; Tan, T.Z.; et al. TRPV4 regulates breast cancer cell extravasation, stiffness and actin cortex. *Sci. Rep.* **2016**, *6*, 27903. [CrossRef]
18. Pinotsis, N.; Zielinska, K.; Babuta, M.; Arolas, J.L.; Kostan, J.; Khan, M.B.; Schreiner, C.; Salmazo, A.; Ciccarelli, L.; Puchinger, M.; et al. Calcium modulates the domain flexibility and function of an α-actinin similar to the ancestral α-actinin. *Proc. Natl. Acad. Sci. USA* **2020**, *117*, 22101–22112. [CrossRef]
19. Oser, M.; Condeelis, J. The cofilin activity cycle in lamellipodia and invadopodia. *J. Cell Biochem.* **2009**, *108*, 1252–1262. [CrossRef]
20. Bravo-Cordero, J.J.; Magalhaes, M.A.; Eddy, R.J.; Hodgson, L.; Condeelis, J. Functions of cofilin in cell locomotion and invasion. *Nat. Rev. Mol. Cell Biol.* **2013**, *14*, 405–415. [CrossRef]
21. Wang, Y.; Shibasaki, F.; Mizuno, K. Calcium signal-induced cofilin dephosphorylation is mediated by Slingshot via calcineurin. *J. Biol. Chem.* **2005**, *280*, 12683–12689. [CrossRef]
22. Padányi, R.; Pászty, K.; Hegedűs, L.; Varga, K.; Papp, B.; Penniston, J.T.; Enyedi, Á. Multifaceted plasma membrane Ca(2+) pumps: From structure to intracellular Ca(2+) handling and cancer. *Biochim. Biophys. Acta* **2016**, *1863*, 1351–1363. [CrossRef]
23. Jeong, J.; Van Houten, J.N.; Dann, P.; Kim, W.; Sullivan, C.; Yu, H.; Liotta, L.; Espina, V.; Stern, D.F.; Friedman, P.A.; et al. PMCA2 regulates HER2 protein kinase localization and signaling and promotes HER2-mediated breast cancer. *Proc. Natl. Acad. Sci. USA* **2016**, *113*, E282–E290. [CrossRef]
24. Rüschoff, J.H.; Brandenburger, T.; Strehler, E.E.; Filoteo, A.G.; Heinmöller, E.; Aumüller, G.; Wilhelm, B. Plasma membrane calcium ATPase expression in human colon multistep carcinogenesis. *Cancer Investig.* **2012**, *30*, 251–257. [CrossRef]
25. Dhanasekaran, S.M.; Barrette, T.R.; Ghosh, D.; Shah, R.; Varambally, S.; Kurachi, K.; Pienta, K.J.; Rubin, M.A.; Chinnaiyan, A.M. Delineation of prognostic biomarkers in prostate cancer. *Nature* **2001**, *412*, 822–826. [CrossRef]
26. Hegedüs, L.; Garay, T.; Molnár, E.; Varga, K.; Bilecz, Á.; Török, S.; Padányi, R.; Pászty, K.; Wolf, M.; Grusch, M.; et al. The plasma membrane Ca^{2+} pump PMCA4b inhibits the migratory and metastatic activity of BRAF mutant melanoma cells. *Int. J. Cancer* **2017**, *140*, 2758–2770. [CrossRef]
27. Naffa, R.; Vogel, L.; Hegedűs, L.; Pászty, K.; Tóth, S.; Kelemen, K.; Singh, N.; Reményi, A.; Kállay, E.; Cserepes, M.; et al. P38 MAPK promotes migration and metastatic activity of BRAF mutant melanoma cells by inducing degradation of PMCA4b. *Cells* **2020**, *9*, 1209. [CrossRef]
28. Hegedüs, L.; Padányi, R.; Molnár, J.; Pászty, K.; Varga, K.; Kenessey, I.; Sárközy, E.; Wolf, M.; Grusch, M.; Hegyi, Z.; et al. Histone deacetylase inhibitor treatment increases the expression of the plasma membrane Ca^{2+} pump PMCA4b and inhibits the migration of melanoma cells independent of ERK. *Front. Oncol.* **2017**, *7*, 95. [CrossRef]
29. Varga, K.; Pászty, K.; Padányi, R.; Hegedűs, L.; Brouland, J.-P.; Papp, B.; Enyedi, A. Histone deacetylase inhibitor- and PMA-induced upregulation of PMCA4b enhances Ca^{2+} clearance from MCF-7 breast cancer cells. *Cell Calcium.* **2014**, *55*, 78–92. [CrossRef]
30. Ribiczey, P.; Tordai, A.; Andrikovics, H.; Filoteo, A.G.; Penniston, J.T.; Enouf, J.; Enyedi, A.; Papp, B.; Kovács, T. Isoform-specific up-regulation of plasma membrane Ca^{2+} ATPase expression during colon and gastric cancer cell differentiation. *Cell Calcium.* **2007**, *42*, 590–605. [CrossRef]
31. Antalffy, G.; Pászty, K.; Varga, K.; Hegedűs, L.; Enyedi, Á.; Padányi, R. A C-terminal di-leucine motif controls plasma membrane expression of PMCA4b. *Biochim. Biophys. Acta* **2013**, *1833*, 2561–2572. [CrossRef]
32. Penniston, J.T.; Padányi, R.; Pászty, K.; Varga, K.; Hegedus, L.; Enyedi, A. Apart from its known function, the plasma membrane Ca^{2+}ATPase can regulate Ca^{2+} signaling by controlling phosphatidylinositol 4,5-bisphosphate levels. *J. Cell Sci.* **2014**, *127*, 784. [CrossRef]
33. Mao, T.; O'Connor, D.H.; Scheuss, V.; Nakai, J.; Svoboda, K. Characterization and subcellular targeting of GCaMP-type genetically-encoded calcium indicators. *PLoS ONE* **2008**, *3*, e1796. [CrossRef]
34. Taylor, M.J.; Perrais, D.; Merrifield, C.J. A high precision survey of the molecular dynamics of mammalian clathrin-mediated endocytosis. *PLoS Biol.* **2011**, *9*, e1000604. [CrossRef]
35. Zhao, Y.; Araki, S.; Wu, J.; Teramoto, T.; Chang, Y.F.; Nakano, M.; Abdelfattah, A.S.; Fujiwara, M.; Ishihara, T.; Nagai, T.; et al. An expanded palette of genetically encoded Ca^{2+} indicators. *Science* **2011**, *333*, 1888–1891. [CrossRef]
36. Traynor, D.; Kay, R.R. Possible roles of the endocytic cycle in cell motility. *J. Cell Sci.* **2007**, *120*, 2318–2327. [CrossRef]
37. Wilson, B.J.; Allen, J.L.; Caswell, P.T. Vesicle trafficking pathways that direct cell migration in 3D matrices and in vivo. *Traffic* **2018**, *19*, 899–909. [CrossRef]
38. Sigismund, S.; Confalonieri, S.; Ciliberto, A.; Polo, S.; Scita, G.; Di Fiore, P.P. Endocytosis and signaling: Cell logistics shape the eukaryotic cell plan. *Physiol. Rev.* **2012**, *92*, 273–366. [CrossRef]
39. Svitkina, T. The Actin cytoskeleton and actin-based motility. *Cold Spring Harb. Perspect. Biol.* **2018**, *10*, a018267. [CrossRef]
40. Franco, S.J.; Rodgers, M.A.; Perrin, B.J.; Han, J.; Bennin, D.A.; Critchley, D.R.; Huttenlocher, A. Calpain-mediated proteolysis of talin regulates adhesion dynamics. *Nat. Cell Biol.* **2004**, *6*, 977–983. [CrossRef]
41. Giannone, G.; Rondé, P.; Gaire, M.; Beaudouin, J.; Haiech, J.; Ellenberg, J.; Takeda, K. Calcium rises locally trigger focal adhesion disassembly and enhance residency of focal adhesion kinase at focal adhesions. *J. Biol. Chem.* **2004**, *279*, 28715–28723. [CrossRef]
42. Pászty, K.; Caride, A.J.; Bajzer, Ž.; Offord, C.P.; Padányi, R.; Hegedűs, L.; Varga, K.; Strehler, E.E.; Enyedi, A. Plasma membrane Ca^{2+}-ATPases can shape the pattern of Ca^{2+} transients induced by store-operated Ca^{2+} entry. *Sci. Signal.* **2015**, *8*, ra19. [CrossRef]
43. Koskinen, M.; Hotulainen, P. Measuring F-actin properties in dendritic spines. *Front. Neuroanat.* **2014**, *8*, 74. [CrossRef]
44. Hartzell, C.A.; Jankowska, K.I.; Burkhardt, J.K.; Lewis, R.S. Calcium influx through CRAC channels controls actin organization and dynamics at the immune synapse. *eLife* **2016**, *5*, e14850. [CrossRef]

45. Yoneda, M.; Nishizaki, T.; Tasaka, K.; Kurachi, H.; Miyake, A.; Murata, Y. Changes in actin network during calcium-induced exocytosis in permeabilized GH3 cells: Calcium directly regulates F-actin disassembly. *J. Endocrinol.* **2000**, *166*, 677–687. [CrossRef]
46. Wales, P.; Schuberth, C.E.; Aufschnaiter, R.; Fels, J.; García-Aguilar, I.; Janning, A.; Dlugos, C.P.; Schäfer-Herte, M.; Klingner, C.; Wälte, M.; et al. Calcium-mediated actin reset (CaAR) mediates acute cell adaptations. *eLife* **2016**, *5*, e19850. [CrossRef]
47. Moon, A.; Drubin, D.G. The ADF/cofilin proteins: Stimulus-responsive modulators of actin dynamics. *Mol. Biol. Cell* **1995**, *6*, 1423–1431. [CrossRef]
48. Tsai, F.C.; Seki, A.; Yang, H.W.; Hayer, A.; Carrasco, S.; Malmersjö, S.; Meyer, T. A polarized Ca^{2+}, diacylglycerol and STIM1 signalling system regulates directed cell migration. *Nat. Cell Biol.* **2014**, *16*, 133–144. [CrossRef]
49. Hsu, M.-Y.; Meier, F.; Herlyn, M. Melanoma development and progression: A conspiracy between tumor and host. *Differentiation* **2002**, *70*, 522–536. [CrossRef]
50. Elsholz, F.; Harteneck, C.; Muller, W.; Friedland, K. Calcium—A central regulator of keratinocyte differentiation in health and disease. *Eur. J. Dermatol.* **2014**, *24*, 650–661. [CrossRef]
51. Kodet, O.; Lacina, L.; Krejčí, E.; Dvořánková, B.; Grim, M.; Štork, J.; Kodetová, D.; Vlček, Č.; Šáchová, J.; Kolář, M.; et al. Melanoma cells influence the differentiation pattern of human epidermal keratinocytes. *Mol. Cancer* **2015**, *14*, 1. [CrossRef]
52. Škalamera, D.; Stevenson, A.J.; Ehmann, A.; Ainger, S.A.; Lanagan, C.; Sturm, R.A.; Gabrielli, B. Melanoma mutations modify melanocyte dynamics in co-culture with keratinocytes or fibroblasts. *J. Cell Sci.* **2019**, *132*, jcs234716. [CrossRef]
53. Parri, M.; Taddei, M.L.; Bianchini, F.; Calorini, L.; Chiarugi, P. EphA2 reexpression prompts invasion of melanoma cells shifting from mesenchymal to amoeboid-like motility style. *Cancer Res.* **2009**, *69*, 2072–2081. [CrossRef]
54. Taddei, M.L.; Giannoni, E.; Morandi, A.; Ippolito, L.; Ramazzotti, M.; Callari, M.; Gandellini, P.; Chiarugi, P. Mesenchymal to amoeboid transition is associated with stem-like features of melanoma cells. *Cell Commun. Signal.* **2014**, *12*, 24. [CrossRef]
55. Friedl, P.; Wolf, K. Plasticity of cell migration: A multiscale tuning model. *J. Cell Biol.* **2010**, *188*, 11–19. [CrossRef]
56. Anand, P.; Harper, A.G.S. Human platelets use a cytosolic Ca^{2+} nanodomain to activate Ca^{2+}-dependent shape change independently of platelet aggregation. *Cell Calcium.* **2020**, *90*, 102248. [CrossRef]
57. Mermelstein, C.S.; Rebello, M.I.; Amaral, L.M.; Costa, M.L. Changes in cell shape, cytoskeletal proteins and adhesion sites of cultured cells after extracellular Ca^{2+} chelation. *Braz. J. Med. Biol. Res.* **2003**, *36*, 1111–1116. [CrossRef]
58. Moore, T.M.; Brough, G.H.; Babal, P.; Kelly, J.J.; Li, M.; Stevens, T. Store-operated calcium entry promotes shape change in pulmonary endothelial cells expressing Trp1. *Am. J. Physiol.* **1998**, *275*, L574–L582. [CrossRef]
59. Tojkander, S.; Gateva, G.; Lappalainen, P. Actin stress fibers-assembly, dynamics and biological roles. *J. Cell Sci.* **2012**, *125*, 1855–1864. [CrossRef]
60. Pellegrin, S.; Mellor, H. Actin stress fibres. *J. Cell Sci.* **2007**, *120*, 3491–3499. [CrossRef]
61. Yang, S.; Zhang, J.J.; Huang, X.Y. Orai1 and STIM1 are critical for breast tumor cell migration and metastasis. *Cancer Cell* **2009**, *15*, 124–134. [CrossRef]
62. Huang, H.K.; Lin, Y.H.; Chang, H.A.; Lai, Y.S.; Chen, Y.C.; Huang, S.C.; Chou, C.Y.; Chiu, W.T. Chemoresistant ovarian cancer enhances its migration abilities by increasing store-operated Ca^{2+} entry-mediated turnover of focal adhesions. *J. Biomed. Sci.* **2020**, *27*, 36. [CrossRef]
63. Go, C.K.; Hooper, R.; Aronson, M.R.; Schultz, B.; Cangoz, T.; Nemani, N.; Zhang, Y.; Madesh, M.; Soboloff, J. The Ca^{2+} export pump PMCA clears near-membrane Ca^{2+} to facilitate store-operated Ca^{2+} entry and NFAT activation. *Sci. Signal.* **2019**, *12*, eaaw2627. [CrossRef]
64. Wang, Y.; Kuramitsu, Y.; Ueno, T.; Suzuki, N.; Yoshino, S.; Iizuka, N.; Zhang, X.; Akada, J.; Oka, M.; Nakamura, K. Proteomic differential display identifies upregulated vinculin as a possible biomarker of pancreatic cancer. *Oncol. Rep.* **2012**, *28*, 1845–1850. [CrossRef]
65. Zhang, Y.; Yan, J.; Xu, H.; Yang, Y.; Li, W.; Wu, H.; Liu, C. Extremely low frequency electromagnetic fields promote mesenchymal stem cell migration by increasing intracellular Ca^{2+} and activating the FAK/Rho GTPases signaling pathways in vitro. *Stem. Cell Res. Ther.* **2018**, *9*, 143. [CrossRef]
66. Li, F.; Abuarab, N.; Sivaprasadarao, A. Reciprocal regulation of actin cytoskeleton remodelling and cell migration by Ca^{2+} and Zn^{2+}: Role of TRPM2 channels. *J. Cell Sci.* **2016**, *129*, 2016–2029. [CrossRef]
67. Lopez-Guerrero, A.M.; Espinosa-Bermejo, N.; Sanchez-Lopez, I.; Macartney, T.; Pascual-Caro, C.; Orantos-Aguilera, Y.; Rodriguez-Ruiz, L.; Perez-Oliva, A.B.; Mulero, V.; Pozo-Guisado, E.; et al. RAC1-dependent ORAI1 translocation to the leading edge supports lamellipodia formation and directional persistence. *Sci. Rep.* **2020**, *10*, 6580. [CrossRef]
68. Ballestrem, C.; Wehrle-Haller, B.; Hinz, B.; Imhof, B.A. Actin-dependent lamellipodia formation and microtubule-dependent tail retraction control-directed cell migration. *Mol. Biol. Cell* **2000**, *11*, 2999–3012. [CrossRef]
69. Zabe, M.; Dean, W.L. Plasma membrane Ca (2+)-ATPase associates with the cytoskeleton in activated platelets through a PDZ-binding domain. *J. Biol. Chem.* **2001**, *276*, 14704–14709. [CrossRef]
70. Dalghi, M.G.; Ferreira-Gomes, M.; Rossi, J.P. Regulation of the plasma membrane calcium ATPases by the actin cytoskeleton. *Biochem. Biophys. Res. Commun.* **2018**, *506*, 347–354. [CrossRef]
71. Lee, S.; Kumar, S. Cofilin is required for polarization of tension in stress fiber networks during migration. *J. Cell Sci.* **2020**, *133*. [CrossRef]
72. Redondo, P.C.; Harper, M.T.; Rosado, J.A.; Sage, S.O. A role for cofilin in the activation of store-operated calcium entry by de novo conformational coupling in human platelets. *Blood* **2006**, *107*, 973–979. [CrossRef]

73. Pla, A.F.; Gkika, D. Emerging role of TRP channels in cell migration: From tumor vascularization to metastasis. *Front. Physiol.* **2013**, *4*, 311. [CrossRef]
74. Ju, R.J.; Stehbens, S.J.; Haass, N.K. The role of melanoma cell-stroma interaction in cell motility, invasion, and metastasis. *Front. Med.* **2018**, *5*, 307. [CrossRef]
75. Varga, K.; Hollósi, A.; Pászty, K.; Hegedűs, L.; Szakács, G.; Tímár, J.; Papp, B.; Enyedi, Á.; Padányi, R. Expression of calcium pumps is differentially regulated by histone deacetylase inhibitors and estrogen receptor alpha in breast cancer cells. *BMC Cancer* **2018**, *18*, 1029. [CrossRef]

cancers

MDPI

Article

Targeting PVT1 Exon 9 Re-Expresses Claudin 4 Protein and Inhibits Migration by Claudin—Low Triple Negative Breast Cancer Cells

Fayola Levine [1] and Olorunseun O. Ogunwobi [1,2,3,*]

1 Department of Biological Sciences, Hunter College of The City University of New York, New York, NY 10065, USA; FAYOLA.LEVINE20@myhunter.cuny.edu
2 The Graduate Center Departments of Biology and Biochemistry, The City University of New York, New York, NY 10016, USA
3 Joan and Sanford I. Weill Department of Medicine, Weill Cornell Medicine, Cornell University, New York, NY 10021, USA
* Correspondence: ogunwobi@genectr.hunter.cuny.edu

Simple Summary: Triple negative breast cancer accounts for 10–15% of all breast cancers. Specific molecular characteristics have led to the identification of six subtypes of triple negative breast cancer, with one in particular being claudin-low. PVT1, a non-protein coding gene, has been demonstrated to play an oncogenic role in various cancers. Specifically, PVT1 exon 9 has been shown to have oncogenic capability. In this study, we aimed to assess the role of PVT1 exon 9 in triple negative breast cancer cells. We have observed that siRNA targeting of PVT1 exon 9 in claudin-low triple negative breast cancer cells resulted in re-expression of claudin 4 protein, and inhibition of migration. These findings indicate that PVT1 exon 9 regulates claudin 4 expression and migration in triple negative breast cancer cells.

Citation: Levine, F.; Ogunwobi, O.O. Targeting PVT1 Exon 9 Re-Expresses Claudin 4 Protein and Inhibits Migration by Claudin—Low Triple Negative Breast Cancer Cells. *Cancers* **2021**, *13*, 1046. https://doi.org/10.3390/cancers13051046

Academic Editors: José I. López and Ildefonso M. de la Fuente

Received: 12 February 2021
Accepted: 24 February 2021
Published: 2 March 2021

Abstract: PVT1 is a long non-coding RNA transcribed from a gene located at the 8q24 chromosomal region that has been implicated in multiple cancers including breast cancer (BC). Amplification of the 8q24 chromosomal region is a common event in BC and is associated with poor clinical outcomes. Claudin-low (CL) triple negative breast cancer (TNBC) is a subtype of BC with a particularly dismal outcome. We assessed PVT1 exon 9 expression in the T47D estrogen receptor positive BC cell line, and in the MDA MB 468 and MDA MB 231 TNBC cell lines, followed by the assessment of the expression of claudins 1, 3, 4 and 7, in MDA MB 468 and MDA MB 231 (TNBC) cells. We found that MDA MB 231 TNBC cells significantly express less claudin 1, 3, 4, and 7 than MDA MB 468 TNBC cells. PVT1 exon 9 is significantly upregulated in MDA MB 231 CL TNBC cells, and significantly downregulated in MDA MB 468 claudin high (CH) TNBC cells, in comparison to T47D estrogen receptor positive BC cells. We then analyzed the functional consequences of siRNA targeting of PVT1 exon 9 expression in the MDA MB 231 CL TNBC cells. Notably, siRNA targeting of PVT1 exon 9 expression in the MDA MB 231 CL TNBC cells led to a significant reduction in migration and the re-expression of claudin 4. Taken together, our data indicate that PVT1 exon 9 regulates claudin 4 expression and migration in CL TNBC cells, and may have clinical implications in CL TNBC.

Keywords: PVT1; triple negative breast cancer; claudin-low

1. Introduction

Breast cancer is extraordinarily common worldwide. In 2018, breast cancer was the fifth leading cause of cancer related deaths globally [1]. In the United States, breast cancer is the second leading cause of cancer mortality in women [2,3] with a 13% lifetime risk of diagnosis and a 2.6% risk of death. An estimated 276,480 new cases of invasive breast cancer will be diagnosed in U.S women, of which close to 42,170 women will die from the disease in 2020 alone [2]. Breast cancer is classified according to the expression of three specific molecular markers; estrogen receptor (ER), progesterone receptor (PR) and human epithelial growth factor receptor 2 (EGFR2/HER2) [4]. Loss of these three receptors characterizes an intrinsic subtype of breast cancer called triple negative breast cancer (TNBC). TNBC accounts for 10–15% of all breast cancer cases and is typically more aggressive with poor clinical outcomes [5–7]. Based on molecular characteristics, TNBC can be further subdivided into additional subtypes. Based on gene expression profile study, six TNBC subtypes have been identified, each of which differs in histopathological features and their response to chemotherapy. These subtypes include basal-like 1 (BL1), basal-like 2 (BL2), immunomodulatory (IM), mesenchymal (M), mesenchymal stem-like/claudin low (MSL/CL), and luminal androgen receptor (LAR) [8]. Claudin–low triple negative breast cancer (CL TNBC) makes up 7–14% of all invasive breast cancers. Moreover, CL TNBC is associated with poor prognosis, and some studies report that they exhibit chemo-resistance [9–11].

Plasmacytoma variant translocation 1 (PVT1) is a long noncoding RNA (lncRNA) that is transcribed from a gene located at the 8q24 chromosomal region and has been demonstrated to play an oncogenic role in multiple cancers including breast cancer [12]. The PVT1 gene is located approximately 53 kb downstream of the oncogene MYC [13], and contains several exons, including exons 1A, 1B, 1C 2, 3A, 3B, 4A, 4B, 5, 6, 7, 8, 9 and 10 [14] (Figure 1). Additionally, PVT1 encodes multiple alternatively spliced lncRNAs as well as six annotated microRNAs [15]. Alternative splicing is a tissue and cell specific mechanism, in which a diverse amount of mRNA isoforms is generated. Aberrant alternative splicing of pre-mRNAs is one of the characteristics of cancer [16]. Previously, we have reported that PVT1 exon 9 is overexpressed in prostate cancer cell lines, especially in aggressively tumorigenic prostate cancer cell lines derived from men of African Ancestry [17] Furthermore, we have observed that PVT1 exon 9 is significantly overexpressed in prostate cancer tissue relative to both normal prostate tissue and benign prostatic hyperplasia [18]. Also, we have demonstrated that PVT1 exon 9 induces malignant transformation and resistance to androgen deprivation therapy in prostate epithelial cells [19]. Although studies have shown that PVT1 splice variants are also overexpressed in breast cancer [20], and play a role in cancer progression [21,22], the underlying mechanisms by which these transcripts promote tumorigenicity is yet to be elucidated. PVT1 amplification is associated with many clinicopathological characteristics in breast cancer, including regulation of apoptosis [12], EMT [23] and metastasis [24]. Moreover, there is evidence that PVT1-derived transcripts also promote breast tumorigenesis [25–27]. Furthermore, PVT1 promotes breast tumorigenicity by modulating transcription factors that have been demonstrated to have oncogenic roles in cancer [28,29]. Further studies are required to elucidate PVT1's role in TNBC, and other cancers.

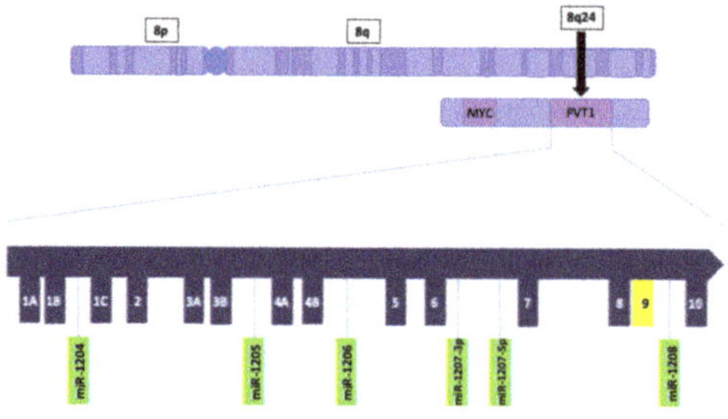

Figure 1. Schematic illustration of PVT1 showing exons and microRNAs. PVT1 is located downstream of the MYC gene on chromosome 8q24. PVT1 exon 9 is highlighted in yellow.

Epithelial and endothelial cell-cell adhesion are mediated through multifunctional complexes known as tight junctions (TJ). The involvement of TJs in cancer biology is associated with dysfunctional signal transduction pathways that regulate cell-cell interactions [30]. Many studies have demonstrated that dysregulation of TJ proteins disrupts normal physiological function, which could lead to pathological consequences such as cancer [31–35]. Claudins (CLDNs) are a family of TJ proteins that consists of 27 members [36]. These tetraspan proteins contain an amino and carboxyl-terminal cytoplasmic domain, as well as two extracellular loops which are critical for maintaining TJ function [37]. CLDNs are tissue- or cell-specific and most tissues or cells express CLDNs in various combinations, or a single CLDN [38]. CLDN functions primarily involve maintaining cellular polarity, signaling [39], maintaining paracellular permeability [40,41] or paracellular channel [36]. Various studies demonstrate compelling evidence of CLDNs and their role in tumorigenicity. Both loss of function, and gain of function, of CLDNs in multiple cancers are well documented [42–49]. Claudin 4 (CLDN4) has been reported to be involved in various biological processes [50–52]. Patients with CLDN4 overexpression develop various clinicopathological characteristics including high tumor grade, poor prognosis and shorter disease-free survival. Additionally, it was reported that there is an association between ER status and CLDN4 expression in which ER- tumors significantly overexpressed CLDN4 [53,54]. Studies have shown that CLDN4 can be a useful prognostic marker in breast cancer [55,56]. Basal-like carcinomas, compared to tumor groups of grades 1–3, overexpressed CLDN4, while tumors of grades 1 and 2, displayed decreased, or absent, expression of CLDN4 [57]. While enhanced expression of CLDN4 in luminal breast cancers was linked to poor clinical outcomes, contrastingly, overexpression of CLDN4 in TNBC was associated with favorable outcomes in which tumors that overexpressed CLDN4 displayed a less aggressive phenotype [53]. Consequently, further research into molecular mechanisms regulating CLDN4 expression in triple negative breast cancer is warranted.

In this study we were interested in assessing the regulatory role of PVT1 exon 9 in CL TNBC cell line MDA MB 231. This has not been previously investigated. Altogether, our study revealed for the first time that targeting PVT1 in MDA MB 231 CL TNBC cells, specifically PVT1 exon 9, inhibits migration and induces re-expression of CLDN 4 in these cells.

2. Results

2.1. Claudins 1, 3, 4 and 7 Are Downregulated in Claudin Low Triple Negative Breast Cancer (TNBC) Cells

To determine the expression profile of claudins in TNBC, we assessed the expression of claudins 1, 3, 4 and 7 in the MDA MB 231 claudin–low TNBC cell line and the MDA MB 468 claudin–high TNBC cell line. We observed that claudins 1, 3, 4 and 7 are significantly downregulated in the MDA MB 231 CL TNBC cell line (Figure 2).

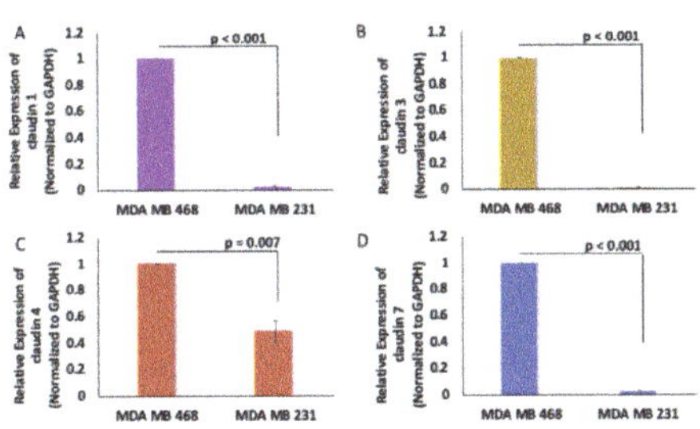

Figure 2. Claudins 1, 3, 4 and 7 expression in MDA MB 468 and MDA MB 231 cells. Claudin 1 (**A**), claudin 3 (**B**), claudin 4 (**C**) and claudin 7 (**D**) expression were assessed using real-time quantitative polymerase chain reaction in MDA MB 231 and MDA MB 468 cell lines; $N = 4$.

2.2. PVT1 Exon 9 Is Upregulated in Caludin- Low TNBC

PVT1 exon 9 is overexpressed in prostate cancer [17] and promotes tumorigenicity by increasing proliferation and migration [19]. To assess the expression of PVT1 exon 9 in TNBC cell lines, we performed total RNA extraction, cDNA synthesis and real-time quantitative polymerase chain reaction (RT-qPCR). We observed that PVT1 exon 9 is significantly overexpressed in the MDA MB 231 CL TNBC cell line when compared to the MDA MB 468 CH TNBC cell line (Figure 3 and Figure S5).

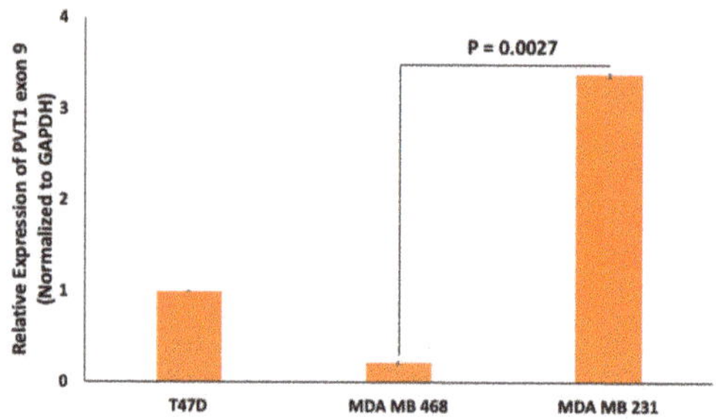

Figure 3. Comparison of PVT1 exon 9 expression in the T47D estrogen receptor positive BC cell line, MDA MB 468 CH TNBC cell line and the MDA MB 231 cell CL TNBC cell line. PVT1 exon 9 expression was assessed using RT-qPCR in the T47D estrogen receptor positive BC cell line, MDA MB 231 CL TNBC cell line and in the MDA MB 468 CH TNBC cell line. Expression was normalized against GAPDH. Data presented are from experiments performed in quadruplicates 6 separate times.

2.3. *PVT1 Exon 9 Regulates Migration in Caludin- Low TNBC Cells*

To determine if PVT1 exon 9 plays a role in the migration of MDA MB 231 CL TNBC cells, we performed wound healing migration assay. Short interfering RNA (SiRNA)-mediated knockdown of PVT1 exon 9 in MDA MB 231 CL TNBC cells significantly decreased migration, when compared to MDA MB 231 CL TNBC cells transfected with control scramble non-targeting SiRNA. Successful knockdown of PVT1 exon 9 in the MDA MB 231 CL TNBC cell line was verified by RT-qPCR. Assessment of the migratory capabilities of cells is based on the rate by which the wound closes. Pictures were taken at 0, 4, 24 and 28 h (Figure S4). We observed that MDA MB 231 CL TNBC cells in which PVT1 exon 9 was knocked down were significantly less migratory than MDA MB 231 CL TNBC control cells in which PVT1 exon 9 was not knocked down (Figure 4). This indicates that PVT1 is involved in regulating the migration of the MDA MB 231 CL TNBC cell line.

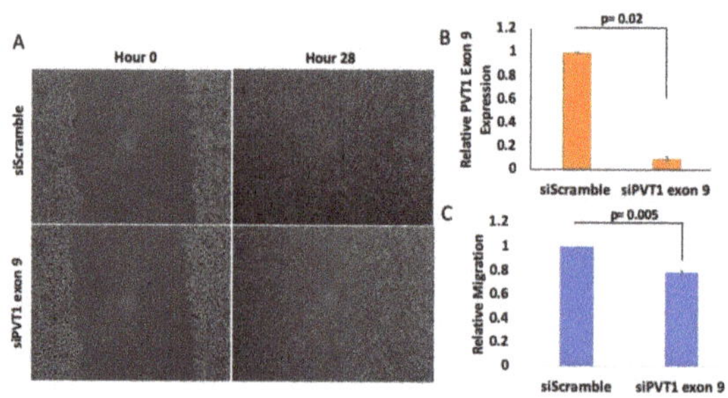

Figure 4. PVT1 regulates migration of MDA MB 231 CL TNBC cells. (**A**) Wound healing migration assays were performed with the MDA MB 231 CL TNBC cell line. MDA MB 231 cells were transfected once confluent. After 24 h, wounds were made and monitored between 0 h and 28 h. Images were taken at 10x magnification using Motic AE30 imaging software. (**B**) Knockdown of PVT1 exon 9 expression in the MDA MB 231 CL TNBC cell line at hour 0. Transfection of SiRNAs that specifically target PVT1 exon 9 was performed. Relative expression of PVT1 exon 9 in MDA MB 231 cells was assessed, based on data from 2 independent experiments. (**C**) Quantification of differences in migration, after 28 h, based on data from 3 independent experiments.

2.4. Targeting PVT1 Exon 9 Induces Re-Expression of Claudin 4 Protein in the Caludin- Low MDA MB 231 TNBC Cell Line

Given that the claudin–low MDA MB 231 TNBC cell line expresses significantly more PVT1 exon 9 than the claudin-high MDA MB 231 TNBC cell line, we hypothesized that PVT1 exon 9 may play a regulatory role in claudin expression in TNBC cells. To determine if PVT1 exon 9 plays a regulatory role in claudin expression, we examined the effect of SiRNA targeting of PVT1 exon 9 on messenger RNA (mRNA) and protein expression of claudins 1, 3, 4, and 7 in the claudin–low MDA MB 231 TNBC cell line. We observed that knockdown of PVT1 exon 9 in the claudin–low MDA MB 231 TNBC cell line does not significantly affect mRNA expression of claudins 1, 3, 4, and 7 when compared to control cells transfected with only control scramble non-targeting siRNAs (Figure 5). Similarly, we did not observe any significant change in protein expression of claudins 1, 3, and 7 when PVT1 exon 9 is knocked down. Interestingly, though, we observed re-expression of CLDN4 protein in the claudin–low MDA MB 231 TNBC cell line when PVT1 exon 9 expression is knocked down (Figure 6). These results suggest that PVT1 exon 9 is regulating claudin 4 protein expression in the claudin–low MDA MB 231 TNBC cell line.

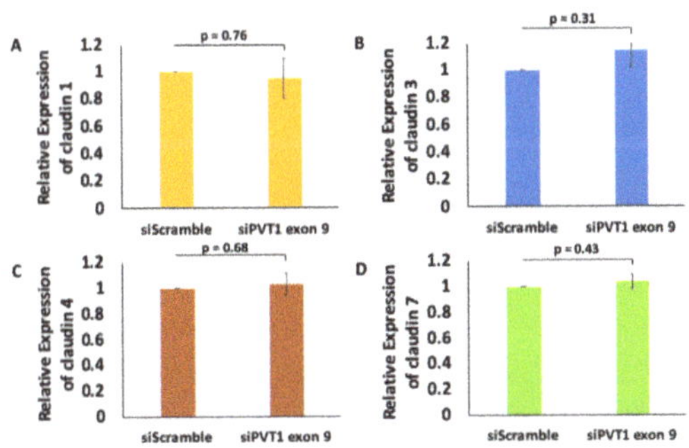

Figure 5. PVT1 exon 9 does not regulate mRNA expression of claudins 1, 3, 4, and 7 in MDA MB 231 cells. Claudin 1 (**A**), claudin 3 (**B**), claudin 4 (**C**) and claudin 7 (**D**) mRNA expression were assessed following knock down of PVT1 exon 9 in MDA MB 231 cell line. Data presented were normalized against GAPDH, and were obtained from three independent experiments.

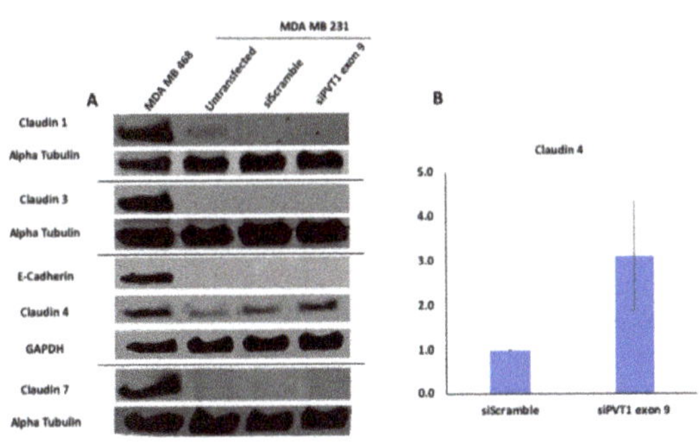

Figure 6. SiRNA targeting of PVT1 exon 9 induces claudin 4 protein re-expression in MDA MB 231 CL TNBC cells. (**A**) MDA MB 231 CL TNBC cells were transfected with PVT1 exon 9 specific siRNAs (siPVT1 exon 9) for 24 h. Western blotting was performed using specific antibodies against claudin 1, claudin 3, claudin 4, claudin 7 and E-Cadherin. SiRNA targeting of PVT1 exon 9 induced claudin 4 protein re-expression in MDA MB 231 CL TNBC cells in comparison to MDA MB 231 CL TNBC cells transfected with only control scramble non-targeting siRNA (siScramble). (**B**) Quantification of relative claudin 4 protein expression normalized to GAPDH protein expression; $N = 2$.

2.5. PVT1 Exon 9 and Epithelial-Mesenchymal Transition (EMT) in Claudin–Low TNBC Cells

Epithelial-mesenchymal transition (EMT) is a critical process that occurs in many types of cancers [52], and PVT1 has been shown to be involved in EMT induction [23]. To investigate if PVT1 exon 9 expression may be regulating EMT CL TNBC cells, western blotting was used to assess the protein expression of EMT markers (vimentin, E-cadherin, fibronectin and caveolin) in the MDA MB 231 CL TNBC cell line after siRNA knockdown of PVT1 exon 9. We observed no changes in EMT markers when PVT1 exon 9 is knocked down. However, our data indicates that EMT markers are more highly expressed in the MDA MB 231 CL TNBC cells in comparison to the MDA MB 468 CH TNBC cell line (Figure 7).

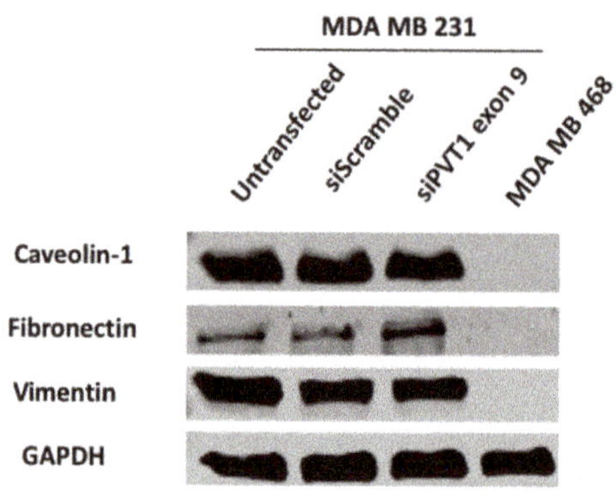

Figure 7. SiRNA targeting of PVT1 exon 9 does not affect EMT in MDA MB 231 CL TNBC cells. MDA MB 231 CL TNBC cells were transfected with PVT1 exon 9 specific siRNAs (siPVT1 exon 9) for 24 h. Western blotting was performed using specific antibodies against vimentin, caveolin and fibronectin. When compared to MDA MB 231 CL TNBC cells transfected with only control scramble non-targeting siRNA (siScramble). siRNA targeting of PVT1 exon 9 did not change the expression of EMT markers in MDA MB 231 CL TNBC cells, except for a slight increase in fibronectin.

3. Discussion

Although much progress has been made in breast cancer management and treatment, patients with TNBC continue to have poor prognosis [4,58,59] with CL TNBC having the worst outcome among the subtypes of BC. Presently, efficient treatment remains unavailable for CL TNBC. Consequently, efforts towards understanding the molecular mechanisms which regulate CLDN expression in TNBC is imperative, as it could potentially uncover novel opportunities for the development of effective therapeutic strategies. Many studies have already demonstrated PVT1's functional role in breast tumorigenesis [24]. However, there is a deficit of studies on the specific mechanisms by which PVT1 plays an important role in breast tumorigenicity.

Differential expression of PVT1 alternatively spliced transcripts in breast cancer have not been previously investigated. The purpose of this study was to investigate the relationship between breast tumorigenesis and PVT1. In this study, we demonstrated that PVT1 may play an important regulatory role in TNBC. Particularly, our data indicates that PVT1 transcripts containing exon 9 may regulate claudin expression in claudin–low MDA MB 231 TNBC cells. Our group previously reported that PVT1 exon 9 was differentially expressed in prostate cancer. More specifically, PVT1 exon 9 was overexpressed in prostate cancer tissue [18]. Furthermore, PVT1 exon 9 expression was reported to be significantly higher in prostate cancer cell lines with an aggressive phenotype [17]. The implications of this study suggest that alternatively spliced transcripts of PVT1, including transcripts containing PVT1 exon 9, may be associated with increased risk of cancer. In a *previous* study, overexpression of PVT1 exon 9 induced malignant transformation, increased cell proliferation, and migration in prostate epithelial cells [19]. These studies established an oncogenic role for PVT1 exon 9 in prostate cancer. However, to our knowledge, the significance of PVT1 exon 9 in breast cancer was not previously investigated.

Our data demonstrated that PVT1 exon 9 was significantly overexpressed in CL MDA MB 231 TNBC cells, and significantly under-expressed in CH MDA MB 468 TNBC cells, when compared to T47D cells. Based on these results, we used CL MDA MB 231 as a model for PVT1 exon 9 overexpression, and CH MDA MB 468 as a model for PVT1 exon 9 under-expression.

Cancer cell migration is a characteristic of metastasis and is associated with poor prognosis in cancer patients [24]. We observed that MDA MB 231 cells transfected with siPVT1 exon 9 were less migratory when compared to cells that were transfected using a scramble control. These results suggest that overexpression of PVT1 exon 9 increases the migratory capacity of the CL MDA MB 231 TNBC cells, and that loss of PVT1 exon 9 expression may have a protective role by making these cells less migratory. A more in-depth study in order to elucidate the underlying mechanisms by which this regulatory process occurs is necessary.

Differential expression of CLDNs are tissue and cell specific. Consequently, their functions are based on their localization and expression pattern. CLDNs 1, 3, 4 and 7 are among the most frequently dysregulated of the CLDN family members [30,60]. Based on other studies, we know that lncRNAs act as critical regulators of gene expression [61]. This is in keeping with our observation that PVT1 exon 9 is overexpressed in CL MDA MB 231 TNBC cells and under expressed in CH MDA MB 468 TNBC cells. We hypothesized that PVT1 may be regulating claudin expression either at a post-transcriptional level, a post-translational level, or indirectly. One way lncRNAs can serve as post transcriptional gene regulators is by forming ribonucleoprotein complexes via interacting directly with various RNA binding proteins (RBPs) to affect mRNA stability [62]. LncRNA PTOV1-AS1 interacts directly with hnRNPK in order to modulate HMOX1 expression [63]. At a post-translational level, lncRNAs can regulate protein stability by slowing down their degradation. LncRNA HOTAIR, for example, inhibited the interaction between AR and E3 ubiquitin ligase MDM2 after binding to AR [64].

Our confirmation of CLDNs 1, 3, 4 and 7 expression in MDA MB 231 cells is in keeping with previous reports. Knockdown of PVT1 exon 9 showed no significant changes in the expression of CLDN mRNA transcripts suggesting that in MDA MB 231 cells PVT1 exon 9 may not be regulating CLDN expression at a transcriptional level. PVT1 may be interacting with CLDN proteins directly, or indirectly, to regulate their expression.

LncRNAs are known to interact with proteins to regulate their stability, or ubiquitination [65]. Our data demonstrate that PVT1 regulates CLDN expression at a post-translational level. Knock down of PVT1 exon 9 did not affect protein expression for CLDNs 1, 3 and 7, however, we did observe re-expression of CLDN4 in MDA MB 231 cells. Since knockdown of PVT1 expression led to an increase in CLDN4 protein expression, it is plausible that PVT1 regulates CLDN4 protein expression via ubiquitination. Protein ubiquitination is one of the most common post-translational modification used to regulate various protein substrates in numerous cellular pathways. LncRNA MEG3, a tumor suppressor reported to play important roles in various malignancies, has been demonstrated to regulate LATS2 by promoting the ubiquitination of EZH2 in gallbladder cancer [66], while lncRNA HOTAIR acts as an inducer of proteolysis by facilitating the ubiquitination of Ataxin-1 and Snurportin-1. Over expression of HOTAIR promotes their rapid degradation [67]. Similarly, PVT1 overexpression may downregulate CLDN4 expression by regulating its ubiquitination. One explanation of how PVT1 could be doing this is by binding directly to CLDN4 thus facilitating its downregulation. However, an RNA immunoprecipitation (RIP) assay in which CLDN4 was used to pull down PVT1 exon 9 transcripts suggests otherwise, as there was no enrichment of PVT1 exon 9 transcript in the IP when compared to the control (Figure S6). Consequently, PVT1 may be regulating CLDN4 protein expression via an as yet undiscovered mechanism such as modulation of a downstream target, crosstalk between ubiquitination mediators, or by utilizing a different molecular mechanism altogether. Additional studies are necessary to further investigate this. Overexpression of CLDN4 is reported to have unfavorable clinical outcomes [57]. Contrastingly, the implications of our results suggest that re-expression of CLDN4 in MDA MB 231 CL TNBC cells is associated with a reduction in migration. This is in keeping with Lin et. al, who reported that loss of CLDN4 promotes EMT, while re-expression, or increased expression, of CLDN4 reduces migration [68]. Moreover, silencing PVT1 exon 9 had no significant effect on cell proliferation, suggesting that PVT1 may not have a regulatory role in cell proliferation (Figure S3).

The epithelial-to-mesenchymal transition (EMT) describes a biological process in which epithelial cells undergo a gradual change becoming more "mesenchymal-like", motile and invasive [69]. Concomitant downregulation of epithelial markers, and upregulation of mesenchymal markers, is characteristic of EMT [70]. Aberrant expression of both CLDNs, and PVT1, induce EMT in many malignancies [71,72]. We assessed the expression of the following EMT markers: E-cadherin, vimentin and fibronectin. Furthermore, we assessed the expression of caveolin-1, an integral membrane protein that participates in several cellular processes including EMT [73]. We observed no changes in the expression of EMT markers when PVT1 exon 9 is knocked down, except for a slight increase in the expression of fibronectin. Fibronectin is a component of the extra cellular matrix whose upregulation is typically associated with increased migration in cancer cells [74,75]. Thus, the slight increase of fibronectin in our study may not be associated with the decrease in migration, but could potentially have implications for other biological processes involving PVT1. All in all, our data suggest that EMT markers are more highly expressed in MDA MB 231 cells in comparison to MDA MB 468 cells. Though this supports that MDA MB 231 CL TNBC cells are more "mesenchymal-like", PVT1 exon 9 does not appear to have a role in this process. Overall, this result suggests that PVT1 may not regulate EMT in TNBC.

Keratin 14 (KRT14), a member of the keratin type I family is overexpressed in breast cancer [76–78]. Cells expressing KRT14 are more migratory [79] and invasive [80,81]. Interestingly, there is evidence which demonstrates that KRT interacts with CLDNs in order to maintain tight junction functionality [82]. Moreover, it was reported that modulation of KRTs affects CLDN expression, cell motility, and invasion in hepatocellular carcinoma [83]. Therefore, in future studies it may be worth investigating PVT1 regulation of CLDN4 in MDA MB 231 CL TNBC cells as part of a potentially novel pathway involving KRT14.

4. Materials and Methods

4.1. Cell Culture

A panel of five breast cancer cell lines were used to assess the expression of PVT1 exon 9. MDA MB 231, MDA MB 468, T47D and MCF-7 were maintained in Dulbecco's Modified Eagle's Medium (DMEM), supplemented with 10% FBS and 0.5% penicillin/streptomycin. Trypsinization of cells occurred using 0.05% trypsin when cells were 70–80% confluent. BT474 was maintained in F12/DMEM (GIBCO, Waltham, MA, USA), supplemented with 10% FBS and 1% penicillin/streptomycin. Trypsinization of cells occurred using 0.25% trypsin when cells were 70–80% confluent. All cell lines were cultured in a 5% CO_2, 37 °C atmosphere.

4.2. Transfection of siRNAs

MDA MB 231 cells were grown in 6-well plates until they have reached 90–100% confluency. According to the manufacturer's instructions, cells were transfected with 10 nM of PVT1 exon 9 siRNA (siPVT1 exon 9) (Forward: 5′ ACCUAUGAGCUUUGAAUAA 3′; Reverse: 5′ UUAUUCAAAGCUCAUAGGU 3′) (Sigma, St. Louis, MO, USA), or a non-targeting scramble control (siScramble) (Forward: 5′ CUCACUACCGUCGACCCCA 3′; Reverse: 5′ UGGGGUCGACGGUAGUGAG 3′) (Sigma) using Lipofectamine RNAiMAX (Thermo Fisher Scientific Inc.; Wilmington, DE, USA). siRNAs and Lipofectamine were diluted in Opti-MEM (Thermo Fisher Scientific Inc.). Following transfection, cells were incubated for 24 h in a 5% CO_2, 37 °C atmosphere before being harvested, or migration assay.

4.3. RNA Extraction and RT-qPCR

At 70–75% confluency, total RNA was extracted from cells in a 60 × 15 mm tissue culture dish, using the RNeasy Mini Kit (Qiagen, Hilden, Germany, cat#: 74104) according to the manufacturers' instructions. RNA concentration was measured using the spectrophotometer NanoDropTM 2000 (Thermo Fisher Scientific, Inc.). cDNA was synthesized from 1 μg of RNA using QuantiTect reverse transcription kit (Qiagen, cat# 205311). Amplification reactions were performed in 25 μL reaction volume using SYBR Green PCR master mix (Life Technologies, Grand Island, NY, USA cat# 4309155), cDNA template and 0.4 μM final concentration for primers. Primers used in this study were composed of the following oligonucleotide sequences listed in Table S1. Using the Quantifect Studio System (Applied Biosystems), relative expression of messenger RNA (mRNA) for each sample was assessed in quadruplicates in at least 3 independent experiments, and quantified via the comparative cycle threshold (ΔΔ Ct) method and normalized to GAPDH mRNA expression.

4.4. Protein Extraction and Immunoblotting

Whole cell extracts were obtained using RIPA lysis buffer (VWR Life Science, Radnor, PA, USA, cat# N653-100ML) supplemented with 1× protease inhibitor and 100 mM of phenyl methylsulfonyl fluoride (PMSF) (Amresco, Solon, OH, USA, cat# M145-5G). Protein concentration was quantified via the Bradford Assay using the Bio-Rad Protein Assay Dye Reagent Concentrate (BioRad, Hercules, CA, USA, cat# 500-0205). For western blot analysis, 50 μg of protein were resolved by 15% SDS-PAGE gels and subsequently transferred onto nitrocellulose membranes. Membranes were blocked in 5% BSA or 5% milk in TBS-T for 1 h at room temperature, incubated overnight at 4 °C in primary antibodies. Next, membranes were washed in 1× TBS-T, incubated with secondary antibodies for 2 h, washes with 1× TBS-T and imaged using the Odyssey CLx imager with infrared fluorescence (LI-COR, Lincoln, NE, USA). The primary antibodies used were Claudin 1 (13050-1-AP), Claudin 4 (16195-1-AP) and Claudin 7 (10118-1-AP) (Proteintech, Rosemont, IL, USA), Claudin 3 (341700; Invitrogen, Waltham, MA, USA), GAPDH (5174S; 1:1000; Cell Signaling, Danvers, MA, USA), α-tubulin (sc-32293; 1:1000; Santa Cruz Biotechnology, Dallas, TX, USA). The secondary antibodies used were anti-mouse (925-32210; 1:15,000; LI-COR) and anti-rabbit (925-32211; 1:15,000; LI-COR). Original blots can be found at Figures S1 and S2.

4.5. Migration Assays

1×10^5 MDA MB 231 cells were grown in a 6-well tissue culture plate until 90–100% confluency. Cells were transfected with siRNA, or siScramble, and incubated at 5% CO_2, 37 °C for 24 h. Wounds were made on the cell monolayer using a sterile 200 μL pipet tip and then washed with 1× PBS and incubated with media containing respective siRNAs. Images of scratched areas were taken at 10× magnification using an AE30 inverted microscope (Motic, Richmond, BC, Canada).

4.6. Cell Viability Assays

10^4 cells were seeded into 96 well plates. At 70% confluency, the cells were transfected with PVT1 exon 9. After 24 h, MTT assays were performed and absorbance measured at 490 nm with a microplate reader (i3 multimode microplate reader, Spectramax, San Jose, CA, USA).

4.7. Crosslinking and RNA Immunoprecipitation (RIP)

RIP experiments were performed using MDA MB 231 cells. Cells were plated in 10 cm tissue culture dishes until 90–100% confluent. Cells were trypsinized and resuspended in culture medium. RNA was crosslinked to proteins by adding formaldehyde drop-wisely to suspended cells at a final concentration of 0.75%. Cells were placed on a shaker at a low speed for 10 min at room temperature. A final concentration of 125 mM of glycine was added to the media and incubated at room temperature on a shaker for 5 min), and then pelleted. Cells were resuspended in 2 mL freshly prepared nuclear isolation buffer (1.28 M sucrose, 40 mM Tris-HCl [pH 7.5], 20 mM MgCl2, 4% Triton X-100), 2 mL 1× PBS and 6 mL nuclease free water, and kept on ice for 20 min with frequent mixing. Cells were pelleted at 2500 *g* for 15 min and resuspended in 1 mL freshly prepared RIP buffer (1861603, Thermo Fisher Scientific Inc.) supplemented with 1× protease inhibitor, 100 U/mL RNase inhibitor (AM2696, Invitrogen, Waltham, MA, USA) and 0.5 mM DTT, for 10 min on ice. Lysate was then sonicated for 3 min (97 amplitude, 30 s on, 30 s off) and centrifuged at 13,000 rpm at 4 °C for 10 min. 40 uL of Dynabeads protein G (10007D, Invitrogen) was washed once with wash buffer and incubated with 10 ug of CLDN4 antibody (16195-1-AP, Proteintech) for 1 h at room temperature. Beads were washed again with wash buffer and incubated with cell lysate for 2 h at 4 °C. Following incubation, beads were washed 3 times with wash buffer and once with 1× PBS. RNA was extracted using Trizol (10296010, Ambion, Austin, TX, USA) as per manufacturers' recommendation. cDNA was synthesized from 150 ng of RNA using QuantiTect reverse transcription kit (Qiagen, cat# 205311). Amplification reactions were performed in 25 µL reaction volume using SYBR Green PCR master mix (Life Technologies, Grand Island, NY, USA cat# 4309155), cDNA template and 0.4 µM final concentration for primers. Using the Quantifect Studio System (Applied Biosystems, Foster City, CA, USA), relative expression of messenger RNA (mRNA) for each sample was assessed in quadruplicates in at least 2 independent experiments, and quantified via the comparative cycle threshold (ΔΔ Ct) method.

4.8. Statistical Analysis

Data from at least three independent experiments were presented as mean ± standard error of the mean (S.E.M). Statistical significance was assessed using a two-tailed Student's *t* test. *p* values less than 0.05 were deemed significant.

5. Conclusions

In conclusion, we demonstrated the potential regulatory role of PVT1 in CL MDA MB 231 TNBC cells by targeting PVT1 exon 9. Knock down of PVT1 exon 9 resulted in the re-expression of CLDN4 protein. Additionally, we demonstrated that overexpression of PVT1 exon 9 is associated with increased migration of MDA MB 231 CL TNBC cells. Our data suggest that PVT1 may have a regulatory role in CL MDA MB 231 TNBC cells by acting as a modulator of CLDN4 protein expression. These data may have implications for prognostication and treatment strategies in CL TNBC.

Supplementary Materials: The following are available online at https://www.mdpi.com/2072-6694/13/5/1046/s1, Figure S1: The following are the corresponding blots to Figure 6. Figure S2: The following are the corresponding blots to Figure 7. Figure S3: PVT1 exon 9 does not regulate cell viability. Figure S4: PVT1 exon 9 regulates migration of MDA MB 231 CL TNBC cells. Figure S5: PVT1 exon 9 is overexpressed in MDA MB 231 cells. Figure S6: PVT1 exon 9 does not bind directly to CLDN4. Table S1: List of primers sequences.

Author Contributions: O.O.O. conceived of the study and supervised the work. O.O.O. designed the studies. F.L. performed all of the experiments and analyzed the data. F.L. wrote the manuscript. O.O.O. revised final version of manuscript. Both authors have read and agreed to the published version of the manuscript.

Funding: This study was supported by the National Cancer Institute Grant numbers U54CA221704 and U54CA221705.

Institutional Review Board Statement: Not applicable.

Informed Consent Statement: Not applicable.

Data Availability Statement: All data generated from this study are contained in the manuscript and supplementary materials.

Acknowledgments: The authors would like to thank Jill Bargonetti for the kind gift of MDA MB 231, MDA MB 468, T47D and MCF-7, and Brian Zeglis for the kind gift of BT474.

Conflicts of Interest: O.O.O. is a co-founder of NucleoBio Inc., a City of New York start-up biotechnology company. F.L. has no conflicts of interest.

References

1. Bray, F.; Ferlay, J.; Soerjomataram, I.; Siegel, R.L.; Torre, L.A.; Jemal, A. Global cancer statistics 2018: GLOBOCAN estimates of incidence and mortality worldwide for 36 cancers in 185 countries. *CA Cancer J. Clin.* **2018**, *68*, 394–424. [CrossRef]
2. American Cancer Society. How Common is Breast Cancer? Available online: https://www.cancer.org/cancer/breast-cancer/about/how-common-is-breast-cancer.html (accessed on 13 May 2020).
3. Siddharth, S.; Sharma, D. Racial Disparity and Triple-Negative Breast Cancer in African-American Women: A Multifaceted Affair between Obesity, Biology, and Socioeconomic Determinants. *Cancers* **2018**, *10*, 514. [CrossRef]
4. Dai, X.; Cheng, H.; Bai, Z.; Li, J. Breast Cancer Cell Line Classification and Its Relevance with Breast Tumor Subtyping. *J. Cancer* **2017**, *8*, 3131–3141. [CrossRef]
5. American Cancer Society. Triple Negative Breast Cancer. Available online: https://www.cancer.org/cancer/breast-cancer/understanding-a-breast-cancer-diagnosis/types-of-breast-cancer/triple-negative.html (accessed on 13 May 2020).
6. Rakha, E.A.; Elsheikh, S.E.; Aleskandarany, M.A.; Habashi, H.O.; Green, A.R.; Powe, D.G.; El-Sayed, M.E.; Benhasouna, A.; Brunet, J.S.; Akslen, L.A.; et al. Triple-negative breast cancer: Distinguishing between basal and nonbasal subtypes. *Clin. Cancer Res.* **2009**, *15*, 2302–2310. [CrossRef]
7. Anders, C.K.; Abramson, V.; Tan, T.; Dent, R. The Evolution of Triple-Negative Breast Cancer: From Biology to Novel Therapeutics. *Am. Soc. Clin. Oncol. Educ. Book* **2016**, *35*, 34–42. [CrossRef]
8. Lehmann, B.D.; Bauer, J.A.; Chen, X.; Sanders, M.E.; Chakravarthy, A.B.; Shyr, Y.; Pietenpol, J.A. Identification of human triple-negative breast cancer subtypes and preclinical models for selection of targeted therapies. *J. Clin. Investig.* **2011**, *121*, 2750–2767. [CrossRef]
9. Wahdan-Alaswad, R.; Harrell, J.C.; Fan, Z.; Edgerton, S.M.; Liu, B.; Thor, A.D. Metformin attenuates transforming growth factor beta (TGF-beta) mediated oncogenesis in mesenchymal stem-like/claudin-low triple negative breast cancer. *Cell Cycle* **2016**, *15*, 1046–1059. [CrossRef] [PubMed]
10. Dias, K.; Dvorkin-Gheva, A.; Hallett, R.M.; Wu, Y.; Hassell, J.; Pond, G.R.; Levine, M.; Whelan, T.; Bane, A.L. Claudin-Low Breast Cancer; Clinical & Pathological Characteristics. *PLoS ONE* **2017**, *12*, e0168669. [CrossRef]
11. Prat, A.; Parker, J.S.; Karginova, O.; Fan, C.; Livasy, C.; Herschkowitz, J.I.; He, X.; Perou, C.M. Phenotypic and molecular characterization of the claudin-low intrinsic subtype of breast cancer. *Breast Cancer Res.* **2010**, *12*, R68. [CrossRef] [PubMed]
12. Guan, Y.; Kuo, W.L.; Stilwell, J.L.; Takano, H.; Lapuk, A.V.; Fridlyand, J.; Mao, J.H.; Yu, M.; Miller, M.A.; Santos, J.L.; et al. Amplification of PVT1 contributes to the pathophysiology of ovarian and breast cancer. *Clin. Cancer Res.* **2007**, *13*, 5745–5755. [CrossRef] [PubMed]
13. Parolia, A.; Cieslik, M.; Chinnaiyan, A.M. Competing for enhancers: PVT1 fine-tunes MYC expression. *Cell Res.* **2018**, *28*, 785–786. [CrossRef] [PubMed]
14. Huppi, K.; Pitt, J.J.; Wahlberg, B.M.; Caplen, N.J. The 8q24 gene desert: An oasis of non-coding transcriptional activity. *Front. Genet.* **2012**, *3*, 69. [CrossRef]
15. Huppi, K.; Volfovsky, N.; Runfola, T.; Jones, T.L.; Mackiewicz, M.; Martin, S.E.; Mushinski, J.F.; Stephens, R.; Caplen, N.J. The identification of microRNAs in a genomically unstable region of human chromosome 8q24. *Mol. Cancer Res.* **2008**, *6*, 212–221. [CrossRef]
16. El Marabti, E.; Younis, I. The Cancer Spliceome: Reprograming of Alternative Splicing in Cancer. *Front. Mol. Biosci.* **2018**, *5*, 80. [CrossRef]
17. Ilboudo, A.; Chouhan, J.; McNeil, B.K.; Osborne, J.R.; Ogunwobi, O.O. PVT1 Exon 9: A Potential Biomarker of Aggressive Prostate Cancer? *Int. J. Environ. Res. Public Health* **2015**, *13*, 12. [CrossRef]
18. Pal, G.; Di, L.; Orunmuyi, A.; Olapade-Olaopa, E.O.; Qiu, W.; Ogunwobi, O.O. Population Differentiation at the PVT1 Gene Locus: Implications for Prostate Cancer. *G3* **2020**, *10*, 2257–2264. [CrossRef]
19. Pal, G.; Huaman, J.; Levine, F.; Orunmuyi, A.; Olapade-Olaopa, E.O.; Onagoruwa, O.T.; Ogunwobi, O.O. Long Noncoding RNA from PVT1 Exon 9 Is Overexpressed in Prostate Cancer and Induces Malignant Transformation and Castration Resistance in Prostate Epithelial Cells. *Genes* **2019**, *10*, 964. [CrossRef]
20. Conte, F.; Fiscon, G.; Chiara, M.; Colombo, T.; Farina, L.; Paci, P. Role of the long non-coding RNA PVT1 in the dysregulation of the ceRNA-ceRNA network in human breast cancer. *PLoS ONE* **2017**, *12*, e0171661. [CrossRef] [PubMed]
21. Yang, T.; Zhou, H.; Liu, P.; Yan, L.; Yao, W.; Chen, K.; Zeng, J.; Li, H.; Hu, J.; Xu, H.; et al. lncRNA PVT1 and its splicing variant function as competing endogenous RNA to regulate clear cell renal cell carcinoma progression. *Oncotarget* **2017**, *8*, 85353–85367. [CrossRef] [PubMed]
22. Paci, P.; Colombo, T.; Farina, L. Computational analysis identifies a sponge interaction network between long non-coding RNAs and messenger RNAs in human breast cancer. *BMC Syst. Biol.* **2014**, *8*, 83. [CrossRef]

23. Wang, L.; Wang, R.; Ye, Z.; Wang, Y.; Li, X.; Chen, W.; Zhang, M.; Cai, C. PVT1 affects EMT and cell proliferation and migration via regulating p21 in triple-negative breast cancer cells cultured with mature adipogenic medium. *Acta Biochim. Biophys. Sin* **2018**, *50*, 1211–1218. [CrossRef]
24. Liu, X.; Bi, L.; Wang, Q.; Wen, M.; Li, C.; Ren, Y.; Jiao, Q.; Mao, J.H.; Wang, C.; Wei, G.; et al. miR-1204 targets VDR to promotes epithelial-mesenchymal transition and metastasis in breast cancer. *Oncogene* **2018**, *37*, 3426–3439. [CrossRef] [PubMed]
25. Pena-Chilet, M.; Martinez, M.T.; Perez-Fidalgo, J.A.; Peiro-Chova, L.; Oltra, S.S.; Tormo, E.; Alonso-Yuste, E.; Martinez-Delgado, B.; Eroles, P.; Climent, J.; et al. MicroRNA profile in very young women with breast cancer. *BMC Cancer* **2014**, *14*, 529. [CrossRef]
26. Hou, X.; Niu, Z.; Liu, L.; Guo, Q.; Li, H.; Yang, X.; Zhang, X. miR-1207-5p regulates the sensitivity of triple-negative breast cancer cells to Taxol treatment via the suppression of LZTS1 expression. *Oncol. Lett.* **2019**, *17*, 990–998. [CrossRef] [PubMed]
27. Yan, C.; Chen, Y.; Kong, W.; Fu, L.; Liu, Y.; Yao, Q.; Yuan, Y. PVT1-derived miR-1207-5p promotes breast cancer cell growth by targeting STAT6. *Cancer Sci.* **2017**, *108*, 868–876. [CrossRef] [PubMed]
28. Wang, Y.; Zhou, J.; Wang, Z.; Wang, P.; Li, S. Upregulation of SOX2 activated LncRNA PVT1 expression promotes breast cancer cell growth and invasion. *Biochem. Biophys. Res. Commun.* **2017**, *493*, 429–436. [CrossRef] [PubMed]
29. Tang, J.; Li, Y.; Sang, Y.; Yu, B.; Lv, D.; Zhang, W.; Feng, H. LncRNA PVT1 regulates triple-negative breast cancer through KLF5/beta-catenin signaling. *Oncogene* **2018**, *37*, 4723–4734. [CrossRef]
30. Singh, A.B.; Sharma, A.; Dhawan, P. Claudin family of proteins and cancer: An overview. *J. Oncol.* **2010**, *2010*, 541957. [CrossRef]
31. Naik, M.U.; Naik, T.U.; Suckow, A.T.; Duncan, M.K.; Naik, U.P. Attenuation of junctional adhesion molecule-A is a contributing factor for breast cancer cell invasion. *Cancer Res.* **2008**, *68*, 2194–2203. [CrossRef]
32. Ivana, B.; Emina, M.; Marijana, M.K.; Irena, J.; Zoran, B.; Radmila, J. High expression of junctional adhesion molecule-A is associated with poor survival in patients with epithelial ovarian cancer. *Int. J. Biol. Mark.* **2019**, *34*, 262–268. [CrossRef] [PubMed]
33. Salehi, P.; Tafvizi, F.; Kamyab Hesari, K. Low Expression of Occludin in the Melanoma Patient. *Iran J. Pathol.* **2019**, *14*, 272–278. [CrossRef]
34. Martin, T.A.; Jordan, N.; Davies, E.L.; Jiang, W.G. Metastasis to Bone in Human Cancer Is Associated with Loss of Occludin Expression. *Anticancer Res.* **2016**, *36*, 1287–1293.
35. Sakamoto, S.; Inoue, H.; Kaneko, M.K.; Ogasawara, S.; Kajikawa, M.; Urano, S.; Ohba, S.I.; Kato, Y.; Kawada, M. Generation and evaluation of a chimeric antibody against coxsackievirus and adenovirus receptor for cancer therapy. *Cancer Sci.* **2019**, *110*, 3595–3602. [CrossRef]
36. Tsukita, S.; Tanaka, H.; Tamura, A. The Claudins: From Tight Junctions to Biological Systems. *Trends Biochem. Sci.* **2019**, *44*, 141–152. [CrossRef]
37. Morita, K.; Furuse, M.; Fujimoto, K.; Tsukita, S. Claudin multigene family encoding four-transmembrane domain protein components of tight junction strands. *Proc. Natl. Acad. Sci. USA* **1999**, *96*, 511–516. [CrossRef] [PubMed]
38. Van Itallie, C.M.; Anderson, J.M. Claudins and epithelial paracellular transport. *Annu. Rev. Physiol.* **2006**, *68*, 403–429. [CrossRef] [PubMed]
39. Krause, G.; Protze, J.; Piontek, J. Assembly and function of claudins: Structure-function relationships based on homology models and crystal structures. *Semin. Cell Dev. Biol.* **2015**, *42*, 3–12. [CrossRef]
40. Matter, K.; Balda, M.S. Signalling to and from tight junctions. *Nat. Rev. Mol. Cell Biol.* **2003**, *4*, 225–236. [CrossRef]
41. Chiba, H.; Osanai, M.; Murata, M.; Kojima, T.; Sawada, N. Transmembrane proteins of tight junctions. *Biochim. Biophys. Acta* **2008**, *1778*, 588–600. [CrossRef] [PubMed]
42. Hewitt, K.J.; Agarwal, R.; Morin, P.J. The claudin gene family: Expression in normal and neoplastic tissues. *BMC Cancer* **2006**, *6*, 186. [CrossRef] [PubMed]
43. Kinugasa, T.; Huo, Q.; Higashi, D.; Shibaguchi, H.; Kuroki, M.; Tanaka, T.; Futami, K.; Yamashita, Y.; Hachimine, K.; Maekawa, S.; et al. Selective up-regulation of claudin-1 and claudin-2 in colorectal cancer. *Anticancer Res.* **2007**, *27*, 3729–3734. [CrossRef]
44. Hichino, A.; Okamoto, M.; Taga, S.; Akizuki, R.; Endo, S.; Matsunaga, T.; Ikari, A. Down-regulation of Claudin-2 Expression and Proliferation by Epigenetic Inhibitors in Human Lung Adenocarcinoma A549 Cells. *J. Biol. Chem.* **2017**, *292*, 2411–2421. [CrossRef]
45. Holczbauer, A.; Gyongyosi, B.; Lotz, G.; Szijarto, A.; Kupcsulik, P.; Schaff, Z.; Kiss, A. Distinct claudin expression profiles of hepatocellular carcinoma and metastatic colorectal and pancreatic carcinomas. *J. Histochem. Cytochem.* **2013**, *61*, 294–305. [CrossRef]
46. Zhang, W.N.; Li, W.; Wang, X.L.; Hu, Z.; Zhu, D.; Ding, W.C.; Liu, D.; Li, K.Z.; Ma, D.; Wang, H. CLDN1 expression in cervical cancer cells is related to tumor invasion and metastasis. *Oncotarget* **2016**, *7*, 87449–87461. [CrossRef] [PubMed]
47. Yang, L.; Sun, X.; Meng, X. Differences in the expression profiles of claudin proteins in human gastric carcinoma compared with nonneoplastic mucosa. *Mol. Med. Rep.* **2018**, *18*, 1271–1278. [CrossRef]
48. Choi, Y.L.; Kim, J.; Kwon, M.J.; Choi, J.S.; Kim, T.J.; Bae, D.S.; Koh, S.S.; In, Y.H.; Park, Y.W.; Kim, S.H.; et al. Expression profile of tight junction protein claudin 3 and claudin 4 in ovarian serous adenocarcinoma with prognostic correlation. *Histol. Histopathol.* **2007**, *22*, 1185–1195. [CrossRef] [PubMed]
49. Hicks, D.A.; Galimanis, C.E.; Webb, P.G.; Spillman, M.A.; Behbakht, K.; Neville, M.C.; Baumgartner, H.K. Claudin-4 activity in ovarian tumor cell apoptosis resistance and migration. *BMC Cancer* **2016**, *16*, 788. [CrossRef] [PubMed]
50. Kwon, M.J.; Kim, S.H.; Jeong, H.M.; Jung, H.S.; Kim, S.S.; Lee, J.E.; Gye, M.C.; Erkin, O.C.; Koh, S.S.; Choi, Y.L.; et al. Claudin-4 overexpression is associated with epigenetic derepression in gastric carcinoma. *Lab. Investig.* **2011**, *91*, 1652–1667. [CrossRef] [PubMed]

51. Michl, P.; Barth, C.; Buchholz, M.; Lerch, M.M.; Rolke, M.; Holzmann, K.H.; Menke, A.; Fensterer, H.; Giehl, K.; Lohr, M.; et al. Claudin-4 expression decreases invasiveness and metastatic potential of pancreatic cancer. *Cancer Res.* **2003**, *63*, 6265–6271.
52. Lin, X.; Shang, X.; Manorek, G.; Howell, S.B. Regulation of the Epithelial-Mesenchymal Transition by Claudin-3 and Claudin-4. *PLoS ONE* **2013**, *8*, e67496. [CrossRef]
53. Kolokytha, P.; Yiannou, P.; Keramopoulos, D.; Kolokythas, A.; Nonni, A.; Patsouris, E.; Pavlakis, K. Claudin-3 and claudin-4: Distinct prognostic significance in triple-negative and luminal breast cancer. *Appl. Immunohistochem. Mol. Morphol.* **2014**, *22*, 125–131. [CrossRef]
54. Lanigan, F.; McKiernan, E.; Brennan, D.J.; Hegarty, S.; Millikan, R.C.; McBryan, J.; Jirstrom, K.; Landberg, G.; Martin, F.; Duffy, M.J.; et al. Increased claudin-4 expression is associated with poor prognosis and high tumour grade in breast cancer. *Int. J. Cancer* **2009**, *124*, 2088–2097. [CrossRef] [PubMed]
55. Szasz, A.M.; Nemeth, Z.; Gyorffy, B.; Micsinai, M.; Krenacs, T.; Baranyai, Z.; Harsanyi, L.; Kiss, A.; Schaff, Z.; Tokes, A.M.; et al. Identification of a claudin-4 and E-cadherin score to predict prognosis in breast cancer. *Cancer Sci.* **2011**, *102*, 2248–2254. [CrossRef] [PubMed]
56. Abd-Elazeem, M.A.; Abd-Elazeem, M.A. Claudin 4 expression in triple-negative breast cancer: Correlation with androgen receptors and Ki-67 expression. *Ann. Diagn. Pathol.* **2015**, *19*, 37–42. [CrossRef] [PubMed]
57. Kulka, J.; Szasz, A.M.; Nemeth, Z.; Madaras, L.; Schaff, Z.; Molnar, I.A.; Tokes, A.M. Expression of tight junction protein claudin-4 in basal-like breast carcinomas. *Pathol. Oncol. Res.* **2009**, *15*, 59–64. [CrossRef]
58. Newman, L.A.; Kaljee, L.M. Health Disparities and Triple-Negative Breast Cancer in African American Women: A Review. *JAMA Surg.* **2017**, *152*, 485–493. [CrossRef]
59. Carey, L.A.; Perou, C.M.; Livasy, C.A.; Dressler, L.G.; Cowan, D.; Conway, K.; Karaca, G.; Troester, M.A.; Tse, C.K.; Edmiston, S.; et al. Race, breast cancer subtypes, and survival in the Carolina Breast Cancer Study. *JAMA* **2006**, *295*, 2492–2502. [CrossRef]
60. Morin, P.J. Claudin proteins in human cancer: Promising new targets for diagnosis and therapy. *Cancer Res.* **2005**, *65*, 9603–9606. [CrossRef]
61. Li, Z.; Zhao, W.; Wang, M.; Zhou, X. The Role of Long Noncoding RNAs in Gene Expression Regulation. *IntechOpen* **2019**. [CrossRef]
62. He, R.Z.; Luo, D.X.; Mo, Y.Y. Emerging roles of lncRNAs in the post-transcriptional regulation in cancer. *Genes Dis.* **2019**, *6*, 6–15. [CrossRef]
63. Shin, C.H.; Ryu, S.; Kim, H.H. hnRNPK-regulated PTOV1-AS1 modulates heme oxygenase-1 expression via miR-1207-5p. *BMB Rep.* **2017**, *50*, 220–225. [CrossRef]
64. Zhang, A.; Zhao, J.C.; Kim, J.; Fong, K.W.; Yang, Y.A.; Chakravarti, D.; Mo, Y.Y.; Yu, J. LncRNA HOTAIR Enhances the Androgen-Receptor-Mediated Transcriptional Program and Drives Castration-Resistant Prostate Cancer. *Cell Rep.* **2015**, *13*, 209–221. [CrossRef]
65. Zhang, X.; Wang, W.; Zhu, W.; Dong, J.; Cheng, Y.; Yin, Z.; Shen, F. Mechanisms and Functions of Long Non-Coding RNAs at Multiple Regulatory Levels. *Int. J. Mol. Sci.* **2019**, *20*, 5573. [CrossRef]
66. Zhang, Y.; Wu, J.; Jing, H.; Huang, G.; Sun, Z.; Xu, S. Long noncoding RNA MEG3 inhibits breast cancer growth via upregulating endoplasmic reticulum stress and activating NF-kappaB and p53. *J. Cell Biochem.* **2019**, *120*, 6789–6797. [CrossRef] [PubMed]
67. Yoon, J.H.; Abdelmohsen, K.; Kim, J.; Yang, X.; Martindale, J.L.; Tominaga-Yamanaka, K.; White, E.J.; Orjalo, A.V.; Rinn, J.L.; Kreft, S.G.; et al. Scaffold function of long non-coding RNA HOTAIR in protein ubiquitination. *Nat. Commun.* **2013**, *4*, 2939. [CrossRef]
68. Shang, X.; Lin, X.; Manorek, G.; Howell, S.B. Claudin-3 and claudin-4 regulate sensitivity to cisplatin by controlling expression of the copper and cisplatin influx transporter CTR1. *Mol. Pharmacol.* **2013**, *83*, 85–94. [CrossRef] [PubMed]
69. Kalluri, R. EMT: When epithelial cells decide to become mesenchymal-like cells. *J. Clin. Investig.* **2009**, *119*, 1417–1419. [CrossRef] [PubMed]
70. Lamouille, S.; Xu, J.; Derynck, R. Molecular mechanisms of epithelial-mesenchymal transition. *Nat. Rev. Mol. Cell Biol.* **2014**, *15*, 178–196. [CrossRef]
71. Katayama, A.; Handa, T.; Komatsu, K.; Togo, M.; Horiguchi, J.; Nishiyama, M.; Oyama, T. Expression patterns of claudins in patients with triple-negative breast cancer are associated with nodal metastasis and worse outcome. *Pathol. Int.* **2017**, *67*, 404–413. [CrossRef]
72. Wu, B.Q.; Jiang, Y.; Zhu, F.; Sun, D.L.; He, X.Z. Long Noncoding RNA PVT1 Promotes EMT and Cell Proliferation and Migration Through Downregulating p21 in Pancreatic Cancer Cells. *Technol. Cancer Res. Treat.* **2017**, *16*, 819–827. [CrossRef]
73. Qian, X.L.; Pan, Y.H.; Huang, Q.Y.; Shi, Y.B.; Huang, Q.Y.; Hu, Z.Z.; Xiong, L.X. Caveolin-1: A multifaceted driver of breast cancer progression and its application in clinical treatment. *OncoTargets Ther.* **2019**, *12*, 1539–1552. [CrossRef] [PubMed]
74. Gopal, S.; Veracini, L.; Grall, D.; Butori, C.; Schaub, S.; Audebert, S.; Camoin, L.; Baudelet, E.; Radwanska, A.; Beghelli-de la Forest Divonne, S.; et al. Fibronectin-guided migration of carcinoma collectives. *Nat. Commun.* **2017**, *8*, 14105. [CrossRef] [PubMed]
75. Yousif, N.G. Fibronectin promotes migration and invasion of ovarian cancer cells through up-regulation of FAK-PI3K/Akt pathway. *Cell Biol. Int.* **2014**, *38*, 85–91. [CrossRef] [PubMed]
76. Kuo, W.H.; Chang, Y.Y.; Lai, L.C.; Tsai, M.H.; Hsiao, C.K.; Chang, K.J.; Chuang, E.Y. Molecular characteristics and metastasis predictor genes of triple-negative breast cancer: A clinical study of triple-negative breast carcinomas. *PLoS ONE* **2012**, *7*, e45831. [CrossRef] [PubMed]

77. Kumar, B.; Prasad, M.; Bhat-Nakshatri, P.; Anjanappa, M.; Kalra, M.; Marino, N.; Storniolo, A.M.; Rao, X.; Liu, S.; Wan, J.; et al. Normal Breast-Derived Epithelial Cells with Luminal and Intrinsic Subtype-Enriched Gene Expression Document Interindividual Differences in Their Differentiation Cascade. *Cancer Res.* **2018**, *78*, 5107–5123. [CrossRef]
78. Ellsworth, R.E.; Seebach, J.; Field, L.A.; Heckman, C.; Kane, J.; Hooke, J.A.; Love, B.; Shriver, C.D. A gene expression signature that defines breast cancer metastases. *Clin. Exp. Metastasis* **2009**, *26*, 205–213. [CrossRef]
79. Xiangying Wang, L.H.; Shan, S.; Sun, Y.; Mao, Y. KRT14 promoting invasion and migration of lung cancer cells through ROCK-1 signaling pathway. *Int. J. Clin. Exp. Pathol.* **2017**, *10*, 795–803.
80. Bilandzic, M.; Rainczuk, A.; Green, E.; Fairweather, N.; Jobling, T.W.; Plebanski, M.; Stephens, A.N. Keratin-14 (KRT14) Positive Leader Cells Mediate Mesothelial Clearance and Invasion by Ovarian Cancer Cells. *Cancers* **2019**, *11*, 1228. [CrossRef]
81. Westcott, J.M.; Prechtl, A.M.; Maine, E.A.; Dang, T.T.; Esparza, M.A.; Sun, H.; Zhou, Y.; Xie, Y.; Pearson, G.W. An epigenetically distinct breast cancer cell subpopulation promotes collective invasion. *J. Clin. Investig.* **2015**, *125*, 1927–1943. [CrossRef]
82. Di Tommaso, T.; Cottle, D.L.; Pearson, H.B.; Schluter, H.; Kaur, P.; Humbert, P.O.; Smyth, I.M. Keratin 76 is required for tight junction function and maintenance of the skin barrier. *PLoS Genet.* **2014**, *10*, e1004706. [CrossRef]
83. Fortier, A.M.; Asselin, E.; Cadrin, M. Keratin 8 and 18 loss in epithelial cancer cells increases collective cell migration and cisplatin sensitivity through claudin1 up-regulation. *J. Biol. Chem.* **2013**, *288*, 11555–11571. [CrossRef] [PubMed]

cancers

MDPI

Review

The Multifaceted Roles of EGFL7 in Cancer and Drug Resistance

Beate Heissig [1,*], Yousef Salama [2], Satoshi Takahashi [3], Ko Okumura [1] and Koichi Hattori [4]

1 Department of Immunological Diagnosis, Graduate School of Medicine, Juntendo University School of Medicine, 2-1-1 Hongo, Bunkyo-Ku, Tokyo 113-8421, Japan; kokumura@juntendo.ac.jp
2 An-Najah Center for Cancer and Stem Cell Research, Faculty of Medicine and Health Sciences, An-Najah National University, P.O. Box 7, Nablus 99900800, Palestine; yousef.ut@najah.edu
3 Division of Molecular Therapy, Center for Stem Cell Biology and Regenerative Medicine, The Institute of Medical Science, The University of Tokyo, 4-6-1 Shirokanedai, Minato-ku, Tokyo 108-8639, Japan; radius@ims.u-tokyo.ac.jp
4 Center for Genome and Regenerative Medicine, Graduate School of Medicine, Juntendo University, 2-1-1 Hongo, Bunkyo-Ku, Tokyo 113-8421, Japan; khattori@juntendo.ac.jp
* Correspondence: heissig@juntendo.ac.jp; Tel.: +81-3813-3111-2158

Simple Summary: Cancer growth and metastasis require interactions with the extracellular matrix (ECM), which is home to many biomolecules that support the formation of new vessels and cancer growth. One of these biomolecules is epidermal growth factor-like protein-7 (EGFL7). EGFL7 alters cellular adhesion to the ECM and migratory behavior of tumor and immune cells contributing to tumor metastasis. EGFL7 is engaged in the formation of new vessels and changes in ECM stiffness. One of its binding partners on the endothelial and cancer cell surface is beta 3 integrin. Beta 3 integrin pathways are under intense investigation in search of new therapies to kill cancer cells. All these properties enable EGFL7 to contribute to drug resistance. In this review, we give insight into recent studies on EGFL7 and its engagement with beta3 integrin, a marker predicting cancer stem cells and drug resistance.

Abstract: Invasion of cancer cells into surrounding tissue and the vasculature is an important step for tumor progression and the establishment of distant metastasis. The extracellular matrix (ECM) is home to many biomolecules that support new vessel formation and cancer growth. Endothelial cells release growth factors such as epidermal growth factor-like protein-7 (EGFL7), which contributes to the formation of the tumor vasculature. The signaling axis formed by EGFL7 and one of its receptors, beta 3 integrin, has emerged as a key mediator in the regulation of tumor metastasis and drug resistance. Here we summarize recent studies on the role of the ECM-linked angiocrine factor EGFL7 in primary tumor growth, neoangiogenesis, tumor metastasis by enhancing epithelial-mesenchymal transition, alterations in ECM rigidity, and drug resistance. We discuss its role in cellular adhesion and migration, vascular leakiness, and the anti-cancer response and provide background on its transcriptional regulation. Finally, we discuss its potential as a drug target as an anti-cancer strategy.

Keywords: beta 3 integrin; integrin; cancer; drug resistance; angiocrine factor; angiogenesis; EGFR; EGFL7; miR-126; adhesion; migration; protease; EMT; FAK; LOX; KLF2; ECM; endothelial cells; cancer; proliferation

Citation: Heissig, B.; Salama, Y.; Takahashi, S.; Okumura, K.; Hattori, K. The Multifaceted Roles of EGFL7 in Cancer and Drug Resistance. *Cancers* **2021**, *13*, 1014. https://doi.org/10.3390/cancers13051014

Academic Editors: José I. López and Ildefonso M. de la Fuente

Received: 8 February 2021
Accepted: 22 February 2021
Published: 1 March 2021

1. Introduction

Tumor growth and metastasis rely on the tumor vascular network for adequate delivery of oxygen and nutrients [1]. Tumor endothelial cells (ECs) are the cellular building blocks of the nutrient-carrying vasculature. During tumor growth, activated ECs expand and form new capillaries in a process called angiogenesis [2]. These capillaries and vessels carry nutrients to hungry cancer cells and ensure proper oxygen delivery.

ECs release so-called angiocrine factors [3], which include the angiogenic factor vascular endothelial growth factor-A (VEGF-A), Jagged1 (Jag1), endothelin, enzymes such as tissue-type plasminogen activator [3], and epidermal growth factor-like protein-7 (EGFL7) [4]. EGFL7 is produced by cancer-associated ECs [4,5] and certain tumor cell types [4,6].

EGFL7 controls intercellular and cell–matrix communication, which are key features of tumor progression and metastasis, by hijacking the receptor tyrosine kinase epidermal growth factor receptor (EGFR), integrin, and Notch signaling pathways [6–8].

EGFL7 modulates cell migration by interacting with extracellular matrix (ECM) sensing integrins [9]. Integrins are a family of cell surface receptors that help cells to interact with the extracellular microenvironment, thereby controlling cell anchorage and movement. Integrins exist as heterodimers with noncovalently linked alpha and beta subunits and link the cytoskeleton with the ECM [10]. Integrins are mechanotransducers and key factors during cell migration and are thereby implicated in many steps of cancer progression, starting with primary tumor development to metastasis, cancer stem cell development, and drug resistance (reviewed in [11]). EGFL7 interacts with two of the most studied integrins in cancer—namely, alphaV:beta 3 (ITGAV:ITGB3) and the alpha5:beta1 integrin (ITGA5:ITGB1).

Integrins bind to a wide range of ECM proteins containing the arginylglycylaspartic acid (RGD)-motif. EGFL7 is one of those ECM proteins with a conserved RGD/Glutamine-Glycine-Asparagine (QGD) motif [12]. The RGD motif is exposed once EGFL7 attaches to the ECM but is hidden in the soluble form of EGFL7 [13]. The ITGAV:ITGB3 integrin can bind to fibronectin, collagen, fibrinogen, thrombospondin, and EGFL7, among others [7]. EGFL7 with its RGD motif competes for binding to ITGAV:ITGB3 integrin with matrix metalloproteinase2 (MMP2), fibronectin, and collagen IV. ITGB3 has important roles in angiogenesis, tumor metastasis, and drug resistance, leading to the development of novel specific RGD-like ligands for use in anti-tumor therapy (reviewed in [14]).

In this review, we introduce the angiocrine factor EGFL7 and one of its receptors (ITGB3) as regulators of angiogenesis and summarize recent knowledge on their involvement in tumor metastasis. We also discuss their involvement in drug resistance in cancer.

2. Epidermal Growth Factor-Like Protein-7

Mouse and human EGFL7 were cloned in 2003 by Soncin [5]. EGFL7 is a molecule that contains an N-terminal signaling sequence, followed by a cysteine-rich Emilin-like (EMI) domain and two epidermal growth factor-like (EGF-like) domains [5] (Figure 1a). The microRNA-126 gene (miR126) is located within intron 7 of the EGFL7 gene. Studies on the effects of miR126 in tumorigenesis are not covered in this review.

The EGFL7 gene locus contains binding sites for the transcription factors Krüppel-like factor 2 (KLF2) [15] and SMAD1/5 [16,17]. EGFL7 expression was upregulated on ECs by the blood-flow-sensitive transcription factor KLF2a after ITGB1-mediated induction [18] and by SMAD transcription factors after the binding of bone morphogenic protein-9 to the transmembrane anaplastic lymphoma kinase 1 receptor [16,17].

Whilst high expression is found during embryonic and neonatal development [19], EGFL7 is downregulated in almost all mature tissues except in the adult mouse lung, with lower expression in the heart, ovary, uterus, and kidneys [20]. EGFL7 expression rises again during vascular injury [21], during pregnancy, in regenerating endothelium following arterial injury, in growth plate injury [12], in atherosclerotic plaques, and in growing tumors, often mainly in tumor ECs [4].

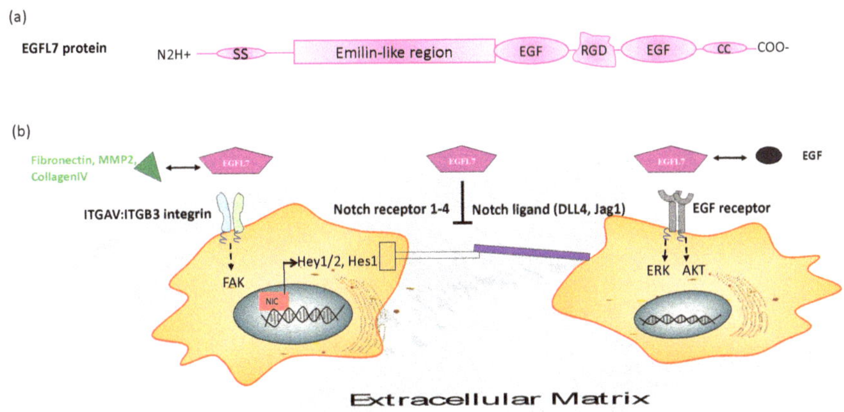

Figure 1. (**a**) Model structure of the EGFL7 protein. EGFL7 contains two EGF-like repeats: an arg-gly-asp integrin-binding motif (RGD) and an Emilin-like region. (**b**) The regulatory network of EGLF7. EGFL7 binds via its RGD domain to ITGAV:ITGB3 integrin and causes among others FAK autophosphorylation. EGFL7 competes for binding to the integrin with the ECM molecules fibronectin, MMP2, and collagen IV. The Emilin-like region of EGFL7 interacts with Notch receptor 1–4 and Notch ligands DLL4 and Jag1 and suppresses Notch signaling, resulting in impaired NIC translocation into the nucleus and reduced Hey1/2 and Hes1 transcription. EGFL7 binding to EGFR results in the activation of the signaling pathways extracellular signal-regulated kinase (ERK) and AKT, among others. EGFL7 competes for binding to EGFR with EGF. Abbreviations: EGF, epidermal growth factor; EGFL7, epidermal growth factor-like protein 7; EGFR, epidermal growth factor receptor; ERK, extracellular signal-regulated kinase; FAK, focal adhesion kinase); MMP2, matrix metalloproteinase 2; DLL4, delta like-protein 4; Jag1, Jagged1; NIC, Notch intracellular domain; ECM, extracellular matrix.

EGFL7 is a 41-kDa secreted signaling factor [4,5] that can be deposited into the ECM. EGFL7 contains a positive C- and a negative N-terminus, enabling the formation of EGFL7 oligomers that are deposited in the ECM in a head-to-tail fashion (Figure 2). It was recently shown that docking of the EGFL7 protein into both fibers and individual aggregates of the EC extracellular space requires the microfibrillar component microfibrillar-associated glycoprotein-1 and fibronectin [17]. The study demonstrated that docking of EGFL7 to the ECM is required for its effects on lysyl oxidase (LOX) activity, but that ECM binding was not necessary to mediate its effect on endothelial adhesion molecule expression or Hairy/enhancer-of-split related with YRPW motif protein 2 (Hey2) expression along the Notch pathway (Figure 1).

EGFL7 facilitates angiogenesis and tumorigenesis. It stimulates the recruitment and proliferation of embryonic or human brain ECs [7,22] and primary mouse embryonic fibroblasts [21]. EGFL7, through its multiple binding partners and cellular receptors (reviewed in [23]), can be found on various cell types, including tumor cells and ECs (Figure 1b). EGFL7 can bind to the NOTCH1–4 extracellular domain through its Emilin-like region [24]. EGFL7 competes with the Notch ligands Jag1 and Jag2 for Notch binding and inhibits Notch signaling (Figure 1b). EGFL7 competes with the Notch ligand Delta-like-4 for Notch4 binding on ECs, while suppressing Notch downstream signals like Hey1 and Hairy/enhancer-of-split 1 (Hes1) and promotes angiogenesis [22]. In acute myeloid leukemia, a hematopoietic blood cell cancer, recombinant EGFL7 inhibited DELTA-like 4-mediated Notch activation while anti-EGFL7 in combination with Dll4 increased Notch activation and induced apoptosis.

EGFL7 can also bind to the EGF receptor on the cell membrane, which results in the activation of the signaling pathways mitogen-activated protein kinase/extracellular signal-regulated kinase (ERK) and phosphatidylinositol-3-kinase/AKT [6,25] (Figure 1b). It was reported that EGFL7 binds to EGFR wildtype but not to the active mutant EGFR variant III, leading to b-catenin activation and upregulation of EGFL7 expression and tumor growth [26]. Binding of EGFL7 to EGFR enhanced cell migration of hepatocellular carcinoma cells and increased intrahepatic and pulmonary metastases in murine liver cancer models but did not alter tumor cell proliferation [27]. On the cellular level, EGFL7–EGFR binding caused phosphorylation of the cytoplasmic protein focal adhesion kinase (FAK) [27]. It is interesting to note that EGFR is expressed on intratumoral vessels but not vessels in non-tumor tissues [28–32], which would suggest that EGFL7 binding to EGFR could drive tumor-angiogenesis. But studies so far indicate that the EGFL7-driven pro-angiogenic effects are mainly mediated by Notch receptor or integrin and not by EGFR signaling [16,33].

EGFL7 enhanced migration of the siman virus 40-mouse microvascular endothelial (SVEC) cell line and resulted in the phosphorylation of ERK1/2, FAK, and STAT5 [12]. EGFL7 treatment of EGFL7-induced SVEC migration was blocked in the presence of RGD peptides, demonstrating the involvement of integrin signaling in EC migration. FAK is activated upon integrin or growth factor receptor signaling, resulting in the autophosphorylation at tyrosine (Y) 397. FAK is a key mediator of integrin signaling through its association with focal adhesion proteins, such as paxillin and talin. The role of FAK as both a cytosol and nuclear protein contributing to cancer progression has been recently reviewed by Murphey et al. [34].

Integrins mediate cell adhesion to the ECM. Adhesions serve as traction points and as signaling centers during cell migration [35]. There is an optimal strength of attachment that allows sufficient adhesion for traction at the cell front and yet allows for efficient release at the rear [36]. Integrin activation in protrusions regulates actin polymerization and myosin II activity through Rho-family GTPases such as Cdc42, Rac1, and RhoA [37]. EGFL7 potentiates EC migration on fibronectin-coated plates through binding to ITGAV:ITGB3 integrin [7], resulting in the activation of the downstream target GTPase Cdc42. EGFL7 cannot directly bind to ITGA5:ITGB1 integrin but enhanced angiogenesis involving this integrin [13,38]. Mechanistically, EGFL7 binding to ITGAV:ITGB3 integrin blocked the endocytosis of fibronectin-associated ITGA5:ITGB1 integrin [13] and ITGAV:ITGB3 and resulted in the upregulation of both integrins on the EC surface, allowing focal adhesion maturation, hydrolysis of Rac1-GFP, and enhanced migration speed of ECs on fibronectin surfaces [39].

EGFL7 controls proliferation in melanoma, hepatocellular carcinoma, and clear cell renal cell carcinoma [40–42] through one of its receptors. Blood cell cancers such as acute myeloid leukemia (AML) or the plasma cell malignancy multiple myeloma (MM) have dysfunctional integrin and Notch signaling [43–45]. EGFL7 caused acute myeloid leukemia (AML) blast proliferation [46]. Anti-EGFL7 blocking antibody through reactivation of Notch signaling in AML cells induced cell differentiation and apoptosis in vitro and in vivo [46]. Our group demonstrated that malignant plasma cells from patients with MM adhere to ECM-deposited EGFL7 and that their cell growth and survival required EGFL7 binding to ITGB3 [45]. These studies demonstrate that EGFL7 could be a potential cancer target.

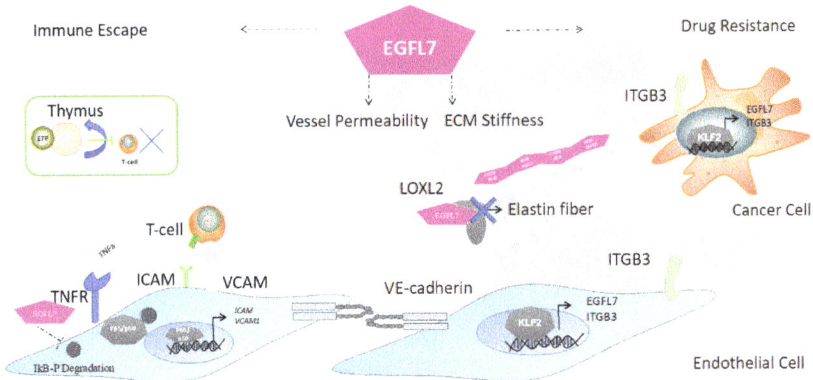

Figure 2. EGFL7 alters tumor growth and metastasis by suppressing the production of immune cells and their recruitment into the growing tumors, vessel permeability, ECM stiffness—all of which contribute to drug resistance. T cell adhesion and rolling and transmigration through the EC are required for T cells to cross the vascular barrier. EGFL7 allows the tumor to escape from the anti-tumor immune response by preventing terminal T cell differentiation in the thymus and inhibiting T cell recruitment via suppression of the adhesion molecules ICAM and VCAM on ECs. EGFL7 prevents adhesion molecule transcription after tumor necrosis factor alpha (TNFa) stimulation by blocking Nuclear factor kappa B (NFkB) signaling. EGFL7 controls ECM stiffness by interacting with LOXL2 so as to mitigate covalent crosslinking of collagen or elastin. ITGB1 integrin on ECs and ITGB3 integrin on cancer cells [45] induce the expression of transcription factor KLF2 which enhances EGFL7 expression, resulting in enhanced cell proliferation. Myelosuppressive drugs such as bortezomib were shown to enhance KLF2-mediated upregulation of ITGB3 and EGFL7. Abbreviations: KLF2, Krüppel-like factor 2; ETP, early thymic progenitor; LOXL2, lysyl oxidase-like 2.

3. EGFL7 Contributes to the Pathological Tumor Vessel Phenotype

As one of the hallmarks of cancer, angiogenesis is necessary for the transition of a small, localized tumor with a diameter of around 1–2 mm into an invasive disease. EGFL7 contributes to the branched and disorganized architecture of tumor endothelium with irregular multi-layered EC lining and an inconsistent smooth muscle and pericyte sheath.

Sprouting angiogenesis refers to the de novo formation of new vessels via the local proliferation of and extension of ECs from the wall of an existing vessel [47]. Cellular components of newly formed vessels include tip cells, which migrate in response to gradients of EGFL7 and VEGF, stalk cells, which proliferate and extend the vessels, and phalanx cells, which are quiescent and support the sprout [48]. For the formation of vessels, EC migration requires the coordinated attachment to and de-attachment from the surrounding ECM [36]. EGFL7 is a component of the interstitial ECM deposited mainly on the basal sides of sprouts at the interface between ECs and interstitial cells. The EGFL7 deposits form a unique ECM coat on the sprout surface. This coat outlines the boundary of a new sprout and marks the migratory path of new ECs [13]. In EGFL7-deficient ECs, the lack of EGFL7-mediated scaffolding of the ECM leaves ECs clueless on where to move. ECs not only attach in the path of the sprout but also attach to the basal sides of other ECs, leading to larger sprouts with multiple layers of ECs on top of each other. The accumulation of ECs hinders appropriate EC movement. Therefore, EGFL7 contributes to the hallmark excessively branched and disorganized architecture of tumor endothelium. All these features from EGFL7 contribute to an unstable vessel wall and promote the vessel leakiness that is characteristic of tumor ECs. Tumor EC-derived EGFL7 promotes glioma growth in experimental glioma models and stimulates tumor vascularization with the generation of mature vessels covered with pericytes and smooth muscle cells [13]. A recent study demonstrated that treatment of ECs with melanoma cell-derived exosome enhanced VE-Cadherin, uPAR, and EGFR upregulation in ECs [49], resulting in a tumor EC phenotype, and through EGFR gave EGFL7 a chance to promote neoangiogenesis.

During the metastatic stage of intravasation, tumor cells gain access to the circulation via vascular or lymphatic vessels, enabled by the transiently increased permeability of the tumor vasculature. Vascular impermeability is achieved with a barrier comprised of ECM components such as the endothelial glycocalyx, the endothelium, basement membrane, and accessory cells such as pericytes and smooth muscle cells. Besides vessel-wrapping, pericytes and smooth muscle cells ensure that no fluid, protein, or immune cell leakage occurs from blood to tissue [50]. Tumor vessels show inconsistent pericyte or smooth muscle cell coverage [51]. EGFL7 reduces EC smooth muscle coverage by blocking platelet-derived growth factor-BB-mediated smooth muscle cell migration [5]. Tumor vessels have irregular cell surface cell walls and show leakiness. EGFL7 knockdown in ECs showed a disturbance of adherens junctions with insufficient phosphorylation of VE-cadherin that contributed to increased EC permeability [33] (Figure 2). EGFL7 knockdown suppresses VEGF-A-mediated angiogenesis, causes the overproduction of endothelial filopodia on the basal side of ECs, and reduces collagen IV deposition at the basal side of ECs. These data demonstrate that EGFL7 controls the actions of one of its angiocrine allies, VEGF. VEGF has been shown to enhance vessel permeability by a cross-activation of ITGB3 and VEGFR2 that directly regulates VE-cadherin [52].

4. EGFL7 Enhances Tumoral Immune Escape

Immune cells, such as T lymphocytes, are part of the body's weapons to fight off tumor cells through their capacity for antigen-directed cytotoxicity. EGFL7 compromises the anti-tumor response on two levels: it impairs the production of terminally differentiated T lymphocytes and it prevents lymphocytes from crossing the vasculature into the tumor bed.

The Flt3/Flt3-ligand pathway is important for early thymic precursors (ETP) expansion [53] and lymphocyte (T and B cell) development. EGFL7 blocks terminal T cell differentiation but expands ETPs in the thymus, the place where T cell precursors reside [54]. Mechanistically, EGFL7-mediated suppression of Notch signaling [55] enhanced the fms-related tyrosine kinase 3 (Flt3) promoter activity in ETPs, causing upregulation of FLT3 on ETPs. EGFL7 also increased the number of thymic ECs, which are a source of Flt3 ligand. Together, EGFL7 stimulates Flt3/Flt3 ligand signaling in ETPs that resulted in the accumulation of immature ETPs and a paucity of circulating T cells and contributes to impaired antigen-directed cytotoxicity against tumor cells.

At the center for the anti-tumor response of immune cells is the ability of immune cells to reach the tumor tissues. Leukocyte immune cell traffic out of the bloodstream into the tumor tissues requires the activation of adhesion molecules. ECs can actively impact T lymphocyte migration by changing the expression of leukocyte adhesion molecules on their surface. The activation of very late antigen-4 (VLA-4; ITGA4:ITGB1, CD49d/CD29) and lymphocyte function-associated antigen 1 (LFA-1,αLβ2, CD11a/CD18) integrin by talin and kindlin allow firm interaction between the immune cell-like T cells or neutrophils and ECs, which express integrin ligands such as intercellular adhesion molecules (ICAMs), vascular cell adhesion molecule 1 (VCAM1), and MAdCAM (reviewed in [56]). These molecules enable T lymphocytes to adhere to ECs (Figure 2), a process that is required for them to cross into the tumor bed. Inhibiting leukocyte migration into the tumor niche at the EC level represents one way for cancer cells to escape anti-tumor host immune responses.

Following the initial observation that EGFL7-overexpressing tumor cells showed enhanced tumor growth and metastasis in immunocompetent but not in immunodeficient mice with an impaired influx of inflammatory cells [57], it was shown that EGFL7 modulates the immune cell recruitment process. Mechanistically, EGFL7 downregulated the endothelial adhesion molecules ICAM1 and VCAM1 on the ECs of 4T1 breast cancer and LLC1 lung adenocarcinoma tumors [57]. Follow-up studies revealed that the treatment of ECs with tumor necrosis factor-alpha (TNFa) repressed VCAM1 or ICAM1 expression and downregulated ICAM1 and VCAM1 transcription via an NFkB dependent mechanism [58]. Because a lack of adhesion factors prevents lymphocytes from binding to ECs, a necessary step for cells to cross the vasculature and enter the tumor bed, EGFL7-induced alteration of adhesive properties endows tumors with the ability to escape immune attack [57] (Figure 2).

5. EGFL7 Regulates ECM Stiffness and EMT

Fibroblasts and tumor cells produce large quantities of the ECM molecules' collagen and fibronectin [59]. The generated highly fibrotic tumor microenvironment causes ECM stiffness that favors tumor progression/metastasis and epithelial–mesenchymal transition (EMT) [60,61]. The lysyl oxidase family comprising LOX and four lysyl oxidase-like proteins (LOXL1–4) (reviewed recently [62]) contributes to ECM stiffness by catalyzing the covalent cross-linking of collagen and elastin. Pro-LOX is synthesized and secreted as a pro-enzyme and, following procollagen and C-proteinase cleavage, gives rise to LOX-PP and LOX. LOX-PP interacts with collagen I, LOXL2, MMP-2, fibromodulin, EGF [63], and as a recent study demonstrated, with EGFL7 [64]. Indeed, EGFL7 binds to all members of the LOX family including LOX and LOXL1–4 [64,65], indicating its importance in the regulation of tumor ECM stiffness. Earlier studies demonstrated that EGFL7 suppresses LOXL2 function [65]. EGFL7 binding to the catalytic LOXL2 domain impaired the conversion of tropoelastin into mature insoluble elastin [65] (Figure 2).

High ECM stiffness promotes EMT and metastasis [60]. In the early stages of metastasis, epithelial cells decrease the expression of cell–cell junction molecules at the primary tumor site and become motile [66,67]. The upregulation of the EMT-associated genes Twist and Snail and the loss of the epithelial marker E-cadherin and the induction of the mesenchymal marker vimentin are typical gene expression patterns during EMT [68]. It was shown that EGFL7 promoted metastasis by triggering EMT in gastric cancer cells with the downregulation of E-cadherin and upregulation of vimentin and Snail [6].

6. EGFL7 Contributes to Drug Resistance

Many cancer patients who receive neoadjuvant or adjuvant treatment to surgery relapse months or years later with a tumor that is resistant to further chemotherapeutic challenge, a feature known as chemoresistance [69]. Despite advances in cancer treatment and the initial tumor reduction after drug treatment, cancer cells develop drug resistance. Resistance to anti-cancer treatment is dependent not only on genetic mutations and epigenetics but also on external factors [70], including cytokines, growth factors, enzymes, glycoproteins, extracellular vesicles, angiocrine factors, and integrins. We showed that irradiation augmented EGFL7 expression in thymic ECs [54]. Drug resistance to the proteosome inhibitor and anti-myeloma bortezomib often occurs in MM patients. Our group reported that treatment with bortezomib, but not other drugs, induced EGFL7 expression in MM cells [45] (Figure 2).

Integrin/ECM interactions mediate cell adhesion-mediated drug resistance, activating a pro-survival and anti-apoptotic program. This escape strategy of tumors is either due to the survival of cells already expressing certain integrins and/or cells capable of inducing integrin gene expression (for review [70]). MM treatment using bortezomib upregulated ITGB3 on MM cells [45]. Our data indicated that targeting EGFL7 using neutralizing antibodies or ITGB3 using an integrin inhibitor could override drug resistance against bortezomib in vitro and using murine in vivo MM models. Mechanistically, we showed that EGFL7-mediated activation of ITGB3 induced the expression of transcription factor KLF2 that further augmented EGFL7 expression in MM cells. Our study demonstrated that an EGFL7-ITGB3-KLF2-EGFL7 amplification loop supports MM cell survival and proliferation in vitro as well as in vivo [45] (Figure 2). In other cell types, such as the human osteosarcoma, where EGFL7 is highly expressed in tumor cells even more so than in ECs, chemotherapy reduced EGFL7 expression [71]. These data suggest that each tumor cell type, dependent on the needs of the ECM niche, develops its own drug escape mechanism.

The existence of tumor cancer stem cells is another tumor and drug escape mechanism. Tumor-initiating cells represent a small fraction within the tumor. These cells have stem cell characteristics with the capacity for self-renewal, they rest in a non-cycling/quiescent state, and retain their ability to differentiate. First shown in neuronal stem cells, EGFL7 can modulate stem cell fate through its liaison with Notch receptors and ligands [24]. We showed that EGFL7 expands immature early thymic progenitor cells [54]. Since many anti-cancer drugs require tumor cell cycling, keeping tumor cells in a quiescent state is just another trick by which EGFL7 enhances drug resistance.

Integrin signaling functions depend on the cytoplasmic protein tyrosine kinase FAK, the major protein found in focal adhesions. While ITGB3-mediated tumor growth was thought to require downstream activation of FAK and cell adhesion, a recent study challenged this doctrine and showed that neither of them was required for c-Src activation and Crk-associated substrate phosphorylation that enhanced tumor growth [72]. Tavora et al. demonstrated that targeting of EC FAK sensitized tumors to DNA-damaging therapy [73]. EC FAK is necessary for the DNA damage-induced NFkB activation required for the cytokine production in ECs. Given that EGFL7 binding to ITGB3 can activate FAK in ECs [7,74], and EC FAK contributes to drug resistance, it is conceivable that blocking EGFL7-ITGB3 can improve chemosensitization to DNA-damaging therapies through the loss of FAK (reviewed in [75]).

7. Conclusions

The ECM molecule EGFL7 is a critical player in the metastatic program, an inhibitor of the anti-cancer immune response escape, and contributes to drug resistance. The effects of EGFL7 on tumor cells and ECs and its diversity of binding receptors enable EGFL7 to impinge on tumor survival strategies.

Proper signals from the ECM maintain stem cell fate. Certain integrins are exclusively abundant in epithelial stem cells, with one of them being ITGAV:ITGB3 integrin, which has been described on lung, breast, and pancreatic tumors with a stem-like phenotype [76]. A deeper understanding of the role of EGFL7 in controlling the fate of tumor-initiating cells and cancer cells and the identification of new ECM binding partners or signaling receptors will open up new avenues for cancer treatment.

EGFL7 unites with main drivers of tumorigenesis from EGFR, ITGB3, Notch ligand, and receptors or LOX family members. Although anti-EGFL7 therapy alone could not sufficiently control tumor growth, the current state of research indicates that combining anti-EGFL7 therapy with other anti-cancer strategies such as chemotherapy might improve the efficacy of conventional anti-cancer strategies.

Author Contributions: Writing—review and editing, B.H., Y.S., S.T. and K.O., and K.H.; visualization, B.H.; supervision, B.H.; project administration, B.H.; funding acquisition, B.H. and K.H. All authors have read and agreed to the published version of the manuscript.

Funding: This work was supported partly by grants from the Japan Society for the Promotion of Science (Kiban C grant no. 19K08857 (B.H.), 17K09941 (K.H.), 19K08858 (S.T.), The Uehara Memorial Foundation (B.H.), a grant from the Nakatani Foundation (K.H.), The Japanese Society of Hematology Research Grant (K.H.), Thermo Life Foundation (K.H.) and Okinaka Memorial Institute for Medical Research (K.H.), and grants from the International Joint Usage/Research Center, the Institute of Medical Science, the University of Tokyo (2019–2058), Japan.

Acknowledgments: The authors would like to dedicate this manuscript to Zena Werb (1945–2020), a pioneer of protease and cancer biology. She was a role model for me as a woman scientist. She has and will continue to be a true inspiration for women and men alike. The authors thank Robert Whittier for proofreading the manuscript.

Conflicts of Interest: The authors declare no conflict of interest.

References

1. Kerbel, R.S. Tumor angiogenesis. *N. Engl. J. Med.* **2008**, *358*, 2039–2049. [CrossRef]
2. Heissig, B.; Ohki-Koizumi, M.; Tashiro, Y.; Gritli, I.; Sato-Kusubata, K.; Hattori, K. New functions of the fibrinolytic system in bone marrow cell-derived angiogenesis. *Int. J. Hematol.* **2012**, *95*, 131–137. [CrossRef]

3. Ohki, M.; Ohki, Y.; Ishihara, M.; Nishida, C.; Tashiro, Y.; Akiyama, H.; Komiyama, H.; Lund, L.R.; Nitta, A.; Yamada, K.; et al. Tissue type plasminogen activator regulates myeloid-cell dependent neoangiogenesis during tissue regeneration. *Blood* **2010**, *115*, 4302–4312. [CrossRef] [PubMed]
4. Parker, L.H.; Schmidt, M.; Jin, S.W.; Gray, A.M.; Beis, D.; Pham, T.; Frantz, G.; Palmieri, S.; Hillan, K.; Stainier, D.Y.; et al. The endothelial-cell-derived secreted factor Egfl7 regulates vascular tube formation. *Nature* **2004**, *428*, 754–758. [CrossRef]
5. Soncin, F.; Mattot, V.; Lionneton, F.; Spruyt, N.; Lepretre, F.; Begue, A.; Stehelin, D. VE-statin, an endothelial repressor of smooth muscle cell migration. *EMBO J.* **2003**, *22*, 5700–5711. [CrossRef] [PubMed]
6. Luo, B.H.; Xiong, F.; Wang, J.P.; Li, J.H.; Zhong, M.; Liu, Q.L.; Luo, G.Q.; Yang, X.J.; Xiao, N.; Xie, B.; et al. Epidermal growth factor-like domain-containing protein 7 (EGFL7) enhances EGF receptor-AKT signaling, epithelial-mesenchymal transition, and metastasis of gastric cancer cells. *PLoS ONE* **2014**, *9*, e99922. [CrossRef] [PubMed]
7. Nikolic, I.; Stankovic, N.D.; Bicker, F.; Meister, J.; Braun, H.; Awwad, K.; Baumgart, J.; Simon, K.; Thal, S.C.; Patra, C.; et al. EGFL7 ligates alphavbeta3 integrin to enhance vessel formation. *Blood* **2013**, *121*, 3041–3050. [CrossRef]
8. LaFoya, B.; Munroe, J.A.; Mia, M.M.; Detweiler, M.A.; Crow, J.J.; Wood, T.; Roth, S.; Sharma, B.; Albig, A.R. Notch: A multi-functional integrating system of microenvironmental signals. *Dev. Biol.* **2016**, *418*, 227–241. [CrossRef]
9. Michael, M.; Parsons, M. New perspectives on integrin-dependent adhesions. *Curr. Opin. Cell Biol.* **2020**, *63*, 31–37. [CrossRef]
10. Hamidi, H.; Ivaska, J. Every step of the way: Integrins in cancer progression and metastasis. *Nat. Rev. Cancer* **2018**, *18*, 533–548. [CrossRef]
11. Su, C.-Y.; Li, J.-Q.; Zhang, L.-L.; Wang, H.; Wang, F.-H.; Tao, Y.-W.; Wang, Y.-Q.; Guo, Q.-R.; Li, J.-J.; Liu, Y.; et al. The Biological Functions and Clinical Applications of Integrins in Cancers. *Front. Pharmacol.* **2020**, *11*, 1435. [CrossRef] [PubMed]
12. Chim, S.M.; Kuek, V.; Chow, S.T.; Lim, B.S.; Tickner, J.; Zhao, J.; Chung, R.; Su, Y.W.; Zhang, G.; Erber, W.; et al. EGFL7 is expressed in bone microenvironment and promotes angiogenesis via ERK, STAT3, and integrin signaling cascades. *J. Cell Physiol.* **2015**, *230*, 82–94. [CrossRef]
13. Dudvarski Stankovic, N.; Bicker, F.; Keller, S.; Jones, D.T.; Harter, P.N.; Kienzle, A.; Gillmann, C.; Arnold, P.; Golebiewska, A.; Keunen, O.; et al. EGFL7 enhances surface expression of integrin alpha5beta1 to promote angiogenesis in malignant brain tumors. *EMBO Mol. Med.* **2018**, *10*, e8420. [CrossRef] [PubMed]
14. Sani, S.; Messe, M.; Fuchs, Q.; Pierrevelcin, M.; Laquerriere, P.; Entz-Werle, N.; Reita, D.; Etienne-Selloum, N.; Bruban, V.; Choulier, L.; et al. Biological Relevance of RGD-Integrin Subtype-Specific Ligands in Cancer. *ChemBioChem* **2020**. [CrossRef]
15. Harris, T.A.; Yamakuchi, M.; Kondo, M.; Oettgen, P.; Lowenstein, C.J. Ets-1 and Ets-2 regulate the expression of miR-126 in endothelial cells. *Arterioscler. Thromb. Vasc. Biol.* **2010**, *30*, 1990–1997. [CrossRef]
16. Richter, A.; Alexdottir, M.S.; Magnus, S.H.; Richter, T.R.; Morikawa, M.; Zwijsen, A.; Valdimarsdottir, G. EGFL7 Mediates BMP9-Induced Sprouting Angiogenesis of Endothelial Cells Derived from Human Embryonic Stem Cells. *Stem Cell Rep.* **2019**, *12*, 1250–1259. [CrossRef]
17. Villain, G.; Lelievre, E.; Broekelmann, T.; Gayet, O.; Havet, C.; Werkmeister, E.; Mecham, R.; Dusetti, N.; Soncin, F.; Mattot, V. MAGP-1 and fibronectin control EGFL7 functions by driving its deposition into distinct endothelial extracellular matrix locations. *FEBS J.* **2018**, *285*, 4394–4412. [CrossRef]
18. Renz, M.; Otten, C.; Faurobert, E.; Rudolph, F.; Zhu, Y.; Boulday, G.; Duchene, J.; Mickoleit, M.; Dietrich, A.C.; Ramspacher, C.; et al. Regulation of β1 integrin-Klf2-mediated angiogenesis by CCM proteins. *Dev. Cell* **2015**, *32*, 181–190. [CrossRef] [PubMed]
19. Nichol, D.; Stuhlmann, H. EGFL7: A unique angiogenic signaling factor in vascular development and disease. *Blood* **2012**, *119*, 1345–1352. [CrossRef]
20. Fitch, M.J.; Campagnolo, L.; Kuhnert, F.; Stuhlmann, H. Egfl7, a novel epidermal growth factor-domain gene expressed in endothelial cells. *Dev. Dyn.* **2004**, *230*, 316–324. [CrossRef] [PubMed]
21. Campagnolo, L.; Leahy, A.; Chitnis, S.; Koschnick, S.; Fitch, M.J.; Fallon, J.T.; Loskutoff, D.; Taubman, M.B.; Stuhlmann, H. EGFL7 is a chemoattractant for endothelial cells and is up-regulated in angiogenesis and arterial injury. *Am. J. Pathol.* **2005**, *167*, 275–284. [CrossRef]
22. Nichol, D.; Shawber, C.; Fitch, M.J.; Bambino, K.; Sharma, A.; Kitajewski, J.; Stuhlmann, H. Impaired angiogenesis and altered Notch signaling in mice overexpressing endothelial Egfl7. *Blood* **2010**, *116*, 6133–6143. [CrossRef] [PubMed]
23. Hong, S.-K.; Lee, H.; Kwon, O.-S.; Song, N.-Y.; Lee, H.-J.; Kang, S.; Kim, J.-H.; Kim, M.; Kim, W.; Cha, H.-J. Large-scale pharmacogenomics based drug discovery for ITGB3 dependent chemoresistance in mesenchymal lung cancer. *Mol. Cancer* **2018**, *17*, 175. [CrossRef]
24. Schmidt, M.H.H.; Bicker, F.; Nikolic, I.; Meister, J.; Babuke, T.; Picuric, S.; Muller-Esterl, W.; Plate, K.H.; Dikic, I. Epidermal growth factor-like domain 7 (EGFL7) modulates Notch signalling and affects neural stem cell renewal. *Nat. Cell Biol.* **2009**, *11*, 873–880. [CrossRef] [PubMed]
25. Huang, C.; Yuan, X.; Wan, Y.; Liu, F.; Chen, X.; Zhan, X.; Li, X. VE-statin/Egfl7 expression in malignant glioma and its relevant molecular network. *Int. J. Clin. Exp. Pathol.* **2014**, *7*, 1022–1031. [PubMed]
26. Wang, F.-Y.-F.; Kang, C.-S.; Wang-Gou, S.-Y.; Huang, C.-H.; Feng, C.-Y.; Li, X.-J. EGFL7 is an intercellular EGFR signal messenger that plays an oncogenic role in glioma. *Cancer Lett.* **2017**, *384*, 9–18. [CrossRef]
27. Wu, F.; Yang, L.Y.; Li, Y.F.; Ou, D.P.; Chen, D.P.; Fan, C. Novel role for epidermal growth factor-like domain 7 in metastasis of human hepatocellular carcinoma. *Hepatology* **2009**, *50*, 1839–1850. [CrossRef] [PubMed]

28. Ellis, L.M. Epidermal growth factor receptor in tumor angiogenesis. *Hematol. Oncol. Clin. N. Am.* **2004**, *18*, 1007–1021. [CrossRef] [PubMed]
29. Hirata, A.; Uehara, H.; Izumi, K.; Naito, S.; Kuwano, M.; Ono, M. Direct inhibition of EGF receptor activation in vascular endothelial cells by gefitinib ('Iressa', ZD1839). *Cancer Sci.* **2004**, *95*, 614–618. [CrossRef] [PubMed]
30. Tabernero, J. The role of VEGF and EGFR inhibition: Implications for combining anti-VEGF and anti-EGFR agents. *Mol. Cancer Res.* **2007**, *5*, 203–220. [CrossRef]
31. Tortora, G.; Ciardiello, F.; Gasparini, G. Combined targeting of EGFR-dependent and VEGF-dependent pathways: Rationale, preclinical studies and clinical applications. *Nat. Clin. Pract. Oncol.* **2008**, *5*, 521–530. [CrossRef]
32. Yen, L.; Benlimame, N.; Nie, Z.R.; Xiao, D.; Wang, T.; Al Moustafa, A.E.; Esumi, H.; Milanini, J.; Hynes, N.E.; Pages, G.; et al. Differential regulation of tumor angiogenesis by distinct ErbB homo- and heterodimers. *Mol. Biol. Cell* **2002**, *13*, 4029–4044. [CrossRef] [PubMed]
33. Usuba, R.; Pauty, J.; Soncin, F.; Matsunaga, Y.T. EGFL7 regulates sprouting angiogenesis and endothelial integrity in a human blood vessel model. *Biomaterials* **2019**, *197*, 305–316. [CrossRef]
34. Murphy, J.M.; Rodriguez, Y.A.R.; Jeong, K.; Ahn, E.-Y.E.; Lim, S.-T.S. Targeting focal adhesion kinase in cancer cells and the tumor microenvironment. *Exp. Mol. Med.* **2020**, *52*, 877–886. [CrossRef] [PubMed]
35. Parsons, J.T.; Horwitz, A.R.; Schwartz, M.A. Cell adhesion: Integrating cytoskeletal dynamics and cellular tension. *Nat. Rev. Mol. Cell Biol.* **2010**, *11*, 633–643. [CrossRef]
36. Devreotes, P.; Horwitz, A.R. Signaling Networks that Regulate Cell Migration. *Cold Spring Harb. Perspect. Biol.* **2015**, *7*, a005959. [CrossRef] [PubMed]
37. Keely, P.J.; Westwick, J.K.; Whitehead, I.P.; Der, C.J.; Parise, L.V. Cdc42 and Rac1 induce integrin-mediated cell motility and invasiveness through PI(3)K. *Nature* **1997**, *390*, 632–636. [CrossRef]
38. White, D.P.; Caswell, P.T.; Norman, J.C. alpha v beta3 and alpha5beta1 integrin recycling pathways dictate downstream Rho kinase signaling to regulate persistent cell migration. *J. Cell Biol.* **2007**, *177*, 515–525. [CrossRef]
39. Jacquemet, G.; Green, D.M.; Bridgewater, R.E.; von Kriegsheim, A.; Humphries, M.J.; Norman, J.C.; Caswell, P.T. RCP-driven α5β1 recycling suppresses Rac and promotes RhoA activity via the RacGAP1-IQGAP1 complex. *J. Cell Biol.* **2013**, *202*, 917–935. [CrossRef]
40. Tang, H.; Xiao, W.R.; Liao, Y.Y.; Li, L.; Xiao, X.; Xu, X.P.; Feng, H. EGFL7 silencing inactivates the Notch signaling pathway; enhancing cell apoptosis and suppressing cell proliferation in human cutaneous melanoma. *Neoplasma* **2019**, *66*, 187–196. [CrossRef]
41. Hu, M.H.; Ma, C.Y.; Wang, X.M.; Ye, C.D.; Zhang, G.X.; Chen, L.; Wang, J.G. MicroRNA-126 inhibits tumor proliferation and angiogenesis of hepatocellular carcinoma by down-regulating EGFL7 expression. *Oncotarget* **2016**, *7*, 66922–66934. [CrossRef]
42. Zhai, W.; Zhu, R.; Ma, J.; Gong, D.; Zhang, H.; Zhang, J.; Chen, Y.; Huang, Y.; Zheng, J.; Xue, W. A positive feed-forward loop between LncRNA-URRCC and EGFL7/P-AKT/FOXO3 signaling promotes proliferation and metastasis of clear cell renal cell carcinoma. *Mol. Cancer* **2019**, *18*, 81. [CrossRef]
43. Dias, S.; Hattori, K.; Heissig, B.; Zhu, Z.; Wu, Y.; Witte, L.; Hicklin, D.J.; Tateno, M.; Bohlen, P.; Moore, M.A.; et al. Inhibition of both paracrine and autocrine VEGF/ VEGFR-2 signaling pathways is essential to induce long-term remission of xenotransplanted human leukemias. *Proc. Natl. Acad. Sci. USA* **2001**, *98*, 10857–10862. [CrossRef] [PubMed]
44. Heissig, B.; Eiamboonsert, S.; Salama, Y.; Shimazu, H.; Dhahri, D.; Munakata, S.; Tashiro, Y.; Hattori, K. Cancer therapy targeting the fibrinolytic system. *Adv. Drug Deliv. Rev.* **2016**, *99*, 172–179. [CrossRef]
45. Salama, Y.; Heida, A.H.; Yokoyama, K.; Takahashi, S.; Hattori, K.; Heissig, B. The EGFL7-ITGB3-KLF2 axis enhances survival of multiple myeloma in preclinical models. *Blood Adv.* **2020**, *4*, 1021–1037. [CrossRef]
46. Bill, M.; Pathmanathan, A.; Karunasiri, M.; Shen, C.; Burke, M.H.; Ranganathan, P.; Papaioannou, D.; Zitzer, N.C.; Snyder, K.; LaRocco, A.; et al. EGFL7 Antagonizes NOTCH Signaling and Represents a Novel Therapeutic Target in Acute Myeloid Leukemia. *Clin. Cancer Res.* **2020**, *26*, 669. [CrossRef]
47. Heissig, B.; Hattori, K.; Friedrich, M.; Rafii, S.; Werb, Z. Angiogenesis: Vascular remodeling of the extracellular matrix involves metalloproteinases. *Curr. Opin. Hematol.* **2003**, *10*, 136–141. [CrossRef] [PubMed]
48. Blanco, R.; Gerhardt, H. VEGF and Notch in tip and stalk cell selection. *Cold Spring Harb. Perspect. Med.* **2013**, *3*, a006569. [CrossRef] [PubMed]
49. Biagioni, A.; Laurenzana, A.; Menicacci, B.; Peppicelli, S.; Andreucci, E.; Bianchini, F.; Guasti, D.; Paoli, P.; Serratì, S.; Mocali, A.; et al. uPAR-expressing melanoma exosomes promote angiogenesis by VE-Cadherin, EGFR and uPAR overexpression and rise of ERK1,2 signaling in endothelial cells. *Cell Mol. Life Sci.* **2020**. [CrossRef] [PubMed]
50. Sarelius, I.H.; Glading, A.J. Control of vascular permeability by adhesion molecules. *Tissue Barriers* **2015**, *3*, e985954. [CrossRef] [PubMed]
51. Tilki, D.; Kilic, N.; Sevinc, S.; Zywietz, F.; Stief, C.G.; Ergun, S. Zone-specific remodeling of tumor blood vessels affects tumor growth. *Cancer* **2007**, *110*, 2347–2362. [CrossRef] [PubMed]
52. Mahabeleshwar, G.H.; Feng, W.; Reddy, K.; Plow, E.F.; Byzova, T.V. Mechanisms of integrin-vascular endothelial growth factor receptor cross-activation in angiogenesis. *Circ. Res.* **2007**, *101*, 570–580. [CrossRef] [PubMed]
53. Kenins, L.; Gill, J.W.; Hollander, G.A.; Wodnar-Filipowicz, A. Flt3 ligand-receptor interaction is important for maintenance of early thymic progenitor numbers in steady-state thymopoiesis. *Eur. J. Immunol.* **2010**, *40*, 81–90. [CrossRef]

54. Salama, Y.; Hattori, K.; Heissig, B. The angiogenic factor Egfl7 alters thymogenesis by activating Flt3 signaling. *Biochem. Biophys. Res. Commun.* **2017**, *490*, 209–216. [CrossRef]
55. Kato, T.; Sakata-Yanagimoto, M.; Nishikii, H.; Ueno, M.; Miyake, Y.; Yokoyama, Y.; Asabe, Y.; Kamada, Y.; Muto, H.; Obara, N.; et al. Hes1 suppresses acute myeloid leukemia development through FLT3 repression. *Leukemia* **2015**, *29*, 576–585. [CrossRef] [PubMed]
56. Harjunpää, H.; Llort Asens, M.; Guenther, C.; Fagerholm, S.C. Cell Adhesion Molecules and Their Roles and Regulation in the Immune and Tumor Microenvironment. *Front. Immunol.* **2019**, *10*, 1078. [CrossRef]
57. Delfortrie, S.; Pinte, S.; Mattot, V.; Samson, C.; Villain, G.; Caetano, B.; Lauridant-Philippin, G.; Baranzelli, M.C.; Bonneterre, J.; Trottein, F.; et al. Egfl7 promotes tumor escape from immunity by repressing endothelial cell activation. *Cancer Res.* **2011**, *71*, 7176–7186. [CrossRef] [PubMed]
58. Pinte, S.; Caetano, B.; Le Bras, A.; Havet, C.; Villain, G.; Dernayka, R.; Duez, C.; Mattot, V.; Soncin, F. Endothelial Cell Activation Is Regulated by Epidermal Growth Factor-like Domain 7 (Egfl7) during Inflammation. *J. Biol. Chem.* **2016**, *291*, 24017–24028. [CrossRef]
59. Naba, A.; Clauser, K.R.; Hoersch, S.; Liu, H.; Carr, S.A.; Hynes, R.O. The matrisome: In silico definition and in vivo characterization by proteomics of normal and tumor extracellular matrices. *Mol. Cell Proteom.* **2012**, *11*, M111.014647. [CrossRef]
60. Fattet, L.; Jung, H.-Y.; Matsumoto, M.W.; Aubol, B.E.; Kumar, A.; Adams, J.A.; Chen, A.C.; Sah, R.L.; Engler, A.J.; Pasquale, E.B.; et al. Matrix Rigidity Controls Epithelial-Mesenchymal Plasticity and Tumor Metastasis via a Mechanoresponsive EPHA2/LYN Complex. *Dev. Cell* **2020**, *54*, 302–316. [CrossRef]
61. Wei, S.C.; Fattet, L.; Tsai, J.H.; Guo, Y.; Pai, V.H.; Majeski, H.E.; Chen, A.C.; Sah, R.L.; Taylor, S.S.; Engler, A.J.; et al. Matrix stiffness drives epithelial–mesenchymal transition and tumour metastasis through a TWIST1–G3BP2 mechanotransduction pathway. *Nat. Cell Biol.* **2015**, *17*, 678–688. [CrossRef] [PubMed]
62. Vallet, S.D.; Ricard-Blum, S. Lysyl oxidases: From enzyme activity to extracellular matrix cross-links. *Essays Biochem.* **2019**, *63*, 349–364. [CrossRef]
63. Vallet, S.D.; Miele, A.E.; Uciechowska-Kaczmarzyk, U.; Liwo, A.; Duclos, B.; Samsonov, S.A.; Ricard-Blum, S. Insights into the structure and dynamics of lysyl oxidase propeptide, a flexible protein with numerous partners. *Sci. Rep.* **2018**, *8*, 11768. [CrossRef] [PubMed]
64. Vallet, S.D.; Berthollier, C.; Salza, R.; Muller, L.; Ricard-Blum, S. The Interactome of Cancer-Related Lysyl Oxidase and Lysyl Oxidase-Like Proteins. *Cancers* **2021**, *13*, 71. [CrossRef] [PubMed]
65. Lelievre, E.; Hinek, A.; Lupu, F.; Buquet, C.; Soncin, F.; Mattot, V. VE-statin/egfl7 regulates vascular elastogenesis by interacting with lysyl oxidases. *EMBO J.* **2008**, *27*, 1658–1670. [CrossRef] [PubMed]
66. Pastushenko, I.; Brisebarre, A.; Sifrim, A.; Fioramonti, M.; Revenco, T.; Boumahdi, S.; Van Keymeulen, A.; Brown, D.; Moers, V.; Lemaire, S.; et al. Identification of the tumour transition states occurring during EMT. *Nature* **2018**, *556*, 463–468. [CrossRef] [PubMed]
67. Yang, J.; Antin, P.; Berx, G.; Blanpain, C.; Brabletz, T.; Bronner, M.; Campbell, K.; Cano, A.; Casanova, J.; Christofori, G.; et al. Guidelines and definitions for research on epithelial–mesenchymal transition. *Nat. Rev. Mol. Cell Biol.* **2020**, *21*, 341–352. [CrossRef]
68. Lambert, A.W.; Pattabiraman, D.R.; Weinberg, R.A. Emerging Biological Principles of Metastasis. *Cell* **2017**, *168*, 670–691. [CrossRef]
69. Dillekås, H.; Rogers, M.S.; Straume, O. Are 90% of deaths from cancer caused by metastases? *Cancer Med.* **2019**, *8*, 5574–5576. [CrossRef]
70. Seguin, L.; Desgrosellier, J.S.; Weis, S.M.; Cheresh, D.A. Integrins and cancer: Regulators of cancer stemness, metastasis, and drug resistance. *Trends Cell Biol.* **2015**, *25*, 234–240. [CrossRef]
71. Liu, Q.; He, H.; Yuan, Y.; Zeng, H.; Wang, Z.; Luo, W. Novel Expression of EGFL7 in Osteosarcoma and Sensitivity to Cisplatin. *Front. Oncol.* **2020**, *10*, 74. [CrossRef]
72. Desgrosellier, J.S.; Barnes, L.A.; Shields, D.J.; Huang, M.; Lau, S.K.; Prévost, N.; Tarin, D.; Shattil, S.J.; Cheresh, D.A. An integrin alpha(v)beta(3)-c-Src oncogenic unit promotes anchorage-independence and tumor progression. *Nat. Med.* **2009**, *15*, 1163–1169. [CrossRef]
73. Tavora, B.; Reynolds, L.E.; Batista, S.; Demircioglu, F.; Fernandez, I.; Lechertier, T.; Lees, D.M.; Wong, P.P.; Alexopoulou, A.; Elia, G.; et al. Endothelial-cell FAK targeting sensitizes tumours to DNA-damaging therapy. *Nature* **2014**, *514*, 112–116. [CrossRef] [PubMed]
74. Schmidt, M.; De Maziere, A.; Smyczek, T.; Gray, A.; Parker, L.; Filvaroff, E.; French, D.; van Dijk, S.; Klumperman, J.; Ye, W. The Role of Egfl7 in Vascular Morphogenesis. In *Vascular Development: Novartis Foundation Symposium*; Novartis Foundation: Basel, Switzerland, 2007; Volume 283, pp. 18–28, discussion 28–36, 238–241.
75. Roy-Luzarraga, M.; Hodivala-Dilke, K. Molecular Pathways: Endothelial Cell FAK-A Target for Cancer Treatment. *Clin. Cancer Res.* **2016**, *22*, 3718–3724. [CrossRef]
76. Seguin, L.; Kato, S.; Franovic, A.; Camargo, M.F.; Lesperance, J.; Elliott, K.C.; Yebra, M.; Mielgo, A.; Lowy, A.M.; Husain, H.; et al. An integrin β_3-KRAS-RalB complex drives tumour stemness and resistance to EGFR inhibition. *Nat. Cell Biol.* **2014**, *16*, 457–468. [CrossRef]

cancers

MDPI

Article

Implications of ABCC4–Mediated cAMP Efflux for CRC Migration

Jakub Kryczka [1], Ewelina Sochacka [1,2], Izabela Papiewska-Pająk [1] and Joanna Boncela [1,*]

[1] Institute of Medical Biology, Polish Academy of Sciences, 93-232 Lodz, Poland; jkryczka@cbm.pan.pl (J.K.); esochacka@cbm.pan.pl (E.S.); ipapiewska-pajak@cbm.pan.pl (I.P.-P.)

[2] Faculty of Biology and Environmental Protection, University of Lodz, 90-236 Lodz, Poland

* Correspondence: jboncela@cbm.pan.pl

Received: 5 October 2020; Accepted: 23 November 2020; Published: 27 November 2020

Simple Summary: Cancer cells have developed a number of mechanisms to overcome anticancer therapy; the active efflux of drugs from cells via multidrug resistance proteins (MRPs) is one of them. MRPs belong to the superfamily of ATP binding cassette (ABC) proteins. It was hypothesized that the inhibition of ABC drug transporter activity during cancer therapy could sensitize drug-resistant tumors and/or improve the initial activity of anticancer agents. We demonstrated that the pharmacological inhibition of ABCC4 increases the migratory rate and invasive protrusion formation in colorectal cancer (CRC). Thus, during the use of ABCC4 inhibitors to reduce chemotherapy resistance or drugs that are potential substrates of ABCC4, the indirect effect on cancer metastasis should be taken into consideration and may be important in selecting a therapy scheme for patients.

Abstract: Colorectal cancer (CRC) presents significant molecular heterogeneity. The cellular plasticity of epithelial to mesenchymal transition (EMT) is one of the key factors responsible for the heterogeneous nature of metastatic CRC. EMT is an important regulator of ATP binding cassette (ABC) protein expression; these proteins are the active transporters of a broad range of endogenous compounds and anticancer drugs. In our previous studies, we performed a transcriptomic and functional analysis of CRC in the early stages of metastasis induced by the overexpression of Snail, the transcription factor involved in EMT initiation. Interestingly, we found a correlation between the Snail expression and ABCC4 (MRP4) protein upregulation. The relationship between epithelial transition and ABCC4 expression and function in CRC has not been previously defined. In the current study, we propose that the ABCC4 expression changes during EMT and may be differentially regulated in various subpopulations of CRC. We confirmed that ABCC4 upregulation is correlated with the phenotype conversion process in CRC. The analysis of Gene Expression Omnibus (GEO) sets showed that the *ABCC4* expression was elevated in CRC patients. The results of a functional study demonstrated that, in CRC, ABCC4 can regulate cell migration in a cyclic nucleotide-dependent manner.

Keywords: ABC transporters; ABCC4 protein; colon cancer; metastasis

1. Introduction

Colorectal cancer (CRC) treatment is challenging due to the heterogeneous nature of cancer, in which prognosis depends on the tumor type and disease stage. Despite progress in diagnosis and therapy, metastasis and chemoresistance are two critical processes for the overall survival of CRC patients [1]. Approximately 50–60% of patients who are diagnosed with CRC will eventually develop metastatic disease. Most often, metastases develop after first-line chemotherapeutic drug and monoclonal antibody treatments for local disease. Over several years, many studies have demonstrated that metastatic CRC presents significant molecular heterogeneity [2,3]. This observation explains the enormous variability noted in regard to treatment outcomes. The cellular plasticity of epithelial to mesenchymal transition (EMT) is one of the key factors responsible for the heterogeneous nature of the metastatic CRC phenotype. EMT is not a binary process [4]. The epithelial cells undergoing EMT give rise to cell populations that may enter into states with various proportions of epithelial and mesenchymal features. Metastatic cells represent various EMT states, from epithelial-like through to mixed epithelial/mesenchymal (hybrid) to a strongly mesenchymal phenotype. Hybrid and mesenchymal cells exhibit increased migratory and invasive features, suggesting a detrimental role of EMT during metastatic dissemination [5,6]. Another complicating factor is that EMT has been linked to additional traits that are not associated with canonical EMT regulation, such as stemness and resistance to anticancer therapeutic drugs [7].

Cancer cells have developed a number of mechanisms to overcome anticancer therapy, and the active efflux of drugs from cells via multidrug resistance proteins (MRPs) is one of them. MRPs belong to the superfamily of ATP-binding cassette (ABC) proteins, active transporters with a broad range of substrate spectra, including anticancer drugs [8]. The human genome contains 48 ABC genes, and they are classified into seven subfamilies (ABCA-ABCG) [9]. Among them, ABCB1, ABCC1, and ABCG2 are highly involved in the acquisition of multidrug resistance (MDR). Increased ABC transporter expression has been correlated with aggressive and invasive cancers. EMT is an important regulator of ABC transporters, and the expression of ABC transporters changes continuously during EMT [10,11]. Mechanistically, it was demonstrated that the promoters of ABC transporters carry several binding sites for EMT-inducing transcription factors, and the overexpression of Twist, Snail, and ZEB increases the promoter activity of ABC transporters [12,13]. In our previous studies, we performed transcriptomic and functional analyses of CRC in the early stages of metastasis induced by the overexpression of Snail, the transcription factor involved in EMT initiation. Our results showed that Snail regulates early phenotype conversion towards a hybrid EMT. Interestingly, we found a correlation between Snail expression and ABCC4 (MRP4) protein upregulation [14]. The relationship between epithelial transition and ABCC4 expression and function in CRC has not been previously defined. In the current study, we propose that ABCC4 expression changes during EMT and may be differentially regulated in various subpopulations of CRC. ABCC4 is able to transport a range of organic anionic compounds out of the cell; thus, most functional studies of ABCC4 have classically focused on its role in cancer chemotherapy [15]. However, the physiological actions of this protein are quite diverse, and drug transport appears not to be the most important evolutionarily conserved function. The efflux of cyclic adenosine monophosphate (cAMP) through ABCC4 has been well documented in various cell types, suggesting that this transporter plays a relevant role in the regulation of cAMP signaling. ABCC4 was shown to modulate the compartmentalization of cAMP signaling in a colon adenocarcinoma cell line (HT29, T84), and ABCC4 inhibition with MK571 compound leads to the accumulation of cAMP at or near the plasma membrane [16]. Notably, the inhibition of ABCC4 function through MK571 or gene knockout was shown to have a direct role in cell migration. The pharmacological inhibition of ABCC4 with MK571 resulted in the intracellular accumulation of cAMP, leading to increased fibroblast migration related to protein kinase A (PKA) activity [17,18]. In the current paper, we show that in CRC, similar to what has been observed in fibroblasts, ABCC4 can regulate cell migration in a cAMP-dependent manner.

The understanding of the mechanistic linkage between two phenomena, ABCC4 transport function and cell migration, would significantly contribute to the improvement of anticancer therapy in CRC. Numerous ABC transporter inhibitors have been developed and tested [19]. The clinical use of ABC transporter inhibitors is still an ongoing challenge, partially due to the intratumor heterogeneity; thus, the evaluation of ABCC4 expression status alone or in combination with other transporters in various CRC subpopulations supported by information on signaling pathways related to ABCC4 transport function may improve the development of patient-tailored therapy.

2. Results

2.1. *ABCC4 Is Overexpressed in CRC*

To identify the expression level of *ABCC4* in CRC, we first analyzed the *ABCC4* expression levels in normal and CRC tissues by a bioinformatics analysis. Microarray data from the public Gene Expression Omnibus (GEO) database (GSE18105, GSE44861, and GSE32323: [20] revealed that *ABCC4* was significantly upregulated in primary tumors compared to normal tissues (Figure 1A). Since we previously observed that ABCC4 expression was upregulated in HT-29 colon cancer cells stably overexpressing Snail and that those cells had transcriptomic profile changes indicating EMT induction [14], we examined the same GEO database to identify mRNA related to EMT, whose expression was changed in our cell line model. Given the previous results, we found that the mRNA levels of the mesenchymal markers fibronectin (Figure 1B) and vimentin (Figure 1C) were elevated and that the mRNA levels of E-cadherin (epithelial marker, Figure 1D) were decreased in the analyzed CRC tissue compared to normal tissue. Among the transcription factors involved in EMT, we found an increase in the expression of *Twist* mRNA in the analyzed CRC patient samples (Figure 1E). Thus, we confirmed that the elevated expression of *ABCC4* in the analyzed CRC data sets was related to changes in phenotypic transition markers. EMT is induced by different stimuli, and TGFβ is its canonical driver. In our previous study, TGFβ was indicated by ingenuity pathway analysis as the most significant upstream regulator of the transcriptomic changes in response to Snail in HT29 cells, and the changes in the expression of the TGFβ signaling pathway components indicated that this pathway was modestly activated [14].

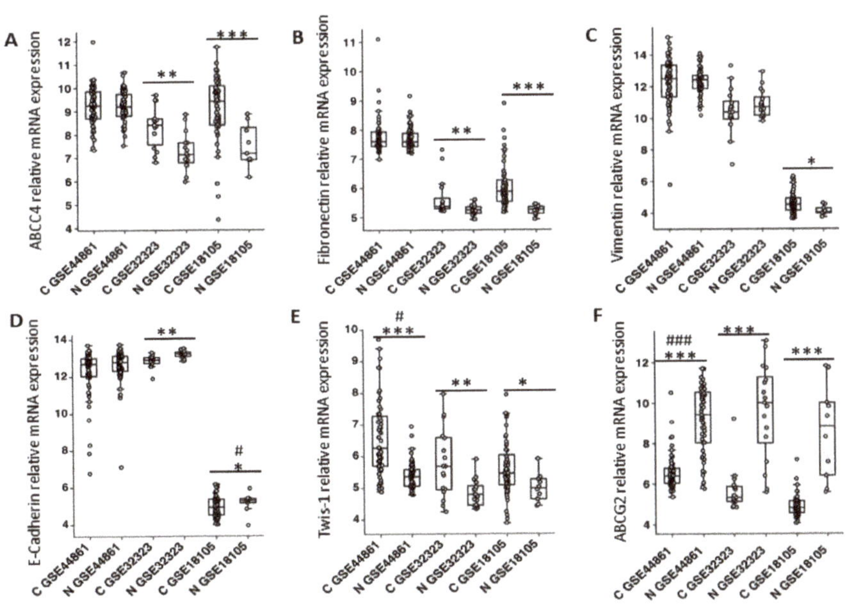

Figure 1. ABCC4 (**A**), ABCG2 (**F**), and EMT marker (**B–E**) mRNA expression in CRC and normal tissue. Microarray data from the public Gene Expression Omnibus (GEO) database (GSE18105, GSE44861, and GSE32323: [20] were analyzed with the respective *n* for c (cancer) and n (normal): GSE18105 $n_c = 57$ $n_n = 10$ (primary tumors only); GSE44861 $n_c = 56$, $n_n = 55$; and GSE32323 $n_c = 17$ $n_n = 17$. Normality test (Shapiro–Wilk) was performed, followed by the Mann-Whitney U test (*)—* $p < 0.05$; ** $p < 0.005$; *** $p < 0.001$, no statistically significant—no indicator. Additionally, all the normally distributed samples were tested using *t*-test (#) # $p < 0.05$; ### $p < 0.001$. Data density distribution is presented in the Supplementary Materials, Figure S2.

To check whether the *ABCC4* expression upregulation is related to the TGFβ signaling pathway in CRC, we analyzed GSE18105 datasets—which represent the statistically most significant differences in EMT markers between normal and cancer cells—for the expression of TGFβ1/2 and their receptors. We observed a significant upregulation of *TGFβ2* and both TGFβ receptors (*TGFβR1* and *TGFβR2*) in tumors compared to normal colon tissue (Figure 2A,D). Furthermore, we noticed a positive correlation between the *ABCC4* expression and *TGFβ2* and both the TGFβ receptors' (*TGFβR1/2*) expression (Figure 2C–F and Figure S1A), and a negative correlation with the *TGFβ1* expression (Figure 2B and Figure S1A). These results confirmed that *ABCC4* expression is related to the TGFβ-induced transcriptomic signature in CRC. To date, in clinical studies ABCG2 has been recognized as the main drug efflux protein in CRC [21]. However, studies comparing the expression of *ABCG2* mRNA in normal colon tissue and tumor tissue showed that primary colon cancer cells exhibit an initial downregulation of *ABCG2* mRNA expression [22]. Our previous results showed that HT29 lines with upregulated Snail expression exhibited an increase in *ABCC4* expression and a decrease in *ABCG2* expression [14]. To validate this observation, we analyzed the same datasets from the GEO database (GSE18105, GSE448, and GSE32323) and found that *ABCC4* was significantly upregulated while *ABCG2* was downregulated in primary tumors compared to normal colon tissues. [20] A representative analysis is presented in Figure 1F. These data may indicate that ABCC4 is a prevalent drug transporter in primary tumors in which ABCG2 is downregulated, and confirm the hypothesis that the changes in the ABC transporter expression might be related to the various stages of CRC progression. Next, to correlate the transcriptomic analysis to the ABCC4 protein function in CRC, we analyzed the ABCC4 protein expression profile and performed a functional study.

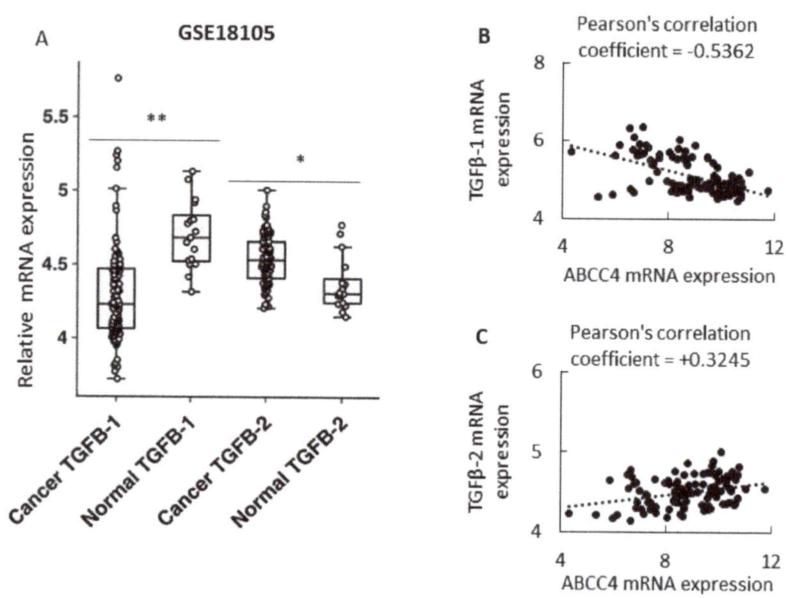

Figure 2. *Cont.*

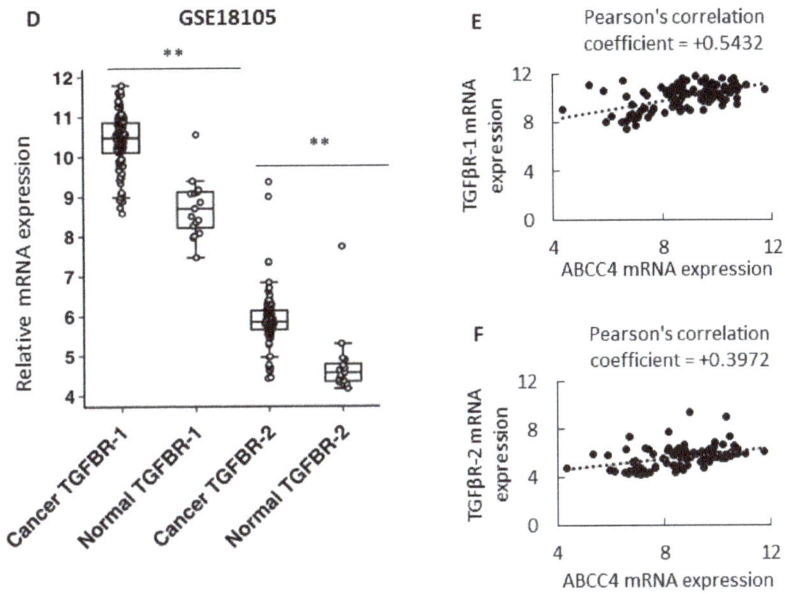

Figure 2. TGFβ pathway and ABCC4 mRNA expression analysis. TGFβ1/2 (**A**) and TGFβR1/2 (**D**) expression analysis in CRC patient tissue. Data obtained from the GSE18105 data set, cancer $n = 94$, normal $n = 16$. Correlation of ABCC4 expression and TGFβ1 (**B**) or TGFβ2 (**C**) or TGFβR1 (**E**) or TGFβR2 (**F**) expression, $n = 110$ [20]. Sample data sets were tested with the Shapiro–Wilk test, presenting a normal distribution, followed by the T-test. Pearson's correlation coefficient (PCC): 0–0.25 no PCC, 0.25–0.5 low PCC, 0.5–0.75 moderate PCC, 0.75-1 strong PCC. Comparison of Pearson's and Spearman's correlation values in Figure S1A, Supplementary Materials. Data density distribution (Figure 2A,D) produced with SinaPlot is presented in Supplementary Materials, Figure S2. * $p < 0.05$; ** $p < 0.005$.

2.2. *ABCC4 Protein Expression in CRC Is Related to Phenotypic Transition*

The ABCC4 protein was identified as an active transporter of cyclic nucleotides and as a mediator of secondary messenger signaling through cAMP in several different cell and tissue types [13]. To determine this function of ABCC4 in CRC, we first analyzed the level of ABCC4 protein in HT29 cells. We observed an increased ABCC4 protein expression in HT29 cells overexpressing Snail (HT29/Snail) compared to control HT29 cells (Figure 3A, Figure S1B, Supplementary Materials). Next, as our previous mRNA analysis of HT29/Snail cells [14] and current bioinformatics analysis of CRC patient samples showed that the upregulation of *ABCC4* accompanied the downregulation of *ABCG2*, we also performed a Western blot analysis of the ABCG2 protein. We observed that this protein was present at lower levels in HT29/Snail cells than in control HT29 cells (Figure 3A and Figure S1B, Supplementary Materials). In our previous study, we found a correlation between Snail expression and the upregulation of *ABCC4* [14]. However, crucial EMT-activating transcription factors, including Snails, ZEB, and Twist, recognize the E-box DNA sequences in the promoter region of the *ABCC4* gene, and the bioinformatic analysis of the *ABCC4* promoter region revealed the presence of 11 E-box sequences [12]; thus, we assume that the *ABCC4* upregulation is correlated with the epithelial reprogramming process rather than with the activity of a single transcription factor during epithelial transition in CRC. To check this hypothesis, we performed an ABCC4 protein expression analysis in various CRC cell lines representing epithelial, intermediate mesenchymal, and strongly mesenchymal phenotypes. All of these cell lines are directly derived from primary colorectal cancers of different clinical stages and differentiation grades [23,24]. Our results showed that cells with an epithelial phenotype (CCD841CoN) expressed less ABCC4 than intermediate or strongly mesenchymal cells (CaCo-2 and Colo-320, respectively) (Figure 3B and Figure S1C, Supplementary Materials), suggesting that the ABCC4 expression is related to the phenotypic status in CRC. The highest expression level was noted for Colo-320 cells. Interestingly, Colo-320 was shown to have the strongest expression of the EMT signature and the highest propensity to local invasion among the analyzed group of cells [25].

2.3. Cellular Localization of ABCC4 in CRC

The cellular localization of ABCC4 regulates cAMP signaling involved in cell migration [18]. Thus, we analyzed the ABCC4 cellular localization in CRC by isolating subcellular fractions and measuring the level of ABCC4 protein, particularly in the outer membrane fraction, in comparison to whole cell extracts (input). We performed cell surface protein biotinylation using EZ-Link Sulfo-NHS-Biotin, and we collected the biotinylated protein fraction with streptavidin agarose. Using Western blot analysis, we detected higher levels of ABCC4 protein in the membrane fraction of HT29 cells overexpressing Snail than in that of the HT29 control cells (Figure 3, Figure S1B, Supplementary Materials). These results indicate that the higher expression of ABCC4 protein determines its membranous localization and transport function in cells that acquire mesenchymal traits.

Generally, the elevated expression of ABC is attributed to drug resistance. Recent data have shown that ABC protein enrichment was present in EVs from drug-resistant cells [26]. We estimated the ABCC4 abundance in extracellular vesicles (EVs) released from HT29 cells. We detected ABCC4 in HT29-derived EVs (Figure 3C, Figure S1D, Supplementary Materials). These EVs were positive for CD63, CD9, and CD81 and negative for cytochrome c and were used in our previous study for mRNA and miRNA analysis [14,27]. Our results showed a higher level of ABCC4 in EVs from two clones of HT29 cells stably overexpressing Snail, suggesting that ABCC4 is packed into CRC EVs and that the level of ABCC4 in EVs correlates with the level of ABCC4 expression in the cells.

2.4. Analysis of Intracellular Accumulation of cAMP

The inhibition of ABCC4 function by MK571, a known ABCC4 inhibitor, has been used in previous works and was shown to increase the intracellular cyclic nucleotide level and have a direct role in mouse fibroblast migration. The effect was more profound, however, on the cAMP level than on the cGMP level, indicating a higher affinity of ABCC4 for cAMP. In our experiments, we noted that treatment with MK571 increased the intracellular level of cAMP in HT29 cells. These results demonstrated that ABCC4 was responsible for cAMP efflux in CRC. However, the effect was more pronounced in HT29 cells that acquired mesenchymal characteristics by Snail overexpression than in control HT29 cells (Figure 3D). We calculated the intracellular level of cAMP after 24 h of incubation with MK571 and used the cAMP competitive test (Cyclic AMP ELISA Kit, #581001, Cayman Chemicals). The concentrations of MK571 used were selected from previous reports and did not affect the cellular viability (data not shown) [28].

2.5. ABCC4 Function Is Necessary for Adequate PKA Activity

Intracellular cAMP has a vast repertoire of effectors; among them, the PKA enzyme family is one of the most studied, and its activity is directly related to serine/tyrosine phosphorylation [29]. Previous reports described a correlation between PKA activity and fibroblast migration [18]. Therefore, we studied whether inhibition of ABCC4 and cAMP efflux might modulate the PKA activity. HT29 cells overexpressing Snail were incubated in the presence or absence of MK571 (20 μM for 0, 1, 5, 30, and 60 min). Next, phosphorylated substrates of PKA (pPKA-Subs) were visualized using Western blot and phospho-(ser/thr) PKA Substrate Antibody (Cell Signaling) (Figure 3e, Figure S1E). The obtained results showed that in HT29 cells with Snail overexpression, short incubation (1 and 5 min) with MK571 resulted in an apparent increase in the phosphorylation of 140 kDa proteins. This effect was further decreased after 30 and 60 min. Additionally, we observed a gradual increase in the phosphorylation of 42 kDa and 140 kDa proteins, Figure 3F. Changes in the phosphorylation of PKA substrates after ABCC4 inhibition were not observed in control HT29 cells (Figure 3E, Figure S1E). This observation suggests that ABCC4 activity is necessary for the early regulation of PKA activity in cells that acquire a mesenchymal phenotype and indicates that in CRC, similar to what was observed in fibroblasts, the inhibition of ABCC4 may increase cell migration.

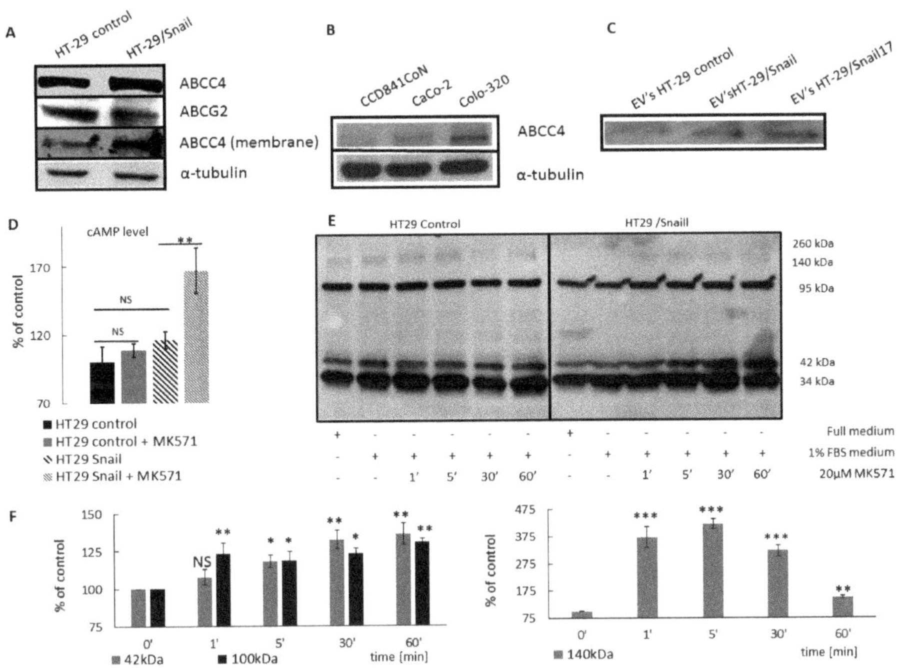

Figure 3. ABCC4 protein expression level in CRC cell lines. Western blot performed in standard reducing SDS PAGE conditions using goat anti ABCC4 (#PA5 18315, Thermo Scientific) and rabbit anti ABCG2 (#ORB 155559 Biorbyt). (**A**) Protein expression level of ABCC4 and ABCG2 in HT-29 stably overexpressing transcription factor Snail (HT29/Snail) and control HT-29. ABCC4 level in the membrane fraction (obtained by biotynylation using EZ-Link Sulfo -NHS-Biotin Thermo Scientific kit) of HT-29 control cells and HT-29 Snail $n = 3$. (**B**) ABCC4 protein expression level in CRC cells in different states of EMT: CCD841CoN (most epithelial), CaCo-2 (moderate EMT), and Colo-320 (most mesenchymal) $n = 3$. (**C**) ABCC4 protein abundance in Extracellular Vesicles (EVs) released from HT-29 control cells and two HT-29 stably overexpressing transcription factor Snail clones (HT-29/Snail and HT-29/Snail17), $n = 2$. (**D**) Intracellular cAMP level measurement. Accumulation of cAMP in HT29 cells was measured using a cAMP competitive kit (#581001 Cayman Chemical). Cells were incubated for 24 h with MK571 20 µM, or untreated ones were assayed according to the manufacturer's protocol. Calculation were conducted using the Cayman data sheet. cAMP concentration of HT29 was set as 100%. *T*-test performed, $n = 5$; * $p < 0.05$; ** $p < 0.005$; *** $p < 0.001$. NS—not statistically significant. (**E**) PKA phosphorylation profile analysis. HT29 Snail cells were seeded on a 6-well plate. Then, 24 h after, full growth medium was changed into starving (FBS free) medium for 24 h. Next, 20uM of MK571 was added to cells for 60, 30, 5, and 1 min. Cells without the starving procedure were used as a positive control, and negative control cells were not treated with MK571. Phosphorylation profile analysis was performed using phospho-(ser/thr) PKA Substrate Antibody #9621 (Cell Signaling Technology). Significant time- (exposure) related impact on the phosphorylation profile was observed for 42 kDa and 95–100 kDa proteins in HT29 Snail cells compared to no time-related changes in control cells, $n = 3$. (**F**) HT-29/Snail PKA phosphorylation profile analyzed with densitometry; statistical significance estimated using *T*-test. * $p < 0.05$; ** $p < 0.005$; *** $p < 0.001$. NS—not statistically significant.

2.6. Analysis of the Migratory Potential of CRC Subtypes Treated with ABCC4 Inhibitor

Given the obtained results, we further evaluated the ability of the cells to migrate in the presence of MK571. First, we performed a scratch (wound healing-like) assay. One of the major advantages of this simple method is that it mimics the migration of cells in vivo [30]. We also decided to test the impact of MK571 on the ability to cross anatomical boundaries using a gelatinolysis assay and transwell invasion assay, performed as described in our previous papers [14,28,31]. The obtained results indicated that MK571 increased the motility of both HT29 and HT29 cells overexpressing Snail (Figure 4A,B). However, the effect of inhibition on cell invasiveness and gelatinolysis activity was detected only in HT29-Snail cells (Figure 4C,D), suggesting that cells with acquired mesenchymal characteristics are more prone to ABCC4 inhibition than cells with epithelial characteristics. To confirm this observation, we investigated the motility of Caco2 cells, which represent an intermediate mesenchymal phenotype [23,24]. We observed that MK571 also increased CaCo2 migration, as detected in the transwell migration assay (Figure 4E). Since the wound healing assay is not recommended for this cell line due to its growth characteristics, we performed this assay using collagen type I-coated 6-well plates (Corning) that increased cell adhesion, preventing spontaneous detachment. MK571-treated CaCo2 cells presented a higher migration rate, as measured by faster wound closure than control cells (Figure 4F).

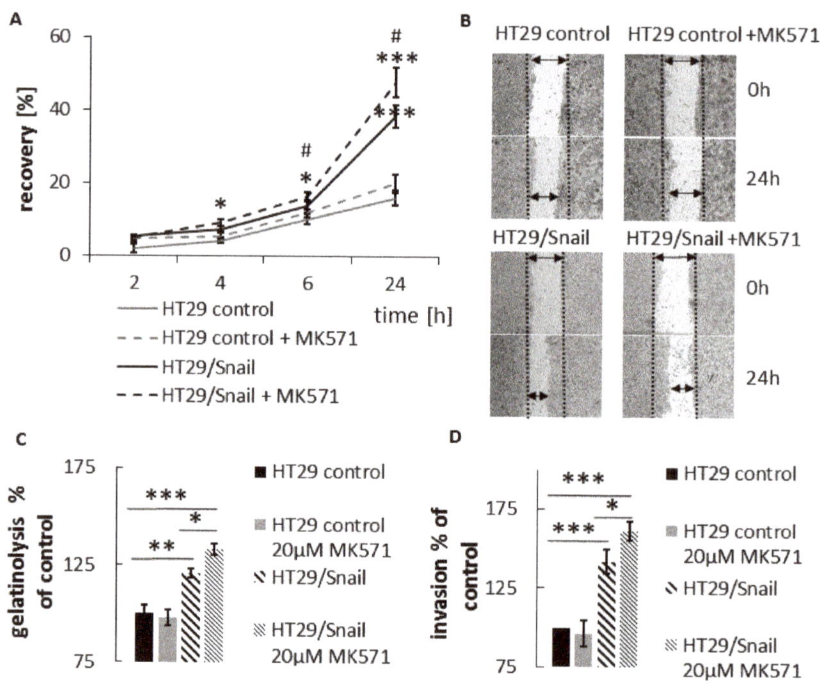

Figure 4. *Cont.*

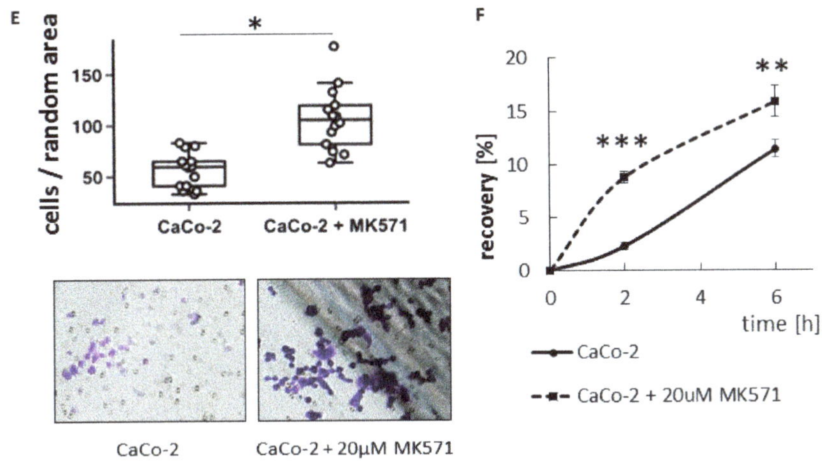

Figure 4. MK571 impact on CRC cells migratory abilities. (**A**) HT29 cells (control or overexpressing Snail) were grown to confluence on 6 well plate, next, wounded across the cell monolayer. New medium containing 20 μM MK571 was added. Wounded area was visualized after 0, 2, 4, 6 and 24 h—and presented in using Nikon Eclipse TE 2000-U microscope (Nikon, Japan) and calculated by ImageJ software [32]. Cell motility was estimated through the quantification of the % of recovery using the equation: R (%) = [1 − (wound area at Tt/wound area at T0)] × 100,where T0 is the wounded area at 0 h and Tt is the wounded area after t; $n = 3$; * HT-29/Snail (w/wo MK571) vs. HT-29 control; # HT-29/Snail MK571 vs. HT-29/Snail * $p < 0.05$; ** $p < 0.005$; *** $p < 0.001$, NS—no statistically significant. (**B**) Representative picture of wound healing assay. (**C**) Gelatinolysis mediated by HT-29 measured by in-situ zymography. The pericellular proteolytic abilities of HT-29 were analyzed by a measure of the increase in FITC fluorescent intensity from digested DQ gelatine relativized to control cells presented as 100%; $n = 3$ (**D**) HT29 cells, transwell assay. Cells incubated for 24 h with MK571 20 μM, or untreated once were seeded on Matrigel coated transwell inserts in the upper chamber in medium supplemented with 0.1% bovine albumin serum (BSA) (w/wo 20 μM MK571). Full medium in lower chamber served as chemoattractant for cell invasion. Membrane were cut out and all cells from membrane were calculated after 6 h of incubation followed by hematoxylin/eosin staining. Number of control HT-29 cells that transmigrate into transwell membrane through 8 μM pores covered with Matrigel was set as 100%, next number of other cells was calculated and presented as % of control. (**E**) CaCo-2 cells, transwell assay. Cells were incubated for 24 h with MK571 20 μM, or untreated once were seeded on un-coated transwell inserts (8 μM pores) in the upper chamber in 2% BSA medium (w/wo 20 μM MK571). Full medium in lower chamber served as chemoattractant for cell migration. Cells were calculated in randomly assigned areas after 3 h of incubation followed by hematoxylin/eosin staining. Interquartile range (Q1–Q3) is shown as gray box with median (Q2) with all data points from all ($n = 3$) experiments overlap on the box plot. (**F**) CaCo-2 cells, wound healing assay. Cells were seeded on collagen coated 6-well plates to full confluence. Next wound was done across cell monolayer, rinsed with phosphate buffer (PBS) PBS. Next fresh medium w/wo 20 μM MK571 was added. Cells were visualized every 2 h and % of wound enclosure was calculated as in A). * $p < 0.05$; ** $p < 0.005$; *** $p < 0.001$, NS—not statistically significant.

To extend this analysis, we more comprehensively evaluated the extent that ABCC4 expression and function correlated with phenotypic transition. We examined the effect of MK571 on endothelial cells upon endothelial to mesenchymal transition (EndoMT). For this purpose, we used HMEC-1 (microvascular endothelial) cells shifted towards the mesenchymal phenotype, which was broadly characterized in our previous study [31]. We investigated the migratory ability of HMEC-1 cells with transient Snail overexpression (Figure 5A–C) or TGFβ stimulation (Figure 5D) in the presence or absence of MK571. To omit any noncanonical impact of TGFβ on EndoMT, HMEC-1 cells treated with TGFβ receptor inhibitor 24 h (SB431542 #1614, Tocris Bioscience, Bristol, UK)) prior to the experiment were used as a control for the TGFβ-mediated EndoMT (as in our previous study [31]). The results showed that ABCC4 inhibition increased the migration of endothelial cells that acquire a mesenchymal phenotype.

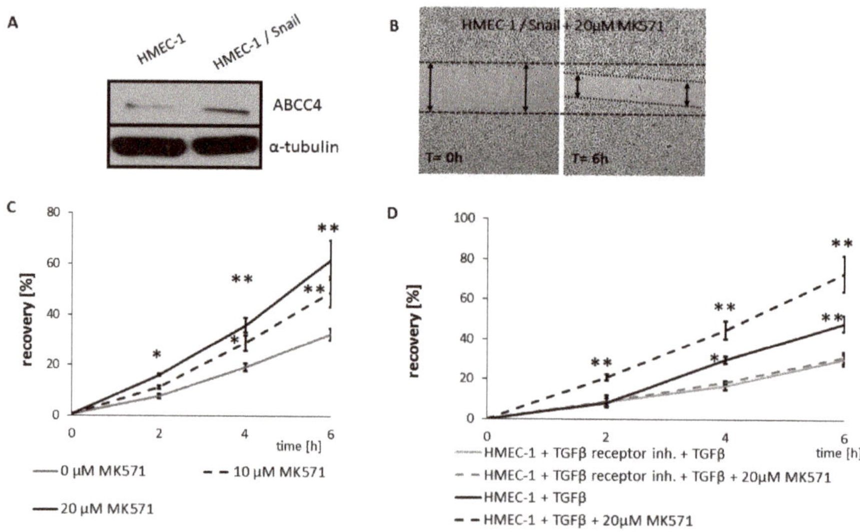

Figure 5. MK571 increased motility of EndoMT undergoing HMEC-1 cells. ABCC4 level in HMEC-1 and HMEC-1 overexpressing Snail cells (HMEC-1/Snail). (**A**) HMEC-1 were grown to confluence on 6-well plate, and transiently transfected with pcDNA/Snail and wounded across monolayer as described in [31]. New medium containing MK571 was added. (**B**) Representative image of HMEC-1 control or HMEC-1/Snail cells in wound healing assay. (**C**) Wounded area was visualized after 0, 2, 4 and 6 h using Nikon Eclipse TE 2000-U microscope (Nikon, Japan) and calculated by ImageJ software [32]. Cell motility was estimated through the quantification of the % of recovery using the equation: R(%) = [1 − (wound area at Tt/wound area at T0)] × 100,where T0 is the wounded area at 0 h and Tt is the wounded area after 2 or 4 h. * $p < 0.05$; ** $p < 0.005$; $n = 3$. (**D**) HMEC-1 treated w/wo TGF-β receptor inhibitor were grown to confluence on 6 well plate, incubated for 48 h with 10ng/mL TGF-β2 in starving condition as described in [31 and wounded. Wounded area was visualized and analyzed as in (**C**).

2.7. Irinotecan Treatment and CRC Migration

Interestingly, irinotecan, a chemotherapeutic drug for CRC, has a high affinity for ABCC4 and was demonstrated (through substrate competition with cAMP) to increase cAMP levels at or near the plasma membrane to levels comparable with the effect of the ABCC family inhibitor MK571 [33]. Irinotecan does not directly inhibit ABCC4 transport and we assume that the endpoint effect of cAMP-mediated signaling may be similar. To test this hypothesis, we investigated whether irinotecan affected the migration of HT29 cells in a manner comparable to that of MK571. First, we established irinotecan cytotoxicity (IC25 and IC50) for HT29/Snail and control HT29 cells (Figure S1E, Supplementary Materials) using a WST-1 assay. The obtained results indicated that both HT29 variants presented similar levels of irinotecan tolerance with IC50 values of approximately 5.5 μM, which corresponds to the literature data [34]. Finally, we tested the impact of irinotecan on migration. Control HT29 cells and HT29 cells overexpressing Snail were seeded on 24-well plates for 24 h to reach confluence. Next, wounds were made across monolayers, the cells were washed with PBS and fresh medium was added w/wo 2.5 μM irinotecan (Figure 6). We decided to use a concentration of 2.5 μM, corresponding to the

IC25, to avoid increased cytotoxic/cytostatic effects in the scratch assay. We noticed that irinotecan enhanced migration of HT29 overexpressing Snail, whereas the migration of control HT29 was not significantly changed. Of note, the most statistically significant increase was observed within the first 8 h of irinotecan supplementation.

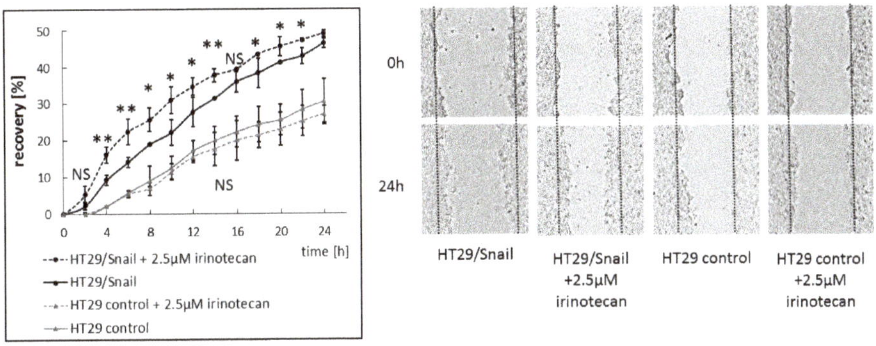

Figure 6. Irinotecan affects CRC migration. HT29 control and HT29/Snail cells were seeded on 24 well plate to confluence for 24 h. Next, wound was done across monolayer and fresh medium was added w/wo 2.5 μM irinotecan (final concentration). Wounded area was visualized after every 2 h by Spark multimode microplate reader (TECAN, Swizerland). Wounded area was calculated by ImageJ software [32]. Cell motility was estimated through the quantification of the % of recovery using the equation: R(%) = [1 − (wound area at Tt/wound area at T0)] × 100,where T0 is the wounded area at 0 h and Tt is the wounded area after 2 or 4 h. * $p < 0.05$; ** $p < 0.005$, NS—no statistically significant (all HT29 control vs HT29 control + 2.5 μM irinotecan were considered NS), $n = 3$.

3. Discussion

During the multistep progression of carcinomas that are initially benign, epithelial cells acquire a few distinctly mesenchymal traits that confer to them the ability to invade adjacent tissues and then disseminate to distant tissues. Much of this phenotypic progression towards increased invasiveness depends on the activation of the EMT [5]. Experimental and clinical studies have shown that EMT is an important regulator of ABC protein expression, the active transporters of a broad range of anticancer drugs and the expression of ABC transporters change continuously during EMT [9,10]. We found a correlation between this phenotypic conversion and ABCC4 protein upregulation in HT29 cells overexpressing Snail; thus, in the current study, we propose that ABCC4 protein expression and function are related to epithelial reprogramming in CRC [14]. To support our hypothesis, we first analyzed the *ABCC4* expression levels in CRC tissue. Our analysis of GEO sets showed that *ABCC4* expression was elevated in CRC patient samples compared to normal colon tissue. Further analysis of the same datasets revealed increased expression of mesenchymal markers and decreased expression of E-cadherin in patient samples. We also found a positive correlation between *ABCC4* expression and *TGFβ1/2* receptors and its ligand *TGFβ2*, which were shown to be involved in epithelial conversion induction in cancers. Clinically, ABC transporters were the first and most studied mechanism of resistance associated with MDR. Interestingly, to date, ABCG2 has been recognized as the main drug efflux protein in CRC [21]. Several studies have shown that ABCG2, through its function in xenobiotic clearance, might play an important role in irinotecan resistance. However, other studies comparing the expression of *ABCG2* mRNA in normal colon tissue and tumor tissue showed a decreased expression in tumor tissue. The latter data suggest that primary colon cancer cells exhibit an initial downregulation of *ABCG2* mRNA expression [22]. Our results clearly indicated that ABCC4 is a prevalent drug transporter in tumors in which ABCG2 is downregulated. We showed that HT29 lines overexpressing Snail, which represent a CRC model in the early stages of phenotype conversion, exhibited an upregulated *ABCC4* expression and concomitant downregulated *ABCG2* expression. These results correspond to microarray data (GEO) from patient samples. We found that *ABCC4* was significantly upregulated while *ABCG2* was downregulated in primary tumors compared to normal colon tissues. These data support the hypothesis that CRC may depend on several drug transporters, specifically regulating their expression and executing their function during cancer progression from primary to metastatic disease. Nevertheless, the ABC transporter mRNA expression may have limited reliability with respect to protein function. A very limited number of studies describing the association between the transcriptional and protein overexpression of ABC transporters in cancers have been published [35]. The correlation between the level of ABC proteins and their transporter function in cancers remains to be proven as well. In CRC, the significance of ABCG2 protein measurement in predicting clinical resistance to irinotecan in patients was examined. ABCG2 protein expression analyzed by IHC showed that ABCG2-positive cells were mainly positioned in the front of the carcinomatous tissue (the invasion front), and strong membranous staining was significantly correlated with a higher Dukes' stage, more lymph nodes, and the presence of distant metastases [36]. However, the role of ABCG2 as a prognostic factor or predictor of irinotecan efficacy in CRC is not well established. The few studies available seem to report discordant results mainly due to the lack of validated assays and standardized reference values for IHC protocols [21,22]. This lack of consistency may also be a result of either cancer heterogeneity and/or an incomplete understanding of the biological role of ABC transporters in cancer progression. In view of the above, the analysis of the mRNA and protein expression levels of specific ABC transporters in relation to their transporter function in various cancer cell subpopulations may have clinical value.

ABCC4 is unique among ABC transporters since its different locations (basolateral/apical membranous versus cytoplasmic) may fundamentally influence its transport function. To date, the mechanism of ABCC4 cellular trafficking has not been elucidated; however, changes in ABCC4 expression led to changes in its localization and function [31–34]. It is widely accepted that ABC transporters can bind their substrates either from the surroundings of the plasma membrane or intracellular vesicles and transport them out of the cell directly to the external milieu [33,37]. Our results showed a higher level of ABCC4 in the plasma membrane fraction and in EVs from two clones of HT29 cells stably overexpressing Snail, suggesting that the level of ABCC4 expression in CRC determines its functional localization in tumor cells.

To further explore the role of the ABCC4 transporter in CRC progression, we confirmed that ABCC4 upregulation is correlated with the phenotype conversion process in CRC. We measured the ABCC4 protein level in CRC cell lines representing various phenotypes from epithelial to intermediate mesenchymal to strongly mesenchymal. All of these cell lines were either directly derived from primary colorectal cancers of different clinical stages and differentiation grades [25]. We believe that all the lines with clearly defined genetic backgrounds—i.e., methylation and epigenetic status—the occurrence (or not) of *KRAS*, *BRAF*, *SMAD4*, and other mutations were the best experimental models to identify the particular ABCC4 expression status in CRC to mimic a patient-specific approach. ABCC4 expression analysis in those lines showed that cells with an epithelial phenotype (CCD841CoN) expressed less ABCC4 than cells with an intermediate or a strongly mesenchymal phenotype (CaCo-2 and Colo-320, respectively). The highest expression of ABCC4 was observed in Colo-320 cells, strongly mesenchymal cells with the strongest expression of the EMT signature and the highest propensity to local invasion in the analyzed group of cells. These results confirmed that the ABCC4 expression is related to phenotypic transition in CRC.

In addition to xenobiotic efflux, ABCC4 was shown to control the export of endogenous signals, such as cyclic nucleotides and prostaglandins, and their cellular concentration; therefore, ABCC4 plays an important role in other processes. Platelet aggregation, retinal neovascularization, CFTR-mediated secretory diarrhea and fibroblast migration are partially related to ABCC4 transport activity [37]. Among the endogenous substrates of ABCC4, cAMP and cGMP play important roles in the signaling pathways at various stages of the cell migration process, either directly or by activating their corresponding kinases. As ABCC4 was shown to have a higher affinity for cAMP, we tried to monitor the effect of ABCC4 inhibition on the intracellular cAMP levels to further evaluate the significance of ABCC4 protein in CRC progression. We noted that treatment with MK571 increased the intracellular level of cAMP in HT29 cells. However, the effect was more pronounced in HT29 cells that mesenchymal characteristics acquired by Snail overexpression.

The cAMP-PKA pathway is the most relevant mechanism to the outcome of ABCC4 expression. Localized increases in cAMP concentration and cAMP-dependent PKA at the leading edge both play pivotal roles in ensuring the polarity of migrating cells [28,29]. The polarized activation of the cAMP-dependent kinase PKA is not only an essential early step for directional cell migration but is also involved in actin polymerization and cytoskeleton dynamics regulation. We observed that inhibition of ABCC4 increased the intracellular level of cAMP and modulated PKA activity and phospho-serine/tyrosine levels in HT29 cells overexpressing Snail. The diversity of PKA substrates permits the regulation of multiple signaling events based on the subcellular localization of PKA [38]. Studies have shown that at the leading edge, PKA activates small GTPases, such as Rac and Cdc42, which are important for lamellipodia and filopodia formation, respectively, during cell migration [39]. Phosphorylation-activated Rac induces the WASP/WAVE-mediated activation of Apr 2/3 and promotes the formation of dendritic actin network-containing lamellipodia [40]. Additionally, the PKA-dependent phosphorylation of VASP can regulate actin polymerization and hence can control protrusion formation during cell migration. [41].

This observation prompted us to validate the effect of ABCC4 inhibition on cell motility. The obtained results confirmed that HT29 cells with acquired mesenchymal characteristics (by Snail overexpression) are more prone to ABCC4 inhibition, which leads to an increase in the migratory and invasive properties of cells. The same effect—i.e., an increase in migration—was shown in the CaCo-2 line, classified as an intermediate mesenchymal phenotype. Thus, our results showed that in CRC, similar to what was observed in fibroblasts, ABCC4 can regulate cell migration in a cyclic nucleotide-dependent manner. However, the intracellular cyclic nucleotide level is controlled not only by the process involving active efflux transport from the cell but also by phosphodiesterase-mediated hydrolysis. In fibroblasts, the inhibition of ABCC4 function through MK571 treatment or gene knockout showed that the intracellular cAMP level was moderately regulated by ABCC4 near the leading edge of the cells, whereas forskolin and PDE inhibitors strongly elevated the cAMP level inside the cells. Therefore, ABCC4 regulates fibroblast migration through the spatial resolution of cAMP signaling and localized PKA activation at the cell front [18]. Similarly, ABCC4 was shown to modulate the compartmentalization of cAMP signaling in a colon adenocarcinoma cell lines (HT29 and T84), and ABCC4 inhibition with MK571 leads to the accumulation of cAMP at or near the plasma membrane. Interestingly, irinotecan, a first- and second-line chemotherapeutic drug for CRC, was also demonstrated to increase the cAMP levels at or near the plasma membrane to levels comparable to the effect of MK571 in mouse intestinal epithelial cells and human CRC cells [33]. This observation raises the question of whether irinotecan increases the migration rate in CRC cells with a specific phenotype and ABCC4 expression level. Since irinotecan has been reported to be a potential substrate for ABCC4, we reasoned that these drugs, through substrate competition with cAMP, may also elevate intracellular levels of cAMP and increase cell migration and ultimately cancer dissemination. Our results confirmed the above assumption: irinotecan increased CRC migration, and the effect was more pronounced in cells with mesenchymal characteristics.

Taken together, our results indicated that the pharmacological inhibition of ABCC4 regulates cAMP signaling and PKA activity and increases the migratory rate and invasive protrusion formation in CRC. Thus, during the use of ABCC4 inhibitors to reduce chemotherapy resistance or drugs that are potential substrates of ABCC4, the indirect effect on cancer metastasis should be taken into consideration and may be important in selecting a therapy scheme for individual patients. However, the involvement of ABCC4 protein in cell migration is ambiguous. The siRNA silencing of ABCC4 in human retinal microvascular endothelial cells (HRECs) enhanced their migration [42], while pharmacological inhibition of ABCC4 activity or downregulation through RNAi in dendritic cells (DCs) resulted in the reduced migration of DCs [43]. The ambiguous role of cyclic nucleotides in cell migration should also be considered. This observation suggests that various intracellular mechanisms may be responsible for ABCC4 involvement in migration and that the ABCC4 function may be cell type-dependent.

As tumor heterogeneity is accepted and heterogeneity seems particularly pronounced in CRC, patient-derived material analysis is required to further increase the translatability of our findings. The single-cell sequencing of normal tissues, primary tumors, circulating tumor cells, and metastases, combined with cellular analyses and functional validations, will reveal the role of ABCC4 protein in the diverse responses of CRC patients to therapy.

4. Materials and Methods

4.1. Patients Samples Analysis.

Microarray profiles and datasets of primary CRC were acquired from the public Gene Expression Omnibus (GEO) databases—National Center for Biotechnology Information (NCBI), U.S. National Library of Medicine 8600 Rockville Pike, Bethesda MD, 20894 USA [20] with the following entries: GSE18105, GSE44861, and GSE32323 (Affymetrix Human Genome U133 Plus 2.0 platform). [20] GSE18105 consisted of 110 samples, GSE44861 consisted of 111 samples, and GSE32323 consisted of 34 samples. For each dataset, samples were divided into two defined groups: colon cancer samples (c) and normal colon samples (n)—and the number of analyzed samples in each group was as follows: GSE18105: $n_c = 94$, $n_n = 16$; GSE44861: $n_c = 56$, $n_n = 55$; GSE32323 $n_c = 17$ $n_n = 18$. Next, the obtained data were analyzed using the same microarray ID for each mRNA in every dataset (e.g., 203196_at for ABCC4 analysis). Data were presented as box charts, with the median and all the data points depicted. Statistical analysis was performed using BioVinci version 1.1.5 developed by BioTuring Inc., San Diego, CA, USA, [44]

4.2. Cell Culture

Colon cancer cell lines and dermal microvascular endothelium were obtained from American Type Culture Collection (Manassas, VA, USA) and cultured: HT29 and HT29/Snail in McCoy's 5A medium (LifeTechnologies, Waltham, MA, USA), COLO-320 in RPMI-1640, CCD 841 CoN and CaCo-2 in Eagle's Minimum Essential Medium (EMEM), HMEC-1 in MCBD-131 medium (Life Technologies, Waltham, MA, USA), all supplemented with 10% FBS (LifeTechnologies, Waltham, MA, USA) and antibiotics—streptomycin and penicillin (Sigma-Aldrich, St. Louis, MO, USA), primocin (Invivogen, San Diego, CA, USA) in a 90–95% humidified atmosphere of 5% CO2. HMEC-1 were additionally supplemented with EGF, hydrocortisone, and L-Glutamine (Sigma-Aldrich, St. Louis, MO, USA). The cells were periodically tested for mycoplasma every 4 weeks using the PlasmoTest (Invivogen, San Diego, CA, USA).

4.3. Western Immunoblotting

Proteins isolated from HT-29 cells were extracted with NP-40 lysis buffer (50 mM Tris, pH 8.0, containing 1% Nonidet-Igepal, 150 mM NaCl, 5 mM EDTA) with the Halt protease inhibitor cocktail (Thermo Scientific, Waltham, MA, USA), and the soluble protein fraction was collected through centrifugation. The protein concentrations in the cell lysates were measured with the BCA method (Pierce/Thermo Scientific, Waltham, MA, USA) and were equalized between samples. The protein extracts were subjected to SDS-PAGE analysis and were electro transferred onto PVDF or nitrocellulose membranes (BioRad, Hercules, CA, USA) followed by immunodetection goat anti human ABCC4 #PA5-18315 (Thermo Fisher Scientific), rabbit anty human ABCG2 #ORB155559 (Biorbyt). The control-mouse rabbit anti-α-tubulin antibody conjugated with HRP (NB100-690H) was obtained from Novus Biologicals (Centennial, CO, USA) and used as a loading control. Detection was performed using secondary HRP-conjugated antibodies (Santa Cruz Biotechnology, Dallas, TX, USA) followed by incubation with an enhanced chemiluminescence kit (Thermo Scientific, Waltham, MA, USA) and development with Kodak BioMax Light Film (Eastman Kodak, Rochester, NY, USA).

4.4. Biotinylation of Cell-Surface Proteins

HT29 cells were seeded on 75 cm^2 bottles. After reaching 80% confluence, they were washed 3× with PBS pH 8.0. Next, 2.5 mL freshly made of 2mM biotin (EZ-Link Sulfo -NHS-Biotin Thermo Scientific, Waltham, MA, USA) solution in pH 8.0 PBS was added for 2 h 4 °C. Next, biotin solution was aspirate and cells were washed 3× with cold TBS pH 7.4 solution, then cells were lysed for 30 min using M-PER™ Mammalian Protein Extraction Reagent #78501 (Thermo Scientific, Waltham, MA, USA), centrifuged (20 min, 4 °C) and supernatant were collected and diluted to 1 mg/mL of total proteins. To 1ml of sample, 100 µL of streptavidin agarose slurry was added and incubated overnight at 4 °C on a rocky platform. Agarose/sample was centrifuged and the pellet was washed 4x with lysis solution. Finally, 80 µL of Laemmli buffer supplemented with 2 βME was added and incubated for 10 min at 95 °C, and next the samples were analyzed by SDS-PAGE and Western blot using Ab anti ABCC4

4.5. cAMP Level Measurement

A cAMP level analysis was performed using the cAMP competitive Kit (#581001 Cayman Chemical, Ann Arbor MA, USA). Cells incubated for 24 h w/wo 20 µM MK571 were treated with 0.1M HCl for 20min in RT and assayed according to the manufacturer's protocol. Calculation was performed using data sheet provided by the Cayman. The cAMP concentration of HT29 control cells was set as 100%, and next all the obtained data were recalculated as the % of control.

4.6. Wound Healing (Scratch) Assay

Cells were seeded on 6-well plate or 24-well pate and were grown to confluence; with a 20 µL pipette tip and rinsed twice with PBS. New medium w/wo tested chemical compound was added. The wounded area was visualized after every 2 h using Nikon Eclipse TE 2000-U microscope (Nikon, Japan) or Spark® multimode microplate reader (TECAN, Switzerland)). Wound area was calculated by ImageJ software. Cell motility was estimated through the quantification of the % of recovery using the equation: R (%) = [1 − (wound area at Tt/wound area at T0)] × 100,where T0 is the wounded area at 0 h and Tt is the wounded area after th.

4.7. Fluorescent Dequenching (DQ) Gelatine Assay

The surface of 24-well plates was coated with 250 µL 0.1 mg/mL DQ gelatine (Life Technologies, Waltham, MA, USA) overnight at 4 °C and then washed 3× with PBS. Then, 1×10^5 cells/well were added for 24 h to earlier prepared DQ gelatine-coated dishes in full medium supplemented w/wo 20 µM MK571. FITC fluorescence generated by the cleavage of DQ gelatine was measured using a Thermo Labsystem Fluoroscan Ascent reader (ThermoFisher Scientific, Waltham, MA, USA) fitted with FITC excitation and emission filters. Data are presented as the percent of increase above background fluorescence (100%) observed in the control HT-29.

4.8. Trans-Well Invasion and Trans-Well Migration Assays

HT-29 control or HT-29/Snail cells were treated with 20 µM of MK571 for 24 h. Then, cells were trypsinized, washed twice with medium, and transferred (2.5×10^4 cells/chamber) to the upper chamber of Nunc™ Cell Culture Inserts (transwell) 8.0 µm pore diameter (#141006) covered with BD Matrigel (2 h, 0.6 mg/mL of Matrigel—75 µL) for 6 h in 0.1% BSA medium—supplemented w/wo MK571. Full medium in lower chamber was used as chemoattractant. Next, the medium and the Matrigel from the top surface of the membrane was removed, invaded cells on the bottom surface of the membrane were washed 2× with PBS, then fixed for 5 min with 96% ethanol at 4 °C. Cells were dyed at RT as follows: 6 min—hematoxylin, 1 min—1% eosin. Finally, membranes were cut out from chambers, placed on microscope glass and number of cells that migrate into the membrane was counted. CaCo2 cells incubated for 24 h with (MK571 20 μM) or untreated once were seeded on un-coated trans-well inserts (8 μM pores) in the upper chamber in 2% BSA medium (w/wo 20 μM MK571). Full medium in the lower chamber served as chemoattractant for cell migration. Cells were calculated in randomly assigned areas after 3 h of incubation followed by hematoxylin/eosin staining. Minimal and maximal cell counts are shown as the lower and upper extremes by respective whiskers.

4.9. PKA Phosphorylation Assay

Cells were seeded on a 6-well plate (5×10^5/well). After 24 h, full growth medium was changed into starving (FBS free) medium for 24 h. Next, 20 μM of MK571 was added to cells for 60, 30, 5, and 1 min. Cells without starving procedure were used as positive control, and cells not treated with MK571 were used as negative control. After washing with PBS, cells were lysed, and SDS/PAGE and Western blot were performed using phospho-(ser/thr) PKA Substrate Antibody #9621 (Cell Signalling, Technology, Danvers, MA, USA).

4.10. WST-1 Cell Viability and Proliferation Assay

A total of 2×10^4 cells per well were seeded on 96-well plate and left for 24 h. Next, 100 μL of fresh medium was added containing irinotecan to a final concentration of 0, 5, 25, 50, 100 μM. Next, after 48 h of incubation 10 μL of WST-1 reagent (ScienCell, Research Lab., Carlsbad, CA, USA) freshly made, #8038 was added for 2 h. Calculation of cell viability was done by OD_{450nm} - OD_{630nm} using the Spark multimode microplate reader.

4.11. Statistical Analysis

All the experiments were repeated at least three times and the results were expressed as mean ± standard deviation (SD). Statistical evaluation was performed using normality test (Shapiro–Wilk) followed by T-Student test (for normally distributed data) or Mann-Whitney U test (for not normally distributed data). Calculation and graphs were performed using BioVinci version 1.1.5 developed by BioTuring Inc., San Diego, CA, USA, [44]. p values < 0.05 were considered statistically significant for all analyses: * $p < 0.05$; ** $p < 0.005$; *** $p < 0.001$, NS—not statistically significant. Pearson's linear correlation and Spearman correlation (Supplementary Materials) analysis were performed to analyze the correlation between TGFβ1/2 and TGFβR1/2 in CRC tissues with Pearson correlation coefficient (PCC): 0–0.25 no PCC, 0.25–0.5 low PCC, 0.5–0.75 moderately PCC, 0.75–1 strong PCC. Data density distribution is presented in Supplementary Materials, Figure S2 was produced using SinaPlot server [45].

The densitometry analysis of WB were performed with $n = 3$ (ABCC4/ABCG2 protein, PKA substrates) or $n = 2$ (EV's analysis) biological replicates. All the functional tests were performed in triplicate with $n = 3$ of biological replicates.

5. Conclusions

During the use of ABCC4 inhibitors to reduce chemotherapy resistance or drugs that are potential substrates of ABCC4, the indirect effect on cancer metastasis should be taken into consideration and may be important in selecting a therapy scheme for individual patients.

Supplementary Materials: The following are available online at http://www.mdpi.com/2072-6694/12/12/3547/s1: Figure S1 (**A**) Comparison of Pearson's and Spearman's correlation values (**B**) Relative densitometry quantification of ABCC4/ABCG2 protein expression level in HT29. (**C**) Relative densitometry quantification of ABCC4 protein expression level in various CRC cells lines. (**D**) Relative densitometry quantification of ABCC4 abundance in HT29 cells EVs. (**E**) Representative films of phosphorylation profile of PKA substrates after ABCC4 inhibition. (**F**) Irinotecan cytotoxicity (IC25 and IC50) for HT29/Snail and control HT29 cells. (**G**) representative images of wound healing of HT-29 control. Figure S2. Data density distribution produced with SinaPlot tools. Figure S3. Original Western Blots.

Author Contributions: J.K. designed the research, carried out all experiments, and prepared the first draft of the manuscript; I.P.-P. participated in HT29 stable clones generation and provided EVs for analysis; E.S. performed CaCo-2 cells migration assay and provided lysates of CRC lines for protein analysis; J.B. designed the research, developed the protocol and HT29 stable clones, provided data evaluation and interpretation, wrote the manuscript. All authors have read and agreed to the published version of the manuscript.

Funding: This work was supported by the grant from the National Science Center, Cracow, Poland, project no. 2018/02/X/NZ3/00144, the Polish-Norwegian Research Program, Project MOMENTO (Pol-Nor/202952/5/2013) and statutory funds to Institute of Medical Biology PAS, Lodz, Poland.

Conflicts of Interest: The authors declare no conflict of interest.

References

1. Henry, J.T.; Johnson, B. Current and evolving biomarkers for precision oncology in the management of metastatic colorectal cancer. *Chin. Clin. Oncol.* **2019**, *8*, 49. [CrossRef] [PubMed]
2. Bray, F.; Ferlay, J.; Soerjomataram, I.; Siegel, R.L.; Torre, L.A.; Jemal, A. Global cancer statistics 2018: GLOBOCAN estimates of incidence and mortality worldwide for 36 cancers in 185 countries. *CA Cancer J. Clin.* **2018**, *68*, 394–424. [CrossRef] [PubMed]
3. Souza e Silva, V.; Chinen, L.T.D.; Abdallah, E.A.; Damascena, A.; Paludo, J.; Chojniak, R.; Dettino, A.L.A.; de Mello, C.A.L.; Alves, V.S.; Fanelli, M.F. Early detection of poor outcome in patients with metastatic colorectal cancer: Tumor kinetics evaluated by circulating tumor cells. *Onco Targets Ther.* **2016**, *9*, 7503–7513. [CrossRef] [PubMed]
4. Jolly, M.K.; Somarelli, J.A.; Sheth, M.; Biddle, A.; Tripathi, S.C.; Armstrong, A.J.; Hanash, S.M.; Bapat, S.A.; Rangarajan, A.; Levine, H. Hybrid epithelial/mesenchymal phenotypes promote metastasis and therapy resistance across carcinomas. *Pharmacol. Ther.* **2019**, *194*, 161–184. [CrossRef] [PubMed]
5. Pastushenko, I.; Blanpain, C. EMT Transition States during Tumor Progression and Metastasis. *Trends Cell Biol.* **2019**, *29*, 212–226. [CrossRef] [PubMed]
6. Gupta, P.B.; Pastushenko, I.; Skibinski, A.; Blanpain, C.; Kuperwasser, C. Phenotypic Plasticity: Driver of Cancer Initiation, Progression, and Therapy Resistance. *Cell Stem Cell* **2019**, *24*, 65–78. [CrossRef] [PubMed]
7. Vu, T.; Datta, P.K. Regulation of EMT in Colorectal Cancer: A Culprit in Metastasis. *Cancers* **2017**, *9*, 171. [CrossRef]
8. El-Awady, R.; Saleh, E.; Hashim, A.; Soliman, N.; Dallah, A.; Elrasheed, A.; Elakraa, G. The Role of Eukaryotic and Prokaryotic ABC Transporter Family in Failure of Chemotherapy. *Front. Pharmacol.* **2016**, *7*, 535. [CrossRef]
9. Wilkens, S. Structure and mechanism of ABC transporters. *F1000Prime Rep.* **2015**, *7*, 14. [CrossRef]
10. Jiang, Z.S.; Sun, Y.Z.; Wang, S.M.; Ruan, J.S. Epithelial-mesenchymal transition: Potential regulator of ABC transporters in tumor progression. *J. Cancer* **2017**, *8*, 2319–2327. [CrossRef]
11. Santamaria, P.G.; Moreno-Bueno, G.; Cano, A. Contribution of Epithelial Plasticity to Therapy Resistance. *J. Clin. Med.* **2019**, *8*, 676. [CrossRef] [PubMed]
12. Saxena, M.; Stephens, M.A.; Pathak, H.; Rangarajan, A. Transcription factors that mediate epithelial–mesenchymal transition lead to multidrug resistance by upregulating ABC transporters. *Cell Death Dis.* **2011**, *2*, e179. [CrossRef] [PubMed]
13. Kryczka, J.; Boncela, J. Cell Migration Related to MDR-Another Impediment to Effective Chemotherapy? *Molecules* **2018**, *23*, 331. [CrossRef]
14. Przygodzka, P.; Papiewska-Pajak, I.; Bogusz, H.; Kryczka, J.; Sobierajska, K.; Kowalska, M.A.; Boncela, J. Neuromedin U is upregulated by Snail at early stages of EMT in HT29 colon cancer cells. *Biochim. Biophys. Acta* **2016**, *1860*, 2445–2453. [CrossRef]
15. Wen, J.; Luo, J.; Huang, W.; Tang, J.; Zhou, H.; Zhang, W. The Pharmacological and Physiological Role of Multidrug-Resistant Protein 4. *J. Pharmacol. Exp. Ther.* **2015**, *354*, 358–375. [CrossRef] [PubMed]
16. Li, C.; Krishnamurthy, P.C.; Penmatsa, H.; Marrs, K.L.; Wang, X.Q.; Zaccolo, M.; Jalink, K.; Li, M.; Nelson, D.J.; Schuetz, J.D.; et al. Spatiotemporal coupling of cAMP transporter to CFTR chloride channel function in the gut epithelia. *Cell* **2007**, *131*, 940–951. [CrossRef]
17. Sinha, C.; Ren, A.; Arora, K.; Moon, C.S.; Yarlagadda, S.; Zhang, W.; Cheepala, S.B.; Schuetz, J.D.; Naren, A.P. Multi-drug Resistance Protein 4 (MRP4)-mediated Regulation of Fibroblast Cell Migration Reflects a Dichotomous Role of Intracellular Cyclic Nucleotides. *J. Biol. Chem.* **2013**, *288*, 3786–3794. [CrossRef]

18. Sinha, C.; Ren, A.; Arora, K.; Moon, C.S.; Yarlagadda, S.; Woodrooffe, K.; Lin, S.; Schuetz, J.D.; Ziady, A.G.; Naren, A.P. PKA and actin play critical roles as downstream effectors in MRP4-mediated regulation of fibroblast migration. *Cell. Signal.* **2015**, *27*, 1345–1355. [CrossRef]
19. Delou, J.; Souza, A.S.O.; Souza, L.C.M.; Borges, H.L. Highlights in Resistance Mechanism Pathways for Combination Therapy. *Cells* **2019**, *8*, 1031. [CrossRef]
20. Geo. GEO2R-GEO-NCBI. 2020. Available online: https://www.ncbi.nlm.nih.gov/geo (accessed on 5 October 2020).
21. Candeil, L.; Gourdier, I.; Peyron, D.; Vezzio, N.; Copois, V.; Bibeau, F.; Orsetti, B.; Scheffer, G.L.; Ychou, M.; Khan, Q.A.; et al. ABCG2 Overexpression in Colon Cancer Cells Resistant to SN38 and in Irinotecan-Treated Metastases. *Int. J. Cancer* **2004**, *109*, 848–854. [CrossRef]
22. Nielsen, D.L.; Palshof, J.A.; Brünner, N.; Stenvang, J.; Viuff, B.M. Implications of ABCG2 Expression on Irinotecan Treatment of Colorectal Cancer Patients: A Review. *Int. J. Mol. Sci.* **2017**, *18*, 1926. [CrossRef] [PubMed]
23. Schlicker, A.; Beran, G.; Chresta, C.M.; McWalter, G.; Pritchard, A.; Weston, S.; Runswick, S.; Davenport, S.; Heathcote, K.; Castro, D.A.; et al. Subtypes of primary colorectal tumors correlate with response to targeted treatment in colorectal cell lines. *BMC Med. Genom.* **2012**, *5*, 66. [CrossRef] [PubMed]
24. Melo, F.D.S.E.; Wang, X.; Jansen, M.; Fessler, E.; Trinh, A.; Rooij, L.P.M.H.d.; Jong, J.H.d.; Boer, O.J.d.; Leersum, R.v.; Bijlsma, M.F.; et al. Poor-prognosis colon cancer is defined by a molecularly distinct subtype and develops from serrated precursor lesions. *Nat. Med.* **2013**, *19*, 614–618. [CrossRef] [PubMed]
25. Christensen, J.; El-Gebali, S.; Natoli, M.; Sengstag, T.; Delorenzi, M.; Bentz, S.; Bouzourene, H.; Rumbo, M.; Felsani, A.; Siissalo, S.; et al. Defining new criteria for selection of cell-based intestinal models using publicly available databases. *BMC Genom.* **2012**, *13*, 274. [CrossRef] [PubMed]
26. Maacha, S.; Bhat, A.A.; Jimenez, L.; Raza, A.; Haris, M.; Uddin, S.; Grivel, J.-C. Extracellular vesicles-mediated intercellular communication: Roles in the tumor microenvironment and anti-cancer drug resistance. *Mol. Cancer* **2019**, *18*, 55. [CrossRef]
27. Przygodzka, P.; Papiewska-Pająk, I.; Bogusz-Koziarska, H.; Sochacka, E.; Boncela, J.; Kowalska, M.A. Regulation of miRNAs by Snail during epithelial-to-mesenchymal transition in HT29 colon cancer cells. *Sci. Rep.* **2019**, *9*, 2165. [CrossRef]
28. Kryczka, J.; Papiewska-Pajak, I.; Kowalska, M.A.; Boncela, J. Cathepsin B Is Upregulated and Mediates ECM Degradation in Colon Adenocarcinoma HT29 Cells Overexpressing Snail. *Cells* **2019**, *8*, 203. [CrossRef]
29. O'Brien, E.D.; Krapf, D.; Cabada, M.O.; Visconti, P.E.; Arranz, S.E. Transmembrane adenylyl cyclase regulates amphibian sperm motility through protein kinase A activation. *Dev. Biol.* **2011**, *350*, 80–88. [CrossRef]
30. Liang, C.C.; Park, A.Y.; Guan, J.L. In vitro scratch assay: A convenient and inexpensive method for analysis of cell migration in vitro. *Nat. Protoc.* **2007**, *2*, 329–333. [CrossRef]
31. Kryczka, J.; Przygodzka, P.; Bogusz, H.; Boncela, J. HMEC-1 adopt the mixed amoeboid-mesenchymal migration type during EndMT. *Eur. J. Cell Biol.* **2017**, *96*, 289–300. [CrossRef]
32. Schneider, C.A.; Rasband, W.S.; Eliceiri, K.W. NIH Image to ImageJ: 25 years of image analysis. *Nat. Methods* **2012**, *9*, 671–675. [CrossRef]
33. Moon, C.; Zhang, W.; Ren, A.; Arora, K.; Sinha, C.; Yarlagadda, S.; Woodrooffe, K.; Schuetz, J.D.; Valasani, K.R.; de Jonge, H.R.; et al. Compartmentalized accumulation of cAMP near complexes of multidrug resistance protein 4 (MRP4) and cystic fibrosis transmembrane conductance regulator (CFTR) contributes to drug-induced diarrhea. *J. Biol. Chem.* **2015**, *290*, 11246–11257. [CrossRef] [PubMed]
34. Pavillard, V.; Agostini, C.; Richard, S.; Charasson, V.; Montaudon, D.; Robert, J. Determinants of the cytotoxicity of irinotecan in two human colorectal tumor cell lines. *Cancer Chemother. Pharmacol.* **2002**, *49*, 329–335.
35. Jensen, N.F.; Stenvang, J.; Beck, M.K.; Hanakova, B.; Belling, K.C.; Do, K.N.; Viuff, B.; Nygard, S.B.; Gupta, R.; Rasmussen, M.H.; et al. Establishment and characterization of models of chemotherapy resistance in colorectal cancer: Towards a predictive signature of chemoresistance. *Mol. Oncol.* **2015**, *9*, 1169–1185. [CrossRef] [PubMed]
36. Giampieri, R.; Scartozzi, M.; Loretelli, C.; Piva, F.; Mandolesi, A.; Lezoche, G.; Del Prete, M.; Bittoni, A.; Faloppi, L.; Bianconi, M.; et al. Cancer stem cell gene profile as predictor of relapse in high risk stage II and stage III, radically resected colon cancer patients. *PLoS ONE* **2013**, *8*, e72843. [CrossRef]
37. Fletcher, J.I.; Haber, M.; Henderson, M.J.; Norris, M.D. ABC transporters in cancer: More than just drug efflux pumps. *Nat. Rev. Cancer* **2010**, *10*, 147–156. [CrossRef]

38. Arora, K.; Sinha, C.; Zhang, W.; Ren, A.; Moon, C.S.; Yarlagadda, S.; Naren, A.P. Compartmentalization of cyclic nucleotide signaling: A question of when, where, and why? *Pflugers Arch.* **2013**, *465*, 1397–1407. [CrossRef]
39. Krause, M.; Dent, E.W.; Bear, J.E.; Loureiro, J.J.; Gertler, F.B. Ena/VASP proteins: Regulators of the actin cytoskeleton and cell migration. *Annu. Rev. Cell Dev. Biol.* **2003**, *19*, 541–564. [CrossRef]
40. Le Clainche, C.; Carlier, M.F. Regulation of actin assembly associated with protrusion and adhesion in cell migration. *Physiol. Rev.* **2008**, *88*, 489–513. [CrossRef]
41. Hara, Y.; Sassi, Y.; Guibert, C.; Gambaryan, N.; Dorfmuller, P.; Eddahibi, S.; Lompre, A.M.; Humbert, M.; Hulot, J.S. Inhibition of MRP4 prevents and reverses pulmonary hypertension in mice. *J. Clin. Investig.* **2011**, *121*, 2888–2897. [CrossRef]
42. Tagami, M.; Kusuhara, S.; Imai, H.; Uemura, A.; Honda, S.; Tsukahara, Y.; Negi, A. MRP4 knockdown enhances migration, suppresses apoptosis, and produces aggregated morphology in human retinal vascular endothelial cells. *Biochem. Biophys. Res. Commun.* **2010**, *400*, 593–598. [CrossRef] [PubMed]
43. Schaletzki, Y.; Kromrey, M.L.; Broderdorf, S.; Hammer, E.; Grube, M.; Hagen, P.; Sucic, S.; Freissmuth, M.; Volker, U.; Greinacher, A.; et al. Several adaptor proteins promote intracellular localisation of the transporter MRP4/ABCC4 in platelets and haematopoietic cells. *Thromb. Haemost.* **2017**, *117*, 105–115. [CrossRef] [PubMed]
44. INCB. BioTuring Browser|BioTuring. 2020. Available online: www.bioturing.com (accessed on 5 October 2020).
45. Sidiropoulos, N.; Sohi, S.H.; Pedersen, T.L.; Porse, B.T.; Winther, O.; Rapin, N.; Bagger, F.O. SinaPlot: An Enhanced Chart for Simple and Truthful Representation of Single Observations Over Multiple Classes. *J. Comput. Gr. Stat.* **2018**, *27*, 673–676. [CrossRef]

cancers

MDPI

Article

Mifepristone Treatment Promotes Testicular Leydig Cell Tumor Progression in Transgenic Mice

Donata Ponikwicka-Tyszko [1,†], Marcin Chrusciel [2,†], Kamila Pulawska [2], Piotr Bernaczyk [3], Maria Sztachelska [1], Peilan Guo [4], Xiangdong Li [5,6], Jorma Toppari [2,7], Ilpo T. Huhtaniemi [2,8], Slawomir Wołczyński [1,6] and Nafis A. Rahman [2,6,*]

1 Department of Biology and Pathology of Human Reproduction, Institute of Animal Reproduction and Food Research, Polish Academy of Sciences, 10-748 Olsztyn, Poland; d.ponikwicka-tyszko@pan.olsztyn.pl (D.P.-T.); m.sztachelska@pan.olsztyn.pl (M.S.); slawomir.wolczynski@umb.edu.pl (S.W.)

2 Institute of Biomedicine, Research Centre for Integrative Physiology and Pharmacology University of Turku, 20520 Turku, Finland; marcin.chrusciel@orionpharma.com (M.C.); kamila.pulawska@utu.fi (K.P.); jortop@utu.fi (J.T.)

3 Department of Medical Pathomorphology, Medical University of Bialystok, 15-269 Bialystok, Poland; piotr.bernaczyk@umb.edu.pl

4 College of Biological Sciences and Technology, Beijing Forestry University, Beijing 100083, China; guopeilan@bjfu.edu.cn

5 State Key Laboratory of the Agro-Biotechnology, College of Horticultural Science, China Agricultural University, Beijing 100193, China; xiangdongli@cau.edu.cn

6 Department of Reproduction and Gynecological Endocrinology, Medical University of Bialystok, 15-276 Bialystok, Poland

7 Department of Pediatrics, Turku University Hospital, 20520 Turku, Finland

8 Institute of Reproductive and Developmental Biology, Imperial College London, London W12 0NN, UK; ilpo.huhtaniemi@imperial.ac.uk

* Correspondence: nafis.rahman@utu.fi

† Equally contributed.

Received: 9 September 2020; Accepted: 2 November 2020; Published: 4 November 2020

check for updates

Simple Summary: Recently, the antiprogestin activity of selective progesterone receptor (PR) modulator mifepristone (MF) has proven unsuccessful as a potential anti-cancer agent in various clinical trials. Herein, we analyzed the effects of MF treatment on Leydig cell tumor (LCT) progression in a transgenic mouse model (inhibin-α promoter-driven SV40 T-antigen), as well as on the proliferation of two Leydig tumor cell lines. MF significantly stimulated the proliferation of LCT in vitro. Similarly, a 1-mo MF or P4 treatment stimulated LCT tumor growth in vivo. Only the abundant membrane Pgrmc1 expression was found in LCTs, but no other classical Pgr or nonclassical membrane PRs. Functional analysis showed that PGRMC1 is required for MF and P4 to stimulate the proliferation and invasiveness of LCTs. Our findings provide novel information that the use of MF as an anti-cancer agent should be considered with caution due to its potential PGRMC1 tumor-promoting pathway activation in cancers.

Abstract: The selective progesterone receptor modulator mifepristone (MF) may act as a potent antiproliferative agent in different steroid-dependent cancers due to its strong antagonistic effect on the nuclear progesterone receptor (PGR). Hereby, we analyzed the effects of MF treatment on Leydig cell tumor (LCT) progression in a transgenic mouse model (inhibin-α promoter-driven SV40 T-antigen), as well as on LCT (BLTK-1 and mLTC-1) cell proliferation. MF significantly stimulated the proliferation of LCT in vitro. Similarly, a 1-mo MF or P4 treatment stimulated LCT tumor growth in vivo. Traceable/absent classical Pgr or nonclassical membrane PRs α, β, γ and Pgrmc2, but abundant membrane Pgrmc1 expression, was found in LCTs. MF did not activate glucocorticoid or androgen receptors in LCTs. Functional analysis showed that PGRMC1 is required for MF and P4 to stimulate the proliferation and invasiveness of LCTs. Accordingly, MF and P4 induced PGRMC1 translocation into the nucleus and thereby stimulated the release of TGFβ1 in LCT cells. MF and P4 treatments upregulated *Tgfbr1*, *Tgfbr2*, and *Alk1* expression and stimulated TGFβ1 release in LCT cells. Our findings provide novel mechanistic insights into the action of MF as a membrane PR agonist that promotes LCT growth through PGRMC1 and the alternative TGFβ1 signaling pathway.

Keywords: leydig cell tumor; mifepristone; progesterone; progesterone receptors; TGFβ; PGRMC1

---

1. Introduction

Mifepristone (MF, RU486), classified as a selective progesterone receptor (PR) modulator (SPRM) shows strong antagonistic activity on the nuclear progesterone receptor (PGR), but depending on different PGR isoforms may also act as an agonist [1]. Recently, the antiprogestin activity of MF has proven unsuccessful as a potential anti-cancer agent in various clinical trials (such as ovarian, breast, nervous system, prostate, ovarian, and bone cancers) [2–7]. On the contrary, MF has been shown to significantly inhibit the growth of cancer cells in vitro with different PGR expression profiles [8]. The actions of progesterone (P4) may be mediated by PGRs in the genomic way, but also through mPRs α, β and γ, as well as PGRMC1 and PGRMC2 in a rapid non-genomic way [9]. The PR type that may be involved in mediating the MF effect in different cancers still remains unknown [8,10–12]. Recently, we have shown that MF and P4 could induce similar agonistic effects in ovarian cancer in the absence of classical PRs. Moreover, we showed MF treatment of ovarian cancer was ineffective due to its agonistic PGRMC1 action that enhanced the tumor growth [13].

Testicular tumors account for 1% of all tumors in males [14], although they are a common malignancy in men between 15 and 35 years of age [15,16]. Approximately up to 3% of all testicular tumors are believed to represent Leydig cell tumors (LCTs) [17], although a recent study showed that LCTs are more frequent than generally believed and associated with male infertility, cryptorchidism, and gynecomastia [18]. LCTs are usually clinically benign, but about 10% of the reported cases reveal a malignant phenotype [19]. LCTs have been shown to secrete steroids that can locally regulate tumor growth [20,21]. Although the expression of PGRs in the male reproductive system has been demonstrated [22] the exact role of P4 in the regulation of testicular function is still poorly understood. Early studies demonstrated expression of nuclear PGRs in rat Leydig cells (LCs) [23,24]. In human testis, PGRs have been detected in LCTs and LCs hyperplasia, as well as in traceable amounts in normal LCs [25–27]. Recently, it has also been reported that P4 with transforming growth factor β1 (TGF-β1) may increase cell proliferation of mouse LCs [28]. Additionally, P4 stimulates steroidogenic acute regulatory protein (StAR) expression in MA-10 cells [29]. In mouse LCTs (mLTC-1) cell line, P4 significantly inhibited luteinizing hormone receptor (LHR) expression and function. Presumably it was through their membrane PR (mPR), as mLTC-1 cells did not express classical PGRs [24,30]. Interestingly, the disruption of α and β PGR isoforms did not affect male fertility [31,32], which may suggest a non-classical P4 pathway activation in LCs. Therefore, further studies are needed to analyze the P4-PR interaction and P4 signaling pathways involved in the regulation of normal and tumorous LC function.

In the present study, we took advantage of P4 and MF treatments in a transgenic mouse model expressing Simian Virus 40 T antigen under the inhibin α promoter (Inhα/Tag) that develops endocrinologically active LCTs by five months of age with 100% penetrance [33,34]. The onset of LCTs in Inha/Tag TG mice corresponded with increased serum levels of P4, decreased gonadotropin concentrations, and an increased number of P4-secreting tumor cells in the gonads [33,35]. For in vitro MF/P4 treatment experiments, we used an immortalized murine LCT cell line (BLTK-1) derived from the Inhα/Tag TG mice and another murine LCT cell line (mLTC-1) [36]. Our goal was to study the molecular mechanisms underlying the MF and P4 action on LCTs and to characterize their nuclear and membrane PR expression profiles, as well as the MF pharmacokinetics in vivo and the MF metabolite effects on LCT proliferation in vitro.

2. Results

2.1. MF Achieves Low μM Serum Concentrations in Inhα/Tag Mouse Serum

Earlier pharmacological studies on MF have shown that MF and its metabolites (N-demethyl, Di-demethyl, and 22-hydroxy MF) achieve only a low μM serum concentrations in humans [37,38]. We evaluated the levels of MF and its metabolites in Inhα/Tag TG mouse serum. The MF pharmacokinetics was measured following a single-dose i.p. injection of 1 mg/kg or 10 mg/kg MF. The peak MF concentrations after 4 h reached 0.024 μM and 0.32 μM following the 1 mg/kg or 10 mg/kg doses, respectively (Figure S1A,B). MF was metabolized by hydroxylation and demethylation. The peaks of hydroxylated MF, mono- and di-demethylated MF were observed at 5 hours after injection and achieved 0.04, 0.009, and 0.015 μM, and 0.35, 0.135, and 0.218 μM for the 1 mg/kg or 10 mg/kg doses, respectively.

2.2. MF and P4 Treatmenst Affect Leydig Tumor Cell Proliferation In Vitro

We analyzed the effects of MF (0.01–25 μM concentration) and P4 (0.003–3 μM concentration) on the proliferation of two independent Leydig tumor cell lines, BLTK-1 and mLTC-1. Low levels, up to 5 μM of MF (Figure 1A, Figure S2A), as well as 0.03 μM of P4 in BLTK-1, and 0.03 μM and 0.3 of P4 in mLTC-1, increased cell proliferation (Figure 1B, Figure S2B), whereas the higher doses of MF, but not of P4, significantly inhibited the cell proliferation in both cell lines. N-demethyl MF at a concentration of 17.5 μM significantly stimulated cell proliferation, whereas the other MF metabolites (Di-demethyl and 22-hydroxy MF) did not affect the BLTK-1 cell proliferation (Figure 1C). We also found that MF treatment did not inhibit the P4-enhanced proliferation of BLTK-1 cells (Figure 1D).

Furthermore, we found that the translocation of the cell death marker protein HMGB1 from the nucleus to the cytoplasm occurred at 17.5 μM MF, but not at 5 μM MF (Figure 1E,F), proving that the lower doses of MF do not induce cell death.

2.3. MF and P4 Stimulate Leydig Cell Tumor Growth In Vivo

To analyze the effects of MF and P4 on tumor growth in Inhα/Tag TG mice we chose 10 mg/kg of MF, corresponding to the dose used in clinical trials and another dose of 1 mg/kg of P4. The Inhα/Tag TG mice treated with these doses of MF and P4 shown increased testis weights (Figure 2A).

Histopathological analyses demonstrated in non-treated LCTs severe cellular atypia, only a few peripheral tubular structures with spermatogenic cells up to elongated spermatids and in some regions rapid tumor growth with necrosis (Figure 2B). The P4- and MF-treated LCTs showed overall destroyed histological morphology with blood-filled cavities, infiltrating lymphocytes, and with almost no normal testicular structures left (Figure 2D,F). We confirmed LCT progression after MF and P4 treatment by increased Ki67-positive cells to 60–80% vs. 40% in the non-treated group (Figure 2C,E,G).

We also analyzed the hormonal profiles after the treatments. P4 (1 mg/kg) treatment significantly decreased LH (Figure 2H) and increased serum P4 levels (Figure 2I). Both treatments increased serum inhibin B level (Figure 2J). Additionally, MF and P4 down-regulated the *Lhcgr* expression level (Figure S3).

2.4. MF and P4 Stimulate Leydig Cell Tumors Proliferation and Invasiveness through Pgrmc1, Independently of Their Glucocorticoid Receptors

We have characterized the expression profile for all PRs in BLTK-1 and mLTC-1 cells as well as Inhα/Tag LCTs (Figure 3A–L, Figure S4, and Table S1). The *Pgr* expression level was low in non-treated LCTs (Figure 3A).

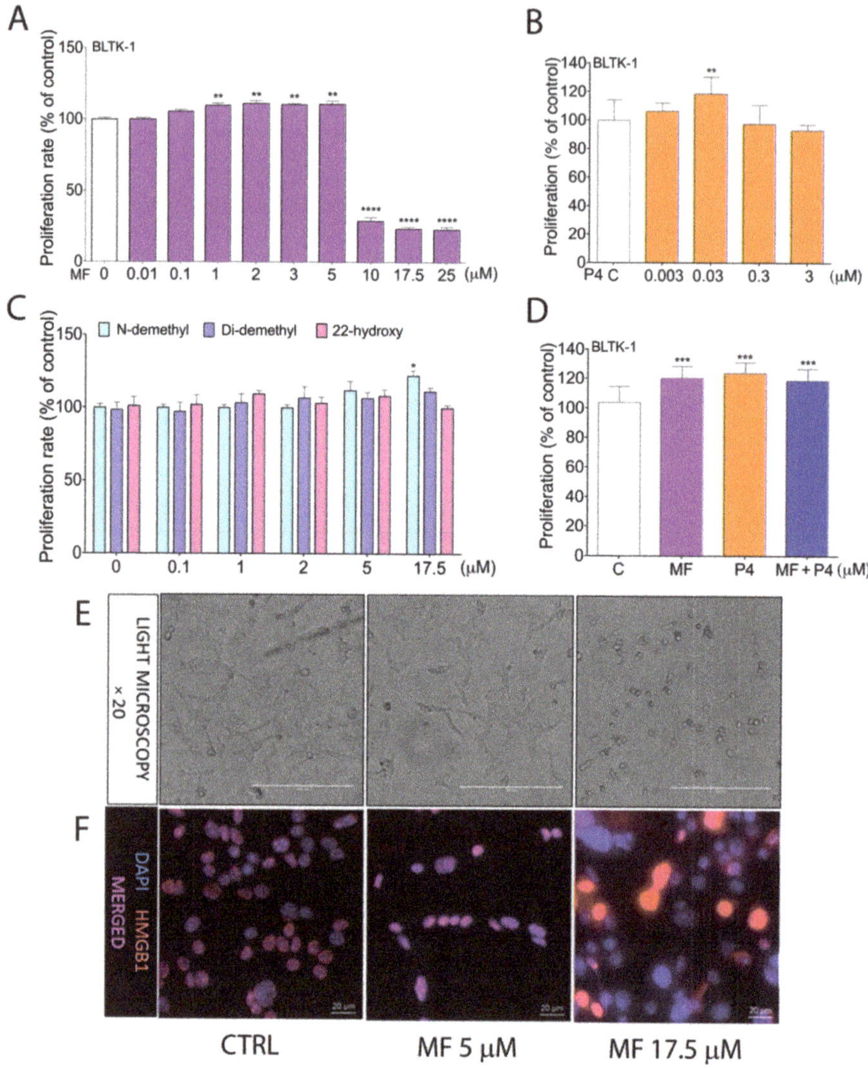

Figure 1. Proliferation of BLTK-1 cells with MF, P4 or MF metabolite treatments. Effects of MF (**A**), P4 (**B**), the 22-hydroxy, N-demethyl and Di-demethyl MF metabolites (**C**) and MF with P4 (**D**) on BLTK-1 cells proliferation after 72 h treatments, measured by MTT and BrdU assay. Light microscopy images of BLTK-1 cells after 5 μM or 17.5 μM MF treatment (**E**). Immunolocalization of HMGB1 protein after 5 μM or 17.5 μM MF treatment of BLTK-1 cells (**F**). The proliferation level of the treated groups is presented as the percentage of control proliferation, considered as 100%. Asterisks indicate significant differences between the control and treated groups (*, $p < 0.05$; **, $p < 0.01$; ***, $p < 0.001$; ****, $p < 0.0001$.). Scale bar, 20 μm. Di-demethyl MF, (11β,17β)-11-(4-Aminophenyl)-17-hydroxy-17-(1-propyn-1-yl)-estra-4,9-dien-3-one; 22-hydroxy MF, (11β,17β)-11-[4-(Dimethylamino)phenyl]-17-hydroxy-17-(3-hydroxy-1-propyn-1-yl)-estra-4,9-dien-3-one; Inhα/Tag mice; transgenic mice expressing the SV40 Taq oncogene under the inhibin α promoter; MF, mifepristone; N-demethyl MF, (11β,17β)-17-Hydroxy-11-[4-(methylamino)phenyl]-17-(1-propyn-1-yl)-estra-4,9-dien-3-one; P4, progesterone.

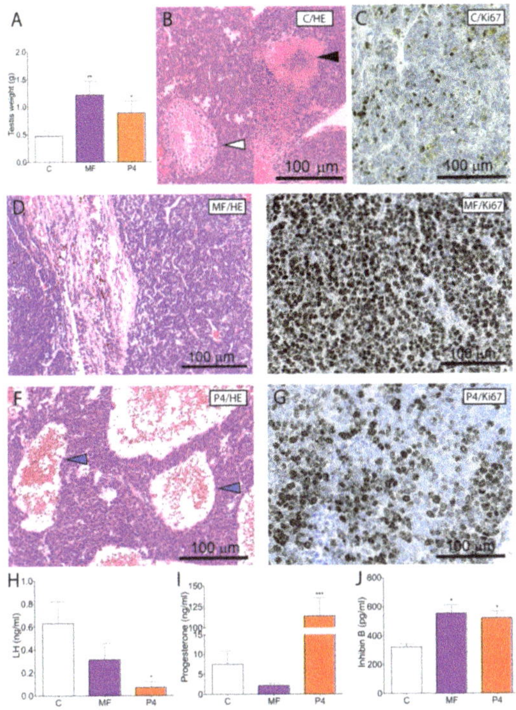

Figure 2. Treatment of Inhα/Tag TG mice presenting Leydig cell tumors and post-treatment hormonal values. Total testicular tumor weights (mean ± SEM) of control, MF-treated and P4-treated Inhα/Tag TG mice (**A**). Analyses of the control histopathology (**B**), control Ki-67 staining (**C**) MF-treated histopathology (**D**), MF-treated Ki-67 staining (**E**) P4-treated histopathology (**F**) and P4-treated Ki-67 staining (**G**) Inhα/Tag TG mice. Serum concentrations (mean ± SEM) of LH (**H**), P4(**I**), and inhibin B (**J**) of the non-treated (vehicle) (control), MF-treated and P4-treated Inhα/Tag TG mice. White arrow heads indicate tubular structure (**B**), black arrow heads necrotic area (**B**), blue arrow heads blood-filled cavities (**F**). Asterisks indicate significant differences between non-treated and treated groups (*, $p < 0.05$; **, $p < 0.01$; ***, $p < 0.001$). Scale bar, 100 μm. C, control; Inhα/Tag TG mice; transgenic mice expressing the SV40 Taq oncogene under the inhibin α promoter; MF, mifepristone; P4, progesterone.

The MF and P4 treatments significantly increased *Pgr* expression in LCTs (Figure 3A) but did not affect the expression of any of the membrane PR (*Paqr5, Paqr7, Paqr8, Pgrmc1, Pgrmc2*) (Figure 3B–F). Immunohistochemical studies demonstrated a weak expression of Pgr (Figure 3G–I) but abundant Pgrmc1 in non-treated, MF-, and P4-treated LCTs (Figure 3J–L). BLTK-1 cells expressed all types of PRs at the mRNA level, however the expression of PGR and mPRγ at the protein level was weak (Figure S4, Table S1). No Pgr and weak mPRγ expression was detected in mLTC-1 cells (Figure S4, Table S1).

To analyze whether PGRMC1 is involved in MF or P4 actions in LCTs, we treated BLTK-1 cells with the PGRMC1 inhibitor AG-205, which inhibited both MF- and P4-stimulated BLTK-1 cells proliferation (Figure 4A).

The PGRMC1 inhibitor cotreatment significantly decreased BLTK-1 cell proliferation compared to the control group (Figure 4A). Moreover, the PGRMC1 inhibition significantly reduced MF- and P4-induced BLTK-1 cell invasion (Figure 4B). We also studied the nuclear translocation of PGRMC1 after the MF and P4 treatments in BLTK-1 cells (Figure 5A–F). MF and P4 treatments induced the translocation of PGRMC1 to the nucleus (Figure 5B,C), whereas PGRMC1 blockage abolished PGRMC1 nuclear translocation in BLTK1 cells (Figure 5D–F).

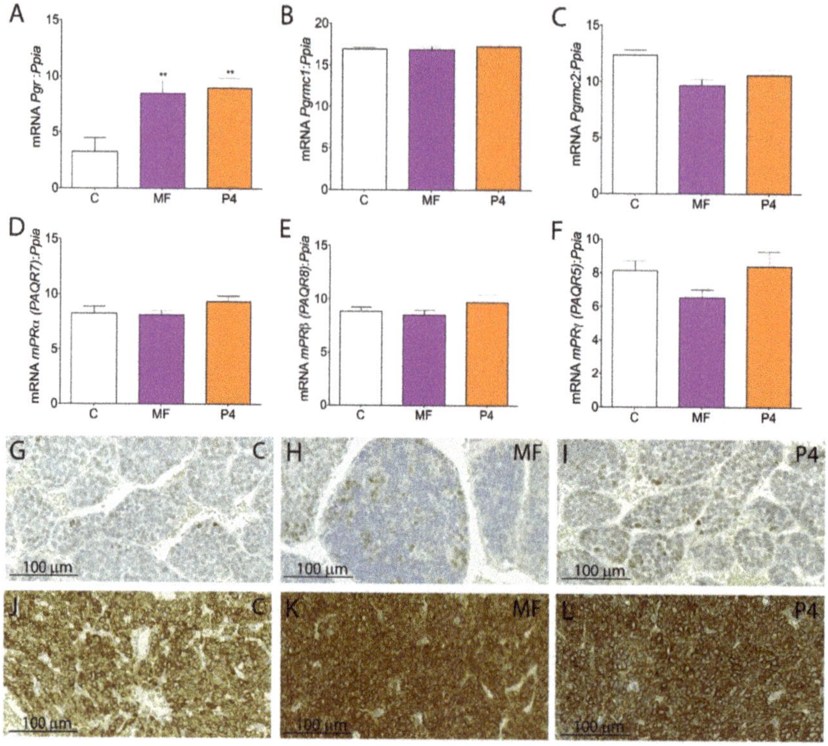

Figure 3. Pgr gene profiling in MF- and P4-treated transgenic Inhα/Tag TG mice and BLTK-1 cells proliferation and invasion. qPCR analysis of *Pgr* (**A**), *Pgrmc1* (**B**), *Pgrmc2* (**C**), *Paqr7* (mPRα) (**D**), *Paqr8* (mPRβ) (**E**), and *Paqr5* (mPRγ) (**F**) expression in the non-, MF- and P4-treated tumors of Inhα/Tag TG mice. Each bar represents the mean ± SEM relative to *Ppia*. Immunohistochemical staining of PGR in the control (**G**), MF-treated (**H**) and P4-treated (**I**) tumors and of PGRMC1 in control (**J**), MF-treated (**K**) and P4-treated (**L**) LCTs of Inhα/Tag TG mice. Scale bar, 100 µm. Asterisks indicate significant differences between the control and treated groups (**, $p < 0.01$). C, control; Inhα/Tag TG mice; transgenic mice expressing the SV40 Taq oncogene under the inhibin α promoter; LCT, Leydig cell tumor; MF, mifepristone; P4, progesterone.

As MF may also bind to glucocorticoid receptors (GR), we assessed the *Gr* and GR-target gene expression levels [39–41] and checked the immunolocalization of Gr after MF treatment in BLTK-1 cells (Figures S5A–F and S6A–D). Neither MF nor P4 treatments affected the expression of *Gr* (Figure S5A,B) or its target genes *Fkbp5*, *Ska2*, *Oct1*, and *Oct2* (Figure S6A–D). Immunocytochemistry demonstrated that Gr did not show nuclear staining after MF or MF with the Gr inhibitor (HSPi90) treatments (Figure S5E–F). A positive control, dexamethasone (DXM), induced Gr nuclear translocation (Figure S5D).

We also checked the MF and P4 treatment effects on androgen receptor (*Ar*) expression level, which was not affected by in vivo treatments in the Inhα/Tag TG mouse LCT or in vitro in BLTK-1 cells (Figure S7A,B).

We finally assessed the Simian Virus 40 T antigen (SV40 Tag) expression levels before and after the MF and P4 treatments in the Inhα/Tag TG mouse LCT and BLTK-1 cells, to exclude the potential interaction of the oncogene with MF and P4 actions in LCTs (Figure S8A,B). The mRNA level of *SV40 Tag* did not change after the MF and P4 treatments in LCTs (Figure S8A), or after the MF, P4, and PGRMC1 inhibitor AG-205 treatments in LCT cells (Figure S8B).

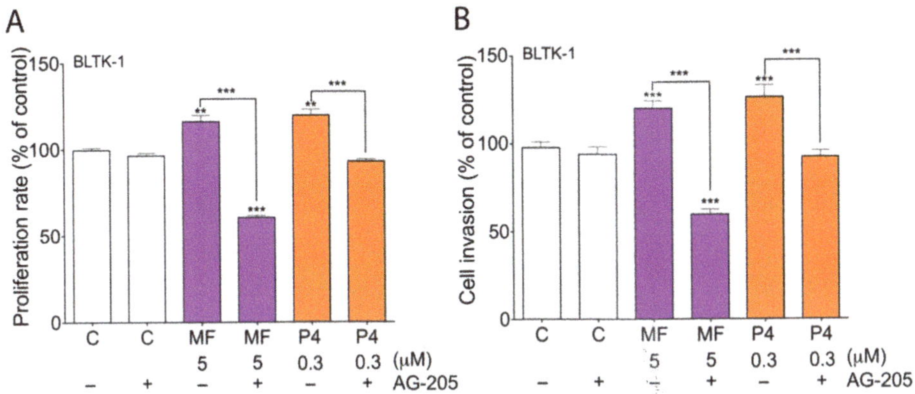

Figure 4. Proliferation of BLTK-1 cells with MF, P4 or PGRMC1 inhibitor treatments. Effects of MF and P4 with or without the AG-205 inhibitor on BLTK-1 cell proliferation after 24 h treatment, measured by BrdU assay (**A**). Cell proliferation of the treated groups is presented as the percentage of the control (considered as 100%). Effects of MF and P4 with or without the AG-205 inhibitor on BLTK-1 cell invasion after 24 h treatment (**B**). Cell invasion of the treated groups is presented as the percentage of the control group (considered as 100%). Asterisks indicate significant differences between the control and treated groups (**, $p < 0.01$; ***, $p < 0.001$). AG-205, PGRMC1 inhibitor; C, control; LCT, Leydig cell tumor; MF, mifepristone; P4, progesterone.

2.5. MF and P4 Promote Tumor Progression through Activation of the TGFβ1 Superfamily Signaling Pathway

As the TGFβ1 superfamily pathway may be involved in cancer progression [42], we characterized the TGFβ1 family member expression profile in LCTs. MF and P4 treatments increased serum TGFβ1 levels in Inhα/Tag TG mice compared with non-treated mice (Figure 6A).

Both the MF and P4 treatments significantly up-regulated *Tgfbr1*, *Tgfbr2*, *Acvrl1*, *Smad2*, and *Smad7* expression level in LCTs (Figure 6B–F). The non-treated group of LCTs showed weak expression of Tgfβr2 (Figure 6G), whereas after MF and P4 treatments, LCTs showed abundant Tgfβr2 staining (Figure 6H,I). Both treatments increased the TGFβ1 release by BLTK-1 cells, whereas PGRMC1 blockage inhibited this effect (Figure 6J). The MF and P4 treatments with recombinant TGFβ1 cotreatment significantly up-regulated the proliferation of BLTK-1 cells and showed an additive effect to MF (Figure 6K).

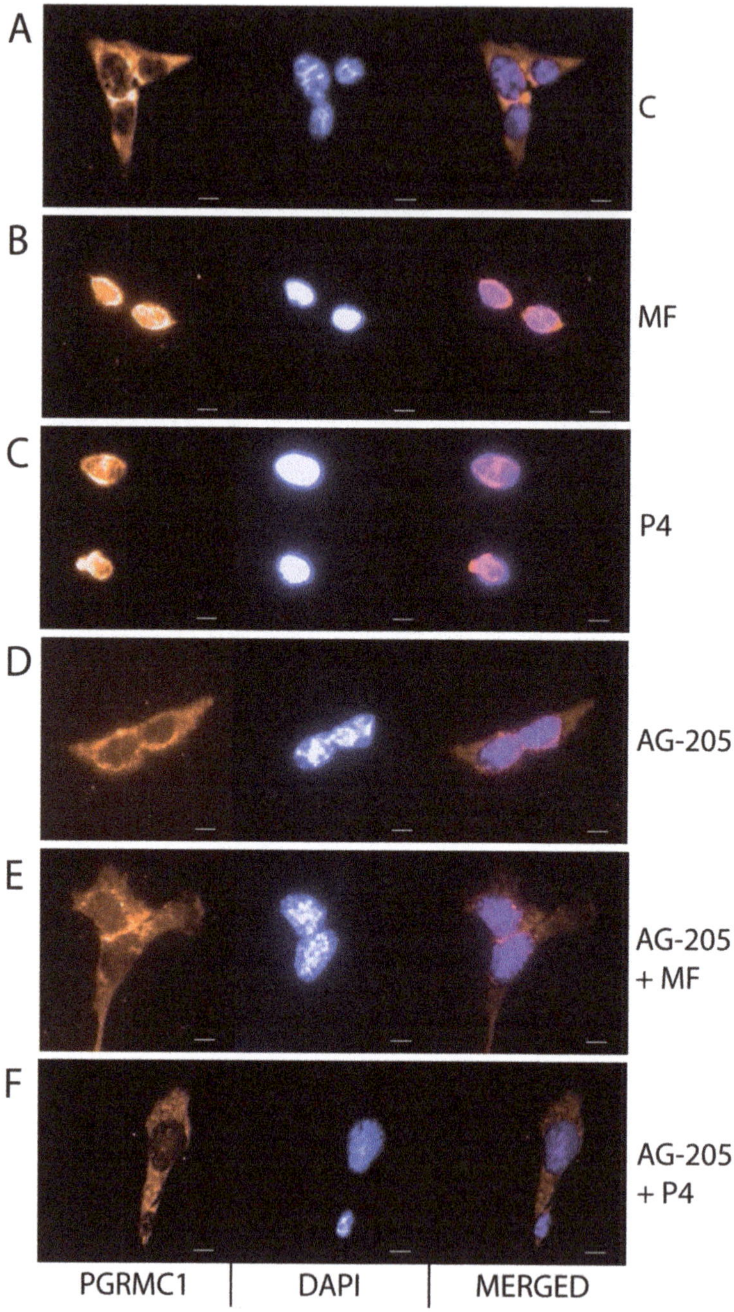

Figure 5. MF and P4 treatments effects on the nuclear translocation of PGRMC1 in BLTK-1 cells. Immunocytochemical localization of PGRMC1 without C (**A**) or with MF (**B**), P4 (**C**), AG-205 (**D**), AG-205 + MF (**E**) and AG-205 + P4 (**F**) in BLTK-1 cells. Scale bar, 20 µm. AG-205, PGRMC1 inhibitor; C, control; MF, mifepristone; P4, progesterone.

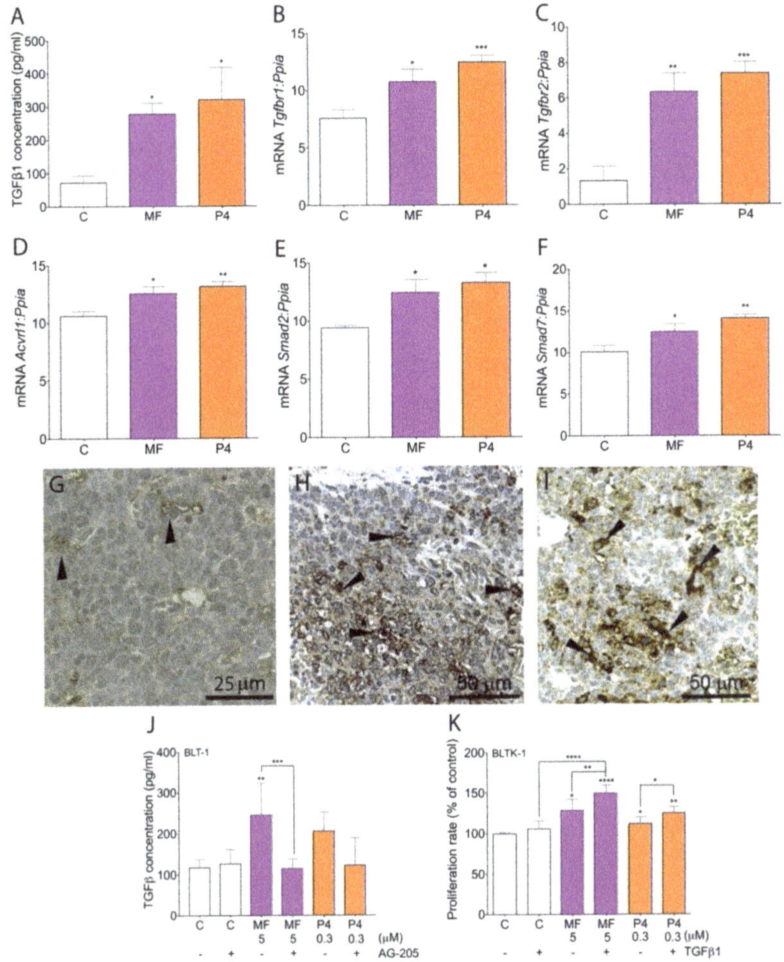

Figure 6. MF and P4 treatments effects on TGFβ1 superfamily signaling pathways. TGFβ1 serum level in the control, MF-treated or P4-treated tumors of Inhα/Tag TG mice (**A**). qPCR analysis of the *Tgfbr1* (**B**), *Tgfbr2* (**C**), *Acvrl1 (Alk1)* (**D**), *Smad2* (**E**), and *Smad7* (**F**) expression levels in the control, MF-treated and P4-treated tumors of Inhα/Tag TG mice. Each bar represents the mean ± SEM relative to *Ppia*. Immunohistochemical staining of TGFβR2 in the C (**G**), MF-treated (**H**) and P4-treated (**I**) tumors of Inhα/Tag TG mice. Black arrows indicate Tgfβr2 positive staining. Scale bar, 25 μm or 50 μm. TGFβ1 level in the control, MF-treated or P4-treated BLTK-1 cells with or without the AG-205 inhibitor (**J**). Effects of recombinant TGFβ1 treatment on MF- or P4-treated BLTK-1 cells (**K**). The proliferation level of the treated groups is presented as the percentage of control proliferation, considered as 100%. Asterisks indicate significant differences between the non-treated control and treated groups (*, $p < 0.05$; **, $p < 0.01$; ***, $p < 0.001$; ****, $p < 0.0001$). AG-205, PGMC1 inhibitor; C, control; Inhα/Tag TG mice; transgenic mice expressing the SV40 Taq oncogene under the inhibin α promoter; MF, mifepristone; ND, non-detectable; P4, progesterone.

3. Discussion

P4 has been shown to be produced and involved in the regulation of LC and LCT function and proliferation [24,30], suggesting that LCTs could be susceptible to antiprogestin treatment. MF has been shown to inhibit in vitro the growth of cancer cells of reproductive and non-reproductive origin, independently of their PGR expression status [8]. However, the clinical MF trial results on several human cancers have been disappointing [3,7]. This discrepancy between the experimental and clinical data prompted further studies to understand the molecular mechanisms underlying the MF action in cancers.

It has been shown that MF in vitro at concentrations of 10–100 μM inhibits cancer cell growth [8,10–12,43]. However, MF at the lower 1 μM concentration did not affect human ovarian epithelial cancer cells proliferation [44]. Our earlier [13] and present results show that MF may exert its effect in a biphasic way. Recently, it has also been shown that the major active MF metabolite, metapristone (N-demethyl mifepristone), inhibited cancer cell proliferation in a dose-dependent manner [45,46]. However, our in vitro experiments demonstrated that metapristone at lower doses may significantly stimulate tumor cell proliferation. Therefore, our data suggest that MF and metapristone at low levels may have a stimulatory effect on tumor cells and may not achieve sufficient tissue levels to inhibit cancer cell progression.

It has been shown that the serum concentrations of MF were not affected by increasing doses of the MF from 200 to 800 mg and reached a maximum of 2.5 μM serum concentration in all patient groups after 24h [47]. Even long-term treatments up to 20 months did not change the serum concentration of MF [48]. The dose of 10 mg/kg of MF in mice corresponds to the highest dose used in human clinical trials [13]. As there were no changes in serum MF levels in humans between the doses of 200–800 mg/day, any changes after higher doses in mice were not either expected. It is highly unlikely to achieve a high level, e.g., 10 μM of MF in the tissues, due to its rapid metabolism and binding to the high-affinity binding protein alpha 1-acid glycoprotein (AAG) [49]. The serum concentration of ~2.5 μM corresponds to the plasma protein AAG binding capacity of MF [49]. Only the unbound drug, i.e., low circulating 2.5 μM of MF concentration, is available for target tissues to exert its pharmacological effects [13,49].

P4 may mediate its signal in a dual mode through genomic and non-genomic ways, although the key mode of PR action on LC and LCT function has not been described [30]. The type of PR that may be involved in mediating the MF effect in different cancers also remains unknown [8]. Expression of the classical Pgr was not observed in mLTC-1 cells [30]. Moreover, the effect of MF on the P4 action on mLCT-1 cells was inconsistent, suggesting that P4 affects LCs independently of the classical P4 signaling pathway [24]. In the present study, we characterized the expression profile of all the nuclear and membrane PRs in LCTs of Inhα/Tag TG mice, BLTK-1, mLTC-1 cells. However, only Pgrmc1 expression was high in LCTs, suggesting its functional role in these tumors. Besides the characterization of PRs, we also checked the GR activation, as MF may also affect the GR and bind to different GRs isoforms (α and β) with high affinity [37]. Recent studies of MF action on GRs are inconsistent, as one of them reported that MF may stimulate GRβ nuclear translocation, but another did not [50,51]. It has also been shown that MF rather inhibits than activates GRs action in LCs and LCTs [52,53]. We were unable to find any connections between MF and nuclear translocation of Gr or Gr-related stimulation of gene expression in BLTK-1 cells. Our data suggest that MF acts independently of GRs activation and rather mediates its action through mPRs in LCTs. Moreover, in MA-10 cells MF significantly stimulated StAR expression at a level comparable with P4 independently of Gr, indicating that the effect may be mediated through the membrane PRs pathway [29]. Our in vitro studies showed that membrane receptor PGRMC1 is required for MF and P4 to stimulate the proliferation and invasiveness of LCTs. These results indicate that MF acts as a selective membrane P4 agonist through PGRMC1 activation in LCTs. In human ovarian cancers, abundant PGRMC1 expression level has also been demonstrated [13,54]. Moreover, PGRMC1 has been involved in ovarian cancer cell invasion [13,55].

The results of this study showed that both MF and P4 may stimulate the alternative tumor-promoting TGFβ1 superfamily signaling pathway in LCTs. TGF-β1 is a member of a large cytokine family involved in many biological processes, including cell proliferation, differentiation, migration, adhesion and survival, in both normal and cancerous cells [56]. In LCs, TGF-β1 has been shown to influence steroidogenesis and regulate cell proliferation [28,57]. We also showed that MF and P4 enhanced *Acvlr1* (*Alk1*), *Smad2*, and *Smad7* expression levels. It has been suggested that TGF-β1 via the ALK1 signaling may lead to epithelial cell proliferation [58,59]. The high expression levels of TGF-β1signaling elements, especially ALK1 has also been shown in patients with LC hyperplasia [27]. Additionally, P4 with TGFβ1 has been considered as the domain factors causing LC hyperplasia/hypertrophy [28]. Morphometric testicular analyses of mice treated with P4 and TGF-β1 revealed increased volume of LCs [28]. Studies also demonstrated enhanced expression of the proliferation marker PCNA in LCs after P4 and TGFβ1 treatments. Additionally, P4 and TGFβ1 treatments reduced the expression level of the proapoptotic gene *Bax* [28]. Our present data revealed that MF and P4 also enhanced the expression of the cell-cycle progression marker Ki-67 in LCTs of Inhα/Tag TG mice. Ki67 is not a cell proliferation marker per sein the sense that it labels cells in S-Phase of the cell cycle. Ki67 labeling can be found in nuclei throughout the cell cycle, usually, except in the cells that are in the G0 phase. Ki67 shows thus that the cells have the capacity to proliferate and are not terminally differentiated. MF treatment increased also TGFβ1 release in BLTK-1 cells. PGRMC1 inhibition significantly reduced this effect, indicating that PGRMC1 is involved in MF and P4 activation of the TGFβ1 signaling pathway in LCTs. Our data suggest that, in MF and P4 action, PGRMC1 may be the key LCT P4 receptor in the tumor-promoting action of TGFβ1.

PGRMC1 expression has also been shown in several cancer cell, like in breast, prostate and lung, emphasizing the translational aspect of such findings [55,60,61]. Recently, it has been demonstrated that PGRMC1 promotes tumorigenesis, cell proliferation, migration, invasion, and antiapoptosis in the same cancer types [55,60–62]. Additionally, higher expression of PGRMC1 may be useful in the prediction of prognosis of breast cancer patients [63]. This issue is becoming even more important, as patients with these cancers are still being recruited for ongoing clinical trials with MF [64–66]. The use of MF as anti-cancer agent should be reconsidered in the light of its potential of tumor promoting action through activation of the PGRMC1 pathway.

4. Materials and Methods

4.1. Experimental Animals

In vivo studies were done on previously characterized Inhα/Tag TG mice [35]. The 5.5 months of age male mice with discernible testicular tumors were randomized into three groups (n = 10 mice/group) and intraperitoneally injected every 2 days either with vehicle (corn oil) or MF (10 mg/kg; Sigma-Aldrich, Saint Louis, MO, USA) or P4 (1 mg/kg; Sigma-Aldrich, Saint Louis, MO, USA) for 1 month. Mice were fed with mouse chow SDS RM-3 (Special Diet Service; E, soy free; Whitham Essex, UK), tap water ad libitum and kept in a specific pathogen-free surrounding and routinely screened for common mouse pathogens. After 30 days of treatments mice were sacrificed, blood and tissue samples were collected. Half of each tumor was fixed in 4% paraformaldehyde and embedded in paraffin for histological and immunohistochemical studies. The second half of the tumor tissue was snap-frozen in liquid nitrogen and stored at −80 °C for RT-PCR analysis. The Ethics Committee for animal experimentation of the University of Turku and the State Provincial Office of Southern Finland approved all animal experiments (Animal Licence number: ESAVI/5757/04.10.07/2017).

4.2. MF Pharmacokinetics

MF pharmacokinetic was analyzed in 6 mo-old Inhα/Tag male mice. Mice were intraperitoneally injected with a single dose of 1 mg/kg of MF (n = 5) and 10 mg/kg of MF (n = 5). Blood samples were collected after 30 min, 4 h, and 8 h, 16 h, 24 h, and 48 h. Concentrations of MF and its metabolites N-demethyl MF, Di-demethyl MF and 22-hydroxyl MF in mouse plasma were determined using high performance liquid chromatography–mass spectrometry (HPLC-MS/MS) after protein precipitation with internal standard alfaxalone. HPLC separation was performed with Agilent 1200 LC system, using a C18 column. Multiple-reaction monitoring with a triple quadrupole mass spectrometer was used for quantitative analyses (AB Sciex 4000 QTrap with Analyst software (v. 1.6.1); MDS Sciex, Ontario, Canada). Standards 22-hydroxy (H948445), Di-demethyl (D439550) and N-demethyl mifepristone (D230950) were bought from Toronto Research Chemicals Inc. (Toronto, Ontario, Canada).

4.3. Cell Cultures

The BLTK-1 cells were cultured in DMEM/F12 medium (GIBCO, Paisley, UK) supplemented with 10% fetal bovine serum (FBS; Biochrom, Berlin, Germany), 100 units/ml penicillin and 100 μg/ml streptomycin (P/S solution; Sigma-Aldrich) at 37 °C in a humidified atmosphere in the presence of 5% CO2. The mLTC-1 cells were cultured in Waymouth's medium (GIBCO, Paisley, UK) supplemented with 10% horse serum (GIBCO, Paisley, UK) and 5% FBS (Biochrom, Berlin, Germany), and P/S solution (Sigma-Aldrich, Saint Louis, MO, USA) at 37 °C in a humidified atmosphere in the presence of 5% CO_2.

4.4. Cell Proliferation

Cell proliferation was analyzed using CellTiter 96®AQueous Non-Radioactive Cell Proliferation Assay (Promega, Madison, WI, USA) and BrdU Cell Proliferation Assay Kit (Cell Signaling Technology, Danvers, MA, USA). BLTK-1 were seeded (10×10^3/well) in culture medium onto 96-well plate and after 16 h treated with vehicle (EtOH 0.05%), MF (0.01; 0.1; 1; 2; 3; 5; 7; 17; 25 μM), P4 (0.003; 0.03; 0.3; 3 μM), a PGRMC1 inhibitor AG-205 (1 μM, Sigma-Aldrich, Saint Louis, MO, USA) and TGFβ1 (10 μM, 240-B; R&D Systems Inc., Minneapolis, MN, USA) in stimulation medium (phenol-free DMEM/F12 with 0.5% charcoal-stripped FBS and P/S solution) for 72 h with MF or P4 for 24 h with AG-205/TGFβ1. The proliferation rate was presented as a percentage of control proliferation considered as 100%. Three independent experiments per cell line were run, each performed in octuplicate wells.

4.5. Cell Invasion

Cell invasion intensity of BLTK-1 cells was assessed using CultreCoat®Cell Invasion Assays (R&D Systems, Minneapolis, MN, USA). Briefly, 2.5×10^4 cells/well were transferred to each of 96-well plate top invasion chamber coated with Basement Membrane Extract (BME). Cells invaded in response to MF, P4 and AG-205 (1 μM) were quantitated using Calcein AM after 24 h of treatment. Three independent experiments were run, each performed in octuplicate wells. Cell invasion intensity of the treated groups was presented as percentage of invasion of control group, considered as 100%.

4.6. Histological and Immunohistochemical Analyses

Mouse testicular tumor tissues were fixed in paraformaldehyde and embedded in paraffin. For histological analysis, 5 μm paraffin sections were stained with hematoxylin-eosin. For immunohistochemistry sections were deparaffinized, hydrated and boiled in 10 mM citric acid buffer (pH 6.0) in retriever for 2.5 h. Tissue sections were incubated with blocking solutions (10% normal goat serum (NGS) with 3% bovine serum albumin (BSA) or only 3% BSA in PBS) for 1 h at room temperature in order to reduce non-specific background staining. Then, sections were incubated overnight at 4 °C with the primary antibodies for PGR (MA5-12658, Thermo Fisher Scientific Inc., Waltham, MA, USA; dilution 1:700), mPRα (ab75508, Abcam, Cambridge, UK; dilution 1:500), mPRβ (ab46534, Abcam; dilution 1:1000), mPRγ (ab79517, Abcam; Cambridge, UK; dilution 1:500), PGRMC1 (PAB20135, Abnova Corporation, Taipei, Taiwan; dilution 1:2000), PGRMC2 (ab125122, Abcam; Cambridge, UK; dilution 1:1000), TGFβRII (sc-220, Santa Cruz Biotechnology, Dallas, TX, USA; dilution 1:700), Ki-67 (Clone TEC-3, Dako, Glostrup, Denmark; dilution 1:500), IgG (ab190475, Abcam; Cambridge, UK; dilution 1:700), IgG2a (ab190463, Abcam; Cambridge, UK; dilution 1:500). After endogenous peroxidase blocking (0.5% H_2O_2 in PBS for 20 min in dark at room temperature) primary antibodies were linked with Envision®anti-mouse or anti-rabbit polymer + HRP (Dako, Glostrup, Denmark) for 30 min at room temperature, only for Ki-67 staining before this step, secondary antibody rabbit anti rat was added (Dako, Glostrup, Denmark; dilution 1:200). The reaction product was visualized using 3′3-diaminobenzidine tetrahydrochloride (DAB, Dako, Glostrup, Denmark).

Three washes were done after each step with PBS with 0.05% Tween (PBS-T). Hematoxylin was used as counterstain and then sections were dehydrated and mounted with Pertex (Histolab Products AB, Spånga, Sweden). Control immunohistochemical stainings of the IgG2a and IgG are shown in Figure S9A–D.

4.7. Immunocytochemistry Analysis

BLTK-1 cells 1–2 × 10^4 cells/well were seeded onto microscope slide coverslips and after 16 h treated with vehicle (0), MF (5 µM, 17.5 µM), vehicle (0), MF (3 µM), DXM (200 nM), MF (3 µM) + HSP90i (50 nM), HSP90i (50 nM) + DXM (200 nM) or vehicle (0), MF (3 µM), P4 (0.3 µM), AG-205 (1 µM), MF (3 µM) + AG-205 (1 µM), and P4 (0.3 µM) + AG-205 (1 µM) in stimulation medium. Cells were fixed in 4% PFA in PBS pH 7.4 for 15 min at room temperature and permeabilized for 10 min in 0.1% Triton X-100. To reduce autofluorescence cells were incubated with 100 mM NH4CL for 10 min. After blocking unspecific binding sites with 3% BSA in PBS with 0.05% Tween 20 for 30 min cells were incubated for 1 h with primary antibodies anti-GR (SC-56851, Santa Cruz Biotechnology, Dallas, TX, USA; dilution 1:400), anti-PGRMC1 (PAB20135, Abnova Corporation; dilution 1:1000) or anti-HMGB1 (ab79823, Abcam, Cambridge, UK; dilution 1:350) diluted in blocking solution. Next, cells were incubated with secondary fluorescent antibody Alexa Fluor 488 goat anti-mouse IgG (ab150113, Abcam, Cambridge, UK; dilution 1:400) or Alexa Fluor 647 donkey anti-rabbit IgG (Life Technologies, Carlsbad, CA, USA; dilution 1:600) for 45 min. To detect cell nuclei, cells were incubated with DAPI for 1 min.

4.8. Real Time RT-PCR

Total RNA from cells and snap-frozen LCTs were prepared using TRIzol extraction method (Invitrogen, Carlsbad, CA). The quantity and quality of isolated RNA was determined by NanoDrop (Thermo Fisher Scientific Inc., Waltham, MA, USA) and gel electrophoresis. Before the reverse transcription (RT) reaction 1 µg of total RNA was incubated for 30 min with DNase I (Invitrogen, Carlsbad, CA) at room temperature. The RT reaction was performed with DyNAmo ^{TM} cDNA Synthesis Kit (Finnzymes, Espoo, Finland) at 37 °C for 1 h in 20 µl. Quantification of investigated genes was performed with FX96™ Real-Time PCR Detection System, Bio Rad using DyNAmo SYBR Green qPCR kit (Finnzymes, Espoo, Finland). Reaction conditions were: initial denaturation at 95 °C for 10 min followed by 40 amplification cycles at 95 °C for 15 s, 56–60 °C at 45 s and 70 °C at 45 s. At the end of the PCR reaction, melting curve was determined to ensure single product amplification. Amplification products were separated on 1.8% agarose gel and stained with ethidium bromide. Expression levels were normalized to the housekeeping gene peptidylprolyl isomerase (*Ppia*). The primer sequences and expected product sizes are shown in Table S2.

4.9. Hormones and TGFβ1 Measurement

Serum levels of LH and FSH were measured by immunofluorometric assays (Delfia; Perkin-Elmer-Wallac, Turku, Finland) as described previously [67,68]. Serum P4 level was measured using Delfia Progesterone Kit (Wallac, Perkin Elmer, Turku, Finland). The intra- and interassay coefficients of variations for these assays were below 10%. Serum level of inhibin B was evaluated by immunoassay Inhibin-B EIA Kit (Sigma-Aldrich, Saint Louis, MO, USA). TGFβ1 level in serum and cell culture supernates was assessed using TGFβ1 Quantikine ELISA Kit (R&D Systems, Minneapolis, MN, USA), following the instructions of the manufacturer.

4.10. Statistical Analysis

Numerical data are presented as mean ± SEM. To analyze statistical significance one-way ANOVA with the post-hoc Bonferroni's multiple comparison post-hoc test with 95% confidence interval was used (GraphPad PRISM v. 7. GraphPad Software Inc., San Diego, CA, USA). Results were considered to be statistically significant at $p < 0.05$ level.

5. Conclusions

In conclusion, based on our results, we suggest that MF in low concentration may act as a membrane PR agonist and activate through PGRMC1 the tumor progression signaling pathway of TGFβ1 superfamily in LCTs. MF may also induce the PGRMC1 nuclear translocation and increase the proliferation and invasion of LCTs. Hence, it is possible that the MF anti-tumor effects observed in many cancer cell lines may not be achievable in vivo in cancer tissues and MF might not be considered as an anti-cancer agent.

Supplementary Materials: The following are available online at http://www.mdpi.com/2072-6694/12/11/3263/s1, Figure S1: Pharmacokinetic analysis of MF metabolism in Inhα/Tag TG mice, Figure S2: Proliferation of mLTC-1 cells with MF or P4 treatments, Figure S3: *Lhcgr* expression level in Leydig cell tumors of Inhα/Tag TG mice, Figure S4: Characteristics of progesterone receptors mRNA levels in BLTK-1 and mLTC-1 cell lines, Figure S5: MF and P4 treatments effects on glucocorticoid receptor, Figure S6: Gr-target genes expression profile in BLTK-1 cells, Figure S7: MF and P4 treatments effects on androgen receptor expression, Figure S8: *SV40 Tag* expression level in Inhα/Tag TG mice and BLTK-1 cells, Figure S9: Isotype negative control staining, Table S1: Characteristics of progesterone receptors mRNA level and immunoreactivity in murine Leydig cell tumor and murine Leydig tumor cell line, Table S2: Primer sequences for RT-qPCR.

Author Contributions: Concept and design: D.P.-T., M.C., S.W., and N.A.R.; Acquisition of data/performed the experiments: D.P.-T., M.C., K.P., P.B., and M.S.; Analysis and interpretation of data: D.P.-T., M.C., P.G., X.L., J.T., I.T.H., S.W., and N.A.R. Writing of the manuscript: D.P.-T., M.C., I.T.H., S.W., and N.A.R. All authors have read and agreed to the published version of the manuscript.

Funding: This work was supported by grants from the National Science Centre, Poland (2013/09/N/NZ5/01831 to D.P.-T.), Medical University of Bialystok Grant (SUB/1/DN/19/005/1104 and SUB/1/DN/20/005/1104 to S.W.), Academy of Finland (254366 to NAR), Moikoinen Cancer Research Foundation (to N.A.R.) and EU PARP Cluster grant (UDA-POIG.05.01.00-005/12-00/NCREMFP to S.W.).

Acknowledgments: We thank Mika Scheinin and Paulina Chrusciel for their help with the pharmacokinetics and Joanna Stelmaszewska with animal maintenance experiment.

Conflicts of Interest: The authors declare no conflict of interest.

Abbreviations

22-hydroxy MF	(11β,17β)-11-[4-(Dimethylamino)phenyl]-17-hydroxy-17-(3-hydroxy-1-propyn-1-yl)-estra-4,9-dien-3-one
AG-205	inhibitor PGRMC1
BLTK-1	Immortalized cell line from Inhα/Tag TG mice
DXM	Dexamethasone
Di-demethyl MF	(11β,17β)-11-(4-Aminophenyl)-17-hydroxy-17-(1-propyn-1-yl)-estra-4,9-dien-3-one
GR	Glucocorticoid receptor
HMGB1	High mobility group box 1 protein
HSP90i	Glucocorticoid receptor inhibitor
Inhα/Tag	Transgenic mice expressing Simian Virus 40 T antigen under inhibin-α promoter
LC	Leydig cell
LCT	Leydig cell tumor
LHR	Luteinizing hormone receptor
MF	Mifepristone
mPR	Membrane progesterone receptor
N-demethyl MF	11β,17β)-17-Hydroxy-11-[4-(methylamino)phenyl]-17-(1-propyn-1-yl)-estra-4,9-dien-3-one
P4	Progesterone
PGR	Nuclear progesterone receptors
PGRMC1	Progesterone receptor membrane component 1
PR	Progesterone receptors
SPRM	Selective progesterone receptor modulator
StAR	Steroidogenic acute regulatory protein
TG	Transgenic
TGF-β1	Transforming growth factor β1

References

1. Meyer, M.E.; Pornon, A.; Ji, J.W.; Bocquel, M.T.; Chambon, P.; Gronemeyer, H. Agonistic and antagonistic activities of RU486 on the functions of the human progesterone receptor. *EMBO J.* **1990**, *9*, 3923–3932. [CrossRef]
2. Perrault, D.; Eisenhauer, E.A.; Pritchard, K.I.; Panasci, L.; Norris, B.; Vandenberg, T.; Fisher, B. Phase II study of the progesterone antagonist mifepristone in patients with untreated metastatic breast carcinoma: A National Cancer Institute of Canada Clinical Trials Group study. *J. Clin. Oncol.* **1996**, *14*, 2709–2712. [CrossRef] [PubMed]
3. Rocereto, T.F.; Saul, H.M.; Aikins, J.A., Jr.; Paulson, J. Phase II study of mifepristone (RU486) in refractory ovarian cancer. *Gynecol. Oncol.* **2000**, *77*, 429–432. [CrossRef] [PubMed]
4. Ho, P.C.; Ng, E.H.Y.; Tang, O.S. Mifepristone: Contraceptive and noncontraceptive uses. *Curr. Opin. Obstet. Gynecol.* **2002**, *14*, 325. [CrossRef] [PubMed]
5. Grunberg, S.M.; Weiss, M.H.; Russell, C.A.; Spitz, I.M.; Ahmadi, J.; Sadun, A.; Sitruk-Ware, R. Long-term administration of mifepristone (RU486): Clinical tolerance during extended treatment of meningioma. *Cancer Investig.* **2006**, *24*, 727–733. [CrossRef] [PubMed]
6. Ramondetta, L.M.; Johnson, A.J.; Sun, C.C.; Atkinson, N.; Smith, J.A.; Jung, M.S.; Broaddus, R.; Iyer, R.B.; Burke, T. Phase 2 trial of mifepristone (RU-486) in advanced or recurrent endometrioid adenocarcinoma or low-grade endometrial stromal sarcoma. *Cancer* **2009**, *115*, 1867–1874. [CrossRef] [PubMed]
7. Rocereto, T.F.; Brady, W.E.; Shahin, M.S.; Hoffman, J.S.; Small, L.; Rotmensch, J.; Mannel, R.S. A phase II evaluation of mifepristone in the treatment of recurrent or persistent epithelial ovarian, fallopian or primary peritoneal cancer: A gynecologic oncology group study. *Gynecol. Oncol.* **2010**, *116*, 332–334. [CrossRef]
8. Tieszen, C.R.; Goyeneche, A.A.; Brandhagen, B.N.; Ortbahn, C.T.; Telleria, C.M. Antiprogestin mifepristone inhibits the growth of cancer cells of reproductive and non-reproductive origin regardless of progesterone receptor expression. *BMC Cancer* **2011**, *11*, 207. [CrossRef] [PubMed]
9. Peluso, J.J. Multiplicity of progesterone's actions and receptors in the mammalian ovary. *Biol. Reprod.* **2006**, *75*, 2–8. [CrossRef] [PubMed]
10. Goyeneche, A.A.; Caron, R.W.; Telleria, C.M. Mifepristone inhibits ovarian cancer cell growth in vitro and in vivo. *Clin. Cancer Res.* **2007**, *13*, 3370–3379. [CrossRef] [PubMed]
11. Freeburg, E.M.; Goyeneche, A.A.; Seidel, E.E.; Telleria, C.M. Resistance to cisplatin does not affect sensitivity of human ovarian cancer cell lines to mifepristone cytotoxicity. *Cancer Cell Int.* **2009**, *9*, 4. [CrossRef]
12. Gamarra-Luques, C.D.; Goyeneche, A.A.; Hapon, M.B.; Telleria, C.M. Mifepristone prevents repopulation of ovarian cancer cells escaping cisplatin-paclitaxel therapy. *BMC Cancer* **2012**, *12*, 200. [CrossRef] [PubMed]
13. Ponikwicka-Tyszko, D.; Chrusciel, M.; Stelmaszewska, J.; Bernaczyk, P.; Chrusciel, P.; Sztachelska, M.; Scheinin, M.; Bidzinski, M.; Szamatowicz, J.; Huhtaniemi, I.T.; et al. Molecular mechanisms underlying mifepristone's agonistic action on ovarian cancer progression. *EBioMedicine* **2019**, *47*, 170–183. [CrossRef]
14. Kinkade, S. Testicular cancer. *Am. Fam. Phys.* **1999**, *59*, 2539–2544, 2549–2550. [PubMed]
15. Rich, M.A.; Keating, M.A. Leydig cell tumors and tumors associated with congenital adrenal hyperplasia. *Urol. Clin. N. Am.* **2000**, *27*, 519–528. [CrossRef]
16. Thomas, J.C.; Ross, J.H.; Kay, R. Stromal testis tumors in children: A report from the prepubertal testis tumor registry. *J. Urol.* **2001**, *166*, 2338–2340. [CrossRef]
17. Hawkins, C.; Miaskowski, C. Testicular cancer: A review. *Oncol. Nurs. Forum* **1996**, *23*, 1203–1211; quiz 1212–1213.
18. Pozza, C.; Pofi, R.; Tenuta, M.; Tarsitano, M.G.; Sbardella, E.; Fattorini, G.; Cantisani, V.; Lenzi, A.; Isidori, A.M.; Gianfrilli, D.; et al. Clinical presentation, management and follow-up of 83 patients with Leydig cell tumors of the testis: A prospective case-cohort study. *Hum. Reprod.* **2019**, *34*, 1389–1403. [CrossRef]
19. McCluggage, W.G.; Shanks, J.H.; Arthur, K.; Banerjee, S.S. Cellular proliferation and nuclear ploidy assessments augment established prognostic factors in predicting malignancy in testicular Leydig cell tumours. *Histopathology* **1998**, *33*, 361–368. [CrossRef]
20. Mostofi, F.K. Proceedings: Testicular tumors. Epidemiologic, etiologic, and pathologic features. *Cancer* **1973**, *32*, 1186–1201. [CrossRef]
21. Al-Agha, O.M.; Axiotis, C.A. An in-depth look at Leydig cell tumor of the testis. *Arch. Pathol. Lab. Med.* **2007**, *131*, 311–317. [CrossRef]

22. Gadkar-Sable, S.; Shah, C.; Rosario, G.; Sachdeva, G.; Puri, C. Progesterone receptors: Various forms and functions in reproductive tissues. *Front. Biosci.* **2005**, *10*, 2118–2130. [CrossRef]
23. Pino, A.M.; Valladares, L.E. Evidence for a Leydig cell progesterone receptor in the rat. *J. Steroid Biochem.* **1988**, *29*, 709–714. [CrossRef]
24. El-Hefnawy, T.; Huhtaniemi, I. Progesterone can participate in down-regulation of the luteinizing hormone receptor gene expression and function in cultured murine Leydig cells. *Mol. Cell Endocrinol.* **1998**, *137*, 127–138. [CrossRef]
25. Due, W.; Dieckmann, K.P.; Loy, V.; Stein, H. Immunohistological determination of oestrogen receptor, progesterone receptor, and intermediate filaments in Leydig cell tumours, Leydig cell hyperplasia, and normal Leydig cells of the human testis. *J. Pathol.* **1989**, *157*, 225–234. [CrossRef]
26. Shah, C.; Modi, D.; Sachdeva, G.; Gadkar, S.; Puri, C. Coexistence of intracellular and membrane-bound progesterone receptors in human testis. *J. Clin. Endocrinol. Metab.* **2005**, *90*, 474–483. [CrossRef] [PubMed]
27. Gonzaález, C.R.; Inserra, P.I.F.; Terradas, C.; Ponzio, R.; Puigdomenech, E.; Levalle, O.; Vitullo, A.D.; Calandra, R.S.; Gonzalez-Calvar, S.I. Altered Expression of Aromatase, Estrogen Receptors and Progesterone Receptors in Human Leydig Cell Hyperplasia. *Adv. Biosci. Biotechnol.* **2015**, *6*, 1–10. [CrossRef]
28. Gonzalez, C.R.; Gonzalez, B.; Rulli, S.B.; Dos Santos, M.L.; Mattos Jardim Costa, G.; Franca, L.R.; Calandra, R.S.; Gonzalez-Calvar, S.I. TGF-beta1 system in Leydig cells. Part II: TGF-beta1 and progesterone, through Smad1/5, are involved in the hyperplasia/hypertrophy of Leydig cells. *J. Reprod. Dev.* **2010**, *56*, 400–404. [CrossRef]
29. Schwarzenbach, H.; Manna, P.R.; Stocco, D.M.; Chakrabarti, G.; Mukhopadhyay, A.K. Stimulatory effect of progesterone on the expression of steroidogenic acute regulatory protein in MA-10 Leydig cells. *Biol. Reprod.* **2003**, *68*, 1054–1063. [CrossRef]
30. El-Hefnawy, T.; Manna, P.R.; Luconi, M.; Baldi, E.; Slotte, J.P.; Huhtaniemi, I. Progesterone action in a murine Leydig tumor cell line (mLTC-1), possibly through a nonclassical receptor type. *Endocrinology* **2000**, *141*, 247–255. [CrossRef]
31. Lydon, J.P.; DeMayo, F.J.; Funk, C.R.; Mani, S.K.; Hughes, A.R.; Montgomery, C.A., Jr.; Shyamala, G.; Conneely, O.M.; O'Malley, B.W. Mice lacking progesterone receptor exhibit pleiotropic reproductive abnormalities. *Genes Dev.* **1995**, *9*, 2266–2278. [CrossRef]
32. Mulac-Jericevic, B.; Lydon, J.P.; DeMayo, F.J.; Conneely, O.M. Defective mammary gland morphogenesis in mice lacking the progesterone receptor B isoform. *Proc. Natl. Acad. Sci USA* **2003**, *100*, 9744–9749. [CrossRef]
33. Kananen Rilianawati, K.; Paukku, T.; Markkula, M.; Rainio, E.M.; Huhtaniemi, I. Suppression of gonadotropins inhibits gonadal tumorigenesis in mice transgenic for the mouse inhibin alpha-subunit promoter/simian virus 40 T-antigen fusion gene. *Endocrinology* **1997**, *138*, 3521–3531. [CrossRef]
34. Rahman, N.A.; Kananen Rilianawati, K.; Paukku, T.; Mikola, M.; Markkula, M.; Hamalainen, T.; Huhtaniemi, I.T. Transgenic mouse models for gonadal tumorigenesis. *Mol. Cell Endocrinol.* **1998**, *145*, 167–174. [CrossRef]
35. Kananen, K.; Markkula, M.; el-Hefnawy, T.; Zhang, F.P.; Paukku, T.; Su, J.G.; Hsueh, A.J.; Huhtaniemi, I. The mouse inhibin alpha-subunit promoter directs SV40 T-antigen to Leydig cells in transgenic mice. *Mol. Cell Endocrinol.* **1996**, *119*, 135–146. [CrossRef]
36. Rebois, R.V. Establishment of gonadotropin-responsive murine leydig tumor cell line. *J. Cell Biol.* **1982**, *94*, 70–76. [CrossRef]
37. Heikinheimo, O.; Kontula, K.; Croxatto, H.; Spitz, I.; Luukkainen, T.; Lahteenmaki, P. Plasma concentrations and receptor binding of RU 486 and its metabolites in humans. *J. Steroid Biochem.* **1987**, *26*, 279–284.
38. Sitruk-Ware, R.; Spitz, I.M. Pharmacological properties of mifepristone: Toxicology and safety in animal and human studies. *Contraception* **2003**, *68*, 409–420. [CrossRef]
39. Rice, L.; Waters, C.E.; Eccles, J.; Garside, H.; Sommer, P.; Kay, P.; Blackhall, F.H.; Zeef, L.; Telfer, B.; Stratford, I.; et al. Identification and functional analysis of SKA2 interaction with the glucocorticoid receptor. *J. Endocrinol.* **2008**, *198*, 499–509. [CrossRef]
40. Prefontaine, G.G.; Lemieux, M.E.; Giffin, W.; Schild-Poulter, C.; Pope, L.; LaCasse, E.; Walker, P.; Hache, R.J. Recruitment of octamer transcription factors to DNA by glucocorticoid receptor. *Mol. Cell Biol.* **1998**, *18*, 3416–3430. [CrossRef]

41. Vermeer, H.; Hendriks-Stegeman, B.I.; van der Burg, B.; van Buul-Offers, S.C.; Jansen, M. Glucocorticoid-induced increase in lymphocytic FKBP51 messenger ribonucleic acid expression: A potential marker for glucocorticoid sensitivity, potency, and bioavailability. *J. Clin. Endocrinol. Metab.* **2003**, *88*, 277–284. [CrossRef]
42. Chou, J.L.; Chen, L.Y.; Lai, H.C.; Chan, M.W. TGF-beta: Friend or foe? The role of TGF-beta/SMAD signaling in epigenetic silencing of ovarian cancer and its implication in epigenetic therapy. *Expert Opin. Ther. Targets* **2010**, *14*, 1213–1223. [CrossRef]
43. Fauvet, R.; Dufournet Etienne, C.; Poncelet, C.; Bringuier, A.F.; Feldmann, G.; Darai, E. Effects of progesterone and anti-progestin (mifepristone) treatment on proliferation and apoptosis of the human ovarian cancer cell line, OVCAR-3. *Oncol. Rep.* **2006**, *15*, 743–748. [CrossRef]
44. Stringer-Reasor, E.M.; Baker, G.M.; Skor, M.N.; Kocherginsky, M.; Lengyel, E.; Fleming, G.F.; Conzen, S.D. Glucocorticoid receptor activation inhibits chemotherapy-induced cell death in high-grade serous ovarian carcinoma. *Gynecol. Oncol.* **2015**, *138*, 656–662. [CrossRef]
45. Yu, S.; Yan, C.; Yang, X.; He, S.; Liu, J.; Qin, C.; Huang, C.; Lu, Y.; Tian, Z.; Jia, L. Pharmacoproteomic analysis reveals that metapristone (RU486 metabolite) intervenes E-cadherin and vimentin to realize cancer metastasis chemoprevention. *Sci. Rep.* **2016**, *6*, 22388. [CrossRef]
46. Wang, J.; Chen, J.; Wan, L.; Shao, J.; Lu, Y.; Zhu, Y.; Ou, M.; Yu, S.; Chen, H.; Jia, L. Synthesis, spectral characterization, and in vitro cellular activities of metapristone, a potential cancer metastatic chemopreventive agent derived from mifepristone (RU486). *AAPS J.* **2014**, *16*, 289–298. [CrossRef]
47. Heikinheimo, O.; Lahteenmaki, P.L.; Koivunen, E.; Shoupe, D.; Croxatto, H.; Luukkainen, T.; Lahteenmaki, P. Metabolism and serum binding of RU 486 in women after various single doses. *Hum. Reprod.* **1987**, *2*, 379–385. [CrossRef] [PubMed]
48. Heikinheimo, O.; Ranta, S.; Grunberg, S.; Spitz, I. Alterations in the pituitary-thyroid and pituitary-adrenal axes—Consequences of long-term mifepristone treatment. *Metabolism* **1997**, *46*, 292–296. [CrossRef]
49. Heikinheimo, O.; Haukkamaa, M.; Lahteenmaki, P. Distribution of RU 486 and its demethylated metabolites in humans. *J. Clin. Endocrinol. Metab.* **1989**, *68*, 270–275. [CrossRef] [PubMed]
50. Lewis-Tuffin, L.J.; Jewell, C.M.; Bienstock, R.J.; Collins, J.B.; Cidlowski, J.A. Human glucocorticoid receptor beta binds RU-486 and is transcriptionally active. *Mol. Cell Biol.* **2007**, *27*, 2266–2282. [CrossRef]
51. Kino, T.E.A. Glucocorticoid receptor (GR) beta has intrinsic, GRalpha-independent transcriptional activity. *Biochem. Biophys Res. Commun.* **2009**, *381*, 671–675. [CrossRef]
52. Hales, D.B.; Payne, A.H. Glucocorticoid-mediated repression of P450scc mRNA and de novo synthesis in cultured Leydig cells. *Endocrinology* **1989**, *124*, 2099–2104. [CrossRef]
53. Panza, S.; Malivindi, R.; Chemi, F.; Rago, V.; Giordano, C.; Barone, I.; Bonofiglio, D.; Gelsomino, L.; Giordano, F.; Ando, S.; et al. Glucocorticoid Receptor as a Potential Target to Decrease Aromatase Expression and Inhibit Leydig Tumor Growth. *Am. J. Pathol.* **2016**, *186*, 1328–1339. [CrossRef]
54. Peluso, J.J.; Liu, X.; Saunders, M.M.; Claffey, K.P.; Phoenix, K. Regulation of ovarian cancer cell viability and sensitivity to cisplatin by progesterone receptor membrane component-1. *J. Clin. Endocrinol. Metab.* **2008**, *93*, 1592–1599. [CrossRef]
55. Ahmed, I.S.; Rohe, H.J.; Twist, K.E.; Mattingly, M.N.; Craven, R.J. Progesterone Receptor Membrane Component 1 (Pgrmc1): A Heme-1 Domain Protein That Promotes Tumorigenesis and Is Inhibited by a Small Molecule. *J. Pharmacol. Exp. Ther.* **2010**, *333*, 564–573. [CrossRef]
56. Bierie, B.; Moses, H.L. TGF-beta and cancer. *Cytokine Growth Factor Rev.* **2006**, *17*, 29–40. [CrossRef] [PubMed]
57. Gonzalez, C.R.; Gonzalez, B.; Rulli, S.B.; Huhtaniemi, I.; Calandra, R.S.; Gonzalez-Calvar, S.I. TGF-beta1 system in Leydig cells. Part I: Effect of hCG and progesterone. *J. Reprod. Dev.* **2010**, *56*, 389–395. [CrossRef]
58. Lebrin, F.; Deckers, M.; Bertolino, P.; Ten Dijke, P. TGF-beta receptor function in the endothelium. *Cardiovasc. Res.* **2005**, *65*, 599–608. [CrossRef]
59. Goumans, M.J.; Valdimarsdottir, G.; Itoh, S.; Rosendahl, A.; Sideras, P.; ten Dijke, P. Balancing the activation state of the endothelium via two distinct TGF-beta type I receptors. *EMBO J.* **2002**, *21*, 1743–1753. [CrossRef]
60. Willibald, M.; Bayer, G.; Stahlhut, V.; Poschmann, G.; Stuhler, K.; Gierke, B.; Pawlak, M.; Seeger, H.; Mueck, A.O.; Niederacher, D.; et al. Progesterone receptor membrane component 1 is phosphorylated upon progestin treatment in breast cancer cells. *Oncotarget* **2017**, *8*, 72480–72493. [CrossRef] [PubMed]

61. Roberto, D.; Selvarajah, S.; Park, P.C.; Berman, D.; Venkateswaran, V. Functional validation of metabolic genes that distinguish Gleason 3 from Gleason 4 prostate cancer foci. *Prostate* **2019**, *79*, 1777–1788. [CrossRef] [PubMed]
62. Peluso, J.J.; Romak, J.; Liu, X. Progesterone receptor membrane component-1 (PGRMC1) is the mediator of progesterone's antiapoptotic action in spontaneously immortalized granulosa cells as revealed by PGRMC1 small interfering ribonucleic acid treatment and functional analysis of PGRMC1 mutations. *Endocrinology* **2008**, *149*, 534–543. [CrossRef]
63. Ruan, X.; Zhang, Y.; Mueck, A.O.; Willibald, M.; Seeger, H.; Fehm, T.; Brucker, S.; Neubauer, H. Increased expression of progesterone receptor membrane component 1 is associated with aggressive phenotype and poor prognosis in ER-positive and negative breast cancer. *Menopause* **2017**, *24*, 203–209. [CrossRef]
64. Study of Oral Mifepristone as Salvage Therapy in Patients With Advanced or Metastatic Non-Small Cell Lung Cancer. Available online: https://ClinicalTrials.gov/show/NCT02642939 (accessed on 27 July 2020).
65. Enzalutamide and Mifepristone in Treating Patients with Metastatic Hormone Resistant Prostate Cancer. Available online: https://ClinicalTrials.gov/show/NCT02012296 (accessed on 27 July 2020).
66. BRCA1/2 and Effect of Mifepristone on the Breast. Available online: https://ClinicalTrials.gov/show/NCT01898312 (accessed on 27 July 2020).
67. Haavisto, A.M.; Pettersson, K.; Bergendahl, M.; Perheentupa, A.; Roser, J.F.; Huhtaniemi, I. A supersensitive immunofluorometric assay for rat luteinizing hormone. *Endocrinology* **1993**, *132*, 1687–1691. [CrossRef]
68. Van Casteren, J.I.; Schoonen, W.G.; Kloosterboer, H.J. Development of time-resolved immunofluorometric assays for rat follicle-stimulating hormone and luteinizing hormone and application on sera of cycling rats. *Biol. Reprod.* **2000**, *62*, 886–894. [CrossRef]

cancers

MDPI

Article

SNAIL Promotes Metastatic Behavior of Rhabdomyosarcoma by Increasing EZRIN and AKT Expression and Regulating MicroRNA Networks

Klaudia Skrzypek [1,2,*], Marta Kot [1,2], Paweł Konieczny [1,2], Artur Nieszporek [1,2], Anna Kusienicka [1,2], Małgorzata Lasota [1,2], Wojciech Bobela [1,2], Urszula Jankowska [3], Sylwia Kędracka-Krok [4] and Marcin Majka [1,2,*]

1 Jagiellonian University Medical College, 31-008 Krakow, Poland; marta.kot@uj.edu.pl (M.K.); pawel.konieczny@outlook.com (P.K.); artur.nieszporek@gmail.com (A.N.); anna.kusienicka@gmail.com (A.K.); malgorzata.lasota@uj.edu.pl (M.L.); wojciech.bobela@gmail.com (W.B.)
2 Department of Transplantation, Institute of Pediatrics, Faculty of Medicine, Jagiellonian University Medical College, 30-663 Krakow, Poland
3 Laboratory of Proteomics and Mass Spectrometry, Malopolska Centre of Biotechnology, Jagiellonian University, 30-387 Krakow, Poland; urszula.jankowska@uj.edu.pl
4 Department of Physical Biochemistry, Faculty of Biochemistry, Biophysics and Biotechnology, Jagiellonian University, 30-387 Krakow, Poland; sylwia.kedracka-krok@uj.edu.pl
* Correspondence: klaudia.skrzypek@uj.edu.pl (K.S.); mmajka@cm-uj.krakow.pl (M.M.); Tel.: +48-12-659-15-93 (K.S. & M.M.)

Received: 19 May 2020; Accepted: 6 July 2020; Published: 11 July 2020

check for updates

Abstract: Rhabdomyosarcoma (RMS) is a predominant soft tissue tumor in children and adolescents. For high-grade RMS with metastatic involvement, the 3-year overall survival rate is only 25 to 30%. Thus, understanding the regulatory mechanisms involved in promoting the metastasis of RMS is important. Here, we demonstrate for the first time that the SNAIL transcription factor regulates the metastatic behavior of RMS both in vitro and in vivo. SNAIL upregulates the protein expression of EZRIN and AKT, known to promote metastatic behavior, by direct interaction with their promoters. Our data suggest that SNAIL promotes RMS cell motility, invasion and chemotaxis towards the prometastatic factors: HGF and SDF-1 by regulating RHO, AKT and GSK3β activity. In addition, miRNA transcriptome analysis revealed that SNAIL-miRNA axis regulates processes associated with actin cytoskeleton reorganization. Our data show a novel role of SNAIL in regulating RMS cell metastasis that may also be important in other mesenchymal tumor types and clearly suggests SNAIL as a promising new target for future RMS therapies.

Keywords: rhabdomyosarcoma (RMS); tumor metastasis; SNAIL transcription factor; microRNA (miRNA); EZRIN; AKT kinase

1. Introduction

Rhabdomyosarcoma (RMS) is a mesenchymal soft tissue tumor that causes death and morbidity predominantly in children and adolescents. Despite general improvement in the 5-year overall survival of pediatric RMS patients, for high grade tumors with metastatic involvement, the overall survival rate at 3 years is only 25–30% [1]. 6% of all cases with RMS display bone marrow (BM) metastasis, what significantly diminishes survival of the patients and makes them more prone to disease relapse and progression [2]. Therefore, understanding of the metastasis mechanisms is highly necessary and may help to find new treatment strategies in future. Important factors regulating the metastatic behavior of different tumor types, including RMS, are hepatocyte growth factor (HGF) and stromal-derived factor-1 (SDF-1) [3–5].

Histological analysis of tumors distinguishes two main subtypes of RMS, embryonal (ERMS) and alveolar (ARMS), with ARMS generally having a significantly worse prognosis [6]. The presence of PAX3-FOXO1 and PAX7/FOXO1 fusion genes and increased levels of MET receptors may cause the increased aggressiveness of ARMS tumors [7]. Mutations affecting the RAS-MAPK and PI3K-AKT pathways also affect RMS development and progression [1]. Among the critical regulators of RMS metastasis is the EZRIN protein, which acts as the actin filament-plasma membrane linker [8]. Crosstalk between different signaling pathways may create an integrated signaling network supporting the metastatic behavior of RMS.

Our previous studies identified the SNAIL family zinc finger 1 (SNAIL or SNAI1) transcription factor as a novel key regulator of RMS growth and differentiation [9,10]. We discovered noncanonical action mechanisms of SNAIL in RMS [9]. Importantly, the SNAIL level has been shown to be elevated in the ARMS tumor subtype, which is usually associated with a worse prognosis [11]. Moreover, the SNAIL level was significantly increased in RMS samples from patients displaying stages 2, 3, and 4 of the disease compared to those from patients with stage 1, suggesting an important role of SNAIL in RMS progression [9]. Furthermore, another member of the SNAIL family of zinc finger transcription factors, SNAI2, was identified by an integrative computational pipeline analysis as potential crucial factor in RMS growth [12].

SNAIL is best known as a regulator of epithelial to mesenchymal transition (EMT) through canonical regulation of the E-cadherin (CDH1) level [13–15]. It belongs to the SNAIL family of zinc finger transcription factors, which consists of 3 members: SNAIL (SNAI1), SLUG (SNAI2) and SMUG (SNAI3). SNAIL may act as a transcriptional repressor or as a gene activator by binding to target E-box sequences (CANNTG) [16] and recruiting histone deacetylases (HDACs) [13]. In addition to regulating protein expression levels, SNAIL is also a crucial regulator of microRNAs expression [17]. SNAIL silencing has been described to effectively suppress the growth and invasiveness of different tumor types [18]. SNAIL degradation is prevented in cancer cells due to stabilization of its level by the USP27X deubiquitinase [19]. SNAIL can also modulate myogenic differentiation by binding to the MYF5 promoter [9].

RMS originates from impaired differentiation of myogenic progenitors or mesenchymal stem cells (MSCs) [6]. Interestingly, SNAIL-deficient MSCs prematurely differentiate into osteoblasts or adipocytes [20], whereas in MSCs with a constitutively activated MET signaling pathway, SNAIL seems to act as a mediator of myogenic differentiation [21]. Moreover, SNAIL expression was demonstrated to be required for sarcomagenesis, as SNAIL controls the tumorigenic potential of MSCs [22]. In ARMS cells, SNAIL silencing completely abolished the growth of human tumor xenotransplants by upregulating myogenic differentiation [9], whereas in ERMS cells, SNAIL was identified as a mediator of the NOTCH pathway [23]. Similarly to SNAIL, myogenic transcription factors that may be important in RMS differentiation, such as MYOD and MYOG, bind to E-box sequences [24]. Furthermore, SNAIL can displace MYOD from E-box sequences that are associated with genes expressed during differentiation and it that way it may regulate RMS growth [9].

The metastatic process is very important for the dissemination of RMS cells and, consequently, for patients' long-term prognoses. In epithelial tumors, SNAIL affects metastasis by regulating EMT [14]. The canonical action mechanism of SNAIL that promotes the metastatic process is the repression of E-cadherin expression by SNAIL binding to the E-cadherin promoter [13]. In addition, during the EMT process, cells acquire migratory properties by reorganization of the actin cytoskeleton and activation of the RhoA GTPase [25]. However, in mesenchymal tumors, different mechanisms might operate, and the mechanistic role of SNAIL in the metastasis of mesenchymal tumors is poorly understood. However, as in epithelial tumors, regulation of the actin cytoskeleton may also be a key process in the metastasis of mesenchymal tumor types.

2. Results

2.1. SNAIL Regulates the Metastatic Behavior of RMS Cells In Vivo and In Vitro

In our previous studies, we stably silenced SNAIL with a mix of three different shRNA variants to study the effects of SNAIL on RMS cell growth and differentiation (Supplementary Figure S1) [9]. In the present study, we used that model to evaluate the effects of SNAIL on the metastatic behavior of RMS cells. RH30 cells were intravenously implanted into immunodeficient NOD-SCID mice. High level of engraftment of both RH30 WT and shCTRL (scrambled shRNA vector) cells was detected in nine out of 30 bone marrow samples per each group, whereas SNAIL silencing in RH30 cells reduced this number to five (Figure 1A). Furthermore, engraftment of SNAIL-deficient (shSNAIL) cells into the lungs in vivo was significantly impaired compared to that of both wild-type (WT) cells and cells modified with scrambled shRNA (shCTRL) (Figure 1B).

The observed impairment in the metastatic potential in vivo could be explained by the diminished invasion and motility of the cells. Indeed, in the invasion assay, SNAIL-deficient cells showed impaired migration through Matrigel (Figure 1C) after stimulation with SDF-1 and HGF, which promote RMS metastasis [4]. In the scratch assay, shSNAIL cells closed the gap in a monolayer slower than control cells (Figure 1D and Supplementary Figure S2). SNAIL silencing also resulted in diminished chemotaxis (Figure 1E) towards both SDF-1 and HGF. Interestingly, as suggested by the literature [26], SDF-1 is expressed in the murine lung (Supplementary Figure S3) and migration towards its gradient may be responsible for the tumor cells engraftment into the lungs in vivo. Expression of the HGF receptor MET was slightly diminished in SNAIL-deficient cells, whereas we did not observe any significant effect of SNAIL on the SDF-1 receptor CXCR4 (Figure 1F). Interestingly, bioinformatic analysis of microarray data deposited in the GEO database revealed a positive correlation between MET and SNAIL expression in 158 RMS tumor samples derived from patients but not between SNAIL and CXCR4 (Figure 1G). RH30 cells were also screened for expression of adhesion receptors that might be responsible for the cells' engraftment in vivo: ICAMs and VLA integrins. VLA family members CD49a, b, c, d, e and f and CD29 were highly expressed in WT cells. Interestingly, SNAIL silencing completely diminished CD49b expression. Moreover, ICAM family receptors, i.e., CD54 (ICAM-1) and CD-102 (ICAM-2) were expressed in WT RH30 cells, whereas their expression in SNAIL-deficient cells was again diminished (Figure 1H). Thus, our data show an important role of SNAIL silencing in diminishing RMS cell metastatic properties by regulating their motility and adhesion properties.

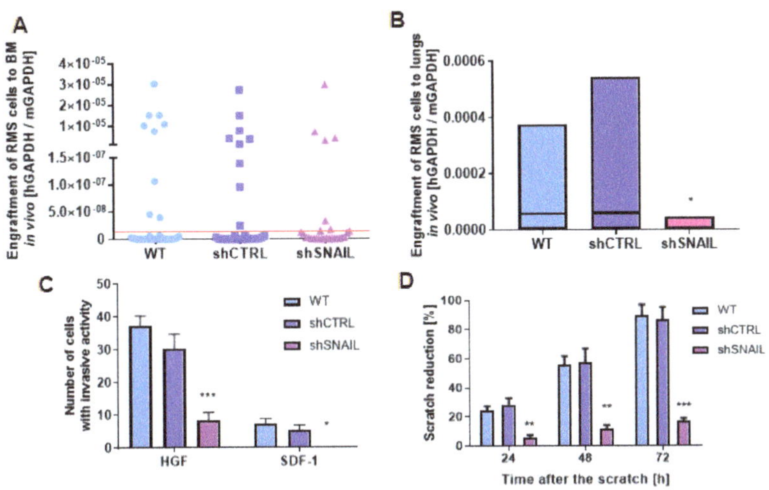

Figure 1. *Cont.*

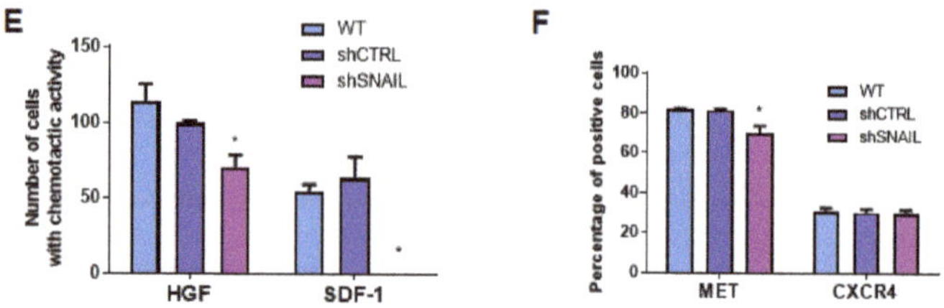

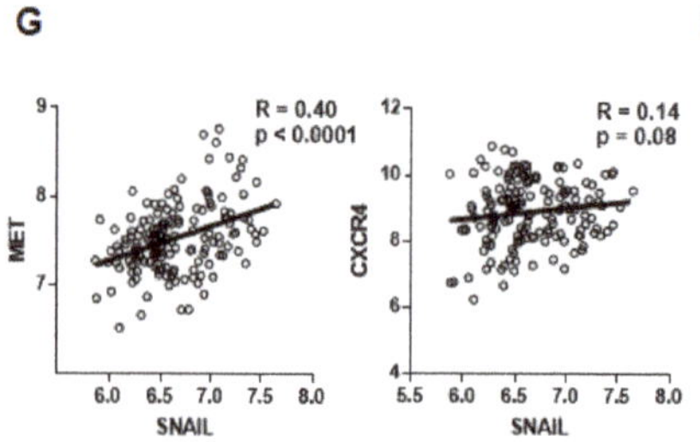

H

Adhesion molecules [% of positive cells]

Marker	WT	shCTRL	shSNAIL
CD29	97,58	99,17	99,24
CD49a	96,24	93,91	99,16
CD49b	85,29	85,75	0,36
CD49c	98,71	99,16	98,08
CD49d	92,22	86,65	99,31
CD49e	99,57	99,69	99,21
CD49f	99,12	97,78	99,31
CD54	96,04	98,53	0,72
CD102	23,72	44,05	1,72

Figure 1. SNAIL regulates the motility of RMS cells in vitro and in vivo. (**A**) Stable SNAIL silencing by transduction with shRNA lentiviral vectors (shSNAIL) tended to inhibit the high level of engraftment of RH30 cells into murine bone marrow 7 days after intravenous implantation compared to that in WT cells and cells transduced with scrambled shRNA vectors (shCTRL). The presence of human cells was detected by qPCR and calculated as the ratio of the human to murine GAPDH levels with the ΔCt method. Data are presented as single points; $n = 30$ bone marrow samples (left and right) from 15 mice; red line discriminates bone marrow samples with high level of engraftment of human RMS cells. (**B**) Stable SNAIL silencing by transduction with shRNA lentiviral vectors (shSNAIL) inhibited the engraftment of RH30 cells into murine lungs 7 days after intravenous implantation compared to that in WT cells and cells transduced with scrambled shRNA vectors (shCTRL). The presence of human cells was detected by qPCR and calculated as the ratio of the human to murine GAPDH levels with the ΔCt method. Data are presented as a floating bar graph (min to max) with line at the mean; $n = 14$–15. (**C**) SNAIL-deficient cells displayed diminished invasion through Matrigel towards HGF (20 ng/mL) and SDF-1 (100 ng/mL), $n = 3$. (**D**) SNAIL deficient RH30 cells closed the gap in a scratch assay slower than control cells, $n = 3$. (**E**) In a chemotaxis assay SNAIL-deficient cells displayed diminished migration towards HGF (20 ng/mL) and SDF-1 (100 ng/mL), $n = 3$ (**F**) SNAIL silencing in RH30 cells diminished MET receptor levels but barely affected CXCR4 levels, as estimated by flow cytometry as the percentage of positive cells, $n = 3$. (**G**) SNAIL levels positively correlated with MET levels and slightly correlated with CXCR4 levels in 158 RMS samples from patients (Pearson correlation). Analysis was performed on data deposited in the GEO database with accession number: GSE92689. (**H**) SNAIL silencing in RH30 cells regulated the expression of adhesion molecules. The results are shown as the percentage of cells labeled with a Lyoplate Screening Panel (flow cytometry). * $p < 0.05$, ** $p < 0.01$, *** $p < 0.001$. Graphical data are presented as the means with SEMs.

2.2. *EZRIN is a Crucial Mediator of the Metastatic Action of SNAIL in RMS Cells*

We evaluated the morphology of RMS cells and found strong reorganization of the actin cytoskeleton in shSNAIL RH30 cells (Figure 2A). Staining with phalloidin has not shown visible differences in fluorescence intensity between the examined lines. Significant differences were visible in the spatial organization of F-actin fibers, what can affect the cell adhesion and migration capacity [27]. In shSNAIL cells, the fibers are arranged parallel to each other, while in the control lines their arrangement is less ordered. Longitudinally oriented stress fibers in shSNAIL cells affect elongated morphology of these cells. Moreover, after SNAIL silencing the formation of membrane protrusions seems to be reduced. It has been proved that membrane protrusions (e.g. filopodia) are a key structure directly involved in cancer cell motility [28,29]. These findings can explain lower migration capacity of shSNAIL cells compared to control lines (WT and shCTRL).

Unbiased shotgun proteomic analysis of the cytoplasmic and nuclear fractions in RH30 cells revealed that SNAIL regulates the expression of several proteins associated with actin cytoskeleton organization including EZRIN (Figure 2B), which has been previously shown to be a key regulator of metastasis in RMS [8]. Diminished expression of EZRIN in shSNAIL RH30 cells was confirmed by Western blotting (Figure 2C). Moreover, when SNAIL was transiently silenced with siRNA in RH41 ARMS cells and RD ERMS cells, we observed downregulation of EZRIN expression (Figure 2D).

The eukaryotic promoter database describes two fragments of the *EZRIN* promoter: *EZR_1* chromosome [NC_000006.12]; strand [-]; position [158819364], and *EZR_2* chromosome [NC_000006.12]; strand [-]; position [158818235] [30]. Bioinformatic analysis demonstrated that the two fragments of the *EZRIN* promoter contain several putative SNAIL binding sites. The chromatin immunoprecipitation (ChIP) assay results revealed that SNAIL binds to one of the *EZRIN* promoter fragments (Figure 2E) and thus may directly regulate its expression. Interestingly, bioinformatic analysis of microarray data for 158 RMS patient samples deposited in the GEO database revealed a positive correlation between SNAIL and EZRIN expression levels (Figure 2F).

A

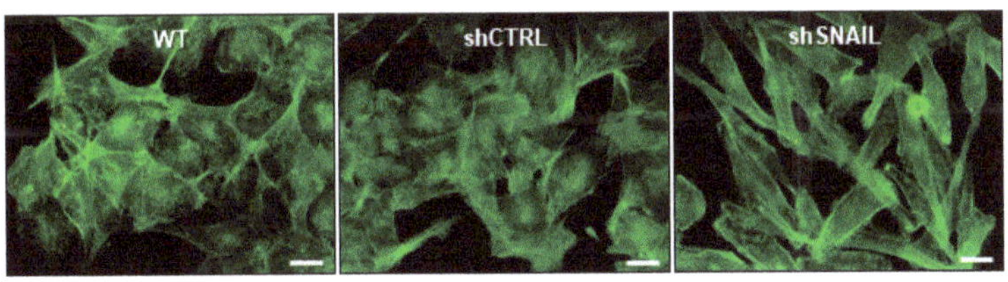

B

Proteins in RH30 shSNAIL cells

Accession	Protein	MW [kDa]	pI	Scores	#Peptides	SC [%]
ACTG_HUMAN	Actin, cytoplasmic 2	41.8	5.2	2281.4	30	63.2
RADI_HUMAN	Radixin	68.5	6.0	255.5	5	10.6
ACTN1_HUMAN	Alpha-actinin-1	103.0	5.1	183.6	3	3.7
ARP3_HUMAN	Actin-related protein 3	47.3	5.5	67.0	2	9.8
DEST_HUMAN	Destrin	18.5	9.2	66.2	2	18.2

Proteins in RH30 control cells

Accession	Protein	MW [kDa]	pI	Scores	#Peptides	SC [%]
ACTB_HUMAN	Actin, cytoplasmic 1	41.7	5.2	1939.2	27	64.0
ACTN4_HUMAN	Alpha-actinin-4	104.8	5.2	261.1	6	9.0
EZRI_HUMAN	Ezrin	69.4	5.9	108.7	3	6.7
GELS_HUMAN	Gelsolin	85.6	5.9	117.5	3	6.1

Number of proteins

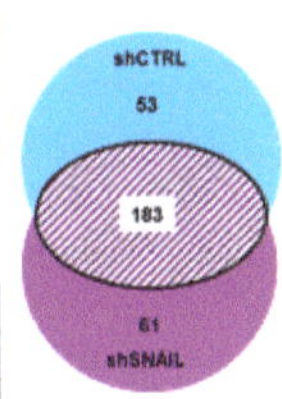

C **D**

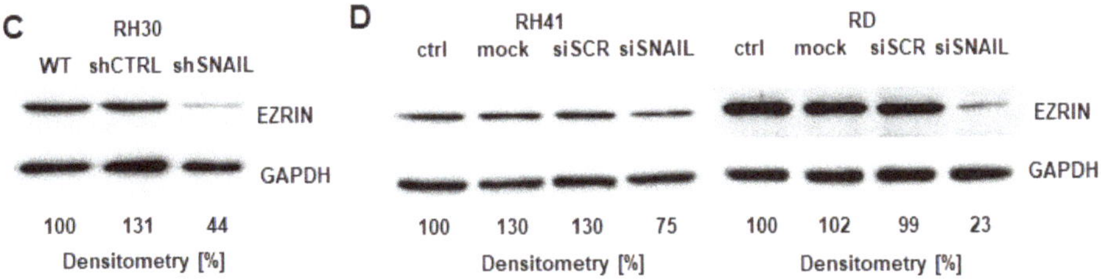

Figure 2. *Cont.*

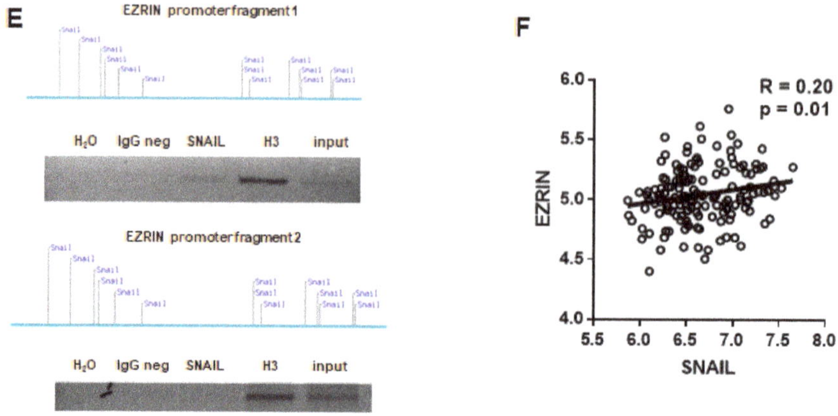

Figure 2. SNAIL regulates EZRIN expression in RMS cells. (**A**) SNAIL silencing in RH30 cells reorganized actin cytoskeleton (staining of RH30 cells with phalloidin conjugated with Alexa Fluor 488). White scale bar represents 50 μm. (**B**) SNAIL regulated proteins associated with the actin cytoskeleton in RH30 cells. Analysis of proteome by shotgun LC-MS/MS and MASCOT software revealed the expression of different proteins associated with the actin cytoskeleton in the cytoplasm of RH30 shSNAIL and shCTRL cells. Proteins for further consideration were identified based on at least 2 peptides. A total of 61 proteins were identified only in shSNAIL cells; 53, only in shCTRL cells; and 183, proteins in both cell types. Here only proteins related to actin cytoskeleton organization are shown. MW – molecular weight; pI –isoelectric point; scores – parameter describing the probability of the correct identification of protein; peptides – number of peptides in the analysis; SC [%]–obtained sequence coverage. (**C**) Stable SNAIL silencing in RH30 cells downregulated EZRIN expression at the protein level (Western blotting). The Western blot results show the representative image of three independent biological experiments. Densitometric analysis evaluated the ratio of EZRIN/GAPDH and was presented as percentage of control. (**D**) Temporal SNAIL silencing in RH41 and RD cells downregulated EZRIN expression at the protein level (Western blotting). The Western blot results show the representative image of three independent biological experiments. Densitometric analysis evaluated the ratio of EZRIN/GAPDH and was presented as percentage of control. (**E**) SNAIL bound to the *EZRIN* promoter in RH30 cells. Two fragments of *EZRIN* promoters (~1000 bp) were screened for putative SNAIL transcription factor binding sites and the results were validated by a ChIP assay. The images show one representative result of the ChIP assay. Proteins bound to DNA were immunoprecipitated with an anti-SNAIL antibody, negative IgG control, and positive histone H3 control. An input DNA served as control for analysis. Two fragments of the *EZRIN* promoter were amplified by PCR and visualized on agarose gels stained with ethidium bromide. (**F**) *SNAIL* levels were positively correlated with *EZRIN* levels in 158 RMS samples from patients (Pearson correlation). The analysis was done on data deposited in GEO database with accession number: GSE92689.

Subsequently, EZRIN and SNAIL were transiently silenced in three different RMS cell lines (ARMS: RH30 and RH41; EMRS: RD) (Supplementary Figure S4A,B) to verify their effects on the motility of these cells. Silencing of either SNAIL or EZRIN diminished the motility of RH30, RH41 and RD cells in a scratch assay (Figure 3A). SNAIL silencing also inhibited chemotaxis towards both HGF and SDF-1, whereas EZRIN silencing inhibited chemotaxis towards HGF in one cell line (RD) and chemotaxis towards SDF-1 in all cell lines (RH30, RH41, RD). However, the effect was less potent than in SNAIL-deficient cells (Figure 3B). Furthermore, to investigate not only effects of SNAIL silencing, but also its overexpression, SMS-CTR ERMS cells, which displayed low basal SNAIL level, were transduced with lentiviral vectors encoding SNAIL, whereas control cells were modified with vector encoding GFP (Figure 3C). SNAIL overexpression in SMS-CTR cell upregulated EZRIN levels

(Figure 4B) and, accordingly, increased motility of the cells in a scratch assay (Figure 3E) and chemotaxis towards SDF-1 and HGF (Figure 3F).

2.3. Mutual Regulation of SNAIL and AKT Kinase Expression in RMS Cells

We performed ChIP-seq analysis of SNAIL binding sites in RMS cells, and the results revealed many significant SNAIL targets in RH30 cells (Supplementary Table S1). SNAIL transcription factor bound to the AKT2 kinase promoter and to its own promoter (Figure 4A and Supplementary Figure S5). These data suggested that SNAIL directly regulates the expression of AKT2 kinase in RMS cells. Additionally, Western blot analysis showed that SNAIL silencing diminished the expression of not only AKT2, but also AKT1 and total AKT (Figure 4B).

Subsequently, we evaluated the mechanism by which SNAIL can be regulated in RMS cells by HGF and SDF-1, the factors determining the metastatic capabilities of these cells. HGF and SDF-1 control GSK3β phosphorylation via the PI3K-AKT signaling pathway and not the MAPK pathway, as demonstrated by simultaneous treatment with HGF or SDF-1 and MET inhibitor, PI3K-AKT signaling inhibitor (LY294002), MAPK inhibitor (UO126) or pertussis toxin (PTX) (Figure 4C,D). LY294002 administration resulted in complete attenuation of AKT phosphorylation, accompanied by a decrease in GSK3β phosphorylation, whereas this effect was not observed after treatment with UO126 (Figure 4C,D). Moreover, PTX, an inhibitor of G-protein-coupled receptors (such as CXCR4) that impedes normal Gα subunit action, almost entirely abolished the stimulatory effect of SDF-1 (Figure 4D).

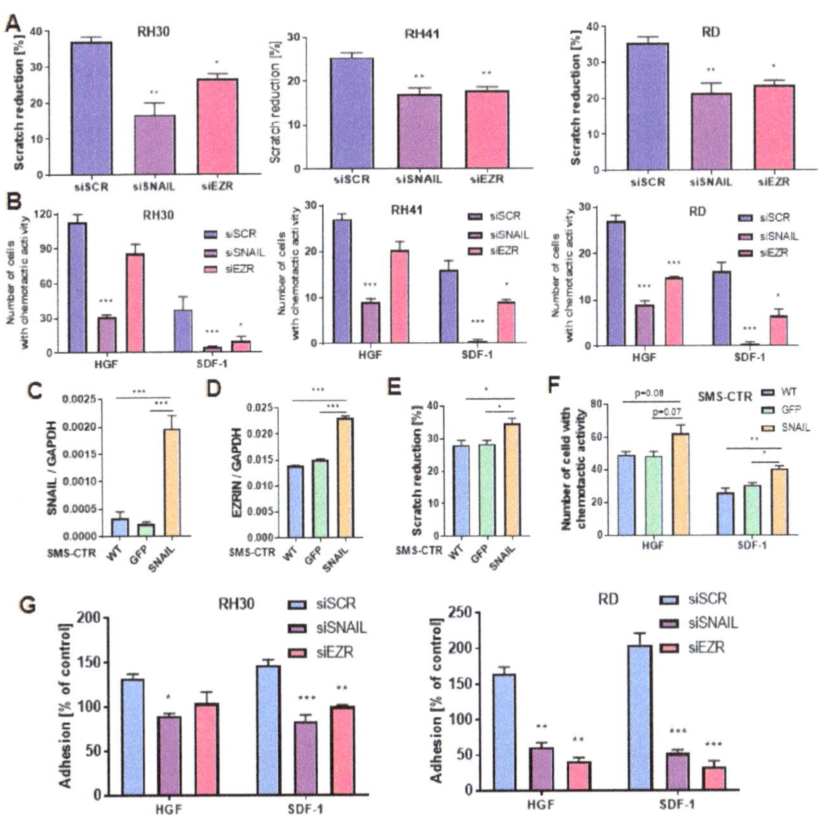

Figure 3. *Cont.*

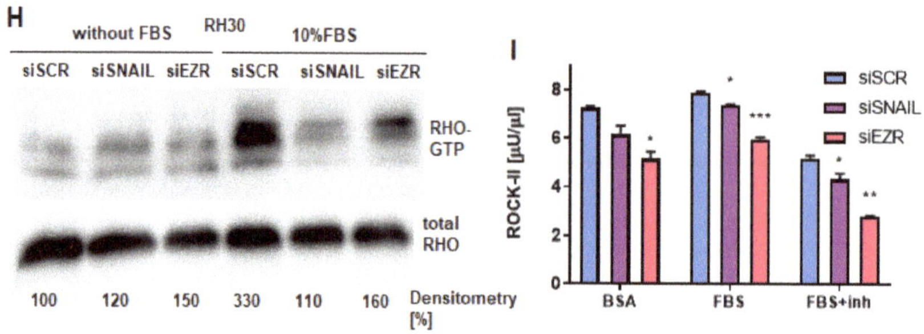

Figure 3. SNAIL and EZRIN are crucial regulators of migration, chemotaxis and adhesion of RMS cells to endothelial cells by affecting RHO activity. (**A**) SNAIL (siSNAIL) and EZRIN (siEZR) silencing with siRNA diminished the motility of RH30, RH41 and RD cells in a scratch assay. Cells were transfected three times every 3–4 days, and the assay was then performed; $n = 3$ (**B**) SNAIL silencing with siRNA inhibited chemotaxis of RH30, RH41 and RD cells towards HGF and SDF-1, whereas EZRIN silencing inhibited chemotaxis towards SDF-1 and HGF; $n = 3$. (**C**) SNAIL was overexpressed in SMS-CTR cells by transduction with lentiviral vectors, whereas control cells were modified with vectors encoding GFP; qPCR results; $n = 2$; WT- wild-type cells. (**D**) SNAIL overexpression in SMS-CTR cells upregulated EZRIN levels; qPCR results; $n = 2$. (**E**) SNAIL overexpression in SMS-CTR cells increased motility of the cells in a scratch assay; $n = 3$. (**F**) SNAIL overexpression in SMS-CTR cells increased chemotaxis towards SDF-1 and HGF; $n = 3$. (**G**) Both SNAIL and EZRIN silencing diminished the adhesion of RH30 and RD cells treated with HGF and SDF-1 to HUVECs pretreated with TNF-α. The graphs show representative results from three independent experiments, and the results were calculated as a percentage of the control. (**H**) Both SNAIL and EZRIN silencing by siRNA in RH30 cells inhibited RHO activation. RHO-GTP was detected by Western blotting. Representative Western blot images from two independent experiments after using the RHO Activation Assay Biochem Kit are shown. Densitometric analysis evaluated the ratio of RHO-GTP/total RHO and was presented as percentage of control. (**I**) SNAIL and EZRIN silencing with siRNA slightly diminished ROCK-II enzyme activity, as evaluated in protein lysates with a ROCK Activity Assay. Two independent experiments were performed with duplicate samples. * $p < 0.05$, ** $p < 0.01$, *** $p < 0.001$. Graphical data are presented as the means ±SEMs.

A

Gene name	Gene.ID	Chr	Start	End	Strand	Peak.Score	Annotation	Distance.to.TSS
AKT2	ENSG00000105221	19	40284853	40285380	+	124	promoter-TSS (ENST00000583859)	-24
SNAI1	ENSG00000124216	20	49982512	49982875	+	188	promoter-TSS (ENST00000244050)	-306

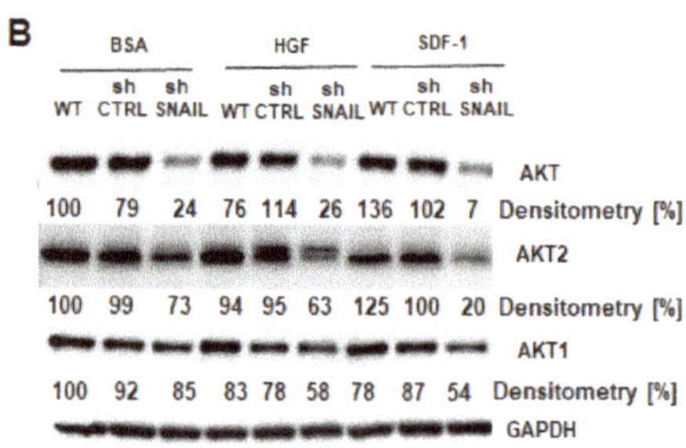

Figure 4. *Cont.*

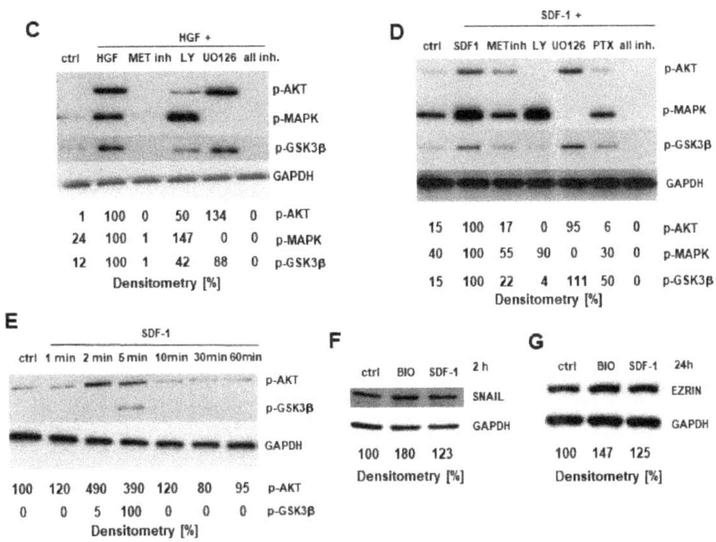

Figure 4. Mutual regulation of SNAIL and AKT expression in RMS cells. (**A**) SNAIL bound to the *AKT2* promoter and its own promoter, as shown by ChIP-seq analysis results in RH30 cells. (**B**) Stable SNAIL silencing in RH30 cells diminished total AKT and AKT2 and slightly decreased AKT1 levels under different treatment conditions (0.5% BSA, 20 ng/mL HGF, 100 ng/mL SDF1 and 10% FBS). Typical Western blot images of two independent biological experiments are shown. Densitometric analysis evaluated the ratio of protein of interest to GAPDH, presented as percentage of control. (**C**) HGF induced GSK3β phosphorylation via AKT kinase signaling pathway and not the MAPK kinase signaling pathway. RH30 cells were treated (20 ng/mL HGF, 5 min) together with the following inhibitors: 5 μM MET inhibitor, 30 μM LY294002 (PI3K-AKT inhibitor) or 30 μM UO126 (MAPK inhibitor). Negative control was starvation medium with 0.5% BSA without stimulants. Phosphorylation of AKT, MAPK and GSK3β was evaluated with GAPDH as loading control. Representative Western blot images of two independent biological experiments are shown. Densitometric analysis evaluated the ratio of protein of interest to GAPDH, presented as percentage of HGF-treated cells. (**D**) SDF-1 induced GSK3β phosphorylation via the AKT kinase signaling pathway and not the MAPK kinase signaling pathway. RH30 cells were treated (100 ng/mL SDF-1, 5 min) together with the following inhibitors: 5 μM MET inhibitor, 30 μM LY294002 (PI3K-AKT inhibitor), 30 μM UO126 (MAPK inhibitor) or 20 μg/mL PTX (signaling inhibitor via the Gα subunit). Negative control was starvation medium with 0.5% BSA and no stimulants. Phosphorylation of AKT, MAPK and GSK3β was evaluated with GAPDH as the loading control. Representative Western blot images of two independent biological experiments are shown. Densitometric analysis evaluated the ratio of protein of interest to GAPDH and was presented as percentage of SDF-1-treated cells. (**E**) SDF-1 induces phosphorylation of AKT and GSK3β in a sequential manner in RH30 cells. RH30 cells were treated with 100 ng/mL SDF-1. Negative control was starvation medium containing 0.5% BSA and no stimulants. Phosphorylation of AKT and GSK3β was evaluated with GAPDH as loading control. Western blot images of two independent biological experiments are shown. Densitometric analysis evaluated the ratio of protein of interest to GAPDH and is presented as % control (p-AKT) or % SDF-1-treated cells (p-GSK3β). (**F**) SDF-1 and BIO (inhibitor of GSK3β) induced expression of SNAIL protein 2 h after treatment. RH30 cells were treated with 100 ng/mL SDF-1 +1 μM BIO for two hours. SNAIL levels were evaluated with GAPDH as loading control. Western blot images of two independent biological experiments are shown. Densitometric analysis evaluated the ratio of SNAIL/GAPDH and was presented as percentage of control. (**G**) SDF-1 and BIO (inhibitor of GSK3β) induced EZRIN protein expression 24 h after treatment. RH30 cells were treated with 100 ng/mL SDF-1 and 1 μM BIO for 24 h. The EZRIN level was evaluated with GAPDH as a loading control. Western blot images of two independent biological experiments are shown. Densitometric analysis evaluated the ratio of genes of interests to GAPDH and was presented as percentage of control.

SDF-1 was demonstrated to induce the phosphorylation of AKT and GSK3β in sequential manner: first, AKT was phosphorylated, followed by GSK3β (Figure 4E). Interestingly, prolonged treatment with SDF-1 or the GSK3β inhibitor BIO resulted in slightly elevated SNAIL levels (Figure 4F) and, subsequently, slightly increased EZRIN level (Figure 4G). Our results demonstrate that SNAIL expression is regulated via the induction of GSK3β phosphorylation by the PI3K-AKT kinase signaling pathway and, moreover, identified SNAIL as an important regulator of AKT and EZRIN expression.

2.4. Inhibition of PI3K-AKT Signaling Diminishes Migration and Chemotaxis of RMS Cells

To indirectly verify whether AKT kinase may be an important mediator of SNAIL-mediated RMS cell motility, WT RH30 cells were treated with LY294002, PI3K-AKT signaling inhibitor. We observed that a 10 μM concentration of the inhibitor partially inhibited AKT kinase phosphorylation, whereas 50 μM almost completely blocked its activation (Figure 5A). Moreover, both 10 and 50 μM LY294002 inhibited migration in a scratch assay (Figure 5B) and chemotaxis towards HGF and SDF-1 (Figure 5C). We discovered that the impaired AKT signaling mimicked the effects of SNAIL downregulation, what indirectly suggests that AKT kinase might be an important mediator of SNAIL action.

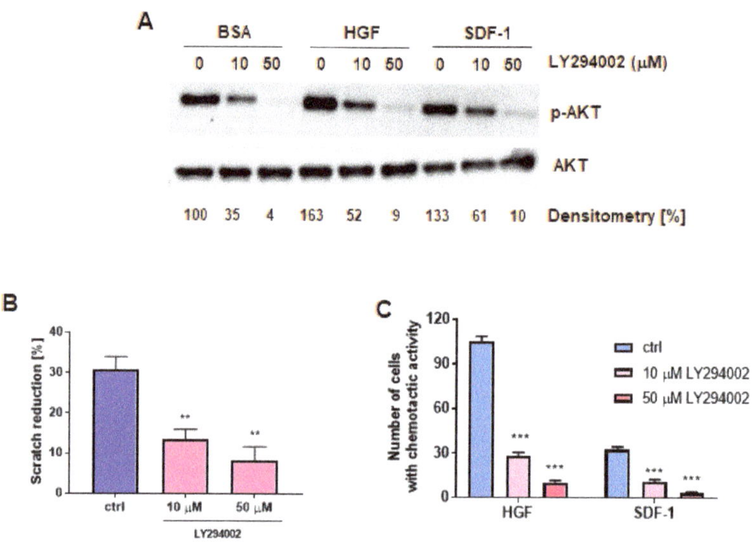

Figure 5. Inhibition of PI3K-AKT signaling diminishes migration and chemotaxis of RMS cells. (**A**) Treatment of RH30 cells with 10 and 50 μM LY294002 inhibited phosphorylation of AKT kinases under different treatments conditions (0.5% BSA, 20 ng/mL HGF, 100 ng/mL SDF-1). Cells were treated with the inhibitor 1 h before further experiments. Representative image of Western blots of two independent biological experiments is shown. Densitometric analysis evaluated the ratio of p-AKT to AKT and was presented as percentage of control. (**B**) Treatment of RH30 cells with 10 and 50 μM LY294002 inhibited the migration of the cells in a scratch assay; $n = 3$. (**C**) Treatment of RH30 cells with 10 and 50 μM LY294002 inhibited chemotaxis of RH30 cells towards 20 ng/mL HGF and 100 ng/mL SDF-1; $n = 3$. * $p < 0.05$, ** $p < 0.01$, *** $p < 0.001$. Graphical data are presented as the means ± SEMs. Alternatively, representative Western blot images of two independent biological experiments are presented.

2.5. The SNAIL-miRNA Axis Regulates the Motility of RMS Cells

In addition to proteins, miRNAs may be the other important mediators of SNAIL action [17]. Therefore, the effect of SNAIL on the miRNA transcriptome was evaluated. Analysis of the miRNA transcriptome by next-generation sequencing showed that SNAIL silencing either upregulated or downregulated different miRNAs (Table 1 and Figure 6A and Supplementary Table S2). Gene ontology (GO) analysis of processes regulated by miRNAs revealed actin cytoskeleton reorganization and differentiation, as well as on HGF-activated receptor activity (Figure 6B), indicating that miRNAs are important mediators of SNAIL action. From the group of SNAIL-dependent miRNAs (Table 1 and Supplementary Table S2), four different candidate miRNAs were selected for further research. The two significantly upregulated and two downregulated miRNAs were selected for further analyses after verification of their levels using qPCR. The candidate miRNAs need to be expressed at a reasonable level, basing on TPM (tags per million) values (Supplementary Table S2).

Table 1. miRNA sequencing results revealed miRNAs that were significantly differentially expressed between RH30 shSNAIL and shCTRL cells. The table shows the log fold change (logFC) between the shSNAIL and shCTRL groups; raw *p*-values, Benjamini-Hochberg FDR corrected *p*-values; $n = 3$.

Names	logFC	*p* Value	FDR
hsa-miR-1269b	−11.78	1.12×10^{-83}	6.95×10^{-82}
hsa-miR-218-5p	−11.07	1.50×10^{-142}	2.10×10^{-140}
hsa-miR-548h-5p	−10.81	5.84×10^{-61}	1.82×10^{-59}
hsa-miR-452-5p	−10.73	3.40×10^{-174}	1.47×10^{-171}
hsa-miR-224-5p	−10.68	1.22×10^{-40}	1.89×10^{-39}
hsa-miR-3148	10.32	9.71×10^{-56}	2.82×10^{-54}
hsa-miR-4652-5p	−9.78	7.14×10^{-43}	1.24×10^{-41}
hsa-miR-1269a	−9.77	6.90×10^{-156}	1.51×10^{-153}
hsa-miR-873-5p	−9.63	5.99×10^{-41}	9.65×10^{-40}
hsa-miR-873-3p	−8.79	1.56×10^{-53}	4.25×10^{-52}
hsa-miR-302a-5p	−8.07	8.53×10^{-45}	1.69×10^{-43}
hsa-miR-105-5p	−6.72	2.21×10^{-61}	7.38×10^{-60}
hsa-miR-139-5p	6.53	2.01×10^{-65}	8.75×10^{-64}
hsa-miR-199b-5p	−6.52	4.51×10^{-94}	3.27×10^{-92}
hsa-miR-95-3p	6.42	2.34×10^{-40}	3.51×10^{-39}
hsa-miR-767-5p	−6.15	8.59×10^{-51}	1.97×10^{-49}
hsa-miR-139-3p	5.93	1.25×10^{-44}	2.37×10^{-43}
hsa-miR-28-5p	5.43	1.15×10^{-94}	1.00×10^{-92}
hsa-miR-143-3p	−5.26	2.17×10^{-53}	5.55×10^{-52}
hsa-miR-28-3p	5.1	3.60×10^{-113}	3.88×10^{-111}
hsa-miR-193a-5p	4.45	6.18×10^{-75}	3.36×10^{-73}
hsa-miR-541-3p	−4.32	4.72×10^{-63}	1.71×10^{-61}
hsa-miR-412-5p	−4.15	4.11×10^{-71}	1.99×10^{-69}
hsa-miR-1197	−3.83	1.45×10^{-50}	3.15×10^{-49}
hsa-miR-200c-3p	−3.77	1.50×10^{-64}	5.95×10^{-63}
hsa-miR-431-5p	−3.76	5.00×10^{-45}	1.04×10^{-43}
hsa-miR-200b-3p	−3.65	3.26×10^{-41}	5.45×10^{-40}
hsa-miR-129-5p	−3.32	2.27×10^{-43}	4.11×10^{-42}
hsa-miR-1180-3p	−3.27	4.16×10^{-51}	1.01×10^{-49}
hsa-miR-486-5p	2.74	9.05×10^{-40}	1.31×10^{-38}

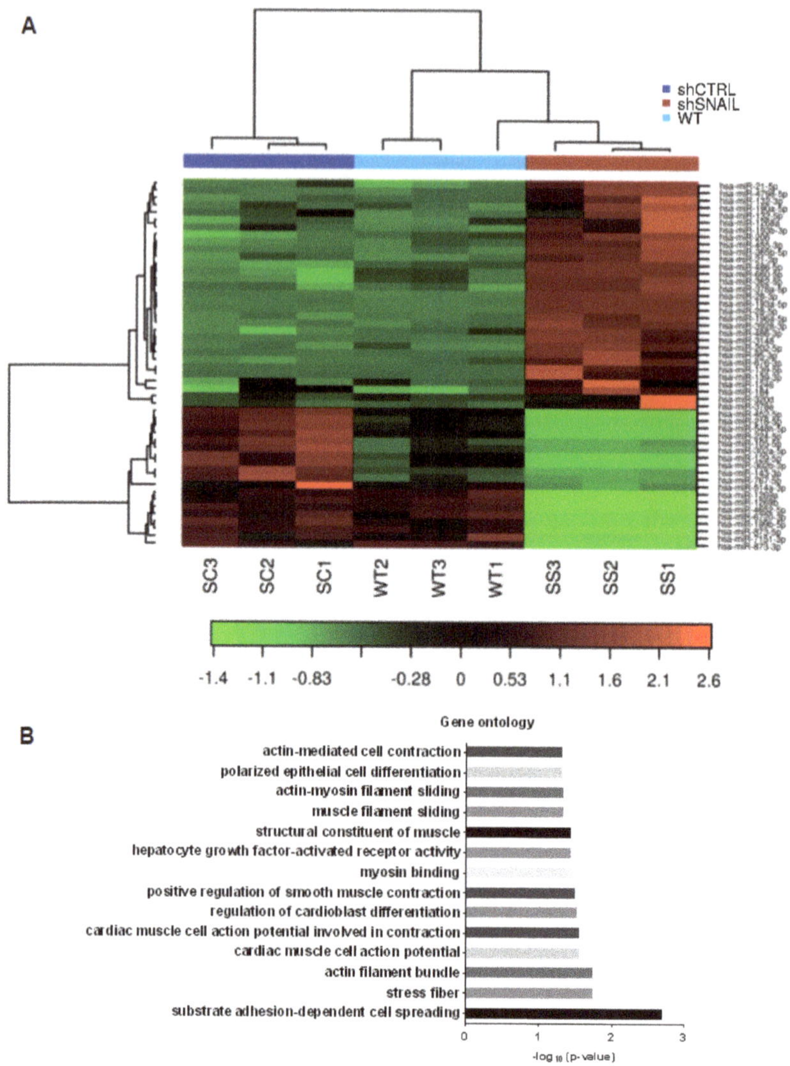

Figure 6. miRNAs are mediators of SNAIL action on cytoskeleton organization. (**A**) miRNA sequencing in RH30 WT, shCTRL (SC) and shSNAIL (SS) cells revealed that SNAIL silencing downregulates and upregulates many different miRNA sequences. These alterations were visualized as a heatmap via unsupervised hierarchical clustering by sample and miRNA. Clustering was performed on all samples, and the top 50 miRNAs with the highest coefficient of variation (CV) were identified based on the trimmed mean of M-values (TMM) normalized counts; $n = 3$. (**B**) GO enrichment analysis using Fisher's test and the 'Elim' method revealed potentially significant biological processes and molecular functions associated with muscle and actin cytoskeleton structure in miRNA sequencing results; RH30 shSNAIL cells vs RH30 shCTRL cells, $p < 0.05$.

Among the selected candidates, miR-28-3p and miR-193a-5p were overexpressed in WT RH30 cells by transfection with miRNA mimics, as they were upregulated in SNAIL-deficient cells, whereas miR-218-5p and miR-452-5p were inhibited with miRNA inhibitors, as they were downregulated in SNAIL-deficient cells. The miRNAs expression levels after transfection were verified by qPCR (Figure 7A). The effect of those miRNAs on cell motility was screened by a scratch assay. miR-28-3p and miR-193a-5p inhibited the migration of RH30 cells the most potently (Figure 7B), indicating that they might be important mediators of SNAIL action.

Subsequently, we evaluated the effect of miRNA mimics and inhibitors on EZRIN levels. Only miR-28-3p downregulated EZRIN at the mRNA (Figure 7C) and at protein levels (Figure 7D). However, bioinformatic analysis did not show any binding sites for miR-28-3p in the *EZRIN* 3′-UTR, so we speculate that miR-28-3p regulates the EZRIN level indirectly. Overexpression of miR-28-3p inhibited chemotaxis towards HGF and SDF-1 (Figure 7E) and diminished the adhesion of RH30 cells to endothelial cells after treatment with HGF and SDF-1 (Figure 7F).

These results indicate that miR-28-3p may be an important mediator of the effects of SNAIL on migration, chemotaxis and adhesion through indirect modulation of EZRIN levels. The proposed mechanism of SNAIL regulation and its action in RMS is presented in graphical abstract.

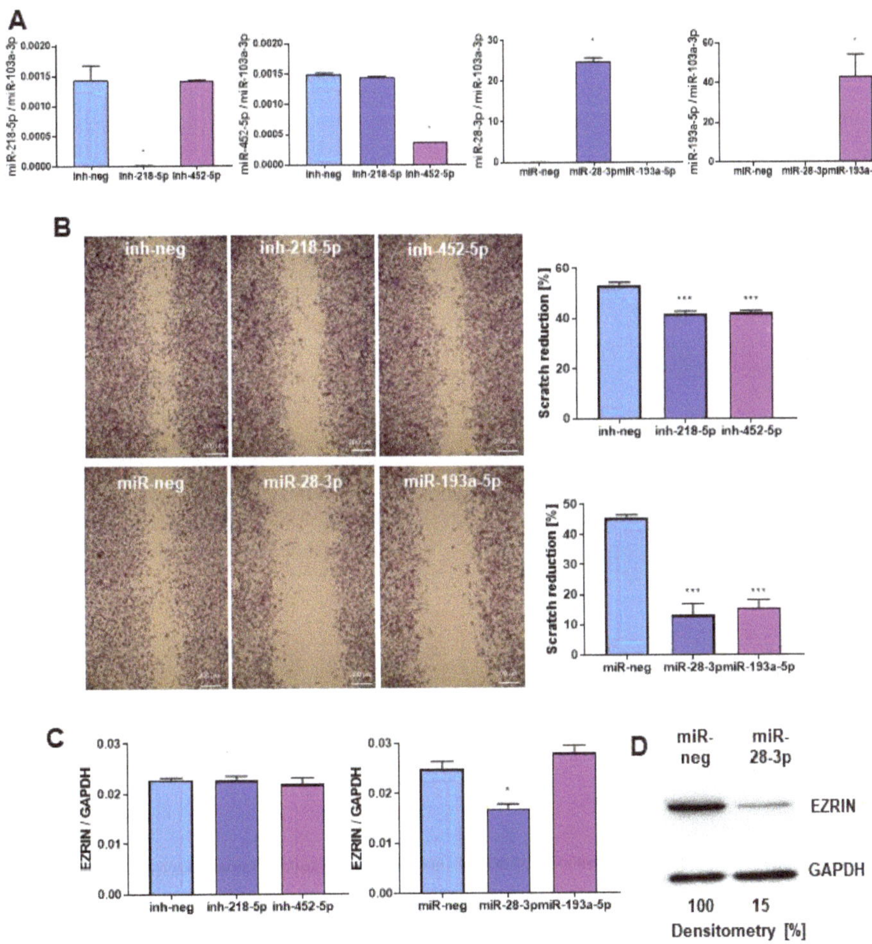

Figure 7. *Cont.*

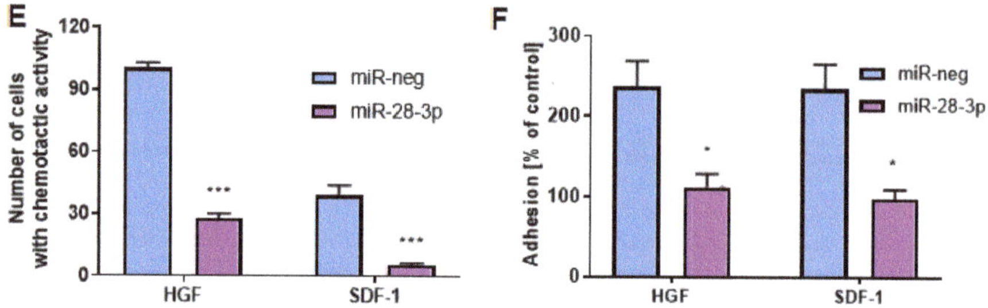

Figure 7. miR-28-3p is a mediator of SNAIL action on RMS cell motility and adhesion to endothelial cells. (**A**) SNAIL-dependent candidates miRNAs were either overexpressed by transfection with miRNA mimics (miR-28-3p and miR-193a-5p) or inhibited with miRNA inhibitors (miR-218-5p and miR-452-5p) in RH30 cells to evaluate whether the candidate miRNAs are important mediators of SNAIL action. The expression levels of miRNAs were evaluated three days after transfection by qPCR with the ΔCt method, and miR-103a-3p was used as a constitutively expressed miRNA control. The results are presented as the means of two independent experiments with duplicate samples. (**B**) Among the four miRNA candidates, miR-28-3p and miR-193a-5p regulated the most potently the motility of RH30 cells in a scratch assay. The results are presented as representative images of three independent experiments 24 h after scratching and graphs presenting the calculated percentage of scratch reduction; $n = 3$. White scale bar represents 200 μm. (**C**) Among the four miRNA candidates, only miR-28-3p was a regulator of *EZRIN* expression at the mRNA level in RH30 cells; $n = 3$. (**D**) miR-28-3p downregulated EZRIN at the protein level in RH30 cells (representative Western blot image of 2 independent biological experiments). (**E**) miR-28-3p overexpression diminished the chemotaxis of RH30 cells towards HGF and SDF-1; $n = 3$. (**F**) miR-28-3p overexpression in RH30 cells diminished the adhesion of RH30 cells treated with HGF and SDF-1 to HUVEC endothelial cells pretreated with TNF-α; $n = 3$. * $p < 0.05$, ** $p < 0.01$, *** $p < 0.001$. Graphical data are presented as the means ±SEMs. Alternatively, representative images are presented.

3. Discussion

This study was undertaken to investigate the role of the SNAIL transcription factor in the regulation of RMS metastasis, since the action mechanism of SNAIL in mesenchymal tumors is not well understood. In our studies, we sought to identify different noncanonical action mechanisms of SNAIL in RMS metastasis that may also be important in other tumor types.

Previously, we showed that SNAIL is a key regulator of ARMS tumor growth and differentiation through functional repression of MYF5 and MYOD [9]. Here, we demonstrated that SNAIL affects RMS metastatic behavior by reorganizing the actin cytoskeleton and regulating intracellular pathways important for tumor cell metastasis. Actin cytoskeleton reorganization is also a key event in the acquisition of migratory properties by epithelial cancer cells undergoing EMT [25]. In addition, SNAIL was identified as a crucial factor involved in PI3K-AKT signaling and the EZRIN-RHO pathway that may be regulated by SDF-1 and HGF, factors that promote tumor cell metastasis.

An important pathway in RMS progression is the PI3K/AKT signaling pathway. Mutational activation of PI3K/AKT signaling has been previously associated with a clinically aggressive RMS subset [31]. We found that blocking this pathway in RMS cells results in decreased motility and chemotaxis towards HGF and SDF-1. Importantly, SNAIL was identified as a regulator of AKT expression by direct binding to the AKT2 promoter. Interestingly, SNAIL has been shown to enhance the binding of AKT2 to the E-cadherin (*CDH1*) promoter in epithelial cells and interference with AKT2 prevents SNAIL-mediated repression of the *CDH1* gene [32]. In addition, the literature suggests that AKT2 activates SNAIL and that a switch from the AKT1 to the AKT2 isoform is required for the induction of SNAIL expression [33]. Based on our results, we postulate the existence of a stimulatory loop between SNAIL and AKT that regulates RMS progression and metastasis. Our studies demonstrated that SDF-1 and HGF, which may stimulate tumor cells to form metastases, increase the SNAIL level by inducing GSK3β phosphorylation via the PI3K-AKT pathway. GSK3β regulation by AKT was previously demonstrated in different cell types [34]. The canonical pathway of AKT activation is initiated by G-protein-coupled receptors, such as CXCR4 and receptor tyrosine kinases, such as MET; subsequently, AKT mediates inhibitory phosphorylation of GSK3β S9 [35]. GSK3β regulates the SNAIL level by phosphorylation at two consensus motifs to affect the function of the protein either by ubiquitination or subcellular localization [36,37]. SNAIL can also bind to its own promoter, forming a feedback mechanism for its transcriptional regulation; this mechanism can be overridden in cells receiving potent stimulation of the PI3K pathway, which can activate SNAIL expression and subsequently induce EMT [38]. Our data confirm the existence of such feedback mechanisms in mesenchymal cells, since our ChIP-seq results showed the binding of SNAIL to its own promoter.

We have shown here for the first time that the EZRIN protein, which has been previously shown to be a key regulator of tumor metastasis [8,39], is a crucial mediator of SNAIL signaling pathways by acting as an actin filament-plasma membrane linker [8]. These results may be extrapolated to RMS patients because we found a positive correlation between *SNAIL* and *EZRIN* expression. EZRIN has been described as a metastatic determinant in many different tumor types [40], including mesenchymal tumors such as osteosarcoma [41].

The literature indicates a link between the EZRIN-dependent metastatic potential and the activity of the small GTPase RHO proteins [8]. Our studies demonstrated that both EZRIN and SNAIL silencing diminished the activity of RHO and its downstream mediator ROCK-II. RHO family GTPases and their downstream effector proteins – ROCK I and II – are often associated with enhanced invasive and metastatic phenotypes, as they are known regulators of the cytoskeleton and cell migration that are frequently overexpressed in different tumor types [42]. In addition, cell motility promoted by RHO signaling through ROCK was shown to require EZRIN localized in the direction of cell movement [43]. In oral cancer cells, SNAIL was previously associated with changes in RHO activity and phosphorylation of EZRIN-RADIXIN-MOESIN family, but no precise mechanism was described [44].

Our studies showed that SNAIL may regulate EZRIN expression by directly binding to its promoter. Via that mechanism SNAIL acts as an activator of EZRIN expression. SNAIL is usually considered as a transcriptional repressor, but it may also act as an activator [16], such as for MMP15 [45]. SNAIL can also potentiate enhancer activation by collaborating with different activators [46].

In this study we showed that in addition to directly regulating EZRIN, SNAIL can indirectly affect its level via miR-28-3p since bioinformatic analysis using TargetScanHuman 7.1 [47] and miRDB [48] did not reveal potential miR-28-3p binding sites in the 3′UTR region of *EZRIN* gene. miR-28-3p was identified as a SNAIL mediator in the motility, chemotaxis and adhesion of RMS cells. The important role of miR-28-3p in RMS was demonstrated for the first time. Previously, the role of miR-28 was suggested to be important in different tumor types, such as non-Hodgkin lymphoma [49] and colorectal cancer [50]. Our research also demonstrated that SNAIL silencing in RMS cells changes the miRNA transcriptome and that SNAIL-miRNA signaling regulates many different processes associated with actin cytoskeleton reorganization and cellular motility and differentiation. In addition to miR-28-3p, many different miRNAs are mediators of SNAIL action. Several studies in the literature describe SNAIL as a regulator of miRNA expression with implications for epithelial tumor progression [15,17], such as the interaction between miR-34 and SNAIL in the regulation of the EMT process [51].

Using an in vivo model, we showed diminished engraftment of SNAIL-deficient RMS cells in murine lungs after intravenous implantation. The metastatic capabilities of RMS cells may be regulated by the SDF-1 gradient. CXCR4–SDF-1 signaling regulates locomotion, chemotaxis, and adhesion [4]. The SDF-1 gradient was identified as an important factor in chemotaxis and adhesion of SNAIL-deficient RMS cells. Interestingly, our data suggest that SNAIL mediated motility of RMS cells is induced by the modulation of EZRIN levels rather than changes in CXCR4 expression. Moreover, SDF-1 may regulate SNAIL expression and thus may promote metastasis. HGF has also been shown to play an important role in RMS metastasis [7,21,52]. Our studies demonstrated that SNAIL regulates the level of the MET receptor, chemotaxis towards HGF and HGF-induced adhesion to endothelial cells. These results may also be extrapolated to patients suffering from RMS due to the correlation between SNAIL and MET receptor levels. Moreover, SDF-1 and HGF may also regulate the SNAIL level, as they induce rapid phosphorylation of AKT kinase and GSK3β [53].

The diminished engraftment of SNAIL-deficient RMS cells in murine lungs may be explained by the inhibited adhesion of RMS cells to activated endothelial cells. This process is affected by SNAIL through its regulation of EZRIN, miR-28-3p levels and adhesion molecules. In this study, we showed that SNAIL silencing completely blocked CD49b and ICAM-1 (CD54) expression and profoundly diminished ICAM-2 (CD102) levels. Integrin adhesion to the ECM provides the traction required for tumor cell invasion [54]. Importantly, CD49b (VLA-2 integrin) expression plays major roles in the post extravasation movement of RMS cells [55].

In conclusion, our data demonstrate a novel mechanism by which the SNAIL transcription factor regulates RMS metastasis. We showed that SNAIL forms an integrated signaling network supporting the progression of RMS and that its expression may be induced by factors attracting tumor cells to form metastases, such as SDF-1 and HGF. Regulatory mechanisms mediated by SNAIL crosstalk between AKT kinase, EZRIN, RHO-GTPase and miRNAs modulate tumor cell motility, adhesion and actin cytoskeleton organization, supporting the progression and metastasis of RMS (see the Graphical Abstract). We believe that these newly discovered mechanisms may also be important in other tumor types.

4. Materials and Methods

4.1. Cell Culture

RMS cell lines (RH30, RH41, RD, SMS-CTR) were kindly provided by Dr. PJ Houghton (Center for Childhood Cancer, Columbus, OH, USA). The cells were cultured in DMEM high-glucose medium (Lonza Group Ltd., Basel, Switzerland) supplemented with 10% fetal bovine serum (FBS, EURx, Gdansk, Poland) and 50 μg/mL gentamicin (Lonza) at 37 °C, 5% CO_2 and 95% humidity.

Human umbilical vein endothelial cells (HUVEC) have been ordered from Becton Dickinson Biosciences. They were cultured in endothelial cell growth medium (PromoCell, Heidelberg, Germany), with endothelial cell growth supplement (PromoCell).

The cell lines were routinely tested for *Mycoplasma* spp. contamination using by MycoAlert™ Mycoplasma Detection Kit (Lonza). RMS cell line authentication was performed by STR profiling using AmpFlSTR SGM PLUS Kit (Applied Biosystems, Foster City, CA, USA) and sequencing apparatus ABI Prism 310 Genetic Analyser (Applied Biosystems) according to the manufacturer's protocol.

4.2. Treatments of the Cells

For evaluation of expression and phosphorylation of kinases, RH30 cells were examined at 70% confluency. They were starved overnight in DMEM medium with 0.5% BSA. Subsequently, they were treated with 20 ng/mL HGF (Peprotech, London, UK) for 10 min, 100 ng/mL SDF-1 for 2–5 min in starving medium and 10% FBS for 30 min. Subsequently, protein was isolated with MPER buffer, as described below.

For experiments with LY294002 inhibitor (Merck, Darmstadt, Germany) cells were pretreated with 10 or 50 μM LY294002 one hour before further treatments with HGF, SDF-1 and FBS. In scratch assay and chemotaxis experiments the inhibitor was added one day before the experiment and it was present for the whole experiment.

The other inhibitors used in the experiments with treatment of the cells with SDF-1 or HGF were following: pertussis toxin (PTX) 20 μg/mL (Sigma-Aldrich, Darmstadt, Germany) was added 1.5 h before stimulation, UO126 30 μM (Merck) was added 1 h before stimulation, MET inhibitor 5 mM (Merck) was added 16 h prior to stimulation, BIO 1 μM was added for different time periods between 2 and 24 h.

4.3. Production of Viral Vectors and Transduction of Cells

RH30 cells were transduced with shRNA Lentiviral Particles targeting SNAIL and control lentiviral particles (Santa Cruz Biotechnology, Santa Cruz, CA, USA), as described previously [9]. shRNA Lentiviral Particles consisted of three different shRNA sequences targeting SNAIL.

Lentiviral particles encoding GFP-P2A-SNAIL (GFP-P2A-SNAIL @pLenti6/UbC) and GFP (GFP@pLenti6/UbC) were produced using the Vira Power Lentiviral Expression System (Invitrogen, Carlsbad, CA, USA), as previously described [9,21,56]. SMS-CTR cells were transduced with lentiviral vectors (at MOI = 10) in the presence of 6 μg/mL polybrene (Sigma-Aldrich). After 72 h the cells were subject to selection with 2.5 μg/mL blasticidin (InvivoGen, Toulouse, France) for 2 weeks.

4.4. Transfection with siRNA

RH30, RH41 and RD cells were transfected with 20 nM siRNA against SNAIL (combination of two Silencer Select siRNA ID variants: s13185 and s13187, Ambion Inc., Austin, TX, USA) or against EZRIN (combination of two Silencer Select siRNA ID variants: s14795 and s14797) or scrambled control siRNA (Silencer Select Negative Control #1 siRNA, cat. 4390844, Ambion) using Lipofectamine RNAiMAX (Invitrogen) transfection reagent according to vendor's protocol. Protein level was verified three days after transfection. Three subsequent transfections were performed every 3–4 days and after that time cells were seeded for the experiments involving scratch assay and chemotaxis.

4.5. Transfection of Cells with miRNA Precursors and Inhibitors

RH30 cells were transfected with 30 nM mirVana™ miRNA Mimic: hsa-miR-28-3p (ID: MC12933), hsa-miR-193a-5p (ID: MC1178), Negative Control 1 or alternatively with mirVana™ miRNA Inhibitor: hsa-miR-218-5p (MH10328), hsa-miR-452-5p (ID: MH12509), Negative Control 1 using using Lipofectamine RNAiMAX (Invitrogen) transfection reagent according to vendor's protocol. RNA was isolated 48–72 h after transfection. The cells were seeded for further experiments 48 h after transfection.

4.6. DNA and RNA Isolation and Reverse Transcription

Total RNA was extracted using the GeneMATRIX Universal RNA/miRNA Purification Kit (EURx) or mirVana miRNA Isolation Kit (Ambion), according to the manufacturer's protocol. Reverse transcription of mRNA was performed using MMLV reverse transcriptase (Promega, Madison, WI, USA) according to the vendor's protocol. Reverse transcription of miRNA was performed using the Universal cDNA Synthesis Kit (Exiqon, Vedbaek, Denmark) or miRCURY LNA RT Kit (Qiagen, Hilden, Germany), according to the manufacturer's protocol.

4.7. Quantitative Real-Time PCR

Gene expression was determined by qRT-PCR analysis using QuantStudio™ 7 Flex Real-Time PCR System (Applied Biosystems), Blank qPCR Master Mix (EURx) and the indicated Taq-Man probes (Applied Biosystems): human: *GAPDH* (Hs99999905_m1), *SNAI1* (Hs00195591_m1) and *EZRIN* (Hs00931653_m1). The mRNA expression level for all of the samples was normalized to the housekeeping gene *GAPDH*, using the $2^{-\Delta Ct}$ method.

For the evaluation of miRNA expression by quantitative real-time PCR, SYBR Green qPCR Master Mix (EURx) with LNA™ PCR primer set (Exiqon) or miRCURY LNA miRNA PCR Assay (Qiagen) for human miR-28-3p, miR-193a-5p, miR-218-5p, miR452-5p and miR-103a-3p were used. The miRNAs expression levels were quantified using the $2^{-\Delta Ct}$ method, using miR-103a-3p as a relative control.

4.8. MicroRNA Sequencing

Library preparation, next generation sequencing and analysis of data were performed at Exiqon Services in Denmark. A total of 1 μg of total RNA was converted into microRNA NGS libraries using NEBNEXT library generation kit (New England Biolabs Inc., Ipswich, MA, USA) according to the manufacturer's instructions. Each individual RNA sample had adaptors ligated to its 3′ and 5′ ends and converted into cDNA. Then the cDNA was pre-amplified with specific primers containing sample specific indexes. After 15 cycle pre-PCR the libraries were purified on QiaQuick columns and the insert efficiency evaluated by Bioanalyzer 2100 instrument on high sensitivity DNA chip (Agilent Inc., Santa Clara, CA, USA) The microRNA cDNA libraries were size fractionated on a LabChip XT (Caliper Inc., Waltham, Massachusetts, USA) and a band representing adaptors and 15–40 bp insert excised using manufacturer's instructions. Samples were then quantified using qPCR and concentration standards. Based on quality of the inserts and the concentration measurements the libraries were pooled in equimolar concentrations (all concentrations of libraries to be pooled are of the same concentration). The library pools were finally quantified again with qPCR and optimal concentration of the library pool used to generate the clusters on the surface of a flow cell before sequencing using v2 sequencing methodology according to the manufacturer instructions (Illumina Inc., San Diego, CA, USA).

Samples were sequenced on the Illumina NextSeq 500 system. The system uses quality score binning enabling a more compact storage of raw sequences. Using only eight levels (Levels: No call, 6, 15, 22, 27, 33, 37, 50) of quality method has been tested and found to virtually loss-less. Average number of reads per sample was 7.5mio. Number of sequencing cycles (read length) was 50 nt, single-end read. The differential expression analysis was done using TMM (the trimmed mean of M-values normalization method) in the EdgeR statistical software package (Bioconductor, http://www.bioconductor.org/). Exiqon Services also performed TPM (tags per million) normalization. MicroRNA NGS data were deposited in Gene Expression Omnibus (GEO) under accession number GSE100114.

4.9. Western Blotting

Total extracts of protein were isolated with M-PER lysing buffer (Pierce, Rockford, IL, USA) as described previously [21], whereas nuclear and cytoplasmic fractions of protein, were isolated using the Nuclear Extract Kit (Active Motif, La Hulpe Belgium) according to the manufacturer's protocol. The protein concentration was measured using the Bio-Rad Protein Assay (Bio-Rad, Hercules, CA, USA) according to the vendor's protocol. Western blotting was performed using the anti-GAPDH rabbit mAb (14C10; #2118; Cell Signaling Technology, Leiden, The Netherlands), the anti-histone H3 (ab1791; Abcam, Cambridge UK), the anti-SNAIL mouse mAb (L70G2; #3895; Cell Signaling), anti-EZRIN (3C12, Santa Cruz Biotechnology), anti-phospho-AKT (Ser473) rabbit mAb (#9272, Cell Signaling, Danvers, MA, USA), anti-AKT rabbit pAb (#9272, Cell Signaling), anti-AKT2 mouse mAb (#5239, Cell Signaling), anti-AKT1 rabbit mAb (#2938T, Cell Signaling), anti-phospho-p44/42 MAPK (9106S, Cell Signaling), anti-phospho-GSK-3β (Ser9) rabbit mAb (#9336, Cell Signaling), anti-GSK-3β rabbit mAb (#9315, Cell Signaling), and secondary anti-rabbit and anti-mouse antibodies conjugated with horseradish peroxidase (HRP, Santa Cruz Biotechnology). Proteins were separated by electrophoresis in a 12% resolving sodium dodecyl sulfate–PAGE gel, and the fractionated proteins were transferred into a PVDF membrane (BioRad). Chemiluminescent signals were developed using (SuperSignal™ West Pico PLUS Chemiluminescent Substrate (Thermo Scientific) and ChemiDoc MP Imaging System (Bio-Rad) or using developing films. Western blot results are presented as representative images of two or three independent biological experiments. Densitometric analysis of each Western blot image was performed using ImageLab software (BioRad). The ratio of the adjusted volume band of the gene of interest to the constitutive gene was evaluated, and subsequently results were presented as percentage of control.

4.10. RHO and ROCK-II Enzymes Activity

RHO protein activity was evaluated using Rho Activation Assay Biochem Kit (Cytoskeleton, Inc., Denver, CO, USA), according to vendor's protocol. For the experiment, RH30 cells were transfected with siRNA for the three times, then starved in 0.5% FBS for 24 h, then 0% FBS for a further 24 h and then treated with 10% FBS 15 min before lysis according to vendor's protocol. The results were analyzed by Western blot.

Rho-associated Kinase (ROCK) Activity Assay (Merck Millipore, Darmstadt, Germany) was used to evaluate ROCK-II enzyme activity in protein lysates, according to vendor's protocol. Inhibitor from the kit was used to evaluate assay specificity.

4.11. Analysis of Subproteomes of Nuclear and Cytoplasmic Fractions

Cytoplasmic and nuclear protein fractions were isolated by the ProteoExtract Kit (Calbiochem, San Diego, CA, USA). Protein samples were diluted with 8 M urea in 50 mM ammonium bicarbonate, reduced with DTT, alkylated with iodoacetamide and digested with trypsin on 30 kDa cut-off filter (Vivacon 500, Sartorius Stedim, Goettingen, Germany) using filter-aided sample preparation (FASP) procedure described previously [57].

Peptides were further analyzed by use of shotgun LC-MS/MS technique using reversed-phase liquid chromatography (RP-LC) system (UltiMate 3000RS LCnanoSystem, Dionex, Sunnyvale, CA, USA) coupled with a quadrupole time-of-flight mass spectrometer (micrOTOF-Q II, Bruker Daltonics, Bremen, Germany).

The RP-LC system consisted of a desalting trap column (75 μm × 2 cm, C_{18} material, 3 μm, 100 Å) and an analytical column (75 μm × 50 cm, C_{18} material 2 μm, 100 Å) with a nanoflow solvent delivery. The LC-MS/MS data were acquired by online analysis of peptides eluted with a 90 min gradient ranging from 2% to 40% acetonitrile in 0.05% formic acid/water at a 300 nl/min flow rate. MS/MS data were obtained by targeting 5 precursor ions in the scan range of m/z 50 to 2500 Da. The raw data were processed by Data Analysis 4.1 (Bruker Daltonics) and searched against Swiss Prot_201407 database with taxonomy restriction to Homo sapiens (20,284 sequences) using MASCOT search engine (v.2.3.0) embedded into ProteinScape 3.0 (Bruker Daltonics). The following parameters were used for the search: trypsin with maximum one missed cleavage, precursor and product ions mass tolerance were respectively ±20 ppm and ±0.05 Da; fixed modification—carbamidomethylation (C); variable modifications—oxidation (M), deamidation (NQ). Proteins below the 1% false discovery rate were considered. The complete analysis was composed of two LC-MS/MS runs. On the basis of the first run, a scheduled precursor list (SPL) was generated and used as an exclusion list during second LC-MS/MS analysis. A compilation of obtained two search results was performed with the ProteinScape 3.0 platform.

The mass spectrometry proteomics data have been deposited to the ProteomeXchange Consortium via the PRIDE [58] partner repository with the dataset identifier PXD006711 and 10.6019/PXD006711. The data include Bruker mass spectrometer output files (.baf) and PRIDE XML files generated from Mascot DAT files with the use of PRIDE Converter 2 [59].

4.12. Bioinformatic Analysis

Ezrin promoter fragments were found in the Eukaryotic promoter database [30]. EPD describes two fragments of *EZRIN* promoter: *EZR_1* 9Chromosome [NC_000006.12]; Strand [-]; Position [158819364]) and *EZR_2* (Chromosome [NC_000006.12]; Strand [-]; Position [158818235]). Two fragments of the promoter of *EZRIN* (~1000 bp) was screened for putative SNAIL transcription factor binding sites using a TF prediction tool called ConSite (http://consite.genereg.net/). The search for putative TF binding sites was performed at 80% cutoff [60]. The results were then compared with other TF prediction tools.

4.13. Chromatin Immunoprecipitation (ChIP) Assay

ChIP was performed using SimpleChip Enzymatic Chromatin IP Kit (Cell Signaling Technology) according to the manufacturer's protocol. For ChIP assays 10 μg of goat antibody against SNAIL (cat. AB-108-C; R&D Systems, Minneapolis, MN, USA), 10 μg of the positive control histone H3 (D2B12 XP Rabbit mAb Chip formulated, Cell Signaling Technology) and the normal goat IgG control

(R&D Systems, cat. AF3639) as a negative control. After immunoprecipitation, the DNA was isolated using spin columns from the kit and eluted in 50 µL Elution Reagent C. PCR was performed with 2 µL of immunoprecipitated material and the products were analyzed on an 1.5% agarose gel, and visualized using a gel documentation system.

The following primers were used to quantify SNAIL binding to the *EZRIN* promoter fragments described in previous section:

EZR_1 fragment forward primer: 5′-GAGGCTAGCACGAGTTAAGCA-3′
EZR_1 fragment reverse primer: 5′-GCACGTTTGTGGCCTCTTTT-3′
EZR_2 fragment forward primer: 5′-GGAGCACACGGAGCACTG-3′
EZR_2 fragment reverse primer: 5′-CGGAGAGAGGCGGAGAAGA-3′

Additionally, Chip-Seq analysis was performed using the Intelliseq sp. z o. o. (Cracow, Poland) company service in cooperation with Novogene Co., Ltd (Cambridge, UK) using Illumina high-throughput sequencer. Quality assessment of the reads was performed using the FastQC tool (v0.11.7). Reads containing adapters were filtered using the Trim Galore tool (v0.5.0). Mapping of reads to the human reference genome GRCh38 from the Ensembl database was done using the BWA-MEM software (v. 2.1.0). The aligned reads were filtered out more than once. Subsequently, the analysis using the MASC2 callpeak tool (v. 2.1.2) with default parameters (minimum FDR 0.05) was performed. The detected transcription factor binding sites were noted using the HOMER annotatePeaks.pl tool (v4.10) and a GTF file from the Ensembl database. The Integrative Genomics Viewer (IGV) was used as a visualization tool for interactive exploration of large, integrated genomic datasets [61]. ChIP-seq data data were deposited in Gene Expression Omnibus (GEO) under accession number GSE152355.

4.14. Scratch Assay

Confluent RMS cells were treated with DMEM medium with 0.5% BSA for 24 h. Subsequently, a scratch was generated with a pipette tip. Starving medium was replaced every day. Photographs were taken after 24, 48 and 72 h and they were analyzed using ImageJ software (National Institutes of Health, Bethesda, MD, USA).

4.15. Chemotaxis and Invasion Assays

Chemotaxis of RMS cells to 20 ng/mL HGF (R&D System) and 100 ng/mL SDF-1 (Peprotech, Rocky Hill, NJ, USA) was evaluated using modified Boyden's chamber with 8 µm pore polycarbonate membrane inserts (Transwell; Corning Life Sciences—PZ HTL SA, Warsaw, Poland), as described previously [21]. 0.5% BSA served as a negative control. Similarly, invasion of RMS cells through growth factor reduced Matrigel invasion inserts (Corning Life Sciences) to 10% FBS, 20 ng/mL HGF, 100 ng/mL SDF-1, 0.5% BSA was also investigated, as described previously [21].

4.16. Immunofluorescent Staining with Phalloidin

RH30 cells were fixed in 4% formaldehyde (POCH, Gliwice, Poland) in PBS, permeabilized in 0.1% TritonX-100 (Sigma-Aldrich), blocked in 1% bovine serum albumin (BSA, Sigma-Aldrich). For visualization of the actin cytoskeleton, the cells were stained with phalloidin conjugated with Alexa Fluor 488 (Life Technologies) according to the manufacturer's protocol. Labeling was assessed by fluorescence microscopy using an Olympus BX51 or IX70 microscope (Olympus Corporation, Tokyo, Japan) and Olympus XC50 camera with cellSens Dimension software (both from Olympus). The images were processed using cellSens Dimension software.

4.17. Flow Cytometry

For evaluation of MET and CXCR4 receptors expression levels RMS cells were stained with monoclonal FITC-labeled anti-human HGFR/c-MET antibody, clone 95106 (R&D), PE-labeled anti-human CXCR4 antibody (Becton Dickinson, Franklin Lakes, NJ, USA) or mouse IgG1 isotype control (R&D) labeled with FITC or PE respectively. The cells were acquired using FACS Canto II cytometer (Becton Dickinson) and analyzed using FACS Diva software (Becton Dickinson), as described previously [21].

The expression level of adhesion molecules was evaluated using Lyoplate technology (Lyoplate Screening Panel, Becton Dickinson) according to the manufacturer's protocol. The cells were acquired by use of Attune Next Flow Cytometer and analyzed using Attune NxT Software v2.2 (Thermo Fisher Scientific, Waltham, MA, USA).

4.18. Adhesion Assay

HUVEC endothelial cells (5×10^4 per well) were seeded in black 96-well plates with a clear bottom (Corning Costar, Amsterdam, The Netherlands) and grown overnight to a confluent monolayer. After stimulation of endothelial cells with TNF-α (50 ng/mL) for 24 h, RMS cells were incubated with 2.5 μM Calcein AM (BD Pharminogen, Franklin Lakes, NJ, USA) for 30 min at 37 °C in cell culture medium, washed, then they rested for 30 min and treated with 100 ng/mL SDF-1 for 15 min or 20 ng/mL HGF for 30 min or they were in control medium without factors. 1×10^4 RMS cells were added to the endothelial monolayers and incubated at 37 °C for 15 min. Plates were washed three times with phosphate–buffered saline to remove unbound cells and the fluorescence was read using a fluorescence plate reader (Spark™ 10M multimode microplate reader, Tecan, Männedorf, Switzerland) with excitation at 495 nm and emission intensity detected at 515 nm. The results were normalized to percentage of the cells in control conditions.

4.19. In Vivo Experiments

Animal experiments were approved by the Local Ethics Committee in Krakow (no 23/2013). To study metastasis, 1×10^6 RH30 cells were implanted intravenously into immunodeficient NOD-SCID mice for 7 days. Each experimental group contained five animals, and all of the experiments were repeated three times. The appearance of RMS cells in the lungs and in the bone marrow samples from the left and right legs was evaluated by real-time PCR using human GAPDH specific primers-probe set (Hs99999905_m1; Applied Biosystems) compared to murine GAPDH (Mm99999915_g1; Applied Biosystems) using ΔC_t method. The level of SDF-1 in murine lungs was evaluated by real-time PCR using murine SDF-1 primers-probe set (Mm00445552_m1; Applied Biosystems) compared to murine GAPDH (Mm99999915_g1; Applied Biosystems) using ΔC_t method.

4.20. Bioinformatical Analysis of Microarray Data from RMS Patients

For gene expression analysis in a group of 158 RMS patients we used data from GEO database, stored under accession number GSE92689 [62]. Background subtraction and data normalization was performed with affy [63] package in R/Bioconductor and the average expression was used for further statistical analysis. Pearson correlation of gene expression was analyzed using GraphPad Prism software (GraphPad Software, San Diego, CA, USA).

4.21. Statistical Analysis

Unless stated otherwise, the results show the mean ± standard error of the mean (SEM) of at least 2 to 4 independent biological experiments, as stated in figure legends (n value). Statistical analysis was performed by one-way analysis of variance (ANOVA) with Tukey or Dunnett's post-test or Student's t-test using GraphPad Prism software. Differences with a p-value less than 0.05 were considered statistically significant.

5. Conclusions

Here, we demonstrate for the first time that SNAIL stimulates the metastatic abilities of RMS cells both in vitro and in vivo. We propose a novel mechanism of its action that may be important in other tumor types. We postulate that, EZRIN, as a protein mediating between plasma membrane and cytoskeleton, PI3K-AKT signaling and miRNAs link SNAIL with the RMS cell migratory machinery.

Supplementary Materials: The following are available online at http://www.mdpi.com/2072-6694/12/7/1870/s1. Table S1. Table presenting SNAIL binding sites in RH30 cells evaluated by ChIP-seq. Table S2. Table presenting the results from next generation sequencing of miRNAs in RH30 WT, shCTRL and shSNAIL cells. Figure S1. SNAIL expression in RH30 cells after transduction with lentiviral vectors encoding shRNA. Figure S2. The effect of SNAIL silencing on motility of RH30 cells in a scratch assay. Figure S3. SDF-1 level in murine lungs. Figure S4. SNAIL and EZRIN levels after transfection of RMS cells with siRNA. Figure S5. Genome browser views of SNAIL binding to AKT2 promoter and its own promoter (ChIP-seq data). Figure S6. Supplementary Western blot images—images of the whole uncropped membranes with ladder.

Author Contributions: K.S. conceived the study, designed, planned and conducted most of the experiments in vitro and in vivo, analyzed and interpreted data, performed the statistical analysis and wrote the manuscript. M.K. participated in the flow cytometry analysis and prepared graphical abstract of the results. P.K. designed primers and performed bioinformatical analysis of data stored in GEO database. A.N. participated in the selected Western blot analysis. A.K. was involved in development of the stable cell line RH30 shSNAIL and evaluation of the MET and CXCR4 receptors levels. M.L. participated in the adhesion assays. W.B. participated in the selected Western blot analysis. U.J. and S.K.-K. analyzed the proteome of the RH30 cells. M.M. conceived the study, designed and coordinated the study, wrote manuscript. All authors have read and agreed to the published version of the manuscript.

Funding: The project was supported by the research grants from the National Science Centre in Poland to M.M.: 2013/09/B/NZ5/00769 and 2018/29/B/NZ5/00915 and to K.S.: 2015/17/D/NZ5/02202. The publication of this article was funded by the Priority Research Area qLife under the program "Excellence Initiative—Research University" at the Jagiellonian University in Krakow (application number 06/IDUB/2019/94).

Acknowledgments: We would like to acknowledge Malgorzata Sekula-Stryjewska for transduction of RH30 cells with lentiviral vectors encoding shRNA and Elzbieta Trzyna for validation of STR profile in RMS cells and technical help with ChIP assay and proteome analysis, Tomasz Adamus for his help in designing viral vectors encoding GFP and SNAIL, and Aleksander Galas for his advice about statistical analysis. KS was a recipient of the Foundation for Polish Science's START scholarship for outstanding young scientists and the Polish Ministry of Science and Higher Education's scholarship for outstanding young scientists.

Conflicts of Interest: The authors declare no competing interests.

References

1. Skapek, S.X.; Ferrari, A.; Gupta, A.A.; Lupo, P.J.; Butler, E.; Shipley, J.; Barr, F.G.; Hawkins, D.S. Rhabdomyosarcoma. *Nat. Rev. Dis. Prim.* **2019**, *5*, 1. [CrossRef] [PubMed]
2. Bailey, K.A.; Wexler, L.H. Pediatric rhabdomyosarcoma with bone marrow metastasis. *Pediatr. Blood Cancer* **2020**. [CrossRef]
3. Jankowski, K.; Kucia, M.; Wysoczynski, M.; Reca, R.; Zhao, D.; Trzyna, E.; Trent, J.; Peiper, S.; Zembala, M.; Ratajczak, J.; et al. Both Hepatocyte Growth Factor (HGF) and Stromal-Derived Factor-1 Regulate the Metastatic Behavior of Human Rhabdomyosarcoma Cells, but only HGF Enhances Their Resistance to Radiochemotherapy. *Cancer Res.* **2003**, *63*, 7926–7935.
4. Libura, J.; Drukala, J.; Majka, M.; Tomescu, O.; Navenot, J.M.; Kucia, M.; Marquez, L.; Peiper, S.C.; Barr, F.G.; Janowska-Wieczorek, A.; et al. CXCR4-SDF-1 signaling is active in rhabdomyosarcoma cells and regulates locomotion, chemotaxis, and adhesion. *Blood* **2002**, *100*, 2597–2606. [CrossRef]
5. Müller, A.; Homey, B.; Soto, H.; Ge, N.; Catron, D.; Buchanan, M.E.; McClanahan, T.; Murphy, E.; Yuan, W.; Wagner, S.N.; et al. Involvement of chemokine receptors in breast cancer metastasis. *Nature* **2001**, *410*, 50–56. [CrossRef]
6. Hettmer, S.; Wagers, A.J. Muscling in: Uncovering the origins of rhabdomyosarcoma. *Nat. Med.* **2010**, *16*, 171–173. [CrossRef]
7. Szewczyk, B.; Skrzypek, K.; Majka, M. Targeting MET Receptor in Rhabdomyosarcoma: Rationale and Progress. *Curr. Drug Targets* **2017**, *18*, 98–107. [CrossRef]

8. Yu, Y.; Khan, J.; Khanna, C.; Helman, L.; Meltzer, P.S.; Merlino, G. Expression profiling identifies the cytoskeletal organizer ezrin and the developmental homeoprotein Six-1 as key metastatic regulators. *Nat. Med.* **2004**, *10*, 175–181. [CrossRef]
9. Skrzypek, K.; Kusienicka, A.; Trzyna, E.; Szewczyk, B.; Ulman, A.; Konieczny, P.; Adamus, T.; Badyra, B.; Kortylewski, M.; Majka, M. SNAIL is a key regulator of alveolar rhabdomyosarcoma tumor growth and differentiation through repression of MYF5 and MYOD function. *Cell Death Dis.* **2018**, *9*, 643. [CrossRef] [PubMed]
10. Ulman, A.; Skrzypek, K.; Konieczny, P.; Mussolino, C.; Cathomen, T.; Majka, M. Genome Editing of the SNAI1 Gene in Rhabdomyosarcoma: A Novel Model for Studies of Its Role. *Cells* **2020**, *9*, 1095. [CrossRef]
11. Püsküllüoglu, M.; Lukasiewicz, E.; Miekus, K.; Jarocha, D.; Majka, M. Differential expression of Snail1 transcription factor and Snail1-related genes in alveolar and embryonal rhabdomyosarcoma subtypes. *Folia Histochem. Cytobiol.* **2010**, *48*, 671–677. [PubMed]
12. Xu, L.; Zheng, Y.; Liu, J.; Rakheja, D.; Singleterry, S.; Laetsch, T.W.; Shern, J.F.; Khan, J.; Triche, T.J.; Hawkins, D.S.; et al. Integrative Bayesian Analysis Identifies Rhabdomyosarcoma Disease Genes. *Cell Rep.* **2018**, *24*, 238–251. [CrossRef] [PubMed]
13. Peinado, H.; Ballestar, E.; Esteller, M.; Cano, A. Snail mediates E-cadherin repression by the recruitment of the Sin3A/histone deacetylase 1 (HDAC1)/HDAC2 complex. *Mol. Cell Biol.* **2004**, *24*, 306–319. [CrossRef] [PubMed]
14. Peinado, H.; Olmeda, D.; Cano, A. Snail, Zeb and bHLH factors in tumour progression: An alliance against the epithelial phenotype? *Nat. Rev. Cancer* **2007**, *7*, 415–428. [CrossRef]
15. Nieszporek, A.; Skrzypek, K.; Adamek, G.; Majka, M. Molecular mechanisms of epithelial to mesenchymal transition in tumor metastasis. *Acta Biochim. Pol.* **2019**, *66*, 509–520. [CrossRef]
16. Barrallo-Gimeno, A.; Nieto, M.A. The Snail genes as inducers of cell movement and survival: Implications in development and cancer. *Development* **2005**, *132*, 3151–3161. [CrossRef]
17. Skrzypek, K.; Majka, M. Interplay among SNAIL Transcription Factor, MicroRNAs, Long Non-Coding RNAs, and Circular RNAs in the Regulation of Tumor Growth and Metastasis. *Cancers* **2020**, *12*, 209. [CrossRef]
18. Olmeda, D.; Jordá, M.; Peinado, H.; Fabra, Á.; Cano, A. Snail silencing effectively suppresses tumour growth and invasiveness. *Oncogene* **2007**, *26*, 1862–1874. [CrossRef]
19. Lambies, G.; Miceli, M.; Martínez-Guillamon, C.; Olivera-Salguero, R.; Peña, R.; Frías, C.-P.; Calderón, I.; Atanassov, B.S.; Dent, S.Y.R.; Arribas, J.; et al. TGFβ-Activated USP27X Deubiquitinase Regulates Cell Migration and Chemoresistance via Stabilization of Snail1. *Cancer Res.* **2019**, *79*, 33–46. [CrossRef]
20. Batlle, R.; Alba-Castellón, L.; Loubat-Casanovas, J.; Armenteros, E.; Francí, C.; Stanisavljevic, J.; Banderas, R.; Martin-Caballero, J.; Bonilla, F.; Baulida, J.; et al. Snail1 controls TGF-β responsiveness and differentiation of mesenchymal stem cells. *Oncogene* **2013**, *32*, 3381–3389. [CrossRef]
21. Skrzypek, K.; Kusienicka, A.; Szewczyk, B.; Adamus, T.; Lukasiewicz, E.; Miekus, K.; Majka, M. Constitutive activation of MET signaling impairs myogenic differentiation of rhabdomyosarcoma and promotes its development and progression. *Oncotarget* **2015**, *6*, 31378–31398. [CrossRef]
22. Alba-Castellón, L.; Batlle, R.; Francí, C.; Fernández-Aceñero, M.J.; Mazzolini, R.; Peña, R.; Loubat, J.; Alameda, F.; Rodríguez, R.; Curto, J.; et al. Snail1 expression is required for sarcomagenesis. *Neoplasia* **2014**, *16*, 413–421. [CrossRef]
23. Ignatius, M.S.; Hayes, M.N.; Lobbardi, R.; Chen, E.Y.; McCarthy, K.M.; Sreenivas, P.; Motala, Z.; Durbin, A.D.; Molodtsov, A.; Reeder, S.; et al. The NOTCH1/SNAIL1/MEF2C Pathway Regulates Growth and Self-Renewal in Embryonal Rhabdomyosarcoma. *Cell Rep.* **2017**, *19*, 2304–2318. [CrossRef] [PubMed]
24. Bentzinger, C.F.; Wang, Y.X.; Rudnicki, M.A. Building muscle: Molecular regulation of myogenesis. *Cold Spring Harb. Perspect. Biol.* **2012**, *4*, a008342. [CrossRef]
25. Shankar, J.; Nabi, I.R. Actin Cytoskeleton Regulation of Epithelial Mesenchymal Transition in Metastatic Cancer Cells. *PLoS ONE* **2015**, *10*, e0119954.
26. Petty, J.M.; Sueblinvong, V.; Lenox, C.C.; Jones, C.C.; Cosgrove, G.P.; Cool, C.D.; Rai, P.R.; Brown, K.K.; Weiss, D.J.; Poynter, M.E.; et al. Pulmonary Stromal-Derived Factor-1 Expression and Effect on Neutrophil Recruitment during Acute Lung Injury. *J. Immunol.* **2007**, *178*, 8148–8157. [CrossRef] [PubMed]
27. Svitkina, T. The actin cytoskeleton and actin-based motility. *Cold Spring Harb. Perspect. Biol.* **2018**, *10*. [CrossRef]

28. Liu, L.; Jiang, H.; Zhao, W.; Meng, Y.; Li, J.; Huang, T.; Sun, J. Cdc42-mediated supracellular cytoskeleton induced cancer cell migration under low shear stress. *Biochem. Biophys. Res. Commun.* **2019**, *519*, 134–140. [CrossRef] [PubMed]
29. Mattila, P.K.; Lappalainen, P. Filopodia: Molecular architecture and cellular functions. *Nat. Rev. Mol. Cell Biol.* **2008**, *9*, 446–454. [CrossRef] [PubMed]
30. Dreos, R.; Ambrosini, G.; Groux, R.; Cavin Périer, R.; Bucher, P. The eukaryotic promoter database in its 30th year: Focus on non-vertebrate organisms. *Nucleic Acids Res.* **2017**, *45*, D51–D55. [CrossRef]
31. Kohsaka, S.; Shukla, N.; Ameur, N.; Ito, T.; Ng, C.K.Y.; Wang, L.; Lim, D.; Marchetti, A.; Viale, A.; Pirun, M.; et al. A recurrent neomorphic mutation in MYOD1 defines a clinically aggressive subset of embryonal rhabdomyosarcoma associated with PI3K-AKT pathway mutations. *Nat. Genet.* **2014**, *46*, 595–600. [CrossRef] [PubMed]
32. Villagrasa, P.; Díaz, V.M.; Viñas-Castells, R.; Peiró, S.; Del Valle-Pérez, B.; Dave, N.; Rodríguez-Asiain, A.; Casal, J.I.; Lizcano, J.M.; Duñach, M.; et al. Akt2 interacts with Snail1 in the E-cadherin promoter. *Oncogene* **2012**, *31*, 4022–4033. [CrossRef] [PubMed]
33. Frías, A.; Lambies, G.; Viñas-Castells, R.; Martínez-Guillamon, C.; Dave, N.; García de Herreros, A.; Díaz, V.M. A Switch in Akt Isoforms Is Required for Notch-Induced Snail1 Expression and Protection from Cell Death. *Mol. Cell. Biol.* **2016**, *36*, 923–940. [CrossRef] [PubMed]
34. Hermida, M.A.; Dinesh Kumar, J.; Leslie, N.R. GSK3 and its interactions with the PI3K/AKT/mTOR signalling network. *Adv. Biol. Regul.* **2017**, *65*, 5–15. [CrossRef]
35. Manning, B.D.; Toker, A. AKT/PKB Signaling: Navigating the Network. *Cell* **2017**, *169*, 381–405. [CrossRef]
36. Zhou, B.P.; Deng, J.; Xia, W.; Xu, J.; Li, Y.M.; Gunduz, M.; Hung, M.-C. Dual regulation of Snail by GSK-3β-mediated phosphorylation in control of epithelial–mesenchymal transition. *Nat. Cell Biol.* **2004**, *6*, 931–940. [CrossRef]
37. Ye, X.; Weinberg, R.A. The SUMO guards for SNAIL. *Oncotarget* **2017**, *8*, 97701–97702. [CrossRef]
38. Peiro, S.; Escrivà, M.; Puig, I.; Barberà, M.J.; Dave, N.; Herranz, N.; Larriba, M.J.; Takkunen, M.; Francí, C.; Muñoz, A.; et al. Snail1 transcriptional repressor binds to its own promoter and controls its expression. *Nucleic Acids Res.* **2006**, *34*, 2077–2084. [CrossRef]
39. Fröse, J.; Chen, M.B.; Hebron, K.E.; Reinhardt, F.; Hajal, C.; Zijlstra, A.; Kamm, R.D.; Weinberg, R.A. Epithelial-Mesenchymal Transition Induces Podocalyxin to Promote Extravasation via Ezrin Signaling. *Cell Rep.* **2018**, *24*, 962–972. [CrossRef]
40. Curto, M.; McClatchey, A.I. Ezrin...a metastatic detERMinant? *Cancer Cell* **2004**, *5*, 113–114. [CrossRef]
41. Ren, L.; Hong, S.H.; Cassavaugh, J.; Osborne, T.; Chou, A.J.; Kim, S.Y.; Gorlick, R.; Hewitt, S.M.; Khanna, C. The actin-cytoskeleton linker protein ezrin is regulated during osteosarcoma metastasis by PKC. *Oncogene* **2009**, *28*, 792–802. [CrossRef] [PubMed]
42. Croft, D.R.; Sahai, E.; Mavria, G.; Li, S.; Tsai, J.; Lee, W.M.F.; Marshall, C.J.; Olson, M.F. Conditional ROCK Activation In vivo Induces Tumor Cell Dissemination and Angiogenesis. *Cancer Res.* **2004**, *64*, 8994–9001. [CrossRef] [PubMed]
43. Sahai, E.; Marshall, C.J. Differing modes of tumour cell invasion have distinct requirements for Rho/ROCK signalling and extracellular proteolysis. *Nat. Cell Biol.* **2003**, *5*, 711–719. [CrossRef] [PubMed]
44. Li, Y.; Zhou, C.-X.; Gao, Y. Snail regulates the motility of oral cancer cells via RhoA/Cdc42/p-ERM pathway. *Biochem. Biophys. Res. Commun.* **2014**, *452*, 490–496. [CrossRef]
45. Tao, G.; Levay, A.K.; Gridley, T.; Lincoln, J. Mmp15 is a direct target of Snai1 during endothelial to mesenchymal transformation and endocardial cushion development. *Dev. Biol.* **2011**, *359*, 209–221. [CrossRef]
46. Rembold, M.; Ciglar, L.; Yáñez-Cuna, J.O.; Zinzen, R.P.; Girardot, C.; Jain, A.; Welte, M.A.; Stark, A.; Leptin, M.; Furlong, E.E.M. A conserved role for Snail as a potentiator of active transcription. *Genes Dev.* **2014**, *28*, 167. [CrossRef]
47. Subhra Das, S.; James, M.; Paul, S.; Chakravorty, N. miRnalyze: An interactive database linking tool to unlock intuitive microRNA regulation of cell signaling pathways. *Database (Oxford)* **2017**, *2017*. [CrossRef]
48. Wong, N.; Wang, X. miRDB: An online resource for microRNA target prediction and functional annotations. *Nucleic Acids Res.* **2015**, *43*, D146–D152. [CrossRef]
49. Bartolomé-Izquierdo, N.; de Yébenes, V.G.; Álvarez-Prado, A.F.; Mur, S.M.; Lopez del Olmo, J.A.; Roa, S.; Vazquez, J.; Ramiro, A.R. miR-28 regulates the germinal center reaction and blocks tumor growth in preclinical models of non-Hodgkin lymphoma. *Blood* **2017**, *129*, 2408–2419. [CrossRef]

50. Almeida, M.I.; Nicoloso, M.S.; Zeng, L.; Ivan, C.; Spizzo, R.; Gafà, R.; Xiao, L.; Zhang, X.; Vannini, I.; Fanini, F.; et al. Strand-Specific miR-28-5p and miR-28-3p Have Distinct Effects in Colorectal Cancer Cells. *Gastroenterology* **2012**, *142*, 886–896.e9. [CrossRef]
51. Hahn, S.; Jackstadt, R.; Siemens, H.; Hünten, S.; Hermeking, H. SNAIL and miR-34a feed-forward regulation of ZNF281/ZBP99 promotes epithelial-mesenchymal transition. *EMBO J.* **2013**, *32*, 3079–3095. [CrossRef] [PubMed]
52. Miekus, K.; Lukasiewicz, E.; Jarocha, D.; Sekula, M.; Drabik, G.; Majka, M. The decreased metastatic potential of rhabdomyosarcoma cells obtained through MET receptor downregulation and the induction of differentiation. *Cell Death Dis.* **2013**, *4*, e459. [CrossRef] [PubMed]
53. Dash, P.R.; Whitley, G.S.J.; Ayling, L.J.; Johnstone, A.P.; Cartwright, J.E. Trophoblast apoptosis is inhibited by hepatocyte growth factor through the Akt and β-catenin mediated up-regulation of inducible nitric oxide synthase. *Cell. Signal.* **2005**, *17*, 571–580. [CrossRef] [PubMed]
54. Desgrosellier, J.S.; Cheresh, D.A. Integrins in cancer: Biological implications and therapeutic opportunities. *Nat. Rev. Cancer* **2010**, *10*, 9–22. [CrossRef]
55. Hangan, D.; Uniyal, S.; Morris, V.L.; MacDonald, I.C.; von Ballestrem, C.; Chau, T.; Schmidt, E.E.; Chambers, A.F.; Groom, A.C.; Chan, B.M. Integrin VLA-2 (alpha2beta1) function in postextravasation movement of human rhabdomyosarcoma RD cells in the liver. *Cancer Res.* **1996**, *56*, 3142–3149.
56. Adamus, T.; Konieczny, P.; Sekuła, M.; Sułkowski, M.; Majka, M. The strategy of fusion genes construction determines efficient expression of introduced transcription factors. *Acta Biochim. Pol.* **2014**, *61*, 773–778. [CrossRef]
57. Skupien-Rabian, B.; Jankowska, U.; Swiderska, B.; Lukasiewicz, S.; Ryszawy, D.; Dziedzicka-Wasylewska, M.; Kedracka-Krok, S. Proteomic and bioinformatic analysis of a nuclear intrinsically disordered proteome. *J. Proteomics* **2015**, *130*, 76–84. [CrossRef]
58. Vizcaíno, J.A.; Csordas, A.; del-Toro, N.; Dianes, J.A.; Griss, J.; Lavidas, I.; Mayer, G.; Perez-Riverol, Y.; Reisinger, F.; Ternent, T.; et al. 2016 update of the PRIDE database and its related tools. *Nucleic Acids Res.* **2016**, *44*, D447–D456. [CrossRef]
59. Cote, R.G.; Griss, J.; Dianes, J.A.; Wang, R.; Wright, J.C.; van den Toorn, H.W.P.; van Breukelen, B.; Heck, A.J.R.; Hulstaert, N.; Martens, L.; et al. The PRoteomics IDEntification (PRIDE) Converter 2 Framework: An Improved Suite of Tools to Facilitate Data Submission to the PRIDE Database and the ProteomeXchange Consortium. *Mol. Cell. Proteomics* **2012**, *11*, 1682–1689. [CrossRef]
60. Sandelin, A.; Wasserman, W.W.; Lenhard, B. ConSite: Web-based prediction of regulatory elements using cross-species comparison. *Nucleic Acids Res.* **2004**, *32*, W249–W252. [CrossRef]
61. Robinson, J.T.; Thorvaldsdóttir, H.; Winckler, W.; Guttman, M.; Lander, E.S.; Getz, G.; Mesirov, J.P. Integrative genomics viewer. *Nat. Biotechnol.* **2011**, *29*, 24–26. [CrossRef] [PubMed]
62. Davicioni, E.; Finckenstein, F.G.; Shahbazian, V.; Buckley, J.D.; Triche, T.J.; Anderson, M.J. Identification of a PAX-FKHR gene expression signature that defines molecular classes and determines the prognosis of alveolar rhabdomyosarcomas. *Cancer Res.* **2006**, *66*, 6936–6946. [CrossRef] [PubMed]
63. Gautier, L.; Cope, L.; Bolstad, B.M.; Irizarry, R.A. Affy–analysis of Affymetrix GeneChip data at the probe level. *Bioinformatics* **2004**, *20*, 307–315. [CrossRef] [PubMed]

cancers

MDPI

Article

Tspan8 Drives Melanoma Dermal Invasion by Promoting ProMMP-9 Activation and Basement Membrane Proteolysis in a Keratinocyte-Dependent Manner

Manale El Kharbili [1,2,†], Muriel Cario [3,4,†], Nicolas Béchetoille [5], Catherine Pain [3], Claude Boucheix [6], Françoise Degoul [7], Ingrid Masse [1,8,*,‡] and Odile Berthier-Vergnes [1,9,‡]

1 Centre de Génétique et de Physiologie Moléculaires et Cellulaires, CNRS UMR5534, Université de Lyon, F-69003 Lyon, France; manale.elkharbili@cuanschutz.edu (M.E.K.); o.vergnes@icloud.com (O.B.-V.)
2 Department of Dermatology, University of Colorado Anschutz Medical Campus, Aurora, CO 80045, USA
3 National Reference Center for Rare Skin Disease, Department of Dermatology, University Hospital, INSERM 1035, F-33000 Bordeaux, France; muriel.cario-andre@u-bordeaux.fr (M.C.); paincatherine4@gmail.com (C.P.)
4 AquiDerm, University Bordeaux, F-33076 Bordeaux, France
5 R&D Department, Gattefossé, F-69800 Saint-Priest, France; nbechetoille@gattefosse.com
6 INSERM U935, Université Paris-Sud, F-94800 Villejuif, France; claude.boucheix@inserm.fr
7 INSERM U1240, Université Clermont Auvergne, Imagerie Moléculaire et Stratégies Théranostiques, F-63000 Clermont Ferrand, France; francoise.degoul@inserm.fr
8 Centre de Recherche en Cancérologie de Lyon, CNRS-UMR5286, INSERM U1052, Université de Lyon, F-69008 Lyon, France
9 US7INSERM /UMS3453 UCBL SFR Santé Lyon-Est, F-69372 Lyon, France
* Correspondence: ingrid.masse@univ-lyon1.fr
† These two authors contributed equally to this paper.
‡ These two authors contributed equally to this paper.

Received: 17 April 2020; Accepted: 16 May 2020; Published: 21 May 2020

check for updates

Abstract: Melanoma is the most aggressive skin cancer with an extremely challenging therapy. The dermal-epidermal junction (DEJ) degradation and subsequent dermal invasion are the earliest steps of melanoma dissemination, but the mechanisms remain elusive. We previously identified Tspan8 as a key actor in melanoma invasiveness. Here, we investigated Tspan8 mechanisms of action during dermal invasion, using a validated skin-reconstruct-model that recapitulates melanoma dermal penetration through an authentic DEJ. We demonstrate that Tspan8 is sufficient to induce melanoma cells' translocation to the dermis. Mechanistically, Tspan8$^+$ melanoma cells cooperate with surrounding keratinocytes within the epidermis to promote keratinocyte-originated proMMP-9 activation process, collagen IV degradation and dermal colonization. This concurs with elevated active MMP-3 and low TIMP-1 levels, known to promote MMP-9 activity. Finally, a specific Tspan8-antibody reduces proMMP-9 activation and dermal invasion. Overall, our results provide new insights into the role of keratinocytes in melanoma dermal colonization through a cooperative mechanism never reported before, and establish for the first time the pro-invasive role of a tetraspanin family member in a cell non-autonomous manner. This work also displays solid arguments for the use of Tspan8-blocking antibodies to impede early melanoma spreading and therefore metastasis.

Keywords: tumor microenvironment; melanoma invasion; melanoma-keratinocytes crosstalk; dermal-epidermal junction; dermis; Tetraspanin 8; MMP-9

1. Introduction

Cutaneous melanoma is the deadliest skin cancer due to its high metastatic propensity and resistance to most conventional and targeted therapies [1]. It usually progresses from an early radial growth phase (RGP) confined within the epidermis to a vertical growth phase (VGP) characterized by dermal invasion, where metastasis risk is high [2]. To date, Breslow thickness remains the most powerful prognostic factor, as long as metastases are not present at the time of diagnosis. The most recent AJCC 8th guidelines introduced mitotic rate as an additional criterion for thinner melanomas, the presence of >1 mitosis/mm^2 predicts poorer outcome [3]. Moreover, the ulceration status used for the sub-classification of thin melanomas [3] emerges as another important histological factor predicting survival [4]. However, such histological features define prognostic groups but not individual patient risk. Indeed, even though the survival rate for thin melanomas is high, some patients develop metastases [3]. Moreover, Werner-Klein et al. [5] recently showed that dissemination occurs shortly after dermal invasion at a median tumor thickness of ~0.5 mm. Therefore, understanding the mechanisms that convert RGP melanoma into VGP is crucial to identify reliable predictive biomarkers and novel therapeutic targets.

Cutaneous melanomas are composed of genotypically and phenotypically distinct subpopulations, dynamically regulated by the selective pressure imposed from the host tumor microenvironment and host immune system [6]. This tumor heterogeneity contributes largely to their strong resistance to standard, targeted and immune therapies [7,8]. Indeed, it appears that cancer/immune cell interactions are informative of resistance to immunotherapy whereas cancer/stromal cell interactions are informative of MAPK inhibitors' resistance [9]. Consistently with the high inter- and intra-tumoral heterogenity of cutaneous melanomas, we have previously defined a subset of melanoma cells expressing strong levels of peanut agglutinin-receptors that possesses a high metastatic frequency [10] and correlates with poor patient survival [11], which simultaneously express Tetraspanin 8 (Tspan8) [12]. Tspan8 belongs to a four-transmembrane-domain protein family called tetraspanins, that organize membrane microdomains via interactions with other tetraspanins and a variety of transmembrane/cytosolic proteins to regulate a wide range of cellular functions, including proliferation, motility, metastasis and angiogenesis [13,14]. Tspan8 is categorized as pro-metastatic in various carcinomas [14] and emerged as an attractive therapeutic target [15,16] and a blood biomarker [17].

We were the first to reveal that Tspan8 expression is sufficient to transform non-invasive melanoma cells into invasive cells [12]. Tspan8 is undetectable at both mRNA and protein in healthy skin, but its expression is acquired by aggressive primary melanomas and lymph node metastases. We also demonstrated that *TSPAN8* is under the transcriptional control of LCMR1 and p53 [18,19] and acts not only by reducing matrix adherence via the β1-integrin/ILK signaling pathway [20], but also by promoting invasion through β-catenin activation [21].

It is accepted that reciprocal stroma–tumor interactions contribute to metastatic progression, especially through the production of matrix degrading enzymes such as MMPs [22,23]. However, the exact mechanisms governing the interplay between melanoma cells and epidermal microenvironment in controlling MMP-dependent invasion have not been studied to date. Here, we address how Tspan8 participates in the dermal–epidermal junction (DEJ) proteolysis during melanoma invasion and whether it contributes to tumor–keratinocyte crosstalk. To this aim, we used 3D-skin reconstructs (SR) with an authentic DEJ, which recapitulate early melanoma stages [24,25]. We found that mere Tspan8 gain of expression is sufficient to promote melanoma invasive behavior and acts by driving proMMP-9 activation leading to DEJ proteolysis. More importantly, we showed that Tspan8 function hinges on the dialog between tumor cells and neighboring keratinocytes. Our work provides strong evidence of the primary involvement of Tspan8 in melanoma–keratinocyte crosstalk leading to efficient DEJ degradation. This is, to our knowledge, the first report demonstrating bidirectional interplay between melanoma cells and epidermal microenvironment to regulate MMP-dependent invasion. This is also the first study characterizing the role of a tetraspanin family member in a cell non-autonomous mechanism that controls basement membrane proteolysis and local invasion.

2. Results

2.1. Tspan8 is Exclusively Expressed in the In Vivo-Selected Highly Metastatic and Invasive Melanoma Subsets

We previously developed an orthotopic rat model for the spontaneous metastasis of human melanoma [10]. This model allowed the selection from a non-aggressive parental cell line of subpopulations with low (NM#1, NM#2, NM#3) or high (M#1, M#2, M#3) lung metastatic potential. Figure 1a depicts a schematic of the selection procedures. M#1, M#2 and M#3 subsets expressed Tspan8 at the mRNA (Figure 1b), protein (Figure 1c), cell-surface (Figure 1d) levels, and displayed a high ability to invade Matrigel (Figure 1e), unlike the parental line and the non-metastatic NM#1, NM#2, NM#3 subsets. These results showed that the parental line is populated by melanoma cells with heterogeneous metastatic phenotypes and that Tspan8 is strongly expressed in the invasive/metastatic subsets.

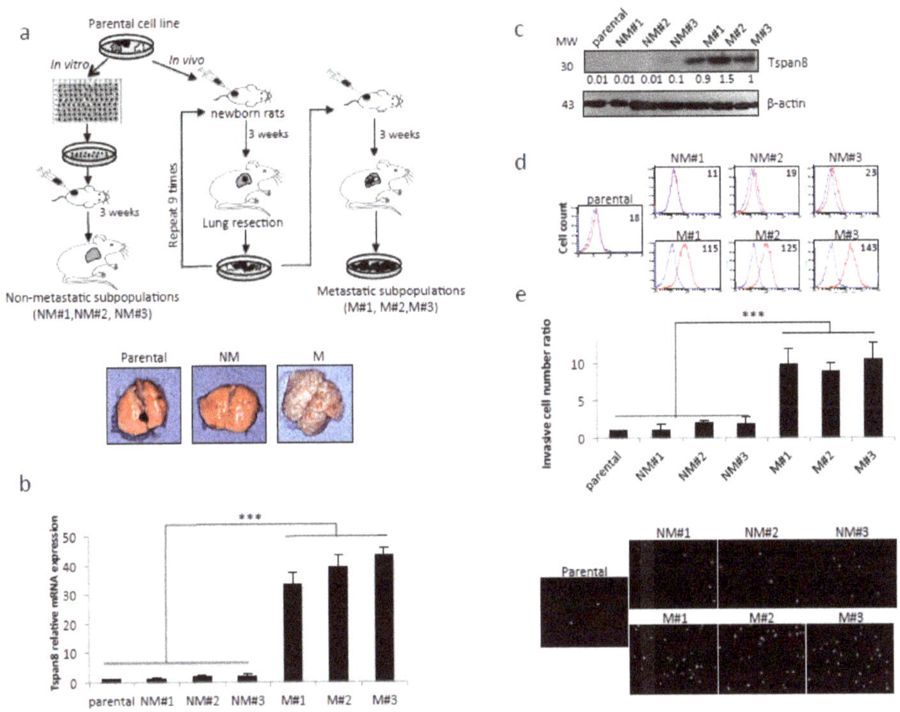

Figure 1. Generation of potent metastatic cell subpopulations expressing the metastatic-associated Tspan8 protein. (**a**) Schematic diagram of the experimental procedure used to sequentially select in an immunosuppressed new-born rat model cell subpopulations with progressively higher metastatic ability from a poorly metastatic melanoma cell line. Lower panel, representative photographs of the rat lungs. (**b**) The parental human M4Be cell line and its derived non metastatic (NM#1-3) and metastatic (M#1-3) subpopulations were examined for *TSPAN8* mRNA levels by QPCR. Expression normalized to GAPDH represented a fold change of control sample (n = 3; ± SD); (**c**) Western blot analysis of Tspan8 expression with β-Actin as loading control and reference for quantification (one representative experiment of three), uncropped western blots figures in Figure S1; (**d**) Tspan8 cell surface expression by flow cytometry analysis. In red, the specific staining and in blue the isotype-matched control antibody (one representative experiment of three). Numbers indicate Mean Fluorescence Intensity (MFI). (**e**) Matrigel invasion assay using transwell chambers. The total number of invasive cells was integrally counted by scanning microscopy and normalized to the value from control parental cell line (n = 3; ± SEM). Representative visual fields are illustrated beneath. ***$p < 0.001$.

2.2. Tspan8 Expression in Melanoma Cells Promotes ProMMP-9 Activation, Collagen IV Degradation and DEJ Crossing

We next determined how Tspan8 expression affects dermal invasion. We used the SR, previously described [24,25] to accurately recapitulate the early steps of melanoma invasion through a preserved 3D architecture of native DEJ. After 21 days of culture, Tspan8$^+$ cells invaded the dermis and formed numerous compact nodules (Figure 2a). By contrast, large nests of Tspan8$^-$ cells were located exclusively in the epidermis, along the DEJ. Evidence of collagen IV dissolution, the major DEJ component, was observed exclusively when Tspan8$^+$ cells were used and integrated within the epidermis (Figure 2b). These data demonstrate that keratinocytes are required for the penetration of Tspan8$^+$ cells across the DEJ, and local degradation of collagen IV, both occurring around day 21.

Collagen IV is known to be primarily degraded by MMP-9 and MMP-2. Therefore, we examined whether Tspan8 regulates MMP-9 and MMP-2 expression and/or activity by using zymographs and ELISA assays on conditioned media harvested from SR. A time-course study revealed that proMMP-9 became active as early as day 10 of culture and drastically increased until reaching active MMP-9 highest amount at the time of collagen IV dissolution (day 21), exclusively in medium from SR containing Tspan8$^+$ cells (Figure 2c,d). Indeed, SR integrating Tspan8$^-$ cells were capable of producing proMMP-9, but unable to generate its active form (Figure 2c,d), in accordance with the intact collagen IV layer (Figure 2b).

Strikingly, melanoma cells were unable to cross the DEJ when keratinocytes were not incorporated, regardless of their Tspan8 expression levels (Figure 2a), even after 5 weeks of culture (not shown). This was consistent with the absence of breaks in collagen IV staining (Figure 2b), low levels of proMMP-9 and absence of active MMP-9 (Figure 2c,d). The MMP-2 proform was detected at very low levels without noticeable active MMP-2 irrespective of Tspan8 expression, at all times and in all tested culture conditions (Figure 2c,e). These data demonstrate that Tspan8 is a key determinant in the activation process of MMP-9, but not of MMP-2, which depends heavily on surrounding keratinocytes.

To further confirm that Tspan8 confers MMP-9-dependent invasive activity, we have generated stable clones expressing ectopic Tspan8 or depleted of endogenous Tspan8. We confirmed the efficiency of Tspan8 expression/silencing at the mRNA (Figure 3a), protein (Figure 3b) and cell-surface (Figure 3c) levels. We observed that non-invasive melanoma cells gained strong invasive properties after the ectopic expression of Tspan8 in matrigel (Figure 3d) and SR (Figure 3e) concomitantly to the production of high levels of active MMP-9 (Figure 3f). Conversely, Tspan8 depletion in invasive cells strongly inhibited matrigel invasion (Figure 3d) and efficiently prevented melanoma cells from crossing the DEJ (Figure 3e), concurrently to a drastic decrease in proMMP-9 activation (Figure 3f). Overall, these data show that, by itself, Tspan8 expression was sufficient to trigger proMMP-9 activation process and collagen IV dissolution, allowing melanoma cells to cross DEJ and invade dermis.

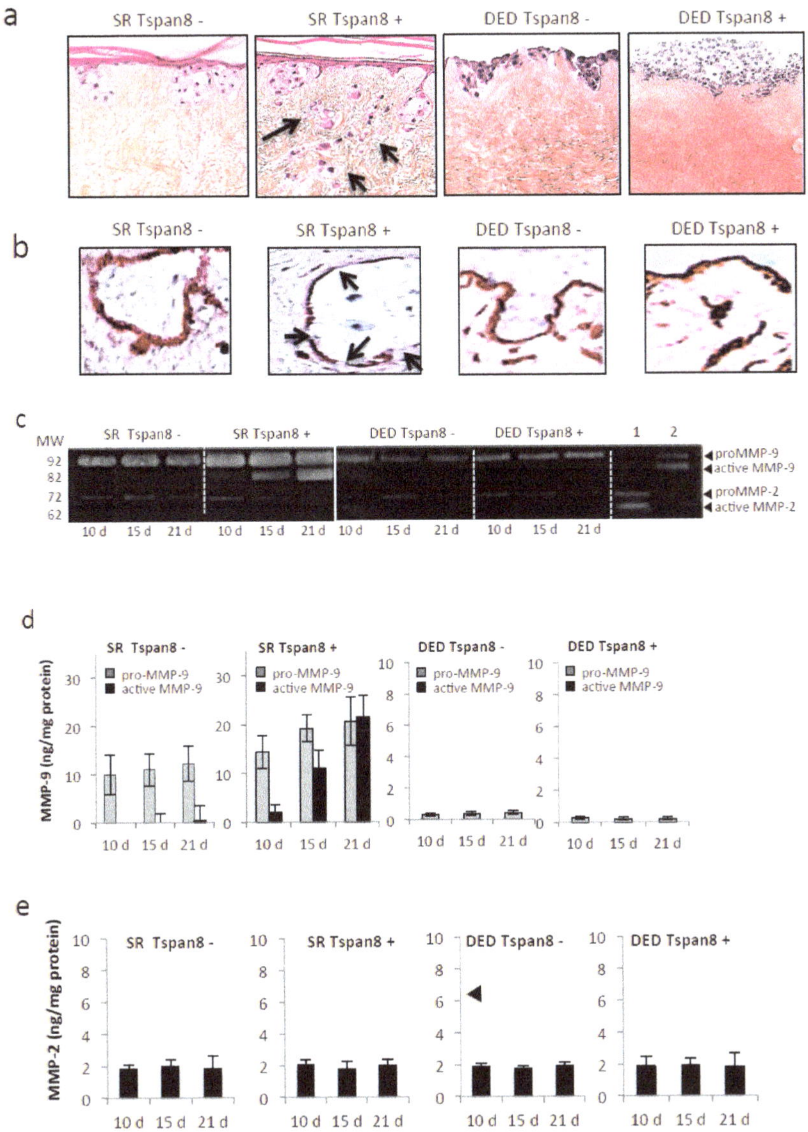

Figure 2. Tspan8-expressing melanoma cells efficiently invade the dermis in human skin reconstructs. Melanoma cells from NM#1 (Tspan8-) and M#1 (Tspan8+) subpopulations were cultured with human keratinocytes (SR) or alone (DED) on acellular dermis. (**a**) Representative photomicrographs of hematoxylin and eosin (H&E)-stained 21-day skin composites (scale bars: 100 μm). Arrows indicate melanoma cells colonizing the dermis (**b**) Representative IHC-staining of collagen IV. Arrows denote collagen IV layer disruptions. (**c**) MMP-9 and MMP-2 activity in gelatin zymography of culture medium from skin composites collected on day 10, 15 and 21. Lane 1, purified MMP-2 standard; lane 2, purified MMP-9 standard both activated with 4-aminophenylmercuric acetate. (**d**,**e**) ELISA quantification of secreted protein levels of proMMP-9, active MMP-9 (d) and MMP-2 (e) (ng/ug total protein) into the composite media. Bars represent the mean ± SD of three separate experiments with 3 ELISA evaluations for each of the 3 independent experiments (n = 9).

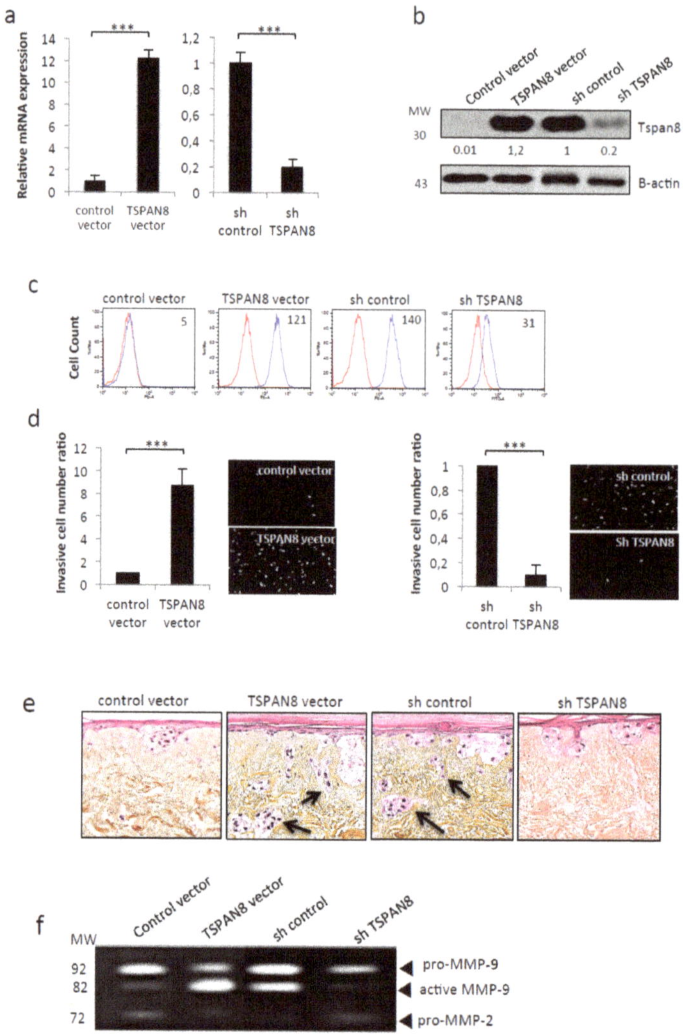

Figure 3. Tspan8 expression in melanoma triggers dermal invasion, concomitantly to MMP-9 activation and collagen IV proteolysis. Non-metastatic stable clones ectopically-expressing Tspan8 (TSPAN8 vector) or not (control vector) Tspan8 (left panel) and metastatic stable clones silenced (shTSPAN8) or not (shcontrol) for Tspan8 (right panel) were subjected to (**a**) QPCR analysis of *TSPAN8* transcripts levels ($n = 3$; mean ± s.d.). (**b**) Western blot analysis of Tspan8 protein levels with β-Actin as loading control. The band intensities were normalized to actin signal (representative experiment of three), uncropped western blots figures in Figure S1 (**c**) Flow cytometry analysis of cell surface Tspan8 expression (representative experiment of three). (**d**) Matrigel cell invasion assay: invading cells were DAPI-stained (right panel) and quantified (left panel). Data are means ± SD with $n = 3$ (*** $p < 0.001$). (**e**) Cells were incorporated into the epidermis of skin reconstructs as described in Materials and Methods. Representative hematoxylin and eosin-stained skin reconstruct were shown. Arrows denote melanoma cells located into the dermis. (**f**) Serum-free conditioned media collected from SR were analyzed by gelatin zymography at 21 days (representative zymogram of 3 independent experiments). Molecular weight (MW) markers are indicated in kDa.

2.3. Tspan8+ Melanoma Cells Require Neighboring Keratinocytes to Promote Dermal Invasion

Tspan8+ cells crossed DEJ exclusively when integrated into an epidermal microenvironment (Figure 2a). We thus investigated whether and how keratinocytes influence melanoma invasion. We developed four different models schematized in Figure 4a. Keratinocytes and Tspan8+ melanoma cells were either cultured alone on de-epidermized dermis (DED) (I and II cultures, respectively) or cocultured, without or with cell–cell contacts (III and IV cultures, respectively). As shown in Figure 4b, Tspan8+ cells were able to penetrate the DEJ only when surrounded by keratinocytes, which coincided with collagen IV breakdown (Figure 4c). Strikingly, when cultured alone or cocultured with keratinocytes without contacts, melanoma cells formed an attached layer along the JDE, several cells thick without noticeable dermal invasion (Figure 4b), nor breaks in collagen IV layer (Figure 4c).

Zymography (Figure 4d), ELISA (Figure 4e) and western blot (Figure 4f) assays revealed that keratinocytes are the major source of proMMP-9 and that active MMP-9 was generated exclusively when Tspan8+ cells were surrounded with keratinocytes (Figure 4d–f). ProMMP-2, detected at low levels in the four types of culture, remained stable throughout the experiment (Figure 4d). Overall, our results indicate that interaction between Tspan8+ melanoma cells and neighboring keratinocytes are essential to drive MMP-9 activation, collagen IV dissolution, and subsequent dermal invasion.

2.4. Tspan8 Expression in Melanoma Cells Surrounded with Keratinocytes Promotes ProMMP-9 Activation by Increasing the Amount of Active MMP-3 and Decreasing TIMP-1 Levels

Since MMP-3 can activate in vitro proMMP-9 but not proMMP-2 [26,27] and because it is the most relevant activator of pro-MMP-9 in vivo [28,29], we wondered whether in our model proMMP-9 activation was MMP-3-dependent. Thus, total MMP-3 levels were measured in the media derived from our four culture models. MMP-3 highest amounts were observed when Tspan8+ melanoma cells were integrated with keratinocytes into the SR, peaking at day 21 (Figure 5a). MMP-3 was undetectable when Tspan8+ cells were cultured alone. MMP-3 activation status was examined by Western blot (Figure 5b) and we observed a 52 kDa band corresponding to the molecular weight of proMMP-3 in all types of cultures, except the culture with melanoma cells alone, indicating that proMMP-3 is generated by keratinocytes and not melanoma cells. However, the 28 kDa band representing the fully activated form of MMP-3 [30] was restricted to co-cultures where Tspan8+ cells were surrounded with keratinocytes (Figure 5b). Importantly, MMP-3 full activation is Tspan8-dependent as SR generated with non-invasive melanoma cells ectopically expressing Tspan8 acquired the property to produce a large amount of fully active MMP-3 (Figure 5c). Concordantly, that property observed in SR containing Tspan8+ melanoma cells was abrogated when Tspan8 expression was silenced (Figure 5c).

It was emphasized that active MMP-3 becomes a potent activator of proMMP-9 in a tumor cell model only when its concentration exceeds that of TIMP-1 [28]. We thus wondered whether active MMP-9, exclusively observed when Tspan8+ melanoma cells were surrounded by keratinocytes, coincided with low levels of TIMP-1. Indeed, of our four culture models, we found that secreted TIMP-1 is at its highest level when Tspan8+ cells are alone and at its lowest when they are in direct contact with keratinocytes (Figure 5d,e). Overall, our data show that Tspan8+ melanoma cells surrounded by keratinocytes maintained lower TIMP-1 level when compared to melanoma cells juxtaposed without contacts with keratinocytes, thus favoring proMMP-9 activation by the fully active MMP-3, in a Tspan8-dependent manner.

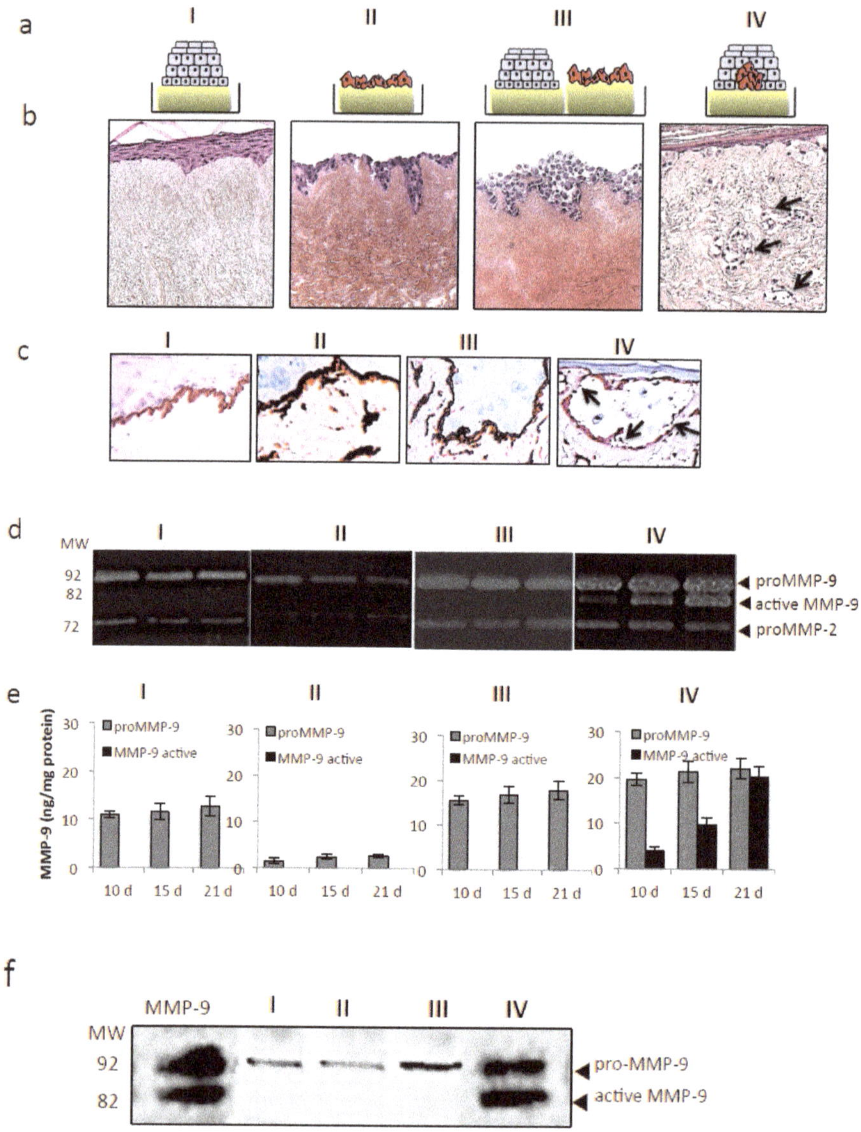

Figure 4. Tspan8-dependent dermal invasion coincides with MMP-9 activity and local dissolution of collagen IV and requires surrounding keratinocytes. (**a**) Schematic drawings of the four different culture conditions. I: SR containing no melanoma cells; II: Tspan8-expressing cells seeded alone on DED (DeEpidermised Dermis); III: SR without melanoma cells juxtaposed with Tspan8+ cells seeded alone on DED; IV: SR containing Tspan8+ cells in contact with keratinocytes. (**b**) Representative H&E staining of skin composites sections. Arrows indicate melanoma cells infiltrating the dermis. (**c**) type IV collagen staining on sections from the four culture conditions described in (**a**). Arrowhead pointed to collagen IV destruction. (**d–f**) Serum-free media from the four culture conditions collected at day 10, 15 and 21 were analyzed for the expression levels of proMMP-9 and active MMP-9 using gelatin zymography (**d**), ELISA (**e**) and western blot (**f**).

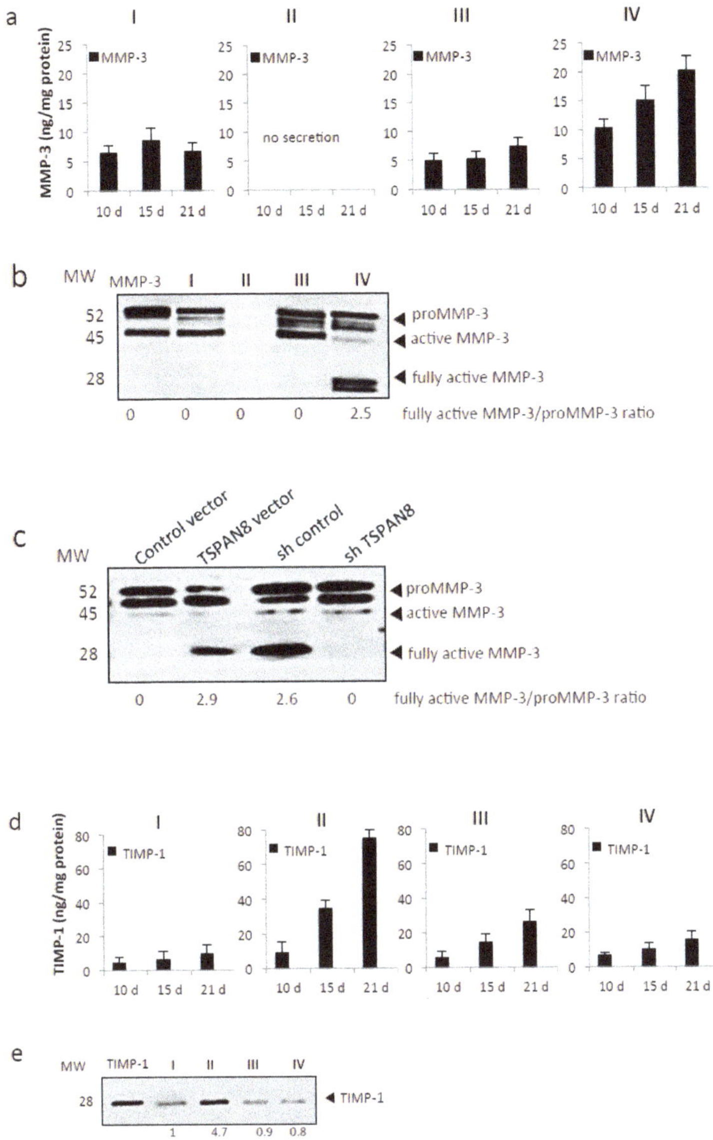

Figure 5. Tspan8-mediated melanoma cell invasion coincides with MMP-3 activation concomitantly with low TIMP-1. (**a**,**b**) The protein levels of total MMP-3 released in the supernatants from each culture model (I, II, II, IV described in Figure 4) were assessed at day 10, 15 and 21 by ELISA (**a**) and western blot (**b**), uncropped western blots figures in Figure S1. (**c**) Immunoblot analysis of MMP-3 in serum-free media harvested from SR integrating non-metastatic NM#1 melanoma cells ectopically expressing Tspan8 (TSPAN8 vector) and their control (control vector) or metastatic M#1 melanoma cells silenced (shTSPAN8) or not (shcontrol) for Tspan8. Equal amounts of total protein were loaded. The intensity value ratio of fully active MMP-3/proMMP-3 was annoted beneath the blot (**d**,**e**). Serum-free media from the 4 culture models collected at day 10, 15 and 21 were analyzed for TIMP-1 content by ELISA (**d**) and western blot (**e**). ELISA results are represented as the mean ± SEM from three independent experiments, each measured in duplicate.

2.5. Keratinocytes are the Main Source of ProMMP-9 and ProMMP-3 Whereas Tspan8+ Melanoma Cells Are the Primary Source of TIMP-1

We next evaluated the cellular source of MMP-9, its activator MMP-3 and its inhibitor TIMP-1. To this end, their respective transcript levels in keratinocytes alone (aK), Tspan8$^+$ melanoma cells alone (bM), keratinocytes cocultured with melanoma cells without contact (cK and cM respectively), and melanoma cells that have penetrated the DEJ (dM) (Figure 6a) were measured by RT-QPCR. *MMP-9* and *MMP-3* were mainly expressed by keratinocytes whereas *TIMP-1* was mainly expressed by Tspan8$^+$ melanoma cells (Figure 6b). The presence of keratinocytes, irrespective of contacts with melanoma cells, slightly augmented the *MMP-9* mRNA levels in keratinocytes and sorely decreased *TIMP-1* transcription in invasive melanoma cells. Surprisingly, *MMP-9* and *MMP-3* transcripts were increased in invading melanoma cells that have penetrated the DEJ, in comparison to melanoma cells still in contact with the DEJ. This was consistent with the immunodetection of MMP-9 and MMP-3 in melanoma cells located into the dermis, but not those localized in the epidermis (Figure 6c). This indicates that Tspan8$^+$ melanoma cells, once invading the dermis, acquire the capability of expressing the precursor forms of MMP-9 and MMP-3, which were previously provided by the keratinocytes when situated in the epidermis.

2.6. Antibody-Specific Blockade of Tspan8 Reduces ProMMP-9 Activation and Melanoma Invasion

We next examined whether a blocking monoclonal anti-Tspan8 antibody, previously shown to be effective in delaying the growth of human colon xenografts [31], could influence melanoma invasion. First, we showed that it allows efficient selective in vivo imaging of Tspan8$^+$ human melanoma xenografts, demonstrating its high target specificity (Figure 7a). When this antibody was added to the culture medium of SR, Tspan8$^+$ M#1 cells grow as clusters in the epidermis without deeply invading the dermis, whereas isotype-matching control Ab-treated SR invaded the dermis by day 10 and progressed deeper by day 20 (Figure 7b). The invasion score confirmed that Tspan8$^+$ cells treated with Tspan8 mAb exhibited minimal invasion (mean score 1.83; $n = 6$) when compared to cells cultured with control mAb (mean score 0.67, $n = 6$) in a statistically significant manner ($p = 0{,}01267$; paired t-test; Figure 7b). These findings were extended to the SKMel28 cell line, broadly used for its ability to invade the dermis of SR [25], and revealed to be Tspan8$^+$ [12]. As depicted in Figure 7c, SKMel28 cells in SR displayed vertically orientated clusters in the upper dermis when control mAb was added. In contrast, tumor nodules remained close to DEJ with less dermal invasion when treated with Tspan8 mAb (Figure 7c). Importantly, Tspan8-mAb treatment correlated with a strong reduction in MMP-9 activation in the SR integrating M#1 and SKMel28 cells (Figure 7d).

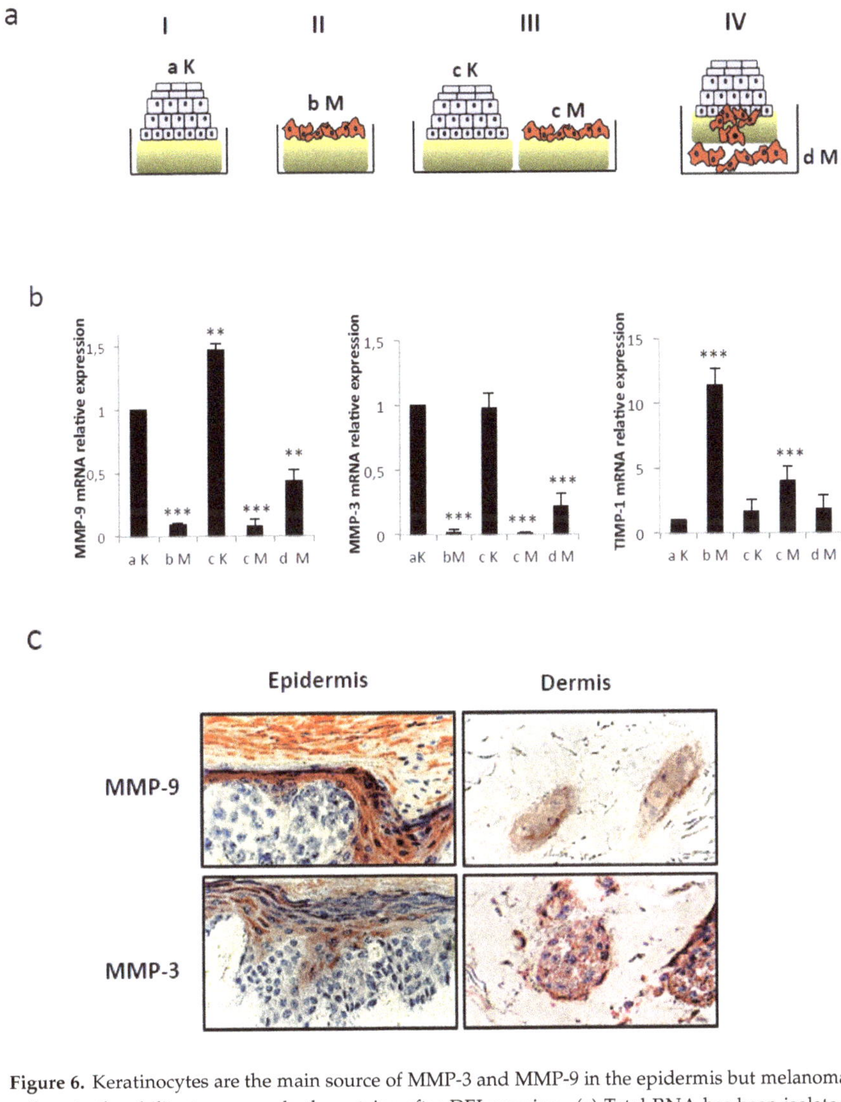

Figure 6. Keratinocytes are the main source of MMP-3 and MMP-9 in the epidermis but melanoma cells gain the ability to express both proteins after DEJ crossing. (**a**) Total RNA has been isolated from keratinocytes and Tspan8+ melanoma cells at day 20 from the four different schematized culture conditions. aK: keratinocytes from culture I; bM: Tspan8+ melanoma cells from culture II; cK and cM: keratinocytes and Tspan8+ melanoma cells from culture III respectively; dM: invading melanoma cells from culture IV. (**b**) QPCR analysis of *MMP-9*, *MMP-3*, and *TIMP-1* transcript expression levels of aK, bM, cK, cM and dM (n =3; ± SD). (**c**) Representative pictures of immunohistochemical staining of MMP-9 and MMP-3 in the epidermis and dermis of SR integrating Tspan8+ melanoma cells (condition IV) at day 20. ** $p < 0.01$, *** $p < 0.001$.

cells were integrated into SR and expressed Tspan8 (Figure S2). These findings are consistent with prior data demonstrating that TIMP-1 overexpression in B16-F10 melanoma cells reduces their invasive capacity [53] and their metastatic potential [54]. Overall, our results strongly suggest that, in the epidermis, Tspan8$^+$ melanoma cells cooperate with surrounding keratinocytes to promote dermal invasion by instigating MMP-3 activation and strongly decreasing TIMP-1 expression, with both events leading to a keratinocyte-originated MMP-9 activation process, and subsequent DEJ penetration. A striking finding is that aggressive cells capable of invading the DEJ to reach the dermis also gain the ability to express MMP-9 and its activator MMP-3 with reduced TIMP-1 levels. This is consistent with previous data conducted in mice, where the selection process for metastatic subclones favors those expressing MMP-9 [40] and those expressing Tspan8 (Figure 1). This profile should allow them to escape the local tissue control, and thus degrade the basal membranes encountered later throughout the metastatic cascade by themselves.

A major discovery in this work is that the ability/inability to cross DEJ is interconvertible and that the switch from one state to another can be accomplished both at the functional and molecular level by simply manipulating Tspan8 expression. Accordingly, a Tspan8-specific antibody efficiently targeting in vivo Tspan8$^+$ melanoma xenografts was able to reduce MMP-9 activity, DEJ breakdown, and dermal invasion. Given that MMPs inhibitors are not highly selective and did not impede a single MMPs function [55,56], it is tempting to speculate that targeting Tspan8 with antibodies might represent an alternative means to specifically block MMP-9 activity, and thereby deeper melanoma invasion of the dermis, the earliest stage before metastatic spreading.

4. Materials and Methods

4.1. Cell lines and Culture

SKMel28 (ATCC, Manassas, VA, USA) and M4Be [10] human melanoma cell lines were derived from lymph node metastasis. Non-metastatic and metastatic subpopulations were selected from immunosuppressed newborn rats that had been subcutaneously injected with M4Be (parental) cells, from lung metastases collected and grown in culture as described previously [57]. Stable clones of human melanoma cells were generated with shRNA-mediated silencing or ectopic overexpression of Tspan8 as described elsewhere [21]. Cells were cultured under standard conditions and tested as mycoplasma-free.

4.2. Matrigel Invasion Assay

Invasion assays were performed in triplicates using BioCoat Matrigel invasion chambers (BD Biosciences) as previously described [21]. Briefly, the cells that migrated to the lower surface of the filter were fixed, stained with DAPI, imaged using an Axiovert 200 (Carl Zeiss Inc., Jena, Germany) equipped with a CoolSNAP HQ camera (Roper Scientific, Lisses, France) and MetaMorph software (MDS Analytical Technologies, Sunnyvale, CA, USA) and then counted on the entire filter using NIH Image J software.

4.3. Invasion Assay in Human Skin Reconstructs

Adult human keratinocytes (4×10^5 cells), mixed or not with human melanoma cells (5.820 cells) at a melanoma/keratinocyte ratio of 1:80, were seeded into a stainless-steel ring deposited on the surface of human dead de-epidermized dermis (DED) squares as previously described [24]. In some experiments, the same respective number of melanoma cells and keratinocytes were seeded alone onto the surface of DED. After 9, 15 and 21 days of incubation at an air-liquid interface, the specimens were collected and embedded in paraffin for hematoxylin and eosin staining or embedded in Tissue-Tek (Miles Inc., Elkhart, IN, USA) for further immunohistochemical staining of type IV collagen (clone CIV 22; Dako, Carpinteria, CA, USA) as described [24]. Four-micrometer vertical sections cut at different levels were subjected to histological and staining evaluation. To test the effect of anti-human Tspan8 antibody, SR were cultured in its constant presence (TS29 clone, 15 μg/mL; [31]) or an isotype-matching control antibody. Dermal invasion was evaluated by a scoring system of 0–2 in a blinded manner: 0 indicated no melanoma cells present in the dermis, 1 invasive melanoma cells were located under JDE, and 2 melanoma cells were observed deep into the dermis. All experiments were done as sixtiplates and were repeated twice for each condition.

4.4. Preparation of Serum-Free Culture Medium

Culture fluids from skin composites were harvested on days 10, 15 and 21. Two days before collection, SR were extensively washed and cultured in serum-free medium. The collected serum-free culture media was centrifuged to remove cellular debris, and concentrated 10-fold in a Centricon ultrafiltration apparatus, containing a polysulfone membrane with an exclusion limit of Mr 10.000 (Millipore, Molshein, France). Protein concentrations were measured by the Bradford method using a commercial kit (Bio-Rad Laboratories, Paris, France). Aliquots with equivalent protein contents were subjected to gelatin zymography, Western blotting, and ELISA assays.

4.5. Gelatin Zymography

The activity of electrophoretically separated gelatinolytic enzymes in the serum-free culture media was analyzed as described previously [58].

4.6. Western Blot Analysis

Western blotting was performed as previously described [12]. Antibodies against MMP-9 (polyclonal antibody, Dako, Trappes. France), MMP-3 (clone 552A4, Oncogene research Product, Boston, Mass) and TIMP-1 (clone 7-6C1; Oncogene Research Products) were used.

Tspan8 was detected using a mouse monoclonal anti-Tspan8 antibody (TS29 clone [12,18–21]. Western blot quantifications were performed using ImageJ software. At least three independent biological replicates were performed.

4.7. Measurement of MMP-9, TIMP-1 and MMP-3

Serum-free culture medium was screened for pro and active MMP-9, total MMP-3 and total TIMP-1 using the commercially available ELISA kits (Amersham Pharmacia Biotech, Saclay, France), following procedures recommended by the manufacturer. All experiments were performed in triplicate from six separate experiments and the results were expressed as ng/mg of total proteins ± SD.

4.8. Real-Time RT-QPCR

Total RNA was extracted using the RNAeasy mini-kit (Qiagen, Germantown, MD, USA), reverse-transcribed into cDNA by PrimeScript™ RT reagent Kit (TaKaRa, Shiga, Japan) and analysed by real-time QPCR using SYBR®Premix ExTaq™II (TaKaRa, Shiga, Japan) on a Mx3000P real-time PCR system (Stratagene, Santa Clara, CA, USA) as described [21]. Results were obtained from at least three independent experiments and normalized to the 18 S rRNA expression level. The primers used are as follows: 18S-F: 5′-CGATGCGGCGGCGTTATT-3′; 18S-R: 5′-CCTGGTCTGTCTCATCCTCCC-3′; TSPAN8-F: 5′-TTGCTTCTGATCCTG CTCCT-3′; TSPAN8-R: 5′-AGGGCCTGCAGGTTCACACCAC-3′; MMP-9-F: 5′-CACTGTCCACCCCTCAGAGC-3′; MMP-9-R: 5′-GCCACTTGTCGGCGATAAGG-3′; MMP-3-F: 5′-GGAAGCTGGACTCCGACACTC-3′; MMP-3-R: 5′-TGGTGTATAATTCACAAT CCTGTATGTAA-3′; TIMP-1-F: 5′-GACGGCCTTCTGCAATTCC-3′; TIMP-1R: 5′-GTATAAGGTGGTCTGGTTGACTTCTG-3′.

4.9. Flow Cytometric Analysis

Cell surface labeling was performed as previously described [12]. Data were collected on a FACSCanto II (BD Biosciences, San Jose, CA, USA) and analyzed using FlowJo Software (Treestar, Ashland, OR, USA).

4.10. Animal Studies

NMR1 Foxn1nu/Foxn1nu female mice (Janvier Labs; Le Genest-Saint-Isle, France) were maintained and used in accordance with the 2010/63/UE directive after approval by the institutional review board C2E2A and the French MESR ministry. Mice were injected subcutaneously with 1.106 Tspan8+ or Tspan8- melanoma cells in their left or in right shoulder, respectively. Radiolabeling of Tspan8 mAB with 111Indium was performed as previously described [59]. The mice were imaged at each timepoint using a γ-camera (γ IMAGER, BIOSPACE Inc., Urbandale, IA 50322, USA) under gaseous anesthesia (Isoflurane, Iso-Vet®1000 mg/g). Removed tumors were weighted and counted using a Wallac 1480 automated calibrated γ-counter (Perkin-Elmer, Waltham, MA, USA).

4.11. Statistical Analysis

Statistical significance was calculated by a two-tailed Student's *t*-test for unpaired samples. Mean differences were considered to be significant when $p < 0.05$.

5. Conclusions

In summary, we report the novel finding that within a human-differentiated epidermis, Tspan8 expression in melanoma cells cooperate with surrounding keratinocytes to promote dermal invasion by instigating keratinocyte-produced MMP-3 activation and decreasing melanoma-derived TIMP-1 levels, leading to keratinocyte-originated MMP-9 activation process, and subsequent DEJ-collagen IV degradation. Furthermore, an anti-Tspan8 monoclonal antibody specifically targeting Tspan8^{+} melanoma xenografts in vivo significantly reduces dermal invasion by strongly impairing proMMP-9 activation process and collagen IV breakdown.

This study is the first to provide evidence for the pro-invasive role of Tspan8 in a cell non-autonomous manner, a mechanism never reported for a tetraspanin family member. This work has important implications since the direct inhibition of MMPs proved disappointing in clinical trials, and therefore targeting Tspan8 might represent a novel alternative and efficient strategy to impede MMP-9 proteolytic activity and greatly reduce metastasis risks.

Supplementary Materials: The following are available online at http://www.mdpi.com/2072-6694/12/5/1297/s1, Figure S1: uncropped western blots figures, Figure S2: Low TIMP-1 levels were exclusively observed when melanoma cells expressed Tspan8 and were integrated into SR. (a,b) Supernatants TIMP-1 protein levels were measured at day 10, 15 and 21 by ELISA in composites consisting of melanoma cells expressing or not Tspan8 cultured on acellular dermis either alone (a) or with human keratinocytes (b). Results are represented as the mean ± SEM from three independent experiments.

Author Contributions: Conceptualization: M.E.K., O.B.-V., I.M.; data curation, M.C., F.D., O.B.-V.; formal analysis, M.E.K., N.B., F.D., O.B.-V.; funding acquisition, O.B.-V., I.M., F.D., C.B.; methodology, M.E.K., M.C., N.B., C.P., F.D.; supervision, O.B.-V.; validation, M.E.K., M.C., I.M., O.B.-V.; writing—original draft preparation, M.E.K., O.B.-V., I.M. All authors have read and agreed to the published version of the manuscript.

Funding: This work was supported by The Ligue against Cancer (Comité Ardèche) and INSERM-Transfer, MelCoMab project.

Acknowledgments: The authors thank Elise Malandain for technical assistance and Cyril Py for immunohistochemical processing, Jean-Michel Chezal for Tspan8 mAb 111Indium labelling, Tiffany Witkowski and Aurélie Maisonial-Besset for radioimaging on mice.

Conflicts of Interest: The authors declare no conflict of interest.

References

1. Arozarena, I.; Wellbrock, C. Phenotype plasticity as enabler of melanoma progression and therapy resistance. *Nat. Rev. Cancer* **2019**, *19*, 377–391. [CrossRef] [PubMed]
2. Shain, A.H.; Bastian, B.C. From melanocytes to melanomas. *Nat. Rev. Cancer* **2016**, *16*, 345–358. [CrossRef]

3. Gershenwald, J.E.; Scolyer, R.A.; Hess, K.R.; Sondak, V.K.; Long, G.V.; Ross, M.I.; Lazar, A.J.; Faries, M.B.; Kirkwood, J.M.; McArthur, G.A.; et al. Melanoma staging: Evidence-based changes in the American Joint Committee on Cancer eighth edition cancer staging manual. *CA Cancer J. Clin.* **2017**, *67*, 472–492. [CrossRef] [PubMed]
4. Roncati, L.; Piscioli, F. AJCC 8th Edition (2017) versus AJCC 7th Edition (2010) in thin melanoma staging. *Neoplasma* **2018**, *65*, 651–655. [CrossRef] [PubMed]
5. Werner-Klein, M.; Scheitler, S.; Hoffmann, M.; Hodak, I.; Dietz, K.; Lehnert, P.; Veronika, N.; Bernhard, P.; Steffi, T.; Christian, W.; et al. Genetic alterations driving metastatic colony formation are acquired outside of the primary tumour in melanoma. *Nat. Commun.* **2018**, *9*, 595–599. [CrossRef] [PubMed]
6. McGranahan, N.; Swanton, C. Clonal Heterogeneity and Tumor Evolution: Past, Present, and the Future. *Cell* **2017**, *168*, 613–628. [CrossRef] [PubMed]
7. Flemming, A. Tumour heterogeneity determines immune response. *Nat. Rev. Immunol.* **2019**, *19*, 662–663. [CrossRef]
8. Dagogo-Jack, I.; Shaw, A.T. Tumour heterogeneity and resistance to cancer therapies. *Nat. Rev. Clin. Oncol.* **2018**, *15*, 81–94. [CrossRef]
9. Fattore, L.; Ruggiero, C.F.; Liguoro, D.; Mancini, R.; Ciliberto, G. Single cell analysis to dissect molecular heterogeneity and disease evolution in metastatic melanoma. *Cell Death Dis.* **2019**, *10*, 827–833. [CrossRef]
10. Zebda, N.; Bailly, M.; Brown, S.; Doré, J.F.; Berthier-Vergnes, O. Expression of PNA-binding sites on specific glycoproteins by human melanoma cells is associated with a high metastatic potential. *J. Cell. Biochem.* **1994**, *54*, 161–173. [CrossRef]
11. Berthier-Vergnes, O.; Zebda, N.; Bailly, M.; Bailly, C.; Dore, J.F.; Thomas, L.; Cochran, A.J. Expression of peanut agglutinin-binding glycoconjugates in primary melanomas with high risk of metastases. *Lancet* **1993**, *341*, 1292. [CrossRef]
12. Berthier-Vergnes, O.; El Kharbili, M.; de la Fouchardiere, A.; Pointecouteau, T.; Verrando, P.; Wierinckx, A.; Lachuer, J.; Le Naour, F.; Lamartine, J. Gene expression profiles of human melanoma cells with different invasive potential reveal TSPAN8 as a novel mediator of invasion. *Br. J. Cancer* **2011**, *104*, 155–165. [CrossRef] [PubMed]
13. Zöller, M. Tetraspanins: Push and pull in suppressing and promoting metastasis. *Nat. Rev. Cancer* **2009**, *9*, 40–55. [CrossRef] [PubMed]
14. Hemler, M.E. Tetraspanin proteins promote multiple cancer stages. *Nat. Rev. Cancer* **2014**, *14*, 49–60. [CrossRef]
15. Bonnet, M.; Maisonial-Besset, A.; Zhu, Y.; Witkowski, T.; Roche, G.; Boucheix, C.; Greco, C.; Degoul, F. Targeting the Tetraspanins with Monoclonal Antibodies in Oncology: Focus on Tspan8/Co-029. *Cancers* **2019**, *11*, 179. [CrossRef]
16. Park, C.S.; Kim, T.K.; Kim, H.G.; Kim, Y.J.; Jeoung, M.H.; Lee, W.R.; Go, N.K.; Heo, K.; Lee, S. Therapeutic targeting of tetraspanin8 in epithelial ovarian cancer invasion and metastasis. *Oncogene* **2016**, *5*, 4540–4548. [CrossRef]
17. Rodia, M.T.; Ugolini, G.; Mattei, G.; Montroni, I.; Zattoni, D.; Ghignone, F.; Veronese, G.; Marisi, G.; Lauriola, M.; Strippoli, P.; et al. Systematic large-scale meta-analysis identifies a panel of two mRNAs as blood biomarkers for colorectal cancer detection. *Oncotarget* **2016**, *7*, 30295–30306. [CrossRef]
18. Agaesse, G.; Barbollat-Boutrand, L.; Sulpice, E.; Bhajun, R.; El Kharbili, M.; Berthier-Vergnes, O.; Degoul, F.; de la Fouchardière, A.; Berger, E.; Voeltzel, T.; et al. A large-scale RNAi screen identifies LCMR1 as a critical regulator of Tspan8-mediated melanoma invasion. *Oncogene* **2017**, *36*, 446–457. [CrossRef]
19. Agaesse, G.; Barbollat-Boutrand, L.; El Kharbili, M.; Berthier-Vergnes, O.; Masse, I. p53 targets TSPAN8 to prevent invasion in melanoma cells. *Oncogenesis* **2017**, *6*, e309. [CrossRef] [PubMed]
20. El Kharbili, M.; Robert, C.; Witkowski, T.; Danty-Berger, E.; Barbollat-Boutrand, L.; Masse, I.; Gadot, N.; de la Fouchardière, A.; McDonald, P.C.; Dedhar, S.; et al. Tetraspanin 8 is a novel regulator of ILK-driven beta1 integrin adhesion and signaling in invasive melanoma cells. *Oncotarget* **2017**, *8*, 17140–17155. [CrossRef] [PubMed]
21. El Kharbili, M.; Agaësse, G.; Barbollat-Boutrand, L.; Pommier, R.M.; de la Fouchardière, A.; Larue, L.; Caramel, J.; Puisieux, A.; Berthier-Vergnes, O.; Masse, I. Tspan8-β-catenin positive feedback loop promotes melanoma invasion. *Oncogene* **2019**, *38*, 3781–3793. [CrossRef]

22. Kessenbrock, K.; Plaks, V.; Werb, Z. Matrix metalloproteinases: Regulators of the tumor microenvironment. *Cell* **2010**, *41*, 52–67. [CrossRef]
23. Kalluri, R. The biology and function of fibroblasts in cancer. *Nat. Rev. Cancer* **2016**, *23*, 582–598. [CrossRef]
24. Béchetoille, N.; Haftek, M.; Staquet, M.J.; Cochran, A.J.; Schmitt, D.; Berthier-Vergnes, O. Penetration of human metastatic melanoma cells through an authentic dermal-epidermal junction is associated with dissolution of native collagen types IV and VII. *Melanoma Res.* **2000**, *10*, 427–434. [CrossRef]
25. Haridas, P.; McGovern, J.A.; McElwain, S.D.L.; Simpson, M.J. Quantitative comparison of the spreading and invasion of radial growth phase and metastatic melanoma cells in a three-dimensional human skin equivalent model. *Peer J.* **2017**, *5*, e3754. [CrossRef] [PubMed]
26. Ogata, Y.; Enghild, J.J.; Nagase, H. Matrix metalloproteinase 3 (stromelysin) activates the precursor for the human matrix metalloproteinase 9. *J. Biol. Chem.* **1992**, *267*, 3581–3584. [PubMed]
27. Okada, Y.; Gonoji, Y.; Naka, K.; Tomita, K.; Nakanishi, I.; Iwata, K.; Yamashita, K.; Hayakawa, T. Matrix metalloproteinase 9 (92-kDa gelatinase/type IV collagenase) from HT 1080 human fibrosarcoma cells. Purification and activation of the precursor and enzymic properties. *J. Biol. Chem.* **1992**, *267*, 21712–21719. [PubMed]
28. Hahn-Dantona, E.; Ramos-DeSimone, N.; Sipley, J.; Nagase, H.; French, D.L.; Quigley, J.P. Activation of proMMP-9 by a plasmin/MMP-3 cascade in a tumor cell model. Regulation by tissue inhibitors of metalloproteinases. *Ann. N. Y. Acad. Sci.* **1999**, *878*, 372–387. [CrossRef]
29. Flores-Pliego, A.; Espejel-Nuñez, A.; Castillo-Castrejon, M.; Meraz-Cruz, N.; Beltran-Montoya, J.; Zaga-Clavellina, V.; Nava-Salazar, S.; Sanchez-Martinez, M.; Vadillo-Ortega, F.; Estrada-Gutierrez, G. Matrix Metalloproteinase-3 (MMP-3) is an endogenous activator of the MMP-9 Secreted by Placental Leukocytes: Implication in Human Labor. *PLoS ONE* **2015**, *10*, e145366. [CrossRef]
30. Sun, S.; Bay-Jensen, A.C.; Karsdal, M.A.; Siebuhr, A.S.; Zheng, Q.; Maksymowych, W.P.; Christiansen, T.G.; Henriksen, K. The active form of MMP-3 is a marker of synovial inflammation and cartilage turnover in inflammatory joint diseases. *BMC Musculoskelet. Disord.* **2014**, *15*, 93–99. [CrossRef]
31. Ailane, N.; Greco, C.; Zhu, Y.; Sala-Valdés, M.; Billard, M.; Casal, I.; Bawa, O.; Opolon, P.; Rubinstein, E.; Boucheix, C. Effect of an anti-human Co-029/tspan8 mouse monoclonal antibody on tumor growth in a nude mouse model. *Front Physiol.* **2014**, *5*, 364. [CrossRef] [PubMed]
32. Eves, P.; Layton, C.; Hedley, S.; Dawson, R.A.; Wagner, M.; Morandini, R.; Ghanem, G.; Mac Neil, S. Characterization of an in vitro model of human melanoma invasion based on reconstructed human skin. *Br. J. Dermatol.* **2000**, *142*, 210–222. [CrossRef] [PubMed]
33. Van Kilsdonk, J.W.; Bergers, M.; Van Kempen, L.C.; Schalkwijk, J.; Swart, G.W. Keratinocytes drive melanoma invasion in a reconstructed skin model. *Melanoma Res.* **2010**, *20*, 372–380. [CrossRef] [PubMed]
34. Huang, H. Matrix Metalloproteinase-9 (MMP-9) as a Cancer Biomarker and MMP-9 Biosensors: Recent Advances. *Sensors* **2018**, *18*, 3249. [CrossRef]
35. Van den Oord, J.J.; Paemen, L.; Opdenakker, G.; de Wolf-Peeters, C. Expression of gelatinase B and the extracellular matrix metalloproteinase inducer EMMPRIN in benign and malignant pigment cell lesions of the skin. *Am. J. Pathol.* **1997**, *151*, 665–670.
36. Hofmann, U.B.; Westphal, J.R.; van Muijen, G.N.P.; Ruiter, D.J. Matrix metalloproteinases in human melanoma. *J. Investig. Dermatol.* **2000**, *115*, 337–344. [CrossRef]
37. Simonetti, O.; Lucarini, G.; Brancorsini, D.; Nita, P.; Bernardini, M.L.; Biagini, G.; Offidani, A. Immunohistochemical expression of vascular endothelial growth factor, matrix metalloproteinase 2, and matrix metalloproteinase 9 in cutaneous melanocytic lesions. *Cancer* **2002**, *95*, 1963–1970. [CrossRef]
38. MacDougall, J.R.; Bani, M.R.; Lin, Y.; Rak, J.; Kerbel, R.S. The 92-kDa gelatinase B is expressed by advanced stage melanoma cells: Suppression by somatic cell hybridization with early stage melanoma cells. *Cancer Res.* **1995**, *55*, 4174–4181.
39. Hofmann, U.B.; Eggert, A.A.; Blass, K.; Bröcker, E.B.; Becker, J.C. Expression of matrix metalloproteinases in the microenvironment of spontaneous and experimental melanoma metastases reflects the requirements for tumor formation. *Cancer Res.* **2003**, *63*, 8221–8225.
40. MacDougall, J.R.; Bani, M.R.; Lin, Y.; Muschel, R.J.; Kerbel, R.S. 'Proteolytic switching': Opposite patterns of regulation of gelatinase B and its inhibitor TIMP-1 during human melanoma progression and consequences of gelatinase B overexpression. *Br. J. Cancer* **1999**, *80*, 504–512. [CrossRef]

41. Itoh, T.; Tanioka, M.; Matsuda, H.; Nishimoto, H.; Yoshioka, T.; Suzuki, R.; Uehira, M. Experimental metastasis is suppressed in MMP-9-deficient mice. *Clin. Exp. Metastasis* **1999**, *17*, 177–181. [CrossRef] [PubMed]
42. Devbhandari, R.P.; Shi, G.M.; Ke, A.W.; Wu, F.Z.; Huang, X.Y.; Wang, X.Y.; Shi, Y.H.; Ding, Z.B.; Xu, Y.; Dai, Z.; et al. Profiling of the tetraspanin CD151 web and conspiracy of CD151/integrin β1 complex in the progression of hepatocellular carcinoma. *PLoS ONE* **2011**, *6*, e24901. [CrossRef] [PubMed]
43. Zhu, J.; Liang, C.; Hua, Y.; Miao, C.; Zhang, J.; Xu, A.; Zhao, K.; Liu, S.; Tian, Y.; Dong, H.; et al. The metastasis suppressor CD82/KAI1 regulates cell migration and invasion via inhibiting TGF-β 1/Smad signaling in renal cell carcinoma. *Oncotarget* **2017**, *8*, 51559–51568. [CrossRef]
44. Sugiura, T.; Berditchevski, F. Function of alpha3beta1-tetraspanin protein complexes in tumor cell invasion. Evidence for the role of the complexes in production of matrix metalloproteinase 2 (MMP-2). *J. Cell. Biol.* **1999**, *146*, 1375–1389. [CrossRef] [PubMed]
45. Saito, Y.; Tachibana, I.; Takeda, Y.; Yamane, H.; He, P.; Suzuki, M.; Minami, S.; Kijima, T.; Yoshida, M.; Kumagai, T.; et al. Absence of CD9 enhances adhesion-dependent morphologic differentiation, survival, and matrix metalloproteinase-2 production in small cell lung cancer cells. *Cancer Res.* **2006**, *66*, 9557–9565. [CrossRef] [PubMed]
46. Herr, M.J.; Kotha, J.; Hagedorn, N.; Smith, B.; Jennings, L.K. Tetraspanin CD9 promotes the invasive phenotype of human fibrosarcoma cells via upregulation of matrix metalloproteinase-9. *PLoS ONE* **2013**, *8*, e67766. [CrossRef]
47. Hong, I.K.; Kim, Y.M.; Jeoung, D.I.; Kim, K.C.; Lee, H. Tetraspanin CD9 induces MMP-2 expression by activating p38 MAPK, JNK and c-Jun pathways in human melanoma cells. *Exp. Mol. Med.* **2005**, *37*, 230–239. [CrossRef]
48. Hong, I.K.; Jin, Y.J.; Byun, H.J.; Jeoung, D.I.; Kim, Y.M.; Lee, H. Homophilic interactions of Tetraspanin CD151 up-regulate motility and matrix metalloproteinase-9 expression of human melanoma cells through adhesion-dependent c-Jun activation signaling pathways. *J. Biol. Chem.* **2006**, *81*, 24279–24292. [CrossRef]
49. Van den Steen, P.E.; Dubois, B.; Nelissen, E.; Rudd, P.M.; Dwek, R.A.; Opdenakker, G. Biochemistry and molecular biology of gelatinase B or matrix metalloproteinase-9 (MMP-9): The next decade. *Crit. Rev. Biochem. Mol. Biol.* **2013**, *48*, 222–272. [CrossRef]
50. Girouard, S.D.; Laga, A.C.; Mihm, M.C.; Scolyer, R.A.; Thompson, J.F.; Zhan, Q.; Widlund, H.R.; Lee, C.W.; Murphy, G.F. SOX2 contributes to melanoma cell invasion. *Lab. Investig.* **2012**, *92*, 362–370. [CrossRef]
51. Nikkola, J.; Vihinen, P.; Vlaykova, T.; Hahka-Kemppinen, M.; Kähäri, V.M.; Pyrhönen, S. High expression levels of collagenase-1 and stromelysin-1 correlate with shorter disease-free survival in human metastatic melanoma. *Int. J. Cancer* **2002**, *97*, 432–438. [CrossRef] [PubMed]
52. Lo Presti, R.; Hopps, E.; Caimi, G. Gelatinases and physical exercise: A systematic review of evidence from human studies. *Medicine* **2017**, *96*, e8072. [CrossRef] [PubMed]
53. Khokha, R.; Zimmer, M.J.; Graham, C.H.; Lala, P.K.; Waterhouse, P. Suppression of invasion by inducible expression of tissue inhibitor of metalloproteinase-1 (TIMP-1) in B16-F10 melanoma cells. *J. Natl. Cancer Inst.* **1992**, *84*, 1017–1022. [CrossRef] [PubMed]
54. Khokha, R.; Zimmer, M.J.; Wilson, S.M.; Chambers, A.F. Up-regulation of TIMP-1 expression in B16-F10 melanoma cells suppresses their metastatic ability in chick embryo. *Clin. Exp. Metastasis* **1992**, *10*, 365–370. [CrossRef] [PubMed]
55. Vandenbroucke, R.E.; Libert, C. Is there new hope for therapeutic matrix metalloproteinase inhibition? *Nat. Rev. Drug Discov.* **2014**, *13*, 904–927. [CrossRef]
56. Winer, A.; Adams, S.; Mignatti, P. Matrix Metalloproteinase Inhibitors in Cancer Therapy: Turning Past Failures into Future Successes. *Mol. Cancer Ther.* **2018**, *17*, 1147–1155. [CrossRef]
57. Bailly, M.; Doré, J.F. Human tumor spontaneous metastasis in immunosuppressed newborn rats. II. Multiple selections of human melanoma metastatic clones and variants. *Int. J. Cancer* **1991**, *49*, 750–757. [CrossRef]
58. Gontier, E.; Cario-André, M.; Vergnes, P.; Bizik, J.; Surlève-Bazeille, J.E.; Taïeb, A. The 'Abtropfung phenomenon' revisited: Dermal nevus cells from congenital nevi cannot activate matrix metalloproteinase 2 (MMP-2). *Pigment Cell Res.* **2003**, *16*, 366–373. [CrossRef]
59. Maisonial-Besset, A.; Witkowski, T.; Navarro-Teulon, I.; Berthier-Vergnes, O.; Fois, G.; Zhu, Y.; Besse, S.; Bawa, O.; Briat, A.; Quintana, M.; et al. Tetraspanin 8 (TSPAN 8) as a potential target for radio-immunotherapy of colorectal cancer. *Oncotarget* **2017**, *8*, 22034–22047. [CrossRef]

cancers

MDPI

Article

Adhesion and Migration Response to Radiation Therapy of Mammary Epithelial and Adenocarcinoma Cells Interacting with Different Stiffness Substrates

Valeria Panzetta [1,2,*], Giuseppe La Verde [3,4], Mariagabriella Pugliese [3,5], Valeria Artiola [5], Cecilia Arrichiello [6], Paolo Muto [6], Marco La Commara [3,4], Paolo A. Netti [1,2] and Sabato Fusco [1,2,*]

1 Centro di Ricerca Interdipartimentale sui Biomateriali, Università degli Studi di Napoli Federico II, Piazzale Tecchio 80, 80125 Napoli, Italy; nettipa@unina.it
2 Centre for Advanced Biomaterial for Health Care, Istituto Italiano di Tecnologia, Largo Barsanti e Matteucci 53, 80125 Napoli, Italy
3 Istituto Nazionale di Fisica Nucleare, INFN sezione di Napoli, Via Cinthia ed. 6, 80126 Napoli, Italy; glaverde@na.infn.it (G.L.V.); pugliese@na.infn.it (M.P.); marco.lacommara@unina.it (M.L.C.)
4 Dipartimento di Farmacia, Università degli Studi di Napoli Federico II, Via Montesano 49, 80131 Napoli, Italy
5 Dipartimento di Fisica "Ettore Pancini", Università degli Studi di Napoli Federico II, Via Cinthia ed. 6, 80126 Napoli, Italy; val.artiola@gmail.com
6 Radiotherapy Unit, Istituto Nazionale Tumori-IRCCS-Fondazione "G. Pascale", Via Semmola, 53, 80131 Naples, Italy; c.arrichiello@istitutotumori.na.it (C.A.); p.muto@istitutotumori.na.it (P.M.)
* Correspondence: valeria.panzetta@unina.it (V.P.); sabfusco@unina.it (S.F.); Tel.: +39-0817682169 (V.P. & S.F.)

Received: 8 April 2020; Accepted: 4 May 2020; Published: 6 May 2020

check for updates

Abstract: The structural and mechanical properties of the microenvironmental context have a profound impact on cancer cell motility, tumor invasion, and metastasis formation. In fact, cells react to their mechanical environment modulating their adhesion, cytoskeleton organization, changes of shape, and, consequently, the dynamics of their motility. In order to elucidate the role of extracellular matrix stiffness as a driving force in cancer cell motility/invasion and the effects of ionizing radiations on these processes, we evaluated adhesion and migration as biophysical properties of two different mammary cell lines, over a range of pathophysiological stiffness (1–13 kPa) in a control condition and after the exposure to two different X-ray doses (2 and 10 Gy, photon beams). We concluded that the microenvironment mimicking the normal mechanics of healthy tissue has a radioprotective role on both cell lines, preventing cell motility and invasion. Supraphysiological extracellular matrix stiffness promoted tumor cell motility instead, but also had a normalizing effect on the response to radiation of tumor cells, lowering their migratory capability. This work lays the foundation for exploiting the extracellular matrix-mediated mechanism underlying the response of healthy and tumor cells to radiation treatments and opens new frontiers in the diagnostic and therapeutic use of radiotherapy.

Keywords: breast cancer; mechanobiology; cell motility; extracellular matrix stiffness; radiotherapy

1. Introduction

Breast cancer is the most common malignancy and the first leading cause of cancer-related death in European women [1]. In particular, women in perimenopausal and postmenopausal age have a higher risk of getting breast cancer [2], and extensive attempts should be made in order to contain breast cancer incidence and mortality. In this context, radiation therapy (RT) is used as adjuvant therapy to prevent tumor recurrence after breast-conserving lumpectomy and mastectomy. RT can induce dramatic consequence for the cells, by stimulating the production of radical and reactive oxygen species that damage the DNA of cancer cells, leading to the formation of lethal chromosome aberrations (double-stranded breaks and/or lesions) and, consequently, killing them or slowing their growth, as widely reported in the literature [3–6]. In any case, the mechanism of the response of the cell cytoskeleton to RT in relationship to the mechanical microenvironment in which cell resides has not yet been elucidated in a comprehensive way. Nowadays, it is widely recognized that cell cytoskeleton and extracellular matrix (ECM) have key roles in the maintenance of the correct functioning of many tissue processes that, if altered, have a determinant contribution in cancer progression. In fact, it has been shown that tumor cells have a less organized and structured cytoskeleton with lower cell mechanical and cyto-adhesive properties compared to their healthy counterparts. Furthermore, the dynamic alteration of the actin cytoskeleton has strong implications on motility, invasion, and metastatic potential of cancer cells [7]. On the other hand, changes in ECM composition and architecture result in a stiffening process of the matrix that activates cell proliferation and a consequent invasion mechanism [8–10]. Recently, several research groups have started to focus their attention on the study of possible impacts of radiation on the cell cytoskeleton and its associated functions. In particular, it has been extensively investigated the way ionizing radiations can influence motility, a prerequisite for the formation of metastasis, and for the invasiveness itself of surviving cancer cells both in vitro and in pre-clinical/clinical experimental studies. The observations made in vitro experiments indicate that the effects of radiation on cell motility depend strongly on the specific cell phenotype and the dose delivered to the cells. So far, it has been reported that irradiation has very different impacts on the motility of four different glioblastoma cell lines, inducing a very strong inhibition of in vitro invasion only on one of them (A-172 cell line) [11]. More recently, Hohmann et al. observed that irradiation leads to changes in motility and to a less invasive phenotype of two glioblastoma cell lines, both associated with an increase of cell mechanical properties and changes in the cytoskeleton structure [12]. Our group reported similar results on healthy and tumor fibroblasts irradiated with 250 keV and 6 MeV [13–15]. A sensitive increase in the mechanical properties of tumor cells, responsible for the enhancement of cell-adhesion and the reduction of migration of tumor cells, was observed. The effects were significant and dose-dependent for the tumor cell line, while healthy fibroblasts resulted in being susceptible principally to high energy X-rays administered at doses higher than 1 Gy. Imaizumi et al. also observed response to X-irradiation strongly dependent on the dose administered to MDA-MB-231 breast cancer cells. If very low doses (0.5 Gy) enhanced cell migration and invasion, higher doses (2 and 10 Gy) suppressed MDA-MB-231 migration in a dose-dependent manner [16]. On the other hand, there are many studies reporting the opposite effect of ionizing radiation on cell motility. In particular, several works demonstrated that X-rays promote migration of neck and head carcinoma cells [17,18], breast cancer cells [19], central nervous system cells [19,20], lung cancer cells [21] and that this enhancement is sometimes accompanied by an increase of cell invasion [19,20], other times by a reduction [17]. Other in vivo experiments suggest that radiations can have a myriad of effects upon the motility and invasion capability of tumor cells. Numerous experimental and clinical studies have evidenced that in addition to a bystander effect that contributes to killing tumor cells, a non-targeted inhibitory effect on distant tumor growth (abscopal effects) exists, mediated by the response of the immune system [22]. However, other clinical trials support the idea that radiations contribute to a higher risk of metastasis through mechanisms mediated by the release of tumor cells into the circulation system or by effects into irradiated non-tumor cells [23]. The differences observed in these works highlight the complexity of the phenomenon of motility and/or invasion after irradiation and suggest wider investigations. To better understand the mechanisms underlying the effects of ionizing radiations on cell motility, here, we propose to consider the role of microenvironment, considering that properties, functions, and healthiness of the tissue are regulated by the close physical crosstalk existing between cells and ECM [24]. When these interactions are impaired, the mechanical integrity of tissue constituents (cells and ECM) changes, triggering cancer formation and progression [8]. Particularly for breast cancer, a large body of evidence has highlighted the existence of a close relationship between ECM stiffening, cancer cell softening and cancer onset, progression and aggression. In fact, on the one hand, breast cancer cells resulted in being significantly more deformable than the non-tumorigenic ones, and this change may facilitate easy migration and invasion of malignant cells during metastasis [25–28]. On the other hand, it has been widely demonstrated that the mechanics of the tumor microenvironment has a central role in the development of the disease. Stowers et al. showed that a protrusive and eventually invasive phenotype of non-transformed epithelial breast cells (MCF10A) can arise from the stiffening of the ECM matrix through the PI3K and Rac1 mechanotransduction pathway. In fact, the authors demonstrated that inhibition of these molecular pathways (Rac1, PI3K, MAPK, FAK, and ROCK) was able to suppress the invasive character of MCF10A cells [9]. Very recently, Panciera et al. observed that only a coordinated interplay between oncogene-mediated transformation and changes in the rigidity of the microenvironment is able to power up the process of tumorigenesis [9]. At the same time, the stiffening of ECM is also relevant to the progression of breast cancers. The increased cross-linking of collagen in breast cancer promotes focal adhesion formation, PI3K activity, as previously mentioned, and breast malignancy [24,29,30]. For this reason, the mechanical evaluation of breast tumor cells and ECM might provide new support to diagnosis and promote new insights in the field of chemo- and radiotherapy and clinical practice. In this work, we focus our attention on breast cancer, not only because it is one of the earliest models used to understand cancer progression and metastasis, but also because it is treated with RT from stages I (tumor dimension up to 2 cm and no lymph nodes involved) to III (tumor spread to lymph nodes or tissue near the breast) to reduce the risk of recurrence after surgery. In particular, we evaluated biophysical properties (strictly correlated to the cytoskeleton integrity, such as adhesion and migration) of two different mammary cell lines. A normal epithelial cell line (MCF10A) and a highly aggressive and invasive adenocarcinoma cell line (MDA-MB-231), on polyacrylamide (PAAm) substrates over a range of pathophysiological stiffness (1–13 kPa) in control conditions and after the exposure to two different doses of X-rays (photon beams), were investigated. Selected doses were 2 and 10 Gy, which represent the daily dose in radiotherapy treatment and the single maximum dose for the treatment of metastasis. Time points of 1 and 3 days (d) after irradiation were chosen based on our previous observations according to which, at these particular time points, healthy and tumor cells exhibited peculiar and different responses to irradiation [13–15]. Shedding light on the effects of X-rays on functions strictly associated with the cytoskeleton architecture and to cell-ECM crosstalk in healthy and tumor cells can open new frontiers in the diagnostic and therapeutic use of RT.

2. Results and Discussion

2.1. Role of Substrate Stiffness on Cell Morphological Features Before and after the Exposure to RT

To test the relevance of the mechanical properties of ECM on both healthy and tumor breast cell behavior and the response to the irradiation, we decided to modulate the mechanical properties of the cell culture substrate using PAAm hydrogels. Their mechanical properties were controlled by adding two different ratios of acrylamide and bis-acrylamide, and the rheological characterization verified that the elastic moduli of the two formulations were equal to 1.3 and 13 kPa, mimicking the values of stiffnesses of healthy breast tissue and malignant biopsies, respectively [29].

It is now widely known that a strong relationship exists between the mechanical properties of the ECM and morphological features, such as cell spreading and nuclear shaping [31–34]. This relationship is strictly related to the myosin-tension generated inside the actin stress fibers necessary for their assembly and their associated focal adhesions, the structures deputed to the sensing of mechanical cues (also identified as mechanosensors). On the one hand, the actin cytoskeleton fosters the cell spreading in a proportional way to the tensional state generated inside it. The cytoskeleton, thanks to the LINC complex (Linker of Nucleoskeleton and Cytoskeleton), can transmit the mechanical forces to the nucleus, changing its shape, distorting the nuclear envelope, and evoking biochemical responses [32]. To evaluate the ability of MCF10A and MDA-MB-231 cells to sense and respond to the reactive forces originating by different mechanical properties of the substrate, we quantified their spreading, intended as the whole-cell area, and the nuclear area on collagen-functionalized PAAm substrates with different stiffness (1.3 and 13 kPa). Cells were plated on the substrates in sparse culture condition in order to avoid or limit the formation of adherens junctions that could impact on the adhesion and, more generally, on the mechanotransduction process [34]. We found that changes in substrate stiffness imposed different degrees of spreading and nuclear area in case of healthy cells, whereas, for metastatic cells, the same morphological features were not affected by the microenvironment stiffness at all (Figure 1a,f,k,p, Figure 2, and Figure S1; Table 1), suggesting compromised mechanosensing machinery. Furthermore, the spreading area of MDA-MB-231 on soft substrate resulted in being higher than that of MCF10A on the same substrate, in agreement with the increased formation and maturation of focal adhesions associated to enhanced intracellular contractility in KRAS-mutated cells cultured on 1 kPa substrate compared to MCF10A cells, as observed by Panciera et al. [10]. Panciera et al. also demonstrated that KRAS-mutated cells, like MDA-MB-231 cells, are characterized by a significant nuclear accumulation of YAP and TAZ, two well-established sensors of mechanotransduction, already when cultured on 1 kPa substrates. On the contrary, MCF10A cells exhibited a YAP/TAZ nuclear/cytoplasmic ratio lower than 1, indicating that YAP and TAZ are principally located inside the cytoplasmic compartment [10]. Consistently, we found higher values of the nuclear area of MDA-MB-231 on a soft substrate compared to MC10A cells.

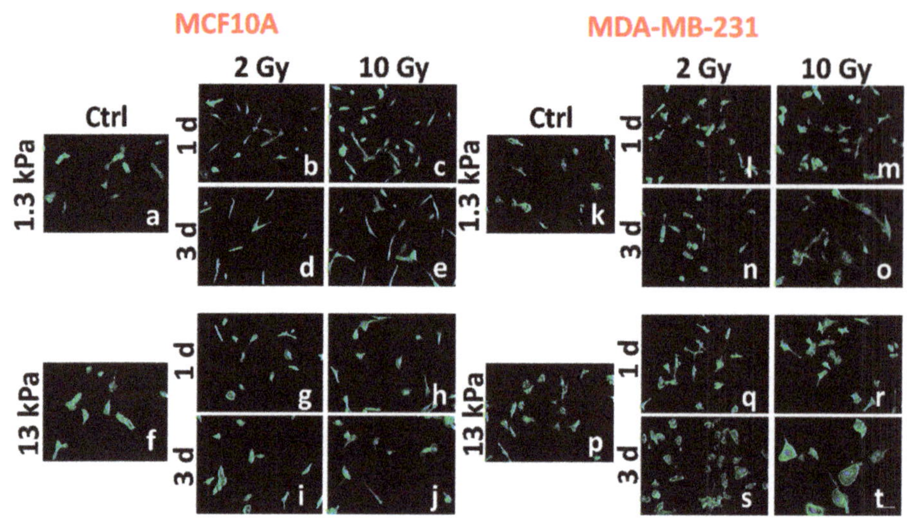

Figure 1. Representative images of two different mammary cell lines, MCF10A and MDA-MB-231, are shown. The spreading of MCF10A and MDA-MB-231 are compared before (**a**,**f**,**k**,**p**) and after RT (radiation therapy) (**b**–**e**,**g**–**j**,**l**–**o**,**q**–**t**). The cells were stained for F-actin (green) and nuclear DNA (blue). Scale bar, 100 μm.

Having characterized the mechanosensing activity of the two cell lines in the control condition, we evaluated the effects of irradiation on cell adhesion at 1 and 3 d after treatment at the two different doses of 2 and 10 Gy. At 1 d after irradiation, MCF10A cells resulted in being less spread compared to cells in the control condition. This effect was relevant for all conditions, even though more significant for cells cultured on a stiff substrate and irradiated with the lower dose (2 Gy; Figure 1a–c,f–h, Figure 2a, Figures S1 and S2; Table 1). Furthermore, the decrease of cell spreading resulted in being not dose-dependent on the soft substrate, whereas it exhibited an inverse dependence on the dose administered in case of cells cultured on the stiff substrate (Table 1). At the same time point, we found that the nuclear areas of irradiated healthy cells cultured on soft substrate increased slightly, but in a significant way and not dose-dependent (Figure 1a–c, Figure 2c, and Figure S1; Table 1). This was an unexpected result if associated with that of the spreading area. It could be explained supposing a protective mechanism, operated by microtubules and intermediate filaments on the nucleus and activated in a physiological environment. In contrast, on the stiff substrate, the nuclear area decreased in agreement with the spreading area, but without dependence on the dose administered (Table 1). At longer times, the effect on the spreading area was maintained only for the lower dose on both substrates, while the initial values were completely restored by the cells irradiated with the higher dose (10 Gy; Figure 1a–c, Figure 2a, and Figure S1; Table 1). The behavior of cells is more complicated if the data on the nuclear area are analyzed. In fact, the higher values were maintained by cells cultured on soft substrates and irradiated with lower dose indicating a more persistent effect of such dosage on cell adhesion, while the nuclear area of cells cultured on stiff substrates returned to its initial value (Figure 1f–j, Figure 2c, and Figure S1; Table 1).

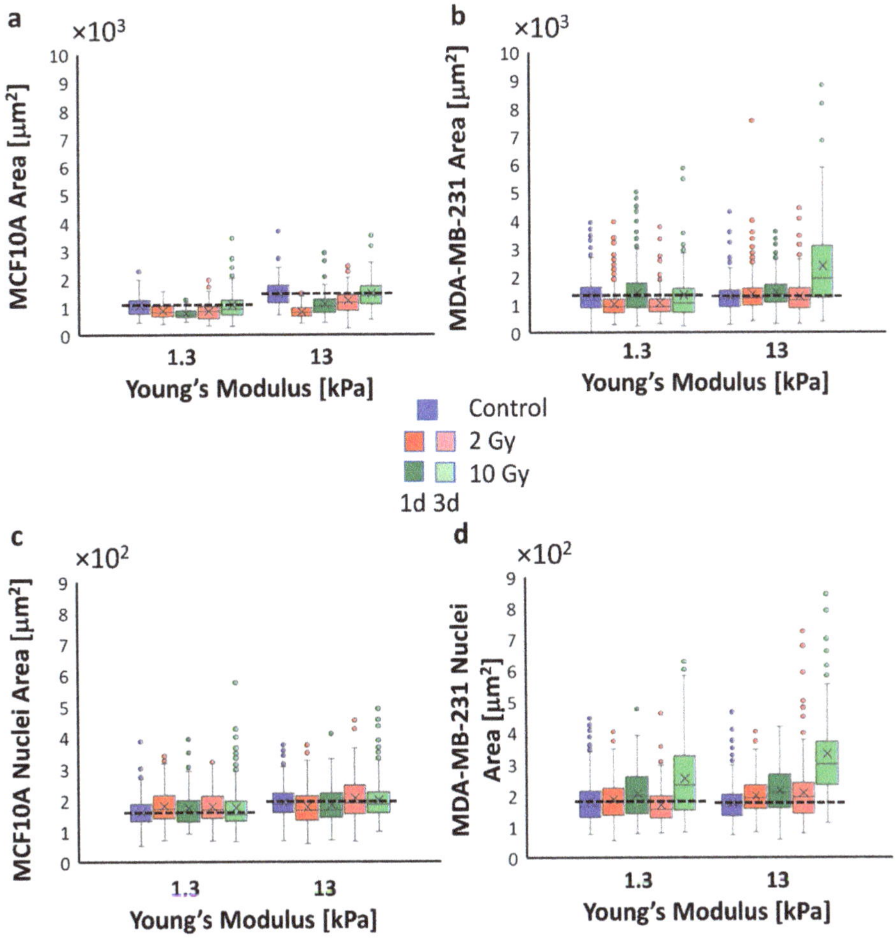

Figure 2. Box plots (mean, median, interquartile range, and outliers) of spreading areas (**a**,**b**) and nuclei areas (**c**,**d**). Spreading areas and nuclei area values were obtained from the analysis of Figure 1 and Figure S1). Dashed lines (**a**–**d**) indicate the mean values of spreading areas and nuclei areas in control conditions. $n > 60$ for cell spreading data, $n > 110$ for nuclear data.

Table 1. Statistical analysis for data of spreading and nuclei area.

			Control	2 Gy				10 Gy			
				1 day		3 days		1 day		3 days	
			13 kPa	1.3 kPa	13 kPa	1.3 kPa	13 kPa	1.3 kPa	13 kPa	1.3 kPa	13 kPa
Control		1.3 kPa	***.NS ***.NS	**.### ***.NS	***.NS **.###	***.### ***.NS	NS.NS ***.###	**.NS NS.##	***.NS NS.###	NS.NS NS.###	***.### ***.###
		13 kPa		***.### **.#	***.# ***.###	***.### **.NS	***.NS NS.###	***.## ***.###	***.### **.###	***.NS ***.###	NS.### NS.###
2 Gy	1 day	1.3 kPa			NS.### NS.#	*.NS NS.###	***.### ***.NS	NS.### NS.#	***.### NS.###	*.## **.###	***.### ***.###
		13 kPa				NS.### NS.###	***.NS ***.NS	NS.NS NS.NS	***.NS NS.NS	***.## NS.###	***.### ***.###
	3 days	1.3 kPa					***.### ***.###	NS.### NS.###	***.### NS.###	***.NS *.###	***.### ***.###
		13 kPa						***.# ***.NS	***.NS **.NS	NS.NS ***.###	***.### **.###
10 Gy	1 day	1.3 kPa							***.## NS.NS	*.## NS.##	***.### ***.###
		13 kPa								***.NS *.NS	***.### **.###
	3 days	1.3 kPa									***.### ***.###

Asterisks (*) refer to spreading area (blue) and nuclei area (red) of MCF10A cell, hash signs (#) to those of MDA-MB-231 cells. ***, ### $p < 0.001$, **, ## $p < 0.01$, *, # $p < 0.05$; NS—not significant.

The behavior of metastatic cells was significantly different. MDA-MB-231 cells resulted in being more sensitive to both doses of irradiation, even though the effects of the RT changed profoundly over time and with doses. In particular, at both time points, metastatic cells cultured on the physiological environment reduced their spreading area when irradiated with the dose of 2 Gy in a similar way to MCF10A cells (Figure 1l–n,q–s, Figure 2b, Figures S1 and S3; Table 1). On the other hand, their adhesion seemed not to be affected by the higher dose, indicating that, in this case, the microenvironment mimicking a healthy tissue mechanics (1.3 kPa) has a sort of protective role on cell properties (Figure 1m–o,r–t, Figure 2b, Figures S1 and S3; Table 1). Cells cultured on stiff substrates showed the opposite behavior. These cells, in fact, increased their spreading area significantly at 1 d in a dose-dependent manner. After 3 d, only cells irradiated with 2 Gy were able to regain their initial features, while for those cells treated with 10 Gy, the spreading area increased to a value 1.9-fold higher than in control condition (Figure 1d and Figure S1; Table 1). The effects of RT on nuclear areas of metastatic cells are reported in Figure 1k–t, Figure 2d, and Figure S1, and Table 1. In particular, the higher dose induced a significant increase of the nuclear area resulting 1.9-fold higher than in control condition (Figure 2d; Table 1), indicating that tumor cells, differently from the healthy ones, were more sensitive to 10 Gy than to 2 Gy dose. The results observed on stiff substrates are partially in agreement with those obtained by our group on MCF10A and MCF7 cells cultured on tissue culture plastics (3 GPa), which responded at the irradiation by, respectively, reducing and increasing their spreading area after irradiation [35]. The results point out the necessity not to neglect the role of mechanical microenvironment in regulating the response of cells to irradiation. The role of substrate stiffness and composition seems to take a regulatory effect on the response of metastatic cells to RT. In fact, Cordes et al. found that substrate, intended as ECM molecules (fibronectin and Matrigel), can have a radioprotective role on glioblastoma cells, which reflects in a significant upregulation of β-integrins by irradiation in its turn correlated with improved β-integrin-mediated adhesion to the substrate [8]. Our data, intersected with those previously obtained on petri-dish [35], indicate a range of stiffness, above 1.3 kPa and below 3 GPa, in which metastatic cells can recover the effect provoked by RT. It is well known, indeed, that ECM of tumoral tissues presents higher mechanical properties due to an increased percentage of aligned collagen fibers [8]. Hence, it is interesting that cells behave differently when cultured on substrates simulating stiffness higher than the physiological ECM of mammary epithelial cells.

Taken together, these results suggest how important is the role of the microenvironment and the necessity to perform further investigations to unravel new mechanisms underlying the response of healthy and tumor cells to RT treatments. In addition to this, observing the behavior of MCF10A and MDA-MB-231 cells after irradiation, it is important to recognize that, even if RT altered the biophysical properties of both cell lines, its impact was less relevant on healthy cells than on tumor cells. This suggests that MCF10A had a stronger ability to preserve or to recover their properties in both mechanical environments by adopting probable mechanisms of protection and repair. Furthermore, after RT tumor cells seem to restore a sort of mechanosensing process having again the ability to recognize the mechanical environment and to respond in a similar way to MCF10A cells.

2.2. Role of Substrate Stiffness on Cell Motility before and after the Exposure to RT

To evaluate MCF10A and MDA-MB-231 migratory behavior in response to substrate stiffness, cell videos were recorded with an interval of 10 min using time-lapse microscopy, and two essential parameters describing the efficiency of cell motility were computed: the mean migration velocity over a 24-h period and directional persistence. Whereas the migration velocity is easy to calculate and interpret, the persistence describes the time a cell employs to change its direction and has been estimated by fitting the mean square displacements (MSDs) over time with the Fürth's formula (see Section 3). As shown in Figure 3u, the velocity of MCF10A decreased significantly from ~0.8 on the soft substrate to ~0.7 μm/min on the stiff substrate (Table 2), consistently with previous reports indicating that the migration velocity of healthy cells presents an inverse proportion to substrate stiffness [31]. This result can be explained by considering that cell motility is a complex process requiring repeated cycles of adhesion to and detachment from the ECM, strictly related to the focal adhesion life cycle (assembly–maturation–disassembly). In this regard, it has been widely demonstrated that when the stiffness increased, healthy cells form bigger stress fibers, contributing a more structured cytoskeleton, and longer focal adhesions with greater assembly/disassembly rate that slows down the cell migration velocity [36]. Moreover, the persistence time resulted in being equal to 0 (Figure 4a,b; Table 3), indicating that the motility of MCF10A cells reflects the behavior of a random Brownian motion on both substrates. On the contrary, the motility of MDA-MB-231 cells increased with the increase of substrate stiffness passing from ~0.7 to ~1 μm/min (Figure 3v, Table 2), consistently with other studies reporting the same effect on the migration of various cancer cells, including pancreatic cancer cells, colorectal cancer cells, breast cancer cells, and so forth [37–41]. What emerged from our experimental results was that MDA-MB-231 cells were more directionally stable on both substrate stiffness compared to healthy cells and also their persistence time was positively correlated to the substrates stiffness, resulting close to 1 and 2 h on soft and stiff substrates, respectively (Figure 4c,d; Table 3). This finding indicated that the ECM stiffening promotes not only motility but also the ability of tumor cells to invade distant sites by increasing the directional persistence, in agreement with observations made in vitro and in vivo experiments. In particular, Levental et al. clearly showed that in breast tumors, the crosslinking of collagen and ECM stiffness regulate the invasive behavior of oncogene pre transformed epithelial cells (H-RAS-transformed MCF10A expressing high levels of activated H-RAS in a similar way to MDA-MB-231 cells) [29]. In addition, the mean migration velocity of MDA-MB-231 cells on the stiffer substrate is significantly higher than that of MCF10A cells, indicating that in the absence of pre-oncogene activity, ECM stiffening does not drive mammary epithelial cell metastasis [10,29]. These two observations, the direct relationship between cell migration velocity and ECM stiffness and the higher velocity of tumor cells compared to that of healthy cells, can be explained by considering that tumor cells have a less organized cytoskeleton with lower mechanical properties, fewer and less-developed focal adhesion than their normal counterpart [42,43]. At the same time, malignant cells exhibit a very high level of focal adhesion kinase (FAK) [44], a nonreceptor tyrosine kinase activated by integrin clustering and involved in the process of disassembly of focal adhesion, and the overexpression and activation of FAK are promoted by the stiffening of the ECM [45]. Consequently, the upregulation of FAK on stiff ECM contributes to increasing the rate of assembly/disassembly of focal adhesions without allowing their maturation and the formation of a robust cytoskeleton but promoting tumor cell invasion (directional persistence) and migration (migration velocity) instead [44].

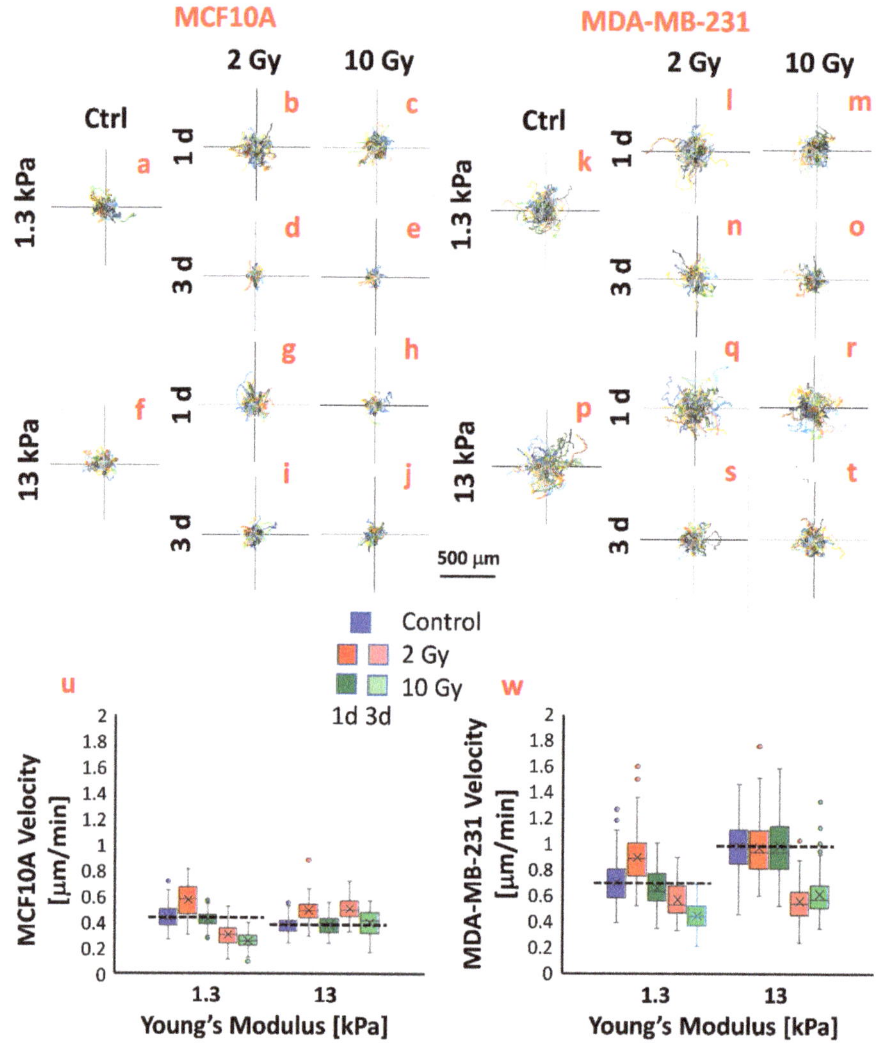

Figure 3. Plot at origin of trajectories of MCF10A and MDA-MB-231 before (**a**,**f**,**k**,**p**) and after RT (**b**–**e**,**g**–**j**,**l**–**o**,**q**–**t**). Trajectories of cells were obtained by manual tracking using ImageJ and Manual Tracking plugin (http://rsweb.nih.gov/ij/). Box plots (mean, median, interquartile range and outliers) of migration velocity of MCF10A (**u**) and MDA-MB-231 (**w**) in all analyzed conditions. Cell velocity was calculated by the trajectories using the formula 1 (see Section 3). Dashed lines (**u**,**w**) indicate the mean values of velocity in control conditions. $n > 58$ for all conditions.

Table 2. Statistical analysis for motility data.

			Control	2 Gy				10 Gy			
				1 day		3 days		1 day		3 days	
			13 kPa	1.3 kPa	13 kPa	1.3 kPa	13 kPa	1.3 kPa	13 kPa	1.3 kPa	13 kPa
Control		1.3 kPa	***.###	***.###	***.###	***.###	***.###	NS.NS	***.###	***.###	***.###
		13 kPa		***.##	***.NS	***.###	***.###	***.###	NS.NS	***.###	***.###
2 Gy	1 day	1.3 kPa			***.#	***.###	***.###	***.###	***.#	***.###	***.###
		13 kPa				***.###	NS.###	***.###	***.NS	***.###	***.###
	3 days	1.3 kPa					***.NS	***.###	***.###	***.###	***.NS
		13 kPa						***.###	***.###	***.###	***.NS
10 Gy	1 day	1.3 kPa							**.###	***.###	***.###
		13 kPa								***.###	***.###
	3 days	1.3 kPa									***.###

Asterisks (*) refer to cell velocity of MCF10A cells, hash signs (#) to those of MDA-MB-231 cells. ***, ### $p < 0.001$, **, ## $p < 0.01$, # $p < 0.05$; NS—not significant.

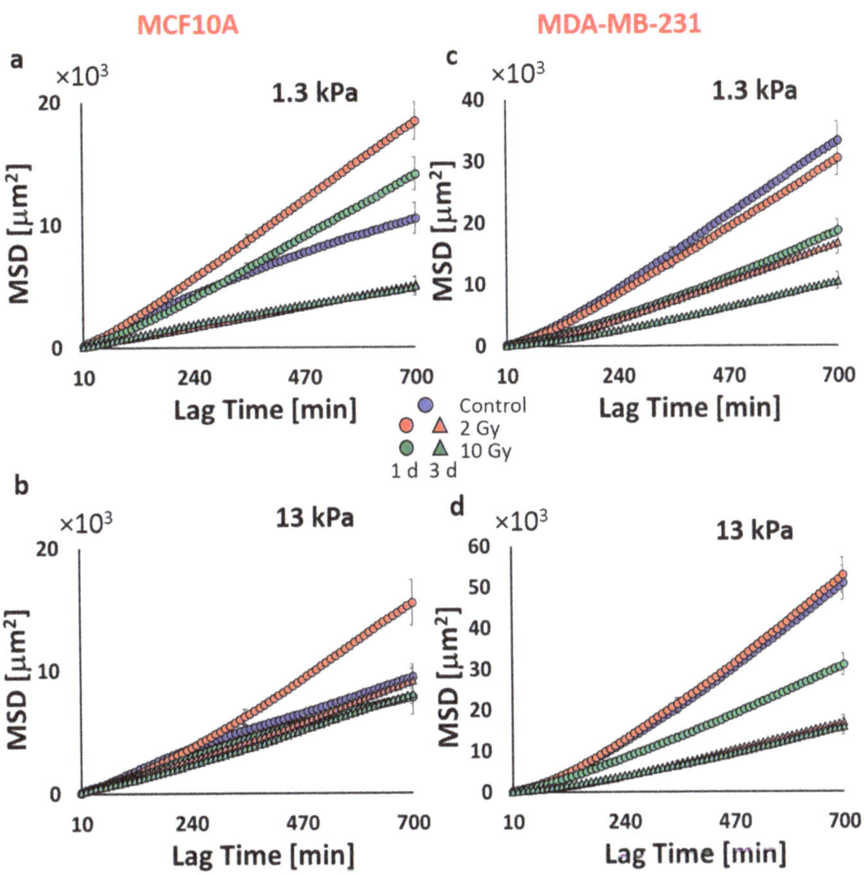

Figure 4. MSDs (mean square displacements) calculated from cell trajectories of MCF10A (**a**,**b**) and MDA-MB-231 (**c**,**d**) by using Formula (2) (see Section 3).

Table 3. Parameters describing the motility of cells.

		Control		2 Gy				10 Gy			
				1 d		3 d		1 d		3 d	
		1.3 kPa	13 kPa	1.3 kPa	13 kPa	1.3 kPa	13 kPa	1.3 kPa	13 kPa	1.3 kPa	13 kPa
MCF10A	D [mm^2/min]	4.0448	3.4916	6.9223	7.8180	1.8238	3.3666	5.4219	2.9704	1.8234	3.0483
	P [min]	0.0090	0.0086	32.2139	113.1808	6.6790	31.9583	49.0114	0.05537	0.03201	43.5825
	R^2	0.9864	0.9937	0.9997	0.9987	0.9993	0.9973	0.9996	0.9939	0.9943	0.9979
MDA-MB-231	D [mm^2/min]	13.0096	21.4693	11.8329	22.1579	6.7391	7.305239	7.594923	12.08702	4.338288	6.487246
	P [min]	59.7990	114.9188	62.1477	112.1086	84.9445	122.0365	97.8991	71.3609	104.0150	100.9757
	R^2	0.9998	0.9995	0.9998	0.9995	0.9998	0.999801	0.998634	0.998922	0.999376	0.998346

Values of diffusion coefficient (D), directional persistence (P), and goodness-of-fit (R^2) obtained by fitting the MSD (mean square displacements) of cells' trajectories to Fürth's formula.

Cell motility analysis showed that irradiation had important effects also on the migratory behavior of healthy and tumor cells. At 1 d after irradiation, MCF10A cells cultured on both physiological and supraphysiological stiffnesses and irradiated with the lower dose responded by increasing their migration velocity (Figure 3u, Table 2), as a consequence of their reduced adhesion (Figure 2a), while the high dose did not seem to affect their motility, even though in this case the adhesion also presented lower values (Figure 2a). As demonstrated by the increase of the persistence time and compared to the control conditions, the motility mode changed sensitively for those conditions in which cell velocity increased (cells irradiated with 2 Gy and 10 Gy and cultured on the stiff substrate; Figure 4a,b; Table 3). At a longer time, cells cultured on soft substrate reduced their velocity in a very significant way, showing a partial concordance with RT effects on the adhesion (Figure 3u, Table 2). In fact, whereas cells irradiated with the higher dose increased their adhesion to the substrate, those irradiated with the lower dose continued to exhibit a reduced adhesion (Figure 2a). At 3 d after irradiation, the persistence of 2 Gy-irradiated cells was lower than those observed at 1 d but higher than those of control cells. This indicates that irradiation can have a strong impact on cell adhesion and motility, both in terms of velocity and directionality (Figure 4a,b; Table 3). According to the results on cell adhesion, the low dose produced effects more durable than the high dose. This, in particular on the stiff substrate, produced a reduction of the spreading area and an enhancement of migration velocity and directional persistence.

MDA-MB-231 cells responded to RT by changing their velocity and persistence time, but the latter remaining in all cases higher than 1 h and retaining the natural propensity of these cells to move directionally. MDA-MB-231 cells on soft substrate responded to RT by increasing their migration velocity when irradiated with 2 Gy, but this effect was completely overturned at 3 d, inducing cells to move slower (Figure 3w, Table 2) and more persistently as a consequence of the reduced adhesion (Figure 4c,d; Table 3). The dose of 10 Gy on cells on the soft substrate produced a time-dependent reduction of migration velocity and, also, in this case, an increase of directional persistence (Figure 3w, Tables 2 and 3). On the other hand, MDA-MB-231 cells cultured on stiff substrate had a very similar response to both irradiation; in fact, their velocity remained unchanged 1 d after irradiation, but the persistence was unperturbed for the low dose and reduced for the high one. At 3 d after irradiation and irrespective of the dose delivered, we measured a significantly reduced migration velocity as a possible consequence of the increased adhesion (Figure 2b). To summarize, the above findings suggest that at 3 d after irradiation, tumor cells exhibited reduced migration, even if their persistence was not substantially changed by dose delivered. This effect was particularly relevant for tumor cells on ECM mimicking tumor microenvironment, which decreased their velocity from ~1 to 0.6 μm/min, approaching the velocity value of healthy cells on stiff ECM (~0.4 μm/min). What appears very clear is the importance of the microenvironment in mediating the cellular response to a physical insult such as photon irradiation. To better elucidate the mechanisms underlying the effect of RT on cell functions in relationship to the mechanics of ECM, it will be necessary to evaluate the effects of radiations on the expression of molecules involved in both processes, first of all, integrins and FAK. In fact, it has been shown that both integrins and FAK regulate biological processes necessary for the pathogenesis of cancer. Paszek et al. suggested that tumor stiffness contributes to the aberrant behavior of epithelial tissue by modulating integrin signaling through FAK [45]. In addition, irradiation can affect the expression of integrins and have serious repercussions on adhesion and invasion. As previously mentioned, Cordes et al. found that ionizing radiation promoted integrin expression in a substrate-dependent way, improving cell adhesion on fibronectin and Matrigel and impairing their invasion ability [11]. Oppositely, Rieken et al. indicated that the overexpression of integrins induced by photons had a promigratory effect on the same tumor cells [46]. On the other side, it will be fundamental also to explore the effects of irradiation on the architecture and mechanical properties of ECM. Our findings, indeed, indicate the importance of including its contribution and crosstalk with cell cytoskeleton architecture in the attempt to reconstruct all the pieces when deciphering cell response to RT treatments.

3. Materials and Methods

3.1. Preparation of Substrate and Mechanical Characterization

Polyacrylamide substrates were prepared as previously reported [47], with minor modifications. Briefly, glass-bottom culture dishes (World Precision Instruments, FD35–100) were silanized with 3-aminopropyltriethoxysilane (Sigma-Aldrich, St. Louis, MO, USA) for 20 min and extensively washed with water. Then, 40% acrylamide and 2% methylene-bis-acrylamide were mixed in phosphate-buffered saline (PBS) solution at two different final concentrations: 4% acrylamide/0.15% methylene-bis-acrylamide and 10% acrylamide/0.1% methylene-bis-acrylamide corresponding to 1.3 and 13 kPa (Young's modulus), respectively. Polymerization was initiated by adding 1/100 total volume of 10% ammonium persulfate and 1/1000 total volume N,N,N′,N′-tetramethylethylenediamide (TEMED, Sigma-Aldrich, T7024, St. Louis, MO, USA). Ten microliters of acrylamide/methylenebis-acrylamide mixture were pipetted on the treated dishes and covered with a 10-mm coverslip. After 20 min, the coverslip was removed and PBS was added to the dish. Before functionalization, the substrates were soaked with a penicillin–streptomycin solution overnight and then exposed to UV light emitted by a germicidal lamp for 1 h. After sterilizations, substrates were functionalized with collagen by using a bifunctional photoreactive crosslinker (sulfosuccinimidyl 6-(4′-azido-2′-nitrophenylamino) hexanoate, sulfo-SANPAH; Fischer Scientific, Loughborough, UK). The sulfo-SANPAH solution was diluted in water at a final concentration of 0.2 mg/mL, placed on PAAm substrates, and exposed to 365 nm UV for 10 min. After washing with PBS, the substrates were incubated with a solution of bovine type I collagen at the final concentration of 50 μg/mL for 2 h at 37 °C. Finally, samples were washed with PBS.

After the process of photoactivation, the sulfo-SANPAH solution was removed, and 0.25 mL solution of PBS and 2% bovine collagen was added to every dish before being incubated at 37 °C for 2 h.

3.2. Cell Culture

The MDA-MB-231 cell line was cultured in Lonza Dulbecco's modified Eagle medium (DMEM/F-12) supplemented with 10% fetal bovine serum (FBS, Gibco, Eggenstein, Germany), 1% L-glutamine (Sigma, St. Louis, MO, USA) and 1% penicillin–streptomycin (Sigma, St. Louis, MO, USA).

The MCF10A cell line was cultured in Lonza mammary epithelium-based medium (MEBM), supplemented with the Mammary Epithelial Cell Growth Medium SingleQuots Kit (MEGM): bovine pituitary extract (BPE), human epidermal growth factor (hEGF) (0.1%), insulin (0.1%), hydrocortisone (0.1%), gentamicin–amphotericin (GA-1000; 0.1%).

3.3. Cell Irradiation

MCF10A and MDA-MB-231 cell lines were exposed to X-rays (photon beams) delivered by the LINAC Synergy Agility (ELEKTA), using a 6MeV energy beam, at the National Cancer Institute "PASCALE" of Naples. Three-dimensional conformal radiation therapy (3D-CRT) treatment plans were realized with Monaco v5.11.03 TPS (treatment planning station) by Elekta to deliver 2 and 10 Gy, with the following setup. Cells were settled in a Petri dish between two solid water phantom slabs (2 and 3 cm each) and exposed at two opposite fields. The dose rate selected was 200 UM/min. Prescribed doses were delivered on a uniform square field of 20×20 cm^2 at the cell level.

The 3D-CRT approach was chosen for the simulation, being broadly used in clinical practice, although breast cancer treatment can be performed by several other radiotherapy techniques, as IMRT (intensity-modulated radiation therapy) and VMAT (volumetric-modulated arc therapy).

The 2 Gy represents the conventional fractional dose delivered in standard treatment, whereas 10 Gy was selected to represent the highest dose/fraction treatment as stereotactic body radiotherapy, even though rarely employed for breast cancer.

3.4. Cell Adhesion Analysis

MCF10A and MDA-MB-231 cells were cultured on PAAm substrates at a final density of 1000 cells/cm^2. Cells were fixed and immunostained for cell spreading and nuclear area in the control condition and 1 and 3 d after irradiation. Cells were washed with PBS and fixed in 4% paraformaldehyde for 20 min, rinsed twice with PBS, permeabilized with 0.1% Triton X100 for 5 min, and blocked with 10% goat serum for 1h. Cells were incubated with Alexa 488 phalloidin at 1:200 dilution. Finally, nuclei were stained with Hoechst 33342. Images of cells were acquired using an Olympus IX81 inverted microscope and a 10× objective. Images were imported into ImageJ software (NIH, Bethesda, MD, USA) for quantification of cell spreading and nuclei area. Individual cells and nuclei were thresholded manually on the basis of phalloidin and nuclei staining, and their spreading and nuclei areas were determined using the "Measure" command in ImageJ.

3.5. Cell Motility

MCF10A and MDA-MB-231 cells were cultured on PAAm substrates at a final density of 1000 cells/cm^2. Cell motility experiments were performed by time-lapse microscopy (Olympus IX81 with 4×). Phase-contrast images were acquired at 10 min interval for 24 h for a total number of 144 frames. Cell positions in each frame were tracked manually using ImageJ and Manual Tracking plugin (http://rsweb.nih.gov/ij/). Migration velocity and MSD were calculated, starting from trajectories using the following formula:

$$v = \sum \frac{\sqrt{(x_{i+1} - x_i)^2 + (y_{i+1} - y_i)^2}}{\Delta T} \quad (1)$$

$$MSD(\tau) = [x(t-\tau) - x(t)]^2 + [y(t-\tau) - y(t)]^2 \quad (2)$$

where x_i e y_i were the coordinates of cell in the i-th frame, ΔT is the time interval between two frames, t is the time, and τ is the lag time.

To estimate diffusion coefficient D and directional persistence P, MSDs curves were fitted with Fürth's Formula [48]:

$$MSD(\tau) = 4D\left(\tau - P\left(1 - e^{-\frac{\tau}{P}}\right)\right) \quad (3)$$

The fitting was done with ordinary nonlinear least-squares regression analysis.

3.6. Statistical Analysis

Statistical comparisons were performed with a Student's unpaired test when data exhibit a normal distribution. Otherwise, a nonparametric Mann-Whitney test was used. P-values of <0.05 denote statistically significant differences.

4. Conclusions

In summary, our findings indicate that ECM mechanics can play a very active role in mediating responses of cells to RT. In particular, RT had significant effects on biophysical properties of MDA-MB-231 cells cultured on stiff ECM mimicking tumor environment, whereas their adhesion and nuclei resulted in being less affected when cells interact with substrate mimicking a physiological mechanical environment. MCF10A features were less affected by RT on both substrates, suggesting a stronger ability of these cells to preserve themselves, at least in terms of spreading and nuclei morphology. On the contrary, the migration velocity of both cell lines was significantly reduced on soft substrate, indicating a sort of radioprotective role of physiological ECM that impaired cell motility and invasion. These preliminary findings, together with further and thorough examinations, can shed light on the ECM-mediated molecular mechanism underlying the response of healthy and tumor cells to radiation treatments, generate data with a higher translational significance, and pave the way for exploring new frontiers in the diagnostic and therapeutic use of RT.

Supplementary Materials: The following are available online at http://www.mdpi.com/2072-6694/12/5/1170/s1, Figure S1: Zoomed pictures of spreading of MCF10A and MDA-MB-231 cells, Figure S2: Percentage distribution of cell spreading area of MCF10A cells before and after RT, Figure S3: Percentage distribution of cell spreading area of MDA-MB-231 cells before and after RT.

Author Contributions: Conceptualization, V.P. and S.F.; methodology, V.P.; software, V.P.; validation, V.P. and S.F.; formal analysis, V.P.; investigation, V.P., G.L.V., and V.A.; resources, M.P., C.A., P.M., and P.A.N.; data curation, V.P., G.L.V., and V.A.; writing—original draft preparation, V.P. and S.F.; writing—review and editing, V.P., G.L.V., M.P., V.A., C.A., P.M., M.L.C., P.A.N., and S.F.; visualization, V.P.; supervision, P.A.N. and S.F.; project administration, V.P. and S.F.; funding acquisition, M.L.C., and P.A.N. All authors have read and agreed to the published version of the manuscript.

Funding: This work was supported by a grant from Regione Campania-POR Campania FESR 2014/2020 "Combattere la resistenza tumorale: piattaforma integrata multidisciplinare per un approccio tecnologico innovativo alle oncoterapie-Campania Oncoterapie" (Project N. B61G18000470007).

Acknowledgments: We thank Elena Leone for her precise proofreading.

Conflicts of Interest: The authors declare no conflict of interest.

References

1. Ferlay, J.; Colombet, M.; Soerjomataram, I.; Dyba, T.; Randi, G.; Bettio, M.; Gavin, A.; Visser, O.; Bray, F. Cancer incidence and mortality patterns in Europe: Estimates for 40 countries and 25 major cancers in 2018. *Eur. J. Cancer* **2018**, *103*, 356–387. [CrossRef] [PubMed]
2. Sun, Y.S.; Zhao, Z.; Yang, Z.N.; Xu, F.; Lu, H.J.; Zhu, Z.Y.; Shi, W.; Jiang, J.; Yao, P.P.; Zhu, H.P. Risk factors and preventions of breast cancer. *Int. J. Biol. Sci.* **2017**, *13*, 1387. [CrossRef] [PubMed]
3. Olive, P.L. The role of DNA single-and double-strand breaks in cell killing by ionizing radiation. *Radiat. Res.* **1998**, *150*, S42–S51. [CrossRef] [PubMed]
4. Lliakis, G. The role of DNA double strand breaks in ionizing radiation-induced killing of eukaryotic cells. *BioEssays* **1991**, *13*, 641–648. [CrossRef]
5. Puck, T.T. Action of radiation on mammalian cells III. Relationship between reproductive death and induction of chromosome anomalies by X-irradiation of euploid human cells in vitro. *Proc. Natl. Acad. Sci. USA* **1958**, *44*, 772. [CrossRef]
6. Iliakis, G.; Wang, Y.A.; Guan, J.; Wang, H. DNA damage checkpoint control in cells exposed to ionizing radiation. *Oncogene* **2003**, *22*, 5834–5847. [CrossRef]
7. Yilmaz, M.; Christofori, G. EMT, the cytoskeleton, and cancer cell invasion. *Cancer Metastasis Rev.* **2009**, *28*, 15–33. [CrossRef]
8. Panzetta, V.; Musella, I.; Rapa, I.; Volante, M.; Netti, P.A.; Fusco, S. Mechanical phenotyping of cells and extracellular matrix as grade and stage markers of lung tumor tissues. *Acta Biomater.* **2017**, *57*, 334–341. [CrossRef]
9. Stowers, R.S.; Allen, S.C.; Sanchez, K.; Davis, C.L.; Ebelt, N.D.; Van Den Berg, C.; Suggs, L.J. Extracellular matrix stiffening induces a malignant phenotypic transition in breast epithelial cells. *Cell Mol. Bioeng.* **2017**, *10*, 114–123. [CrossRef]
10. Panciera, T.; Citron, A.; Di Biagio, D.; Battilana, G.; Gandin, A.; Giulitti, S.; Forcato, M.; Biccoato, S.; Panzetta, V.; Fusco, S.; et al. Reprogramming normal cells into tumour precursors requires ECM stiffness and oncogene-mediated changes of cell mechanical properties. *Nat. Mater.* **2020**, 1–10. [CrossRef]

11. Cordes, N.; Hansmeier, B.; Beinke, C.; Meineke, V.; Van Beuningen, D. Irradiation differentially affects substratum-dependent survival, adhesion, and invasion of glioblastoma cell lines. *Br. J. Cancer* **2003**, *89*, 2122–2132. [CrossRef] [PubMed]
12. Hohmann, T.; Grabiec, U.; Vogel, C.; Ghadban, C.; Ensminger, S.; Bache, M.; Vordermark, D.; Dehghani, F. The impact of non-lethal single-dose radiation on tumor invasion and cytoskeletal properties. *Int. J. Mol.* **2017**, *18*, 2001. [CrossRef] [PubMed]
13. Panzetta, V.; De Menna, M.; Bucci, D.; Giovannini, V.; Pugliese, M.; Quarto, M.; Fusco, S.; Netti, P.A. X-ray irradiation affects morphology, proliferation and migration rate of healthy and cancer cells. *J. Mech. Med. Biol.* **2015**, *15*, 1540022. [CrossRef]
14. Panzetta, V.; Musella, I.; Pugliese, M.; Piccolo, C.; Pasqua, G.; Netti, P.A.; Fusco, S. Effects of high energy X-rays on cell morphology and functions. In Proceedings of the 2017 IEEE 5th Portuguese Meeting on Bioengineering (ENBENG), Coimbra, Portugal, 16–18 February 2017; pp. 1–4.
15. Panzetta, V.; Pugliese, M.G.; Musella, I.; De Menna, M.; Netti, P.A.; Fusco, S. A biophysical analysis to assess X-ray sensitivity of healthy and tumour cells. *Radiat. Prot. Dosim.* **2019**, *183*, 116–120. [CrossRef] [PubMed]
16. Imaizumi, H.; Sato, K.; Nishihara, A.; Minami, K.; Koizumi, M.; Matsuura, N.; Hieda, M. X-ray-enhanced cancer cell migration requires the linker of nucleoskeleton and cytoskeleton complex. *Cancer Sci.* **2018**, *109*, 1158–1165. [CrossRef]
17. Beck, C.; Piontek, G.; Haug, A.; Bas, M.; Knopf, A.; Stark, T.; MartinMißlbeck, M.; Rudelius, M.; Reiter, R.; Brandstetter, M.; et al. The kallikrein–kinin-system in head and neck squamous cell carcinoma (HNSCC) and its role in tumour survival, invasion, migration and response to radiotherapy. *Oral Oncol.* **2012**, *48*, 1208–1219. [CrossRef]
18. Pickhard, A.C.; Margraf, J.; Knopf, A.; Stark, T.; Piontek, G.; Beck, C.; Boulesteix, A.L.; Scherer, E.Q.; Pigorsch, S.; Schelegel, J.; et al. Inhibition of radiation induced migration of human head and neck squamous cell carcinoma cells by blocking of EGF receptor pathways. *BMC Cancer* **2011**, *11*, 388. [CrossRef]
19. De Bacco, F.; Luraghi, P.; Medico, E.; Reato, G.; Girolami, F.; Perera, T.; Gabriele, P.; Comoglio, P.M.; Boccaccio, C. Induction of MET by ionizing radiation and its role in radioresistance and invasive growth of cancer. *J. Natl. Cancer Inst.* **2011**, *103*, 645–661. [CrossRef]
20. Badiga, A.V.; Chetty, C.; Kesanakurti, D.; Are, D.; Gujrati, M.; Klopfenstein, J.D.; Dinh, D.D.; Rao, J.S. MMP-2 siRNA inhibits radiation-enhanced invasiveness in glioma cells. *PLoS ONE* **2011**, *6*, e20614. [CrossRef]
21. Jung, J.W.; Hwang, S.Y.; Hwang, J.S.; Oh, E.S.; Park, S.; Han, I.O. Ionising radiation induces changes associated with epithelial-mesenchymal transdifferentiation and increased cell motility of A549 lung epithelial cells. *Eur. J. Cancer* **2007**, *43*, 1214–1224. [CrossRef]
22. Torok, J.A.; Salama, J.K. Combining immunotherapy and radiotherapy for the STAR treatment. *Nat. Rev. Clin. Oncol.* **2019**, *16*, 666–667. [CrossRef] [PubMed]
23. Blyth, B.J.; Cole, A.J.; MacManus, M.P.; Martin, O.A. Radiation therapy-induced metastasis: Radiobiology and clinical implications. *Cli. Exp. Metastas.* **2018**, *35*, 223–236. [CrossRef] [PubMed]
24. Huang, S.; Ingber, D.E. Cell tension, matrix mechanics, and cancer development. *Cancer Cell* **2005**, *8*, 175–176. [CrossRef] [PubMed]
25. Plodinec, M.; Loparic, M.; Monnier, C.A.; Obermann, E.C.; Zanetti-Dallenbach, R.; Oertle, P.; Hyotyla, J.T.; Aebi, U.; Bentires-Alj, M.; Lim, R.Y.; et al. The nanomechanical signature of breast cancer. *Nat. Nanotechnol.* **2012**, *7*, 757. [CrossRef] [PubMed]
26. Li, Q.S.; Lee, G.Y.; Ong, C.N.; Lim, C.T. AFM indentation study of breast cancer cells. *Biochem. Biophys. Res. Commun.* **2008**, *374*, 609–613. [CrossRef]
27. Baker, E.L.; Lu, J.; Yu, D.; Bonnecaze, R.T.; Zaman, M.H. Cancer cell stiffness: Integrated roles of three-dimensional matrix stiffness and transforming potential. *Biophys. J.* **2010**, *99*, 2048–2057. [CrossRef]
28. Nikkhah, M.; Strobl, J.S.; De Vita, R.; Agah, M. The cytoskeletal organization of breast carcinoma and fibroblast cells inside three dimensional (3-D) isotropic silicon microstructures. *Biomaterials* **2010**, *31*, 4552–4561. [CrossRef]
29. Levental, K.R.; Yu, H.; Kass, L.; Lakins, J.N.; Egeblad, M.; Erler, J.T.; Fong, S.F.; Csiszar, K.; Giaccia, A.; Weninger, W.; et al. Matrix crosslinking forces tumor progression by enhancing integrin signaling. *Cell* **2009**, *139*, 891–906. [CrossRef]

30. Acerbi, I.; Cassereau, L.; Dean, I.; Shi, Q.; Au, A.; Park, C.; Chen, Y.Y.; Liphardt, J.; Hwang, E.S.; Weaver, V.M. Human breast cancer invasion and aggression correlates with ECM stiffening and immune cell infiltration. *Integr. Biol.* **2015**, *7*, 1120–1134. [CrossRef]
31. Fusco, S.; Panzetta, V.; Embrione, V.; Netti, P.A. Crosstalk between focal adhesions and material mechanical properties governs cell mechanics and functions. *Acta Biomater.* **2015**, *23*, 63–71. [CrossRef]
32. Alam, S.; Lovett, D.B.; Dickinson, R.B.; Roux, K.J.; Lele, T.P. Nuclear forces and cell mechanosensing. In *Progress in Molecular Biology and Translational Science*; Engler, A.J., Kumar, S., Eds.; Academic Press: Cambridge, MA, USA, 2014; Volume 126, pp. 205–215.
33. Panzetta, V.; Fusco, S.; Netti, P.A. Cell mechanosensing is regulated by substrate strain energy rather than stiffness. *Proc. Natl. Acad. Sci. USA* **2019**, *116*, 22004–22013. [CrossRef] [PubMed]
34. Dupont, S.; Morsut, L.; Aragona, M.; Enzo, E.; Giulitti, S.; Cordenonsi, M.; Zanconato, F.; Le Digabel, J.; Forcato, M.; Bicciato, S.; et al. Role of YAP/TAZ in mechanotransduction. *Nature* **2011**, *474*, 179–183. [CrossRef] [PubMed]
35. Panzetta, V.; Pugliese, M.; di Gennaro, G.; Federico, C.; Arrichiello, C.; Muto, P.; Netti, P.A.; Fusco, S. X-rays affect cytoskeleton assembly and nanoparticle uptake: Preliminary results. *IL NUOVO CIMENTO* **2018**, *100*, 41.
36. Fusco, S.; Panzetta, V.; Netti, P.A. Mechanosensing of substrate stiffness regulates focal adhesions dynamics in cell. *Meccanica* **2017**, *52*, 3389–3398. [CrossRef]
37. Kraning-Rush, C.M.; Reinhart-King, C.A. Controlling matrix stiffness and topography for the study of tumor cell migration. *Cell Adhes. Migr.* **2012**, *6*, 274–279. [CrossRef]
38. Baker, A.M.; Bird, D.; Lang, G.; Cox, T.R.; Erler, J.T. Lysyl oxidase enzymatic function increases stiffness to drive colorectal cancer progression through FAK. *Oncogene* **2013**, *32*, 1863–1868. [CrossRef]
39. Haage, A.; Schneider, I.C. Cellular contractility and extracellular matrix stiffness regulate matrix metalloproteinase activity in pancreatic cancer cells. *FASEB J.* **2014**, *28*, 3589–3599. [CrossRef]
40. Peela, N.; Sam, F.S.; Christenson, W.; Truong, D.; Watson, A.W.; Mouneimne, G.; Ros, R.; Nikkhah, M. A three dimensional micropatterned tumor model for breast cancer cell migration studies. *Biomaterials* **2016**, *81*, 72–83. [CrossRef]
41. Lin, F.; Zhang, H.; Huang, J.; Xiong, C. Substrate stiffness coupling TGF-β1 modulates migration and traction force of MDA-MB-231 human breast cancer cells in vitro. *ACS Biomater. Sci. Eng.* **2018**, *4*, 1337–1345. [CrossRef]
42. Kornberg, L.J. Focal adhesion kinase and its potential involvement in tumor invasion and metastasis. *Head Neck-J. Sci. Spec. Head Neck* **1998**, *20*, 745–752. [CrossRef]
43. Cance, W.G.; Harris, J.E.; Iacocca, M.V.; Roche, E.; Yang, X.; Chang, J.; Simkins, S.; Xu, L. Immunohistochemical analyses of focal adhesion kinase expression in benign and malignant human breast and colon tissues: Correlation with preinvasive and invasive phenotypes. *Clin. Cancer Res.* **2000**, *6*, 2417–2423. [PubMed]
44. Provenzano, P.P.; Inman, D.R.; Eliceiri, K.W.; Keely, P.J. Matrix density-induced mechanoregulation of breast cell phenotype, signaling and gene expression through a FAK–ERK linkage. *Oncogene* **2009**, *28*, 4326–4343. [CrossRef] [PubMed]
45. Paszek, M.J.; Zahir, N.; Johnson, K.R.; Lakins, J.N.; Rozenberg, G.I.; Gefen, A.; Reinhart-King, C.A.; Margulies, S.S.; Dembo, M.; Boettiger, D.; et al. Tensional homeostasis and the malignant phenotype. *Cancer Cell* **2005**, *8*, 241–254. [CrossRef] [PubMed]
46. Rieken, S.; Habermehl, D.; Wuerth, L.; Brons, S.; Mohr, A.; Lindel, K.; Weber, K.; Haberer, T.; Debus, J.; Combs, S.E. Carbon ion irradiation inhibits glioma cell migration through downregulation of integrin expression. *Int. J. Radiat.* **2012**, *83*, 394–399. [CrossRef] [PubMed]
47. Panzetta, V.; Guarnieri, D.; Paciello, A.; Della Sala, F.; Muscetti, O.; Raiola, L.; Netti, P.A.; Fusco, S. ECM mechano-sensing regulates cytoskeleton assembly and receptor-mediated endocytosis of nanoparticles. *ACS Biomater. Sci. Eng.* **2017**, *3*, 1586–1594. [CrossRef]
48. Fürth, R. Die brownsche bewegung bei berücksichtigung einer persistenz der bewegungsrichtung. mit anwendungen auf die bewegung lebender infusorien. *Z. Für Phys.* **1920**, *2*, 244–256. [CrossRef]

cancers

MDPI

Review

Cell Motility and Cancer

Ildefonso M. De la Fuente [1,2,*] and José I. López [3,*]

1 Department of Nutrition, CEBAS-CSIC Institute, Espinardo University Campus, 30100 Murcia, Spain
2 Department of Mathematics, Faculty of Science and Technology, University of the Basque Country, 48940 Leioa, Spain
3 Department of Pathology, Cruces University Hospital, Biocruces-Bizkaia Health Research Institute, 48903 Barakaldo, Spain
* Correspondence: mtpmadei@ehu.eus (I.M.D.l.F.); joseignacio.lopez@osakidetza.eus (J.I.L.)

Received: 16 July 2020; Accepted: 3 August 2020; Published: 5 August 2020

check for updates

Abstract: Cell migration is an essential systemic behavior, tightly regulated, of all living cells endowed with directional motility that is involved in the major developmental stages of all complex organisms such as morphogenesis, embryogenesis, organogenesis, adult tissue remodeling, wound healing, immunological cell activities, angiogenesis, tissue repair, cell differentiation, tissue regeneration as well as in a myriad of pathological conditions. However, how cells efficiently regulate their locomotion movements is still unclear. Since migration is also a crucial issue in cancer development, the goal of this narrative is to show the connection between basic findings in cell locomotion of unicellular eukaryotic organisms and the regulatory mechanisms of cell migration necessary for tumor invasion and metastases. More specifically, the review focuses on three main issues, (i) the regulation of the locomotion system in unicellular eukaryotic organisms and human cells, (ii) how the nucleus does not significantly affect the migratory trajectories of cells in two-dimension (2D) surfaces and (iii) the conditioned behavior detected in single cells as a primitive form of learning and adaptation to different contexts during cell migration. New findings in the control of cell motility both in unicellular organisms and mammalian cells open up a new framework in the understanding of the complex processes involved in systemic cellular locomotion and adaptation of a wide spectrum of diseases with high impact in the society such as cancer.

Keywords: cell motility; migration; conditioned behavior; learning; cancer; invasion; metastasis

1. Introduction

Cell migration is essential for a plethora of fundamental biological processes and human pathologies such as cancer. The molecular and biochemical mechanisms through which individual cells move have been extensively studied. However, the principles that govern cell motility at a systemic level are still largely unknown. This narrative reviews seminal aspects of cell motility and its application to cancer, in particular, the usefulness of analyzing systemic properties of unicellular eukaryotic organisms to understand cancer cell migration. The importance of a systemic approach to the external stimuli involved in cellular locomotion have provided important findings, such as the limited role of the nucleus in cell motility on two-dimension (2D) surfaces and the emergence of a new behavior by which the cells do learn and develop an associative memory to respond to the environmental changes during cell migration.

Cancer is a health problem of major concern and a leading cause of death in Western societies. Local invasion and metastatic seed in distant territories are complex biological processes that impact negatively in patient prognosis. These issues are receiving much attention in the last years [1]. A better knowledge of the systemic mechanisms underlying cell motility is necessary to advance in the development of efficient therapies to improve cancer prognosis.

2. Simple Organism Models Are Necessary to Understand Human Cell Behavior

Locomotion is a crucial ability to survive for many unicellular eukaryotic cells and the translation of the information obtained from these single organisms to human cancer and other diseases is a milestone in modern medicine. In general terms, motility -understood as cell displacement- has been well conserved over hundreds of millions of years of evolution from unicellular eukaryotic organisms to human cells. For this reason, many studies have analyzed cell motility in non-mammalian model organisms, like amoebae, worms, flies and others. Elegant studies with *Batrachochytrium dendrobatidis* [2–4], *Dictyostelium discoideum* [5–8], *Caenorhabditis elegans* [9–11] and *Drosophila melanogaster* [12–16], among others, have provided useful information to understand the fundamental mechanisms of cell motility.

As it has been recently reviewed [5,6], *D. discoideum* is an important model to analyze cell locomotion, chemotaxis and many other cell characteristics, in part because this social amoeba evolves from a unicellular to a multicellular stage during its cell cycle. *D. discoideum* has a very well developed phagocytic ability and defense mechanisms against potential pathogens making this amoeba a good mimicker of macrophages and other mammalian cells with motile properties. On the other hand, several genes in its genome are homologous to some disease genes in humans [6]. This fact makes once more this microorganism a model to analyze the mechanisms of action involved in several human diseases such as inherited Parkinson's disease [7] or cancer [8].

Basal membrane disruption is the first step in tumor invasion and has been analyzed in simple multicellular organisms like *C. elegans* [9], where the anchor cell during its larval development breaks basement membranes during morphogenesis. Another example of the crucial information obtained from this nematode and translated to cancer are the advances in the knowledge of the molecular mechanisms of apoptosis [10]. In fact, *C. elegans* has emerged as a simple animal model for systematic dissection of the molecular basis of tumorigenesis, focusing on the well-established processes of apoptosis and autophagy [11].

More complex multicellular organisms such as *D. melanogaster* show a model of metastatic potential through the development of several mutations conferring different potentialities to increase cell migration. Such studies are improving the understanding of some fundamental processes in cancer, local invasion and metastases included, for example the discovering of the Hedgehog and WNT pathways [12]. Strikingly, mammalian and *D. melanogaster* intestines share many similarities [13]. For these reasons, this fly has been also a model to understand the development of collective cell migration and metastases through epithelial-mesenchymal transition (EMT) processes driven by the transcription factor Snail [14]. Indeed, this organism has been used for the analysis of possible therapeutic routes in cancer [15,16]. Models of glioblastoma and rhabdomyosarcoma developed in *D. melanogaster* have allowed a better knowledge of the genomic alterations underlying neoplasms. The inhibition of the RET oncogenic activity to treat multiple endocrine neoplasia by newly designed chemicals [15] is another good example of the applicability of these studies in *D. melanogaster*.

3. The Locomotion System in Unicellular Eukaryotic Organisms and Human Cells

Cellular migration is controlled by complex molecular and metabolic networks. These networks shape an intricate interplay of multiple components amongst which the cytoskeleton, a large number of adhesion proteins, varied signaling processes and sophisticated biochemical regulatory networks are fundamentally included.

The cytoskeleton of eukaryotic cells (the main part of the locomotion system) is a dynamic structure formed by three main components—actin microfilaments, microtubules and intermediate filaments, all of them interacting in complex dynamic networks. Cell migration in amoeboid organisms is largely dependent on regulatory mechanisms of the actin cytoskeleton, in which integrins, Rac small GTPases and many post-translational modifications such as Arp2 phosphorylation, adhesion associated proteins including talin, paxillin and vinculin and numerous intracellular signaling molecules, take part modifying the systemic migratory behavior of cells [5].

Most of the spatial-temporally regulated actin dynamics in human cells, for instance leukocytes, share great similarity to amoeboid unicellular organisms [17]. However, there are some differences between them, for example, with respect to the Cdc42 protein, a component that plays a prominent role in the directed actin dynamics of leukocytes [18]. This protein is a small GTPase of the Rho family that is present in a variety of organisms, from yeasts to mammals. Cdc42 regulates the signaling pathways that control important cellular functions including cell migration, endocytosis and cell cycle progression. Recently, Cdc42 has been directly implicated in cancer progression. Actually, Cdc42 is overexpressed in lung, colorectal, breast and testicular cancer, as well as in melanoma [19]. Albeit some variations among different cell types may occur, the basic structures of the cytoskeleton are similar in many unicellular eukaryotic organisms and human cells. These similarities are supported by highly conserved gene products and numerous metabolic processes in most eukaryotic organisms endowed with directional motility [20].

Cell-autonomous polarity is also required for adequate directionality motion and optimal chemoattractant reception [21]. There are a lot of molecular processes that have been implicated in the intrinsic polarity status of cells. One of the best studied is the receptor/G protein network, which detects the external gradients (by means of the chemoattractant receptors) and transmits the external sensorial gradient to the signal transduction system. These processes amplify the directional bias and spreads the asymmetric molecular information to the cytoskeleton system which subsequently generates a protrusive force with specific polarized cell movements according to the prevailing molecular composition in the external cellular medium. A notable difference found between unicellular eukaryotic organisms and human cells is the variable richness in the repertoire of receptors and ligands controlling directed migration in different cells. For example, while only a few chemo-attractants have been identified in *D. discoideum*, human leukocytes respond to numerous molecules, as for instance PAF (Platelet-activating factor), LTB4, C5a, Interleukin-8 (IL8) and growth factors such as IGF-1, EGF, PDGF and TGF-β [22].

On the other hand, GTPases of the Ras superfamily act as enzymatic central mechanism to control a wide range of essential metabolic pathways such as actin cytoskeletal integrity, cell adhesion and cell migration in all eukaryotic cells endowed with motile abilities. In amoeboid unicellular organisms, as for instance in *D. discoideum,* Ras GTPase activity is directly implicated in cell locomotion and signal transduction, where it transfers the input from the receptor/G protein network to several metabolic activities, including PI3K/PIP3, Rap1, cGMP/Myosin II and TORC2/PKB pathways. In mammalian cells GTPases of the Ras superfamily also regulate cell proliferation, differentiation, migration and apoptosis. Roughly 60 types of Ras GTPases have been identified. In leukocytes, Ras GPTase has been involved in PI3K/PIP3 and MAPK processes [5]. *D. discoideum* exhibits several members of the Ras GTPases enzymes, belonging to 14 Ras family genes with 5 characterized isoforms which share similarities with mammalian H-Ras (proto-oncogene involved in the development of several types of cancer) and K-Ras (proto-oncogene involved in the Warburg effect of cancer cells) [23]. Mutations in this Ras family of proto-oncogenes are very common in human cells, being found in 20% to 30% of all tumors [24]. Ras GTPases are highly conserved between *D. discoideum* and mammalian cells and there is a basic similarity in the overall organization of the signal transduction networks in amoeboid unicellular organisms and human cells [5].

Another important cytoskeletal remodeling is the activity of PAKs enzymes which are found in all eukaryotic cells. These groups of enzymes (p21-activated kinases) are serine/threonine protein kinases effectors of the Rho family of GTPases, which are responsible in the direct regulation of cell migration, chemotaxis, cell polarity, plasticity and signaling [25]. Three PAK families of genes have been identified in *D. discoideum* and six isoforms of PAKs are expressed in human cells [25], which are implicated in a variety of processes including cytoskeletal dynamics, cell migration, cell cycle, mitosis, apoptosis, angiogenesis, tumorigenesis and metastasis [26]. In fact, PAKs are frequently up-regulated in human diseases, including various types of cancers [27]. However, it is worth noting that mammalian PAKs activities are still not completely understood. Most of our knowledge about PAK functions has been derived from approaches in unicellular eukaryotic organisms and many of these functions are similar to those seen in human cells. Such studies have demonstrated that the basic structure and functions of PAKs are conserved across practically most eukaryotic cells.

Chemotaxis and cell adhesion are also controlled by Rap1 (Ras-proximate-1 or Ras-related protein 1), another small GTPase, which acts as a molecular switch essential for effective signal transduction being involved in important cellular functions as substratum adhesion, cell motility, apoptosis, cytoskeleton remodeling, motility and intracellular vesicular transport [28]. This enzyme was originally discovered in budding yeast as a telomere-binding protein that is activated in response to a range of stimuli through a number of second-messenger molecules, such as diacylglycerol, cAMP and Ca^{2+} [29]. Rap1 is rapidly activated in response to chemoattractant stimulation regulating the cytoskeletal structure and the adhesion processes in *D. discoideum* [30]. The mechanisms by which Rap1 controls cytoskeletal reorganizations in this unicellular eukaryotic organism are still under investigation. In human cells, however, Rap1 controls cell spreading by mediating the functions of integrins and regulating cell adhesion through the interaction and regulation of adaptor proteins. Specifically, this protein is an important mediator of adhesion, polarity and migration in leukocytes [31]. In addition, Rap1 also plays many roles during cell invasion and metastasis in different human cancers [32]. Different studies have shown that Rap1 is very highly conserved in amoeboid unicellular organisms and human cells [5].

Cell motility requires a complex orchestrated spatial-temporal regulation of thousands of biomolecules which shape complex dynamic networks that are not well understood. This complexity justifies why cell migration still remains a fundamental unresolved problem in contemporary biology with crucial implications in a wide spectrum of diseases, such as cancer. Anyway, several studies performed in unicellular eukaryotic cells have shown important advances in this area and numerous investigations have shown that unicellular organisms provide an excellent experimental model system to understand the precise role of many molecules and metabolic processes in cellular migration. In fact, most of our knowledge about molecular functions on cell motility has been derived from different approaches in lower organisms and many of these functions are similar to that seen in Metazoans and in particular in human cells.

Despite the differences between lower eukaryotes and higher organisms, the fundamental networks' architecture, as well as the principles of the systemic complex orchestration and many individual regulatory modules of cell metabolism has been remarkably conserved during evolution.

4. External Stimuli, Migration and Cancer

The influence of external stimuli during cell migration has been classically analyzed by physiologists. In fact, galvanotaxis and chemotaxis have been documented in unicellular eukaryotic organisms as far as more than one hundred years ago [33]. In addition, other dynamic forces of cell guidance such as haptotaxis [34,35], barotaxis [36,37], durotaxis [38–46], topotaxis [47,48] and plithotaxis [49–51] have been described latterly as additional conditioning factors of cell migration.

Galvanotaxis (electrotaxis), that is, the ability of simple organisms to predictably react in an electric field, has been an issue of research in biology for decades. *Amoeba proteus*, for instance, which has served for more than one hundred years as a cellular model to study cell migration and cytoskeletal function [52], exhibits robust galvanotaxis [53]. The molecular processes that govern cell behavior under galvanotactic conditions are not well understood. However, it is known that different mechanisms are involved in this behavior, for example, the bidirectional traffic of Ca^{2+} through the cell membranes, the sequential events of actin polymerization/depolymerization and the actomyosin contractility [54–56].

The effect of galvanotaxis has also been analyzed in tumor cells [54]. Changes in the concentration of Ca^{2+} may influence cell adhesion, a characteristic that in cancer is related to local invasion and metastases. However, it is interesting to note that cancer cells may behave very differently from their normal counterparts under similar conditions. For example, Wang et al. [57] demonstrated that human lens epithelial cells under galvanotaxis conditions migrate, as expected, to the cathode but their transformed malignant counterparts did it to the anode. Even more, this unexpected migration to the anode of malignant cells seems to be also related to the level of tumor aggressiveness. In this sense, the experiments of Frazer et al. [58] show that a highly aggressive metastatic human breast cancer cell line migrated to the anode whereas a non-metastatic human breast cancer cell line did it to the cathode. This paradoxical effect also affects the microenvironmental compartment since the directional traffic of intravasation/extravasation of tumor-infiltrating lymphocytes may be also governed by galvanotaxis [59]. These experiments and others, show that cells react in different ways depending on their malignancy and the degree of aggressiveness, opening new opportunities for cancer research in vivo.

Chemotaxis is another essential process implicated in cell migration. For instance, chemokines-mediated chemotaxis, a phenomenon that stimulates local invasion and cell migration in malignant tumors [1], has also been previously detected in bacteria, amoebae and other unicellular eukaryotic organisms [1,5,60]. Several types of individual and collective cell migration have been described during Zebrafish morphogenesis through intercellular signaling [1]. Tumor cells at the border of infiltration display a collective behavior via E-cadherin between tumor cells themselves and with the microenvironment [61]. In this sense, macrophages, lymphocytes and neutrophils are modulated by chemokines in pituitary neuroendocrine tumors [62]. Also, tumor-associated fibroblasts increase local tumor aggressiveness favoring the development of metastases, as it has been reported in different tumors [63–65]. Cell migration due to chemotaxis can also be observed in processes related to tumor development, as it has been demonstrated with T-cells serving as the basis for the recently developed immune therapies [66]. In this sense, the PD-1/PD-L1 axis blockage is a promising therapeutic tool that is being successfully used in several types of malignant tumors, including kidney [67], breast [68], lung [69] and bladder [70] cancers, as well as malignant melanoma [71]. Chemotaxis can be associated to the so-called haptotaxis, which refers to the directional motility induced by a gradient of cell adhesion [34] and differs from chemotaxis in the nature of the chemoattractant molecule. So, the chemoattractant is soluble in chemotaxis and insoluble (linked to the extracellular matrix) in haptotaxis. A recent work has demonstrated fibronectin mediated haptotaxis driving directional movements in breast cancer cells during metastatic progression [35].

On the other hand, the capacity of migration following hydraulic gradients in the absence of chemical cues is called barotaxis. Several experiments have demonstrated that cells confined to bifurcating channels select the channel of lower hydraulic resistance to orientate and migrate [36,37]. Interestingly, barotaxis does not need chemical attractors, although both barotaxis and chemotaxis may cooperate influencing cell migration, for example in neutrophils [36] and dendritic cells [37], under specific circumstances. In a system of competition between chemotaxis with the peptide fMLP against barotaxis, Prentice-Mott et al. [36] showed that large cells with high asymmetrical capacities did not respond to chemotactic stimulus and directed their movements to low-resistance channels, whereas neutrophils (small cells with lesser asymmetric potential) could successfully overcome high hydraulic pressures to reach the chemotactic stimulus. Besides, barotaxis may guide dendritic cells to reach the lymph nodes to initiate the immune response regulating micropinocytosis by means of modification in the actomyosin cytoskeleton [37].

In addition, durotaxis refers to the capacity of cells to migrate following matrix stiffness cues and was originally described in fibroblasts [38]. Contact guidance, that is, the capacity of cells to follow extracellular matrix fiber orientation is the context in which durotaxis develops [39]. This cellular systemic property can be observed in many unicellular and multicellular organisms like *D. discoideum* [40] and *C. elegans* [41]. *C. elegans*, for example, is able to detect, adapt and migrate to stiffer regions in the environment by means of undulatory waves in a dorsal-ventral plane that can be modified in shape and speed depending on the changing surrounding parameters [42]. Durotaxis is involved in embryogenesis, organogenesis, inflammation, tissue repair and other physiological processes. Vascular smooth muscle cells and other mesenchymal cells, for example, show a directional motility tendency generated by increasing gradients of extracellular stromal rigidity [43,44]. Isenberg et al. [43] have provided evidence of cellular adaptations in smooth muscle cells to durotactic gradients performed on polyacrylamide gels and hypothesize with connections between chemotactic and durotactic phenomenological responses. Durotaxis appears also in cancer, for instance, tumor cell lines from glioblastoma, breast carcinoma and mesenchymal fibrosarcoma do exhibit this behavior [45]. Here, all tumor cells show a similar pattern to move towards high-stiffness gradients. The stiffening of the stroma occurring in desmoplastic tumors may offer a via of durotactic escape to cells under selective pressures, like hypoxia, for example in colorectal cancer, where extracellular matrix changes with collagen overexpression, pathological collagen crosslinking and fiber arrangement take place [46]. This way, durotaxis could contribute with chemotaxis and other external stimuli to local invasion and metastases in many tumors.

Besides, topotaxis has been recently described as the capacity of cell to mediate their migration following the density of extracellular matrix fibers [47]. Since this cell property depends of the specific stiffness of the cytoskeleton, different migratory behaviors have been observed between benign and malignant counterparts of the same cell. The loss of PTEN in aggressive variants of malignant melanoma modifies cell stiffness making them softer then switching the topotactic polarity and migration [47]. By contrast, topotaxis of melanoma cells with preserved PTEN expression migrate to areas with denser extracellular matrix making aggressive spread more difficult. A similar process has been demonstrated in cutaneous fibroblasts during wound repair [48].

Finally, plithotaxis explain driving forces for collective cell migration in cellular monolayers in vitro [49]. This collective cell behavior takes place in many different contexts with endothelial, fibroblasts and epithelial cells, for example, during morphogenesis of complex branched organs like lungs or kidneys, in the process of wound repair [50] and also in the collective invasion of carcinoma cells [51]. Two different cell-to-cell junction stresses working combined are main actors in the underlying mechanism governing this collective cell behavior—the normal stress or forces working perpendicular to the surface that can be tensile or compressive and the shear stress or forces applied parallel to the tangent of the surface [49]. This mechanism allows groups of cells to move collectively following chemical or physical gradients. How individual cells move inside a collectivity is explained, among other processes, by plithotaxis, an emergent dynamic property necessary to understand complex biological processes involving millions of cells like those participating in tumor invasion and metastases.

5. The Role of the Nucleus in Cell Migration

One of the central issues in cellular migration is the role of the nucleus in the regulation of the locomotion system. The nucleus has been classically considered to be a key structure in cellular migration but its exact role is being understood only very recently. Graham et al. [72] have observed in human enucleated cells (cytoplasts) that the migratory abilities of these cells in 2D surfaces do not depend on the presence of the nucleus. However, cells do require the physical presence of the nucleus to move properly in three-dimensional (3D) spaces. In 3D contexts, the nucleus acts just as a necessary clutch to regulate the adaptative locomotive responses to their mechanical environment in a permanent interplay with the cytoskeleton to which this cellular structure is intimately connected [72,73].

An independent group has verified similar findings to those of Graham et al. [72] in a quantitative analysis in which the authors analyzed the movement trajectories of enucleated and non-enucleated *Amoeba proteus* sp. on flat 2D surfaces using advanced non-linear physical-mathematical tools [74] (Figure 1). The study had been previously deposited in bioRxiv.org by 2017 [75] and represents the first quantitative analysis of cell migration with enucleated cells.

To characterize the movements of cells and cytoplasts from a mathematical perspective, these authors analyzed first the relative move-step fluctuation along their migratory trajectories by applying the root mean square fluctuation (rmsf). This approach is a classical method in Statistical Mechanics based on Gibbs and Einstein's studies [76,77] that has been latterly developed and widely applied to quantify different time-series. The obtained results showed that both cells and cytoplasts displayed migration trajectories characterized by non-trivial long-range positive correlations (Figure 1A,B). Strong correlations over periods of about 41.5 min on average were found in all the analyzed cells and cytoplasts, which corresponded to non-trivial dependencies of the past movements lasting approximately 1245 move-steps. Therefore, each cellular move-step at a given point is strongly influenced by its previous trajectory. This dynamic memory (non-trivial correlations) represents a key characteristic of the movements during cell migration. Besides, this analysis indicated that the move-step fluctuations of all amoebas presented scale-invariance properties related to the increment of the move-step length [74].

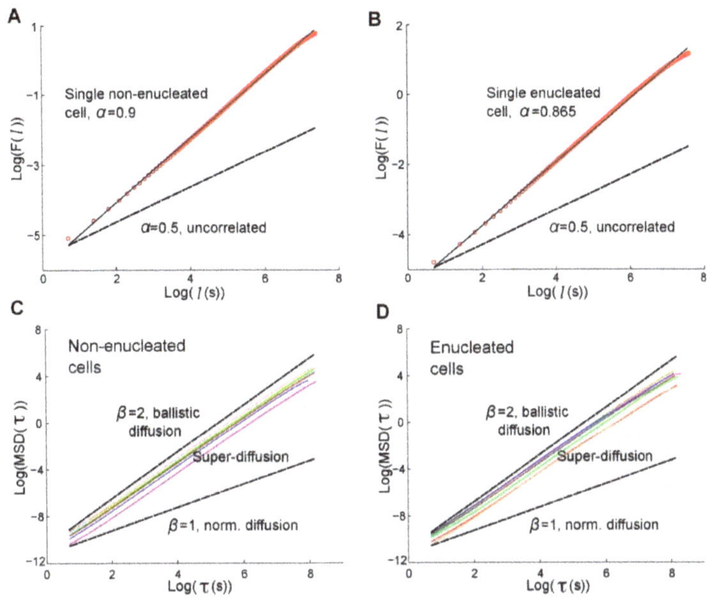

Figure 1. Root mean square fluctuation and Mean Square Displacement (MSD) of the trajectories of non-enucleated and enucleated amoebas. (Part of this figure has been reported previously by the authors in ref. [74].) Log-log plot of *rms* fluctuation F vs. l step for a prototype non-enucleated cell (**A**) and a prototype enucleated cell (**B**). The slope for the non-enucleated cell was $\alpha = 0.9$, while for the enucleated it was $\alpha = 0.865$, indicating non-trivial positive long-term correlations in both cases. In (**C**) and (**D**), log-log plots of MSD against the time interval τ, for 8 prototypic non-enucleated and 8 enucleated cells, respectively. $\beta = 1$ indicates normal diffusion while $\beta = 2$ indicates ballistic diffusion. The grey region defines the area of super-diffusion, in which all experimental slopes are contained. The fact that $\tau_{max} = 1/4$th of the data length, implies that super-diffusion holds in large scales.

Next, the Mean Square Displacement (MSD) was calculated to quantify the amount of space explored over time by the amoebas and the overall migration efficiency. This method was also proposed by Albert Einstein in his work concerning Brownian motion [78]. This approach showed that the migratory trajectories of enucleated and non-enucleated amoebae cells were associated with a non-linear dependence of MSD with time, known as anomalous diffusion, which typically occurs in complex systems with long–range correlated phenomena. Therefore, a super-diffusion process governed all the efficient migration trajectories of cells and cytoplasts (Figure 1B,C). This analysis was furtherly validated by an alternative approach, the renormalization group operator (RGO) developed by Kenneth Wilson, who established the Theory of the Renormalization Group in 1971 [79].

Finally, to quantify some kinematic properties of the cell locomotion trajectories, the directionality ratio (DR), the average speed (AS) and the total distance traveled (TD) of the amoebas were analyzed and no significant differences were observed between cells and cytoplasts [74]. This quantitative analysis showed that both cells and cytoplasts display a kind of dynamic migration structure characterized by highly organized data sequences, non-trivial long-range positive correlations, persistent dynamics with trend-reinforcing behavior, super-diffusion and move-step fluctuations with scale-invariant properties [74].

The systemic locomotion movements of cells and cytoplasts change continuously since all trajectories display random magnitudes that vary over time. These stochastic movements shape a dynamic migration structure whose defining characteristics are preserved. Such a dynamic migration structure characterizes the mathematical way in which the locomotion movements occur and so the move-steps are efficiently organized. Since the cytoplasts preserved the dynamic properties in their migration movements similarly as intact cells, the obtained results quantitatively confirmed that the nucleus does not significantly affect the systemic movements of amoebas in 2D environments [74]. This conclusion, obtained from a mathematical and computational perspective, agrees with the results previously reported using exclusively biological techniques [72].

From a molecular point of view, the amoebas' locomotion is controlled by complex metabolic networks, which operate as non-linear systems with dynamics far from equilibrium [80]. These biochemical networks involve an intricate interplay of multiple components of the cell migration machinery, including the actin cytoskeleton, ion channels, transporters and regulatory proteins such as the Arp2/3 complex or the ADF/cofilin family proteins [5,81]. As a consequence of the efficient self-regulatory activity of the metabolic networks, each amoeba seemed to be endowed with the ability to orientate its movement toward specific goals in the external environment, thereby developing efficient foraging strategies even in conditions of sparse resources when there is limited or no information as to where food is located. In accordance to the aforementioned studies, the enucleated amoeba's behaviors herein observed in 2D environments may be explained by the singular self-regulated properties of the cellular metabolic life [80].

From the studies by Graham et al. [72] and de la Fuente et al. [74,75] performed in 2D surfaces, it should be taken into account that the cytoplasts generated and analyzed in these experiments are smaller than intact cell counterparts. Under these conditions, the adhesion surface is smaller and weaker in cytoplasts thus resulting in a smaller spread area, lower adhesion strength and lower total strain energy on them. In addition, the structural geometrical organization, the cytoplasmic rigidity and density and the contractility of the actomyosin cytoskeleton are also dramatically modified in the cytoplasts. Without the nucleus, the cytoplasm is very deformable and the loss of contractility of the cytoskeleton hinders the optimal travelling speed and the adequate structural functions of cytoplasts to move properly in 3D environments.

These and other observations reveal the critical role of the nucleus for developing appropriate mechanical responses and for regulating both contractility and mechano-sensitivity [73]. Specifically, the physical presence, position and material properties of the nucleus, fundamentally those related with its connections with the cytoskeleton, are essential for a broad range of cell functions. These functions include intracellular nuclear movement, cell polarization, chromatin organization, cellular mechano-sensing and mechano-transduction signaling. Eukaryotic cells require the presence of the nucleus as a necessary component of the molecular clutch involved in the regulation of their mechanical responses to the environment. The physical properties of the nucleus strongly connected with the cytoskeleton allow and guarantee a proper cell migration when the environment displays mechanical complexities, as it happens in 3D conditions [72,73].

6. Conditioned Behavior in Single Cells

In continuation of this study and following the Pavlov's methodological approach with dogs [82], the same group of investigators observed that two different unicellular organisms (*Amoeba proteus* sp.

and *Metamoeba leningradensis* sp.) showed associative learning behaviors [83], which can be essential for adequate and efficient cellular migration (Figure 2). To analyze such conditioned behavior in amoebae they used an electric field as a conditioned stimulus and a specific chemotactic peptide as a non-conditioned stimulus. The migratory trajectories of more than 700 amoebae under different experimental conditions were studied. The results showed that, through the association of stimuli, these unicellular organisms were able to learn and forget new behaviors as time passes [83]. This phenomenon can be considered as a rudimentary form of associative memory and is also crucial to govern properly cell migration.

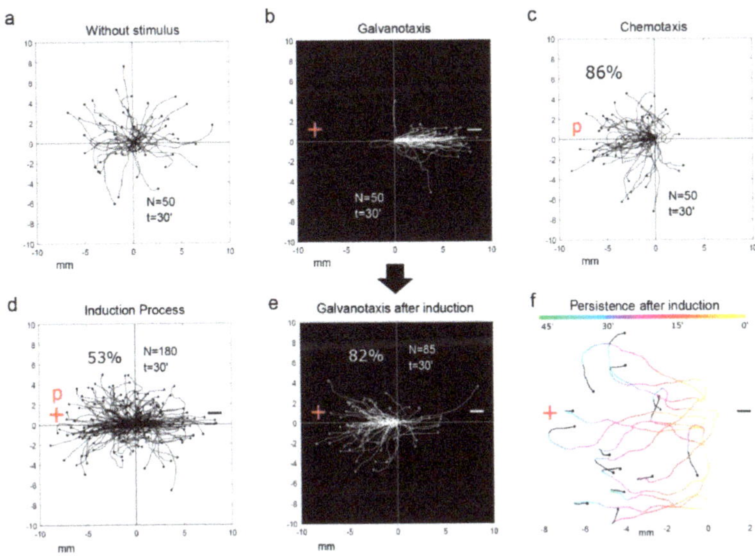

Figure 2. Experimental evidences of conditioned behavior in *Amoeba proteus*. (Part of this figure has been reported previously by the authors in ref. [83].) (**a**) Without stimulus, the cells practically explored all the directions of the experimentation chamber. (**b**) Under galvanotaxis, practically all the amoebae migrated towards the cathode. (**c**) Under chemotaxis, 86% of the cells migrated towards the chemotactic gradient. (**d**) Under galvanotaxis and chemotaxis simultaneously, 53% of the amoebae moved towards the anode-peptide (induced cells). (**e**) After induction process, the cells were placed in Chalkley's medium without any stimulus for 5 min and then they were exposed to galvanotaxis for 30 min, 82% of the inducted cells presented lasting directionality towards the anode where the chemotactic peptide was absent. (**f**) Migratory trajectories of 15 amoebae under galvanotaxis, that previously acquired the conditioned behavior after induction process, lost gradually the persistence towards the anode and turned back to the cathode (times ranging from 27 to 44 min). The colors of the trajectories represent the duration of the conditioned behavior. "*N*" total number of cells, "*t*" time of galvanotaxis or chemotaxis. "*p*" chemotactic peptide (nFMLP), "+" anode, "-" cathode. In (**a**)–(**e**), both the x and y axis show the distance in mm and the initial location of each cell has been placed at the center of the diagram.

The consecutive steps of this study [83] are summarized as follows:

(1) Cell locomotion in the absence of stimuli exhibited a random directional distribution in which amoebae and metamoebae explored practically all the directions of the experimental chamber (Figure 2a),

(2) Amoebae and metamoebae showed an unequivocal systemic response consisting in the migration to the cathode when they were exposed to a strong direct electric field of about 300–600 mV/mm (galvanotaxis, Figure 2b),

(3) The response of both organisms was studied during biochemical guidance by exposing them to an nFMLP peptide gradient placed in the anode of a specific set-up. In these experimental conditions, most of the exposed cells migrated towards the peptide in the anode showing stochastic movements with robust directionality (chemotaxis, Figure 2c),

(4) Cells were exposed simultaneously to the galvanotactic and chemotactic stimuli for 30 min (induction process). For such a purpose, the cathode was placed on the right of the set-up and the anode with the nFMLP peptide solution on the left (Figure 2d). The results showed that roughly half of the amoebae and metamoebae migrated towards the anode where the peptide was placed, while the reminders did it to the cathode,

(5) To verify if the cells that moved to the anode during the induction process (Figure 2d) exhibited some degree of persistence in their migratory behavior, those cells that had previously migrated to the anode-peptide in the fourth step were exposed a second time (30 min) to a single electric field without the peptide. Under these experimental conditions, the analysis of the individual trajectories showed that most cells did migrate to the anode where the peptide was absent (Figure 2e). This evidence corroborated that a new locomotion pattern had appeared in amoebae and metamoebae (Figure 2d) (note that without the induction process practically all the cells migrated to the cathode, (Figure 2c) and after the induction process the cells modified their behavior going to the anode instead to the cathode).

This step-wise experiment showed that some amoebae seemed to associate the anode with the food (the peptide) when the cells were exposed to a stimulus related to their nourishment (the specific peptide nFMLP placed in the anode) and this exposition was simultaneously accompanied by an electric field (induction process, Figure 2d) [83].

After the induction process, most of the conditioned *Amoeba proteus* sp. and *Metamoeba leningradensis* sp. ran to the anode where the peptide was absent, modifying their systemic conduct, behaving against their known tendency to move to the cathode (Figure 2b) and developing a new persistent pattern of cell locomotion characterized by movements towards the anode (Figure 2e).

Strikingly, this conditioned behavior persisted for relatively long periods ranging from 20 to 95 min (Figure 2f). The quantitative analysis of these results emphasized that it was extremely unlikely to obtain them by chance ($p = 10^{-19}$; $Z = 8.878$, Wilcoxon rank-sum test).

Pavlov described four fundamental types of persistent behavior provoked by two stimuli [82]. The experiment of cellular conditioning summarized here was based in one of them, the so-called *simultaneous conditioning*, in which both stimuli are applied at the same time. This finding in which individual cells can generate migratory conditioned patterns guiding their systemic locomotion movements has never been reported so far.

These experiments with unicellular organisms were the consequence of previous physical-mathematical analyses with complex metabolic networks published in 2013 [84], where using advanced tools of Statistic Mechanics and techniques of Artificial Intelligence it was verified from a computational viewpoint that metabolic networks were governed by Hopfield-like dynamics showing associative memory behavior. This quantitative study demonstrated for the first time that an associative memory was also possible in unicellular organisms. Such type of memory could be the manifestation of the emergent properties underlying the complex dynamics of the systemic metabolic networks corresponding to an epigenetic type of cellular memory [80].

7. Concluding Remarks

This narrative reviews fundamental issues related to cell motility and cancer. For such purpose, we revisit the usefulness of analyzing unicellular eukaryotic cells to improve current understanding in several crucial points—(i) the regulation of cell migration in mammalian cells, (ii) the importance of systemic approaches in these studies, (iii) the role of the nucleus in cell displacements, (iv) the emergence of a new behavior by which cells do learn and develop a primitive form of associative memory to respond to the environmental changes during migration and (v) the implications of some important processes related to the control of cell locomotion in cancer. The spectrum of external stimuli and the analysis of their corresponding systemic migratory behaviors (galvanotaxis, chemotaxis, haptotaxis, barotaxis, durotaxis, topotaxis and plithotaxis) in standard physiological conditions and in disease have been briefly reviewed as an example of translational research with potential clinical applicability.

Some special points deserve further special comments, for example, the role of the nucleus in cell migration in diverse environments, fundamentally in 3D spaces. As it has been pointed out in the chapter 5 of this review, several studies have already shown that motility patterns in enucleated cells does not significantly differ from the observed in normal cells on 2D surfaces [72,74]. In this sense, the works of De la Fuente et al. [74,75] with eukaryotic unicellular organisms and of Graham et al. [72] with mammalian cells, demonstrating that cell migration does not depend on the nucleus in 2D scenarios represents a great advance in the understanding of systemic control of cell motility.

However, it is necessary to underline the crucial role of the nucleus during cell locomotion in 3D environments, for instance, cell migration during embryogenesis, wound repair and cancer is performed in 3D spaces in a context of a very different spectrum of extracellular matrix (ECM) proteins that conditionate cell movements. In this particular setting the nucleus acts as a fundamental mechanosensory structure in a continuous interplay with the cytoplasmic architecture.

On the other hand, the comparative analysis of cell motility at the molecular level between unicellular organisms and human cells has been a matter of translational study mentioned in a specific chapter of this review. Similarities and differences have been detected in the analysis, although in general terms, crucial enzymatic routes underlying cell migration have been well preserved during evolution. Numerous studies with eukaryotic unicellular organisms have provided excellent experimental models to understand the precise role of many molecules and metabolic processes involved in cell migration. Interestingly, the translation of these findings at the molecular level from unicellular organisms to cancer cell biology has contributed to unveil new therapeutic opportunities for patients.

Finally, the associative conditioning with long term persistence detected experimentally in unicellular eukaryotic organisms has also been highlighted. In these experiments, amoebas were able to develop a conditioned behavior during cell migration when they were exposed to two different simultaneous stimuli in a similar way as Pavlov's dogs did more than one hundred years ago.

The new studies in the control of cell motility both in unicellular organisms and mammalian cells define a new framework in the understanding of the mechanisms underlying the complex systemic behavior involved both in the cellular migration and in the adaptive capacity of cells to the external medium. Such findings constitute a significant advance in the comprehension of the biological processes involved in critical issues for human life like embryogenesis, tissue repair and carcinogenesis.

Author Contributions: I.M.D.l.F. and J.I.L. conceived, designed and wrote the manuscript. Both authors have read and agreed to the published version of the manuscript.

Funding: This study received no external funding.

Conflicts of Interest: The authors declare no conflict of interest.

References

1. Stuelten, C.H.; Parent, C.A.; Montell, D.J. Cell motility in cancer invasion and metastasis: Insights from simple model organisms. *Nat. Rev. Cancer* **2018**, *18*, 296–312. [CrossRef]
2. Titus, M.A.; Goodson, H.V. An evolutionary perspective on cell migration: Digging for the roots of amoeboid motility. *J. Cell Biol.* **2017**, *216*, 1509–1511. [CrossRef]
3. Fritz-Laylin, L.K.; Lord, S.J.; Mullins, R.D. WASP and SCAR are evolutionarily conserved in actin-filled pseudopod-based motility. *J. Cell Biol.* **2017**, *216*, 1673–1688. [CrossRef] [PubMed]
4. Fritz-Laylin, L.K.; Riel-Mehan, M.; Chen, B.; Lord, S.J.; Goddard, T.D.; Ferrin, T.E.; Nicholson-Dykstra, S.; Higgs, H.; Johnson, G.T.; Betzig, E.; et al. Actin-based protrusions of migrating neutrophils are intrinsically lamellar and facilitate direction changes. *eLife* **2017**, *6*, e26990. [CrossRef] [PubMed]

5. Artemenko, Y.; Lampert, T.J.; Devreotes, P.N. Moving towards a paradigm: Common mechanisms of chemotactic signaling in Dictylostelium and mammalian leukocytes. *Cell Mol. Life Sci.* **2014**, *71*, 3711–3747. [CrossRef] [PubMed]
6. Batsios, P.; Ishikawa-Ankerhold, H.; Roth, H.; Schleicher, M.; Wong, C.; Müller-Taubenberger, A. Ate1-mediated posttranslational arginylation affects substrate adhesion and cell migration in Dictyostelium discoideum. *Mol. Biol. Cell* **2019**, *30*, 453–456. [CrossRef]
7. Gilsbach, B.K.; Ho, F.Y.; Vetter, I.R.; Van Haastert, P.J.; Wittinghofer, A.; Kortholt, A. Roco kinase structures give insights into the mechanism of Parkinson disease-related leucine-rich-repeat kinase 2 mutations. *Proc. Natl. Acad. Sci. USA* **2012**, *109*, 10322–10327. [CrossRef]
8. Chida, J.; Araki, H.; Maeda, Y. Specific growth suppression of human cancer cells by targeted delivery of Dictyostelium mitochondrial ribosomal protein S4. *Cancer Cell Int.* **2014**, *14*, 56. [CrossRef]
9. Sherwood, D.R. Cell invasion through basement membranes: An anchor of understanding. *Trends Cell. Biol.* **2006**, *16*, 250–256. [CrossRef] [PubMed]
10. Ellis, H.M.; Horvitz, H.R. Genetic control of programmed cell death in the nematode *C. elegans*. *Cell* **1986**, *44*, 817–829. [CrossRef]
11. Pinkston-Gosse, J.; Kenyon, C. DAF-16/FOXO targets genes that regulate tumor growth in Caenorhabditis elegans. *Nat. Genet.* **2007**, *39*, 1403–1409. [CrossRef] [PubMed]
12. Nüsslein-Volhard, C.; Wieschaus, E. Mutations affecting segment number and polarity in Drosophila. *Nature* **1980**, *287*, 795–801. [CrossRef] [PubMed]
13. Martorell, O.; Merlos-Suarez, A.; Campbell, K.; Barriga, F.M.; Christov, C.P.; Miguel-Aliaga, I.; Batlle, E.; Casanova, J.; Casali, A. Conserved mechanisms of tumorigenesis in the Drosophila adult midgut. *PLoS ONE* **2014**, *9*, e88413. [CrossRef] [PubMed]
14. Campbell, K.; Rossi, F.; Adams, J.; Pitsidianaki, I.; Barriga, F.M.; García-Gerique, L.; Batlle, E.; Casanova, J.; Casali, A. Collective cell migration and metastases induced by an epithelial-to-mesenchymal transition in *Drosophila* intestinal tumors. *Nat. Commun.* **2019**, *10*, 2311. [CrossRef] [PubMed]
15. Caussinus, E.; Gonzalez, C. Induction of tumor growth by altered stem-cell asymmetric division in Drosophila melanogaster. *Nature Genet.* **2005**, *37*, 1125–1129. [CrossRef] [PubMed]
16. Sonoshita, M.; Cagan, R.L. Modeling human cancers in Drosophila. *Curr. Top. Dev. Biol.* **2017**, *121*, 287–309.
17. Lammermann, T.; Sixt, M. Mechanical modes of 'amoeboid' cell migration. *Curr. Opin. Cell Biol.* **2009**, *21*, 636–644. [CrossRef]
18. Kumar, S.; Xu, J.; Perkins, C.; Guo, F.; Snapper, S.; Finkelman, F.D.; Zheng, Y.; Filippi, M.D. Cdc42 regulates neutrophil migration via crosstalk between WASp, CD11b, and microtubules. *Blood* **2012**, *120*, 3563–3574. [CrossRef]
19. Stengel, K.; Zheng, Y. Cdc42 in oncogenic transformation, invasion, and tumorigenesis. *Cell. Signal.* **2011**, *23*, 1415–1423. [CrossRef]
20. Creed, S.J.; Desouza, M.; Bamburg, J.R.; Gunning, P.; Stehn, J. Tropomyosin isoforms 3 promotes the formation of filopodia by regulating the recruitment of actin-binding proteins to actin filaments. *Exp. Cell Res.* **2011**, *317*, 249–261. [CrossRef]
21. Bornens, M. Cell Polarity: Having and Making Sense of Direction-On the Evolutionary Significance of the Primary Cilium/Centrosome Organ in Metazoa. *Open Biol.* **2018**, *8*, 180052. [CrossRef] [PubMed]
22. Viola, A.; Luster, A.D. Chemokines and their receptors: Drug targets in immunity and inflammation. *Annu. Rev. Pharmacol. Toxicol.* **2008**, *48*, 171–197. [CrossRef] [PubMed]
23. Sasaki, A.T.; Chun, C.; Takeda, K.; Firtel, R.A. Localized Ras signaling at the leading edge regulates PI3K, cell polarity, and directional cell movement. *J. Cell Biol.* **2004**, *167*, 505–518. [CrossRef] [PubMed]
24. Bos, J.L. *ras* oncogenes in human cancer: A review. *Cancer Res.* **1989**, *49*, 4682–4689.
25. Kumar, A.; Molli, P.R.; Pakala, S.B.; Bui Nguyen, T.M.; Rayala, S.K.; Kumar, R. PAK thread from amoeba to mammals. *J. Cell. Biochem.* **2009**, *107*, 579–585. [CrossRef]
26. Eswaran, J.; Soundararajan, M.; Knapp, S. Targeting group II PAKs in cancer and metastasis. *Cancer Metastasis Rev.* **2009**, *28*, 209–217. [CrossRef]
27. Chan, P.M.; Manser, E. PAKs in human disease. *Prog. Mol. Biol. Transl. Sci.* **2012**, *106*, 171–187.
28. Jaśkiewicz, A.; Pająk, B.; Orzechowski, A. The Many Faces of Rap1 GTPase. *Int. J. Mol. Sci.* **2018**, *19*, 2848. [CrossRef]
29. Takai, Y.; Sasaki, T.; Matozaki, T. Small GTP-binding proteins. *Physiol. Rev.* **2001**, *81*, 153–208. [CrossRef]

30. Jeon, T.J.; Lee, D.J.; Merlot, S.; Weeks, G.; Firtel, R.A. Rap1 controls cell adhesion and cell motility through the regulation of myosin II. *J. Cell. Biol.* **2007**, *176*, 1021–1033. [CrossRef]
31. Sawant, K.; Chen, Y.; Kotian, N.; Preuss, K.M.; McDonald, J.A. Rap1 GTPase promotes coordinated collective cell migration in vivo. *Mol. Biol. Cell.* **2018**, *29*, 2656–2673. [CrossRef] [PubMed]
32. Zhang, Y.L.; Wang, R.C.; Cheng, K.; Ring, B.Z.; Su, L. Roles of Rap1 signaling in tumor cell migration and invasion. *Cancer Biol. Med.* **2017**, *14*, 90–99. [PubMed]
33. Dale, H.H. Galvanotaxis and chemotaxis of ciliate infusoria: Part 1. *J. Physiol* **1901**, *25*, 291–361. [CrossRef] [PubMed]
34. Roy, J.; Mazzaferri, J.; Filep, J.G.; Costantino, S. A haptotaxis assay for neutrophils using optical patterning and a high-content approach. *Sci. Rep.* **2017**, *7*, 2869. [CrossRef] [PubMed]
35. Oudin, M.J.; Jonas, O.; Kosciuk, T.; Broye, L.C.; Guido, B.C.; Wyckoft, J.; Riquelme, D.; Lamar, J.M.; Asokan, S.B.; Whittaker, C.; et al. Tumor cell-driven extracellular matrix remodeling drives haptotaxis during metastatic progression. *Cancer Discov.* **2016**, *6*, 516–531. [CrossRef]
36. Prentice-Mott, H.V.; Chang, C.H.; Mahadevan, L.; Mitchison, T.J.; Irimia, D.; Shah, J.V. Biased migration of confined neutrophil-like cells in asymmetric hydraulic environments. *PNAS* **2013**, *110*, 21006–21011. [CrossRef]
37. Moreau, H.D.; Blanch-Mercader, C.; Attia, R.; Maurin, M.; Alraies, Z.; Sanséau, D.; Malbec, O.; Delgado, M.G.; Bousso, P.; Joanny, J.F.; et al. Macropinocytosis overcomes directional bias in dendritic cells due to hydraulic resistance and facilitates space exploration. *Dev. Cell* **2019**, *49*, 171–188. [CrossRef]
38. Lo, C.M.; Wang, H.B.; Dembo, M.; Wang, Y.L. Cell movement is guided by the rigidity of the substrate. *Biophys. J.* **2000**, *79*, 144–152. [CrossRef]
39. Feng, J.; Levine, H.; Mao, X.; Sander, L.M. Cell motility, contact guidance, and durotaxis. *Soft Matter* **2019**, *15*, 4856–4864. [CrossRef]
40. Buenemann, M.; Levine, H.; Rappel, W.J.; Sander, L.M. The role of cell contraction and adhesion in *Dictyostelium* motility. *Biophys. J.* **2010**, *99*, 50–58. [CrossRef]
41. Parida, L.; Padmanabhan, V. Durotaxis in nematode *Caenorhabditis elegans*. *Biophys. J.* **2016**, *111*, 666–674. [CrossRef] [PubMed]
42. Parida, L.; Neogi, S.; Padmanabhan, V. Effect of temperature pre-exposure on the locomotion and chemotaxis of *C. elegans*. *PLoS ONE* **2014**, *9*, e111342. [CrossRef] [PubMed]
43. Isenberg, B.C.; DiMilla, P.A.; Walker, M.; Kim, S.; Womg, J.Y. Vascular smooth muscle cell durotaxis depends on substrate stiffness gradient strength. *Biophys. J.* **2009**, *97*, 1313–1322. [CrossRef] [PubMed]
44. Vincent, L.G.; Choi, I.S.; Alonso-Latorre, B.; Del Alamo, J.C.; Engler, A.J. Mesenchymal stem cell durotaxis depends on substrate stiffness gradient strength. *Biotechnol. J.* **2013**, *8*, 472–484. [CrossRef]
45. DuChez, B.J.; Doyle, A.D.; Dimitriadis, E.K.; Yamada, K.M. Durotaxis by human cancer cells. *Biophys. J.* **2019**, *116*, 670–683. [CrossRef]
46. Liu, C.; Pei, H.; Tan, F. Matrix stiffness and colorectal cancer. *OncoTargets Ther.* **2020**, *13*, 2747–2755. [CrossRef] [PubMed]
47. Park, J.S.; Kim, D.H.; Levchenko, A. Topotaxis: A new mechanism of directed cell migration in topographic ECM gradients. *Biophys. J.* **2018**, *114*, 1257–1263. [CrossRef]
48. Kim, D.H.; Han, K.; Gupta, K.; Kwon, K.W.; Suh, K.Y.; Levchenko, A. Mechanosensitivity of fibroblast cell shape and movement to anisotropic substratum topography gradients. *Biomaterials* **2009**, *30*, 5433–5444. [CrossRef]
49. Trepat, X.; Fredberg, J.J. Plithotaxis and emergent dynamics in collective cellular migration. *Trends Cell Biol.* **2011**, *21*, 638–646. [CrossRef]
50. Kippenberger, S.; Bernd, A.; Loitsch, S.; Guschel, M.; Muller, J.; Bereiter-Hahn, J.; Kaufmann, R. Signaling of mechanical stretch in human keratinocytes via MAP kinases. *J. Investig. Dermatol.* **2000**, *114*, 408–412. [CrossRef]
51. Nabeshima, K.; Inoue, T.; Shimao, Y.; Kataoka, H.; Koono, M. Cohort migration of carcinoma cells: Differentiated colorectal carcinoma cells move as coherent cell clusters or sheets. *Histol. Histopathol.* **1999**, *14*, 1183–1197. [PubMed]
52. Jeon, K.W. The large, free-living amoebae: Wonderful cells for biological studies. *J. Eukaryot. Microbiol.* **1995**, *42*, 1–7. [CrossRef] [PubMed]

53. Korohoda, W.; Mycielska, M.; Janda, E.; Madeja, Z. Immediate and long-term galvanotactic responses of *Amoeba proteus* to electric fields. *Cell Motil. Cytoskeleton.* **2000**, *45*, 10–26. [CrossRef]
54. Mycielska, M.E.; Djamgoz, M.B.A. Cellular mechanisms of direct-current electric field affects: Galvanotaxis and metastatic disease. *J. Cell Sci.* **2004**, *117*, 1631–1639. [CrossRef]
55. Borys, P. The role of passive calcium influx through the cell membrane in galvanotaxis. *Cell. Mol. Biol. Lett.* **2013**, *18*, 187–199. [CrossRef]
56. Pollard, T.D. Regulation of actin filament assembly by Arp2/3 complex and formins. *Annu. Rev. Biophys. Biomol. Struct.* **2007**, *36*, 451–477. [CrossRef]
57. Wang, E.; Zhan, M.; Forrester, J.V.; McCraig, C.D. Reorientation and faster directed migration of lens epithelial cells in a physiological electric field. *Exp. Eye Res.* **2000**, *71*, 91–98. [CrossRef]
58. Frazer, S.P.; Diss, J.K.J.; Mycielska, M.E.; Coombes, R.C.; Djamgoz, M.B.A. Voltage-gated sodium channel expression in human breast cancer: Possible functional role in metastasis. *Breast Cancer Res. Treat.* **2002**, *76*, S142.
59. Cahalan, M.D.; Wulff, H.; Chandy, K.G. Molecular properties and physiological roles of ion channels in the immune system. *J. Clin. Immunol.* **2001**, *21*, 235–252. [CrossRef]
60. Shu, L.; Zhang, B.; Queller, D.C.; Strassmann, J.E. Burkholderia bacteria use chemotaxis to find social amoeba Dictyostelium discoideum hosts. *ISME J.* **2018**, *12*, 1977–1993. [CrossRef]
61. Hirata, E.; Sahai, E. Tumor microenvironment and differential responses to therapy. *Cold Spring Harb. Perspect. Med.* **2017**, *7*, a026781. [CrossRef] [PubMed]
62. Marques, P.; Barry, S.; Carlsen, E.; Collier, D.; Ronaldson, A.; Awad, S.; Dorward, N.; Grieve, J.; Mendoza, N.; Muquit, S.; et al. Chemokines modulate the tumour microenvironment in pituitary neuroendocrine tumours. *Acta Neuropathol. Commun.* **2019**, *7*, 172. [CrossRef] [PubMed]
63. Calvete, J.; Larrinaga, G.; Errarte, P.; Martín, A.M.; Dotor, A.; Esquinas, C.; Nunes-Xavier, C.E.; Pulido, R.; López, J.I.; Angulo, J.C. The coexpression of fibroblast activation protein (FAP) and basal-type markers (CK5/6 and CD44) predicts prognosis in high-grade invasive urothelial carcinoma of the bladder. *Hum. Pathol.* **2019**, *91*, 61–68. [CrossRef] [PubMed]
64. Errarte, P.; Larrinaga, G.; López, J.I. The role of cancer-associated fibroblasts in renal cell carcinoma. An example of tumor modulation through tumor/non-tumor cell interactions. *J. Adv. Res.* **2019**, *21*, 103–108. [CrossRef]
65. Xiang, H.; Ramil, C.P.; Hai, J.; Zhang, C.; Wang, H.; Watkins, A.A.; Afshar, R.; Georgiev, P.; Sze, M.A.; Song, X.S.; et al. Cancer-associated fibroblasts promote immunosuppression by inducing ROS-generating monocytic MDSCs in lung squamous cell carcinoma. *Cancer Immunol. Res.* **2020**. [CrossRef]
66. Angulo, J.C.; Shapiro, O. The changing therapeutic landscape of metastatic renal cancer. *Cancers* **2019**, *11*, E1227. [CrossRef]
67. Nunes-Xavier, C.E.; Angulo, J.C.; Pulido, R.; López, J.I. A critical insight into the clinical translation of PD-1/PD-L1 blockade therapy in clear cell renal cell carcinoma. *Curr. Urol. Rep.* **2019**, *20*, 1. [CrossRef]
68. Xinran, W.; Liu, Y. PD-L1 expression in tumor infiltrated lymphocytes predicts survival in triple-negative breast cancer. *Pathol. Res. Pract.* **2020**, *216*, 152802.
69. Evans, M.; O'Sullivan, B.; Hughes, F.; Mullis, T.; Smith, M.; Trim, N.; Taniere, P. The clinicopathological and molecular associations of PD-L1 expression in non-small cell lung cancer: Analysis of a series of 10,005 cases tested with the 22C3 assay. *Pathol. Oncol. Res.* **2018**. [CrossRef]
70. Morsch, R.; Rose, M.; Maurer, A.; Cassataro, M.A.; Braunschweig, T.; Knüchel, R.; Vögeli, T.A.; Ecke, T.; Eckstein, M.; Weyerer, V.; et al. Therapeutic implications of PD-L1 expression in bladder cancer with squamous differentiation. *BMC Cancer* **2020**, *20*, 230. [CrossRef]
71. Wu, H.; Xia, L.; Jia, D.; Zou, H.; Jin, G.; Qian, W.; Xu, H.; Li, T. PD-L1^{+} regulatory B cells act as a T-cell suppressor in a PD-L1-dependent manner in melanoma patients with bone metastasis. *Mol. Immunol.* **2020**, *119*, 83–91. [CrossRef]
72. Graham, D.M.; Andersen, T.; Sharek, L.; Uzer, G.; Rothenberg, K.; Hoffman, B.D.; Rubin, J.; Balland, M.; Bear, J.E.; Burridge, K. Enucleated cells reveal differential roles of the nucleus in cell migration, polarity, and mechanotransduction. *J. Cell Biol.* **2018**, *217*, 895–914. [CrossRef] [PubMed]
73. Hawkins, R.J. Do migrating cells need a nucleus? *J. Cell Biol.* **2018**, *217*, 799–801. [CrossRef] [PubMed]

74. De la Fuente, I.M.; Bringas, C.; Malaina, I.; Regner, B.; Pérez-Samartin, A.; Boyano, M.D.; Fedetz, M.; López, J.I.; Pérez-Yarza, G.; Cortés, J.M.; et al. The nucleus does not significantly affect the migratory trajectories of amoeba in two-dimensional environments. *Sci. Rep.* **2019**, *9*, 16369. [CrossRef] [PubMed]
75. Bringas, C.; Malaina, I.; Pérez-Samartín, A.; Boyano, M.D.; Fedetz, M.; Pérez-Yarza, G.; Cortés, J.M.; de la Fuente, I.M. Long-term memory in the migration movements of enucleated *Amoeba Proteus*. *BioRxiv* **2017**. [CrossRef]
76. Viswanathan, G.M.; Afanasyev, V.; Buldyrev, S.V.; Murphy, E.J.; Prince, P.A.; Stanley, H.E. Lévy flight search patterns of wandering albatrosses. *Nature* **1996**, *381*, 413–415. [CrossRef]
77. Ivanov, P.C.; Amaral, L.A.; Goldberger, A.L.; Havlin, S.; Rosenblum, M.G.; Strzuik, Z.R.; Stanley, H.E. Multifractality in human heartbeat dynamics. *Nature* **1999**, *399*, 461–465. [CrossRef] [PubMed]
78. Einstein, A. Über die von der molekularkinetischen Theorie der wärme geforderte Bewegung von in ruhenden Flüssigkeiten suspendierten Teilchen. *Ann. Phys.* **1905**, *322*, 549–560. [CrossRef]
79. Wilson, K. Renormalization group and critical phenomena II: Phase space cell analysis of critical behavior. *Phys. Rev. B* **1971**, *4*, 3184. [CrossRef]
80. De la Fuente, I.M. Elements of the cellular metabolic structure. *Front. Mol. Biosci.* **2015**, *2*, 16. [CrossRef]
81. Senoo, H.; Cai, H.; Wang, Y.; Sesaki, H.; Iijima, M. The novel RacE-binding protein GflB sharpens Ras activity at the leading edge of migrating cells. *Mol. Biol. Cell* **2016**, *27*, 1596–1605. [CrossRef] [PubMed]
82. Pavlov, I.P. *Conditioned Reflexes: An Investigation of the Physiological Activity of the Cerebral Cortex*; Oxford University Press: Oxford, UK, 1927.
83. De la Fuente, I.M.; Bringas, C.; Malaina, I.; Fedetz, M.; Carrasco-Pujante, J.; Morales, M.; Knafo, S.; Martinez, L.; Pérez-Samartín, A.; López, J.I.; et al. Evidence of conditioned behavior in amoebae. *Nat. Commun.* **2019**, *10*, 3690. [CrossRef] [PubMed]
84. De la Fuente, I.M.; Cortés, J.M.; Pelta, D.A.; Veguillas, J. Attractor metabolic networks. *PLoS ONE* **2013**, *8*, e58284. [CrossRef] [PubMed]

MDPI
St. Alban-Anlage 66
4052 Basel
Switzerland
Tel. +41 61 683 77 34
Fax +41 61 302 89 18
www.mdpi.com

Cancers Editorial Office
E-mail: cancers@mdpi.com
www.mdpi.com/journal/cancers